Rocky Mountain
Field Guide

Rocky Mountain Field Guide

A TRAILSIDE NATURAL HISTORY

DANIEL MATHEWS

MOUNTAINEERS
BOOKS

MOUNTAINEERS BOOKS is dedicated to the exploration, preservation, and enjoyment of outdoor and wilderness areas.

1001 SW Klickitat Way, Suite 201, Seattle, WA 98134
800-553-4453, www.mountaineersbooks.org

Printed in China

First edition, 2024

Design and layout: Jen Grable
Cartographer: Erin Greb
Portions of this guide first appeared in different form in the author's 2003 self-published book, *Rocky Mountain Natural History: Grand Teton to Jasper*.
All photographs by the author, unless credited otherwise in Photo Credits on p. 540.
Cover photographs, clockwise from upper left: *Townsend's warbler* (Photo by Michael Ashbee), *pika* (Photo by Ann Schonlau, NPS), *northern leopard frog* (Photo by US Fish and Wildlife Service), *Wyoming paintbrush near Crested Butte, Colorado* (Photo by Daniel Mathews); spine: *northern saw-whet owl* (Photo by Michael Ashbee)
Frontispiece: *Aspens over Canyon Creek, Colorado* (Photo by Daniel Mathews)

Library of Congress Cataloging-in-Publication Data
Names: Mathews, Daniel, 1948– author
Title: Rocky Mountain field guide : a trailside natural history / Daniel Mathews
Description: Seattle, WA : Mountaineers Books, 2024 | Includes bibliographical references and index | Summary: "An engaging, comprehensive field guide to the natural wonders of the Rockies"— Provided by publisher
Identifiers: LCCN 2024013094 (print) | LCCN 2024013095 (ebook) | ISBN 9781680516111 (paperback) | ISBN 9781680516128 (epub)
Subjects: LCSH: Mountain ecology—Rocky Mountains. | Natural areas—Rocky Mountains—Identification. | Natural history—Rocky Mountains. | Mountain plants—Rocky Mountains—Identification. | Mountain animals—Rocky Mountains—Identification. | Field guides.
Classification: LCC QH104.5.R6 M36 2024 (print) | LCC QH104.5.R6 (ebook) | DDC 557.8—dc23/eng/20240412
LC record available at https://lccn.loc.gov/2024013094
LC ebook record available at https://lccn.loc.gov/2024013095k

Mountaineers Books titles may be purchased for corporate, educational, or other promotional sales, and our authors are available for a wide range of events. For information on special discounts or booking an author, contact our customer service at 800-553-4453 or mbooks@mountaineersbooks.org.

Printed on FSC-certified materials

MIX
Paper | Supporting responsible forestry
FSC® C188448

ISBN (paperback): 978-1-68051-611-1
ISBN (ebook): 978-1-68051-612-8

An independent nonprofit publisher since 1960

This book is for
the fourlegged people
the standing people
the crawling people
the swimming people
the sitting people
and the flying people

that people walking
with them may know
and honor them.

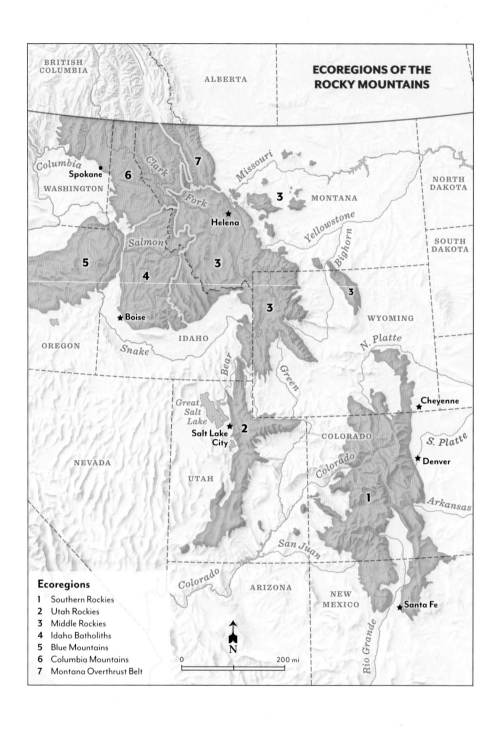

ECOREGIONS OF THE
ROCKY MOUNTAINS

Ecoregions

1 Southern Rockies
2 Utah Rockies
3 Middle Rockies
4 Idaho Batholiths
5 Blue Mountains
6 Columbia Mountains
7 Montana Overthrust Belt

0   200 mi

N

Contents

Introduction: A Natural History of the Rocky Mountains 9
How to Use This Guide 10
Abbreviations and Symbols 16

1   A Grand Landscape 18

2   Weather, Climate, and Fire 32

3   Conifers 50

4   Flowering Trees and Shrubs 79

5   Flowering Herbs 125

6   Ferns, Clubmosses, and Horsetails 244

7   Mosses and Liverworts 253

8   Fungi 262

9   Lichens 286

10   Mammals 305

11   Birds 366

12   Reptiles 424

13   Amphibians 431

14   Fishes 436

15   Insects 449

16   Other Creatures 492

17   Geology 496

Acknowledgments 539
Photo Credits 540
Glossary 544
Sources and Recommended Reading 551
Index 556

The Rockies are a vast, intricate, and discontinuous concatenation of subranges. They comprise a portion of the hemisphere-long swath of mountain ranges and plateaus called the Cordillera ("mountain chain" in Spanish, an exaggeration, since it's missing some links). Its major mountain-building episodes within the Lower 48 took place between 89 million and 45 million years ago—more recently than mountain-building in eastern North America, but longer ago than in the Pacific Northwest. And the Rockies aren't done building. Extensive parts rose to their present elevations within the last six million years, and many got remodeled by either ice or volcanism within the last two million. Some subranges are still rising rapidly (by geologic standards).

Thanks to its grand scale and its sparse human population, this region has the best chance in temperate North America of sustaining an intact ecosystem. But that's far from assured: it requires maintaining—even expanding—the current range of our large carnivores, because even though they are few in number, they are keystone species that affect everything else.

Zoom out on satellite imagery of the Rocky Mountains, and you can easily see their limits: the Rockies are the dark-green forests—plus some smallish forest-enclosed parks and tundra—surrounded by the paler colors of grasslands, shrublands, semideserts, and cities. This field guide focused on the mountains skips most species that live almost exclusively in croplands or deserts, or at elevations below the lowest conifers.

There's a simple reason why the mountains are greener than the plains. Forests and lush meadows demand a wetter climate than grasslands and steppes do, and mountains provide that, both by inducing rain and snow and by being cooler (thanks to higher elevations), which slows the rates of evaporation and snowmelt. So much depends on that cool mountain air.

This fresh incarnation of what began a couple of decades ago as *Rocky Mountain Natural History* can go with you to places where Wikipedia cannot and provide pleasures beyond simple information. I have curated the tastiest morsels for your reading pleasure.

OPPOSITE: Wind River Country, *painted by Albert Bierstadt in 1860, part of the Charles A. Bayly Collection with the Denver Art Museum*

Each species' account begins with common and scientific names. The word "also," when present, introduces names you may see for this species in other sources. An identifying description follows, concluding with likely habitat, sometimes season, and range within the US Rocky Mountains. ("A" means the range touches all eight states; "A–" means all but one or two peripheral states.) When a genus-level description is followed by two or more species names, each with their own paragraph, the genus-level description applies to all of these species, while the species-level descrip-

tions spotlight traits for narrowing your identification (ID) down to the species. A few traits, like having five petals, may be found in the subhead at the bottom of odd-numbered pages.

Elements of these ID paragraphs include: preferred common name, preferred scientific name, stressed syllable, origin of scientific name, a dash for suffixes, page number with a bio of the person it is named for, other common name, other scientific name, ID characteristics, typical habitat, range within Rockies (touches all eight states), and family name. It looks like this:

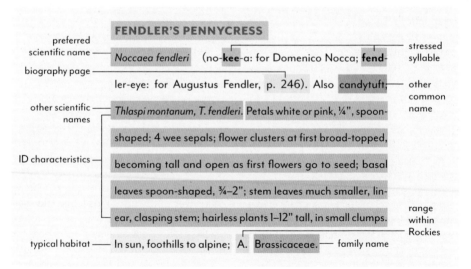

preferred scientific name

biography page

other scientific names

ID characteristics

typical habitat

FENDLER'S PENNYCRESS

Noccaea fendleri (no-**kee**-a: for Domenico Nocca; **fend**-ler-eye: for Augustus Fendler, p. 246). Also candytuft; *Thlaspi montanum, T. fendleri.* Petals white or pink, ¼", spoon-shaped; 4 wee sepals; flower clusters at first broad-topped, becoming tall and open as first flowers go to seed; basal leaves spoon-shaped, ¾–2"; stem leaves much smaller, linear, clasping stem; hairless plants 1–12" tall, in small clumps. In sun, foothills to alpine; A. Brassicaceae.

stressed syllable

other common name

range within Rockies

family name

IN THE FIRST half of the book—the plants and the fungi—the subheads at the bottom of odd-numbered pages recap the nested key (outlined in detail below), which uses con-

spicuous characteristics: evergreen leaves versus deciduous, alternate leaves versus opposite, regular flowers versus irregular (bilaterally symmetrical), flower petals in threes versus fours versus fives, and so on. If any of these characteristics are new concepts to you, they should be easy to learn. Grouping plants this way keeps similar plants together, making it easy for you to compare similar things. Since that's roughly the same idea that got taxonomy started, you'll often find members of a taxonomic family clustered together.

In the second half of the book—the animals and rocks—subheads in the running footers indicate taxonomic groupings directly (e.g., rodents, beetles, sedimentary rocks).

SCIENTIFIC NAMES

For more than a century, we writers of field guides preached that you readers should learn Latin names because, in contrast to common names, they are used consistently from one region or nation or book or decade to another. We lied! Sorry. Or maybe it used to be true.

For example, our mink has been through four scientific names in the last thirty years, while its common name remained "mink." Well, American mink now.

Scientific names are in turmoil. One reason is decentralization. Taxonomic judgments used to be made by a few senior taxonomists at a few major institutions examining animal skeletons or pressed, dried herbarium specimens. Today's taxonomists have more time and more specimens, and often have seen the live organism in its native habitat, where differences may be more visible. A second and bigger reason is that they now have DNA sequencing.

There's a divergence between two main functions of scientific names. The original intent of Linnaean binomials (the genus-species combo) was to standardize the naming of organisms so that scientists could communicate about them with minimal confusion.

Second, Linnaeus sorted organisms into ranked, nested categories based on their degree of similarity. This higher taxonomy served Western science for a century before Darwin published *On the Origin of Species* in 1859. At that point, scientists supposed that Linnaeus's ranked, branching categories might also represent a family tree of evolutionary descent—a second function for scientific names. It was a marriage of convenience, and like others it often proves inconvenient. Molecular biology offers a level of accuracy in drawing the family tree that Darwin never dreamed of, and taxonomists have leapt at the opportunity. Most agree that scientific names should reflect evolutionary descent, but also that the Linnaean-ranked branches are a poor fit for the real tree of life.

To the other parties in the forced marriage (field biologists and field-guide writers and users), it looks like the original function of Linnaean names is ill-served: names are well on their way to being too unstable to facilitate communication. Thanks to the rapid spread of information via the modern internet, online databases often adopt name changes soon after they are published in an obscure journal. In several cases (*Stipa, Tamias, Mustela*), authorities split a long-established name, revert to it 10 or 15 years later, then split it again in a different way.

Some revisions are based solely on DNA analysis, but others consider a range of clues related to evolution. Looking at plants, for example, field biologists may consider insects that coevolved with plants as pollinators, as well as fungal partners and unusual chemical compounds the plant produces. They take a broader view than Linnaeus did when he classified plants by their flower structure. Analyses based solely on DNA may take over in the future, which would make sense and simplify the process—if DNA is to be the last word on evolutionary descent.

Taxonomists are aware of their customers' discomfort. Some try to reassure us: "This is a necessary but temporary phenomenon to correct our unnatural classifications." *Um, when exactly will this phenomenon be done?* From others I hear that the pace of revision

Nested Key Characters and Subsections

Conifers
with bunched needles, p. 50
with single needles, p. 63
with tiny, scalelike leaves, p. 74
that grow exclusively as low shrubs, p. 77

Flowering Trees and Shrubs
deciduous, p. 79
- alternate leaves, and catkins
- opposite leaves
- alternate leaves, flowers not catkins
  › 5 petals, 10+ stamens
  › 5 petals, 5 stamens
  › 4 or 5 petals/sepals, 8 or
    10 stamens
  › composite or irregular flowers

broadleaf evergreens, p. 114
- shrubs taller than 1 foot
- less than 1 foot tall (including cacti)

Flowering Herbs
grasslike herbs, p. 125
- with round, hollow stems (grasses+)
  › bearded grasses (awns at least
    1" long)
  › with narrow flower spikes
  › with ± awnless florets in open
    panicles
  › with short-awned florets in open
    panicles
- with triangular stems (sedges)
- with round, pith-filled stems (rushes)

showy monocots, p. 137
- with 4 or many stamens
- with 6 stamens (lilies)
- with 3 stamens (irises)
- with irregular flowers (orchids)

dicots, p. 153
- without green parts
- with composite flower heads
  › with disk and ray flowers
    (daisylike)
  › without rays (thistlelike)
  › without disk flowers
    (dandelionlike)

- flowers without petals or sepals
- strongly irregular flowers
- regular flowers
  › with 5 petals
    › 5 fused petals, 5 stamens
    › 5 separate petals, 5 stamens
    › 5 separate petals, 10 stamens
    › 15 or more stamens
    › 2 sepals
    › petals/sepals total 5
  › with 4 petals
  › with 3, 6, or several petals or
    sepals

Ferns, Clubmosses, and Horsetails
Ferns, p. 244
- evergreen
- deciduous

Clubmosses and spikemosses, p. 250
Horsetails, p. 251

Mosses and Liverworts
Mosses, p. 253
- upright, fruiting at the tip
- fruiting from midstem

Liverworts, p. 260

Fungi
Mushrooms, p. 266
- with stem, cap, and gills
- with cap and gills but no stem
- with wrinkled undersurface
- with fine teeth under cap
- with a layer of close-packed tubes
  under cap
- with distinct cap without an
  underside

Fungi without cap-and-stem form, p. 278

Lichens
Crust lichens, p. 288
Dust lichens, p. 290
Leaf lichens, p. 291
Shrub lichens, p. 297

Mammals
Eulipotyphla (shrews and moles), p. 306
Chiroptera (bats), p. 307
Lagomorpha (rabbits), p. 308
Rodentia (rodents), p. 313
Carnivora (carnivores), p. 333
Artiodactyla (ungulates and cetaceans),
   p. 352

Birds
Anseriformes (waterfowl), p. 368
Galliformes (chickenlike fowl), p. 372
Podicipediformes (grebes), p. 374
Columbiformes (pigeons), p. 375
Caprimulgiformes (nightjars), p. 375
Apodiformes (swifts and hummingbirds),
   p. 376
Gruiformes (cranes), p. 379
Charadriiformes (shorebirds), p. 380
Gaviiformes (loons), p. 382
Pelecaniformes (herons), p. 383
Cathartiformes (New World vultures),
   p. 383
Accipitriformes (hawks and eagles), p. 385
Strigiformes (owls), p. 390
Coraciiformes (kingfishers), p. 394
Piciformes (woodpeckers), p. 394
Falconiformes (falcons), p. 397
Passeriformes (songbirds), p. 399

Reptiles
Sauria (lizards), p. 424
Serpentes (snakes), p. 426
Testudines (turtles), p. 430

Amphibians
Caudata (salamanders), p. 432
Anura (frogs and toads), p. 433

Fishes
No subsections, p. 436

Insects
Diptera (flies), p. 449
Hemiptera (true bugs), p. 456
Hymenoptera (bees and wasps), p. 457
Coleoptera (beetles), p. 460
Miscellaneous orders, p. 465
Odonata (dragonflies and damselflies),
   p. 467
Lepidoptera (butterflies and moths), p. 471

Other Creatures
Includes extremophiles, algae that live in
snow, giardia, and wood ticks, p. 492

Geology
Sedimentary rocks, p. 521
Intrusive igneous rocks, p. 528
Volcanic igneous rocks, p. 532
Metamorphic rocks, p. 534

will only accelerate as technology improves. Some taxonomists' papers describe our pleas for stability almost as a threat to civilization—"Folk taxonomy weakening 21st-century science" or "'Stable taxonomy' obscures attempts to reconstruct Earth's history" and "weakens the ability to map biodiversity." (Not all scientists studying evolutionary descent are so worked up over naming: some see naming as a distraction, and prefer to publish relationship diagrams without bothering with names.)

Recent taxonomic revisions split a species (or a genus) far more often than they lump two together. In the age-old war between "splitters" and "lumpers," splitters are ascendant. Reflecting that, many current definitions of "species" call it the "smallest" unit, suggesting that varieties and subspecies should all be raised to species rank.

Conservation politics are a hidden hand tipping the scales in favor of splitting. When you name more species, you're seeing more biodiversity, or so they say. More concretely, if you name newly diagnosed taxa at the level of species rather than subspecies, then there are more endangered species, and stronger legal grounds for protecting them. Using that logic, some splitters accuse their critics of shilling for anti-conservation interests. Those critics counter that the rapid proliferation of endangered species (by splitting) will overwhelm species-protection programs and budgets (an understatement if there ever was one) and undermine public trust. Field botanists doing the actual work of conservation will tell you that their hard work is made significantly harder by taxonomic instability. Oregon's top natural heritage botanist said, sighing, "For a while, in the eighties, I knew the names of all the Oregon plants. I don't know when I gave up trying . . . it's just impossible."

Many newly described species are cryptic, meaning that they are identifiable only by genetic analysis—not by anything visible, not even under a microscope.

As a field-guide writer, I think that my readers are interested in the natural history of things we can see out there. It is harder to care about cryptic species. In *Manual of Montana Vascular Plants*, Peter Lesica took the bold step of ignoring cryptic species "as they can't be recognized in the field."

But step back and consider the real world: Nature has never heard of "species." The web of life does not break down into discreet species. If species A and B are related, that means they have a common ancestor. Was the common ancestor species A or B or some other species, X? At what moment did it transform from X into A and B? Necessarily, there was a time when species X had two or more populations on the cusp of becoming A and B. Some such transitions were relatively brief, but some undoubtedly stretched out over millennia.

Our present day is just one snapshot—actually an exceptionally fast-changing moment—in the history of evolution. A portion of today's organisms are in the throes of speciating. While some pairs of related species are just what we want species to be—discreet and non-interbreeding—many others appear to be discreet in some locales, but where they meet up, they act like they're one species: they interbreed and their characteristics intergrade. Some large and diverse genera, like lupines, hold dozens of species, almost any two of which may hybridize where they get the chance. Yet other species pairs are the reverse: they appear to be identical but do not interbreed. Those are cryptic species.

"Family trees" of relationship are also something of an imaginary construct. Evolution can be reticulate—a net, rather than a tree—because lines of descent can converge as well as diverge. Genes can drift from branch to branch.

If plant and vertebrate animal "species" are a fairly amorphous concept, fungal and one-celled species are even more so. Their breeding is more often asexual, and even when it is sexual it often works differently than sex in plants and animals. Bacteria form biofilms that could alternatively be treated as multicelled creatures. Fungi that don't look at all related may be the same species in two different life phases, or taking two forms for reasons we can only guess at. Concepts of the species (and of the individual) that work tolerably well for higher animals and plants don't fit fungi very well. Some mycologists estimate there are 1.5 million species of fungi in the world, of which 5 percent have been named—but doing so may be ill-advised. In a dead-serious article titled "Against the Naming of Fungi," a well-published mycology professor wrote, "It may be more fruitful to abandon the notion of fungal species pending further basic research."

As the scientific project of figuring out evolutionary lineages races ahead thanks to technological advances, the marriage between that project and Linnaeus's project of standardizing names stumbles.

With taxonomic instability and cryptic species proliferating, we "users" of names may find common names more useful than scientific names; we may want to leave the marriage. Mushroom pickers, buyers, and eaters have done so: they call their treasures black morels, yellow morels, and fuzzy-foot morels; they don't send samples off for a costly DNA scan so that they can eat a *Morchella sextelata* specifically. "Folk-taxonomy" common names may increasingly match up with "species groups" like black morels, rather than with single species. (Official common-name checklists handed down by committees don't help much, since they typically insist that common names correspond one-to-one with scientific names.)

But I still like Latin names. At least they're more international than common names. The code regulating them anchors them to particular specimens—something common names can never offer. Biologists continue to use them.

Note that in this book I use the word "also" (short for "also known as") to present other names you may run across for each organism—either common or scientific. The other names are not always synonyms in the strict sense. For example, in many cases, the "also" taxon formerly included populations in the Rockies but then the taxon was split, so the two are not synonymous: the "also" name is still a valid name, but for a species now circumscribed as living outside our range.

Where my translation of a name ends in "—" (an em dash), the name is a root word plus an untranslated suffix or two.

COMMON NAMES

I also like real common names—the ones that came up on their own, through the vernacular. I hate to see common names coined purely for the sake of avoiding italics. If you invented the name, how common can it be? In this book, I include a common name with every species. I do not coin any new ones. In a few cases, I perpetuate a name that charms me, even if I only heard it from my neighbors or saw it in a single, faded book.

Arrowleaf balsamroot

For birds, reptiles, amphibians, and fishes, I follow standard, committee-revised checklists of common names that are broadly accepted, at least in the United States and Canada. (Exception: I eschew two checklist idiosyncrasies—always capitalizing bird and herp names, and compounding snake names like "gophersnake.") The checklists and online floras that I follow are listed on p. 555.

Unfortunately, many common names don't describe their species, but instead assign surnames of individuals, nearly all of them white males and many embodying biases of their day that many people today find offensive. The American Ornithological Society decided in 2023 to fix that: all American bird common names that honor a human will soon be changed, with no attempt to discriminate offensive from virtuous humans. The new name will describe the bird, or perhaps be an Indigenous name for it. (The change won't affect scientific names, nor common names of plant and other species.)

Compared to the situation with animals, plant names are the Wild West. My chief references for plant common names are regional online floras and books—and my own taste. I hate calling a plant a "false"

Abbreviations and Symbols

The following abbreviations and symbols are used in the description paragraphs:

' feet

" inches

° degrees Fahrenheit; degrees of latitude north of the equator

+ or more; and more

± more or less

× by (as in length by width); also lens magnification power; also, in a scientific name, indicates a hybrid

♀ female(s)

♂ male(s)

A found in all eight Rocky Mountain states: CO, WY, MT, ID, NM, UT, WA, OR

A– found in five core states, but missing from either NM, WA, or OR

northerly found widely in MT, ID, WA, OR, and the northern two-thirds of the WY Rockies, but not farther south

southerly found widely in CO, NM, UT, and the southern third of the WY Rockies, but not farther north

alp/subalp in both the alpine and the subalpine zones

avg average

c central

CD Continental Divide

diam diameter (for trees, diameter at 54" above ground)

e east(ern)

ec east central

elev(s) elevation(s)

esp especially

exc except

ID identify; identification; Idaho

incl including/includes

mtn(s) mountain(s)

n north(ern)

nw northwest(ern)

PNW Pacific Northwest

ref refers

RM Rocky Mountain(s)

s south(ern)

sc south central

se southeast(ern)

spp. species plural: any and all species within a genus

ssp. subspecies

sw southwest(ern)

w west(ern)

ws wingspan: the measurement across outspread wings

something, especially when the plant is not a ringer for the one it is said to falsify. But I do use a name starting with "false" if it is the only name in wide circulation.

I have no objection to vernacular names that originated long ago as taxonomic falsehoods. For example, hemlock, the tree, was named for hemlock, the poisonous parsley, just because of a supposedly similar fragrance. But if you ban misnomers, what else are you going to call this tree? And if you can call a tree a hemlock, why would you stop calling a flower a brodiaea just because tax-

onomists decided that, though related, it no longer belongs in genus *Brodiaea*?

If plant names are the Wild West, fungal names are in outer space. Fungal families are so fluid that I decided not to name them in this book. Fungal morphology is turning out, in the light of DNA analysis, to be a stunningly poor predictor of lineage.

Few lichens have common names in vernacular use. However, the excellent tome *Lichens of North America* picked sensible names, which I follow. Unanimous common names for lichens would be a fine thing, as the outlook for their scientific names is dim. The scientific name is that of the fungal partner, not the whole lichen. With different photosynthetic partners, the same fungal species can produce rather different lichens.

PRONOUNCING SCIENTIFIC LATIN

Pronunciations of genus and species names are provided in this guide simply to make Latin names more approachable. I devised no airtight phonetic system; my intent is simply to break each name into units (not always syllables) that you would be unlikely to mispronounce. If you want to pronounce them some other way, feel free. Where I omit the pronunciation and translation of a genus or species, it's either the same genus as the preceding entry or obviously similar to its English translation, like *densa* or *americanus.*

Biologists are far from uniform in their pronunciations. There is an American style and a Continental style. Dr. William Weber in *Colorado Flora* argued that Americans should adopt the Continental style so that taxonomic Latin can be more of an international language. Unfortunately, the two styles are different enough that Americans who adopt Continental pronunciation will find themselves misunderstood during their discussions with other Americans. That said, I wistfully admire the Continental style's consistency: the five vowels are always "ah, eh (or ay), ee, oh, oo," the *ae* diphthong is "eye," *c* is always "k," and *t* has a crisp sound even in *-atius* ("ah-tee-oos"). In contrast, American style Americanizes vowel sounds, both long and short, but sticks to Greek or Latin rules on most consonants and syllable stressing. An initial consonant *x* is pronounced "z," final *es* is "eez," *ch* is "k," *j* is "y," and *th* is always soft as in "thin," never hard as in "then."

Syllable stressing causes difficulty and variation within the American style. The Latin rule says the second-to-last syllable is stressed if its vowel is long, is a diphthong (vowel pair), or is followed by two consonants or by *x* or *y* before the next vowel. Otherwise, the third-to-last syllable gets the stress. Thus *-ophila* is "**ah**-fill-a," but *-ophylla,* thanks to the double *l,* is "o-**fill**-a." When unsure whether the vowel is long, I consult *Webster's Third New International Dictionary,* Gray's *New Manual of Botany* (1908), Jepson's *Manual of the Flowering Plants of California* (1925), or Robbins's *Birds of North America* (1966).

I depart from Latin rules for a few names that have entered the English language: we stress the third-to-last syllables in *Anemone* and *Penstemon.* Native Latin speakers would have stressed the second to last.

In the many cases of proper names with Latin endings tacked on, I try, up to a point, to respect the way the person whose name it was would have pronounced it. For example, *jeffreyi* obviously starts with a "j" sound rather than the "y" sound of the Latin *j.* Sometimes the honoree's pronunciation is just too counterintuitive for us. The surname Menzies is "**ming**-iss" in Scotland, and Douglas is "**doo**-glus," but the scientific names based on them are pronounced American style on these shores.

For the *-oides* ending, I hear "**oy**-deez" today, rather than the "oh-**eye**-deez" I once learned in the classroom of the legendary Leo Hitchcock. I still often hear "ee-eye" for the *-ii* ending, which could nicely be elided into just "ee." Plant families end in *-aceae* with the *a* stressed, and here again most of us streamline, calling the Pinaceae "pie-**nay**-see"; Dr. Hitchcock said "pie-**nay**-seh-ee." Animal families end in *-idae,* with the third-to-last syllable stressed: Felidae is "**fee**-lid-ee." Bird orders end in *-formes,* "**for**-meez." Insect orders end in *-ptera,* with the *p* pronounced and stressed, as in "**dip**-ter-a."

A Grand Landscape

The Rocky Mountains between the Canadian border and Santa Fe, New Mexico, constitute the range of this guide. (See overview map on p. 6.) I'll often refer to our range fondly as "here." It comprises the seven ecoregions that I describe in this chapter. This version of ecoregion boundaries is based on those delineated by the Environmental Protection Agency (EPA).

THE SOUTHERN ROCKIES

Approach Colorado by land from any direction, and you'll have to climb. It's the mid-continent high point. Other Rocky Mountain states have greater local relief in places, but since Colorado's lowlands are high, its mountains are the highest. No other state has a Rocky Mountain reaching 14,000 feet above sea level, and Colorado has fifty-three of them. There are more square miles of flowery alpine tundra here and, of course, more vertical feet of ski runs.

Those fifty-three fourteeners are split up among distinct ranges within the Southern Rockies. Rising directly above the Great Plains and above most of Colorado's humankind, the Front Range is most conspicuous. It boasts the single most imposing sheer cliff—the east face of Longs Peak—and the two Fourteener summits (Mount Blue Sky and Pikes Peak) that see the most people, thanks to paved roads. It has five fourteeners in all, maximum 14,278 feet.

Colorado's higher central spine, the Sawatch Range, includes fifteen relatively broad-topped fourteeners, including the top dog, Mount Elbert (14,440 feet). The Sawatch and Front Ranges are the two biggest exposures of granitic rocks in the US Rockies. Both are Laramide ranges; they played leading roles in our geologic main event, the Laramide orogeny, rising several vertical miles as huge blocks of "basement" rocks. (The orogeny and other geology topics this chapter touches on are explored in detail in chapter 17.)

Northward, the Front Range forks into Wyoming. The east branch is the modest Laramie Range where the Laramide orogeny got its name. The branch west of Laramie is the Medicine Bow Range, whose highest peaks (12,013 feet) are called the Snowy Range because their white, 2.3-billion-year-old quartzites look like year-round snow in the distance.

A subsequent big event, the Rio Grande Rift, split the huge Sawatch block in two, dropping a valley between two north-south faults. The western fault raised the Sawatch while its eastern counterpart raised the Mosquito Range, with its five fourteeners. The Arkansas River originates in the northern section of the rift, west of the Mosquitoes.

The rift's wider southern section holds the Rio Grande. On the rift's east side, faults raise a narrow range 210 miles long: the Sangre de Cristo Mountains. Ten of its peaks are fourteeners. At 14,351 feet, Blanca Peak has only 90 feet to go to beat Mount Elbert, and it may yet get there, as the Sangre de Cristo

Hallett Peak in Colorado's Rocky Mountain National Park

Fault still raises it today. The steep western face of the range includes ancient basement granites, but a maze of faults has broken the basement block into subranges and left much younger sedimentary rocks on the surface across a majority of the Sangre de Cristos.

The Spanish Peaks (13,626 feet) stand defiantly alone, rising from the plains just east of the Sangre de Cristos. They look like volcanoes but are actually a pair of magma intrusions that may have fed a volcano around the time the rift first opened. Sharply defined ridges flanking the peaks reveal a swarm of over 100 dikes radiating from the intrusions.

The Sangre de Cristos extend far into New Mexico and feature that state's ski areas and its high point, Wheeler Peak (13,161 feet). Most geographers consider the range—and the Rocky Mountains as a whole—to terminate at Glorieta Pass, east of Santa Fe.

Twelve fourteeners punctuate a broad, complex range, the San Juan Mountains, covering much of southwest Colorado. The ruggedest San Juan peaks are the Needle Mountains and the Sneffels Range, where a Laramide core of ancient basement rock is augmented by later intrusions. (Mount Sneffels is named after two Snæ Fells, or "snow mountains," on Iceland and the Isle of Man, by way of a silly character in a Jules Verne tale.) In contrast, more than half of the San Juans is volcanic rock from the era of supervolcanoes and giant calderas (p. 509). The West Elk Mountains, a part of the same volcanic complex, rise to the north, across the Gunnison River.

The Elk Mountains proper, northeast of the West Elks, are entirely different. With five fourteeners topping out at 14,265 feet, they are Colorado's finest effort at sculpting sedimentary and metasedimentary rocks. The core area, between Aspen and Crested Butte, comprises the aptly named Maroon Formation. The Elks have granitic rocks as well. In Pyramid Peak, they boast Colorado's peakiest peak (number one in calculated omnidirectional steepness and relief) and its hardest Fourteener to climb, some say.

Both volcanism and granitic intrusions of the Laramide age and younger are scattered across the Colorado Rockies and are central to Colorado's settlement history, because they delivered valuable metals to near the surface. Most Colorado mountain towns began as mining towns.

Some Laramide ranges here lack fourteeners. The Gore (Nuchu) Range is kind of a northern extension of the Mosquito Range, with greater steepness and relief thanks to faults raising it on both sides. That uplift resumes still farther north as the Park Range and reaches Wyoming under the name Sierra Madre. West of these is the broad White River Plateau, which never re-rose high enough to remove its post-Laramide sedimentary cover. Instead, it was capped with flat basalt flows around 15 million years ago. This basalt underlies the Flat Tops and their famous cliff-flanked Devil's Causeway.

Subdivisions of the Southern Rockies ecoregion distinguish between areas of sedimentary, volcanic, and crystalline (granitic or metamorphic) rock. Bedrock does matter, especially to a lot of small plants. Yet despite the diversity of distinct ranges and their bedrock, your basic impressions of plant life will mostly cohere: rather dry stands of ponderosa pine and Douglas-fir at the lower elevations, and Engelmann spruce and subalpine fir at high elevations, all subject to local overwhelm by either lodgepole pine or quaking aspen. (Those six are major species almost throughout our range, not just in the Southern Rockies.) The region's southern half has less lodgepole pine, and more piñon-juniper woodland at its lowest margins. Still higher are the high-elevation pines, the vast subalpine flower meadows, willow thickets, and then tiny-flowered alpine tundra. Brushland is scattered at all elevations

Patterns on Tundra

Alpine areas develop curious small-scale landforms. Most result from permafrost, frozen soil whose year-round average temperature is below 32°F. Above permafrost usually lies a thawed layer in summer; it gets thinner or may freeze up completely in fall and winter. While plants do grow in this thaw zone, the permafrost layer alters conditions by restricting groundwater, plant roots, and mycorrhizae to the thaw zone. With its drainage blocked, the habitable layer may stay sloppy-wet for much of the summer.

While Arctic regions have vast continuous permafrost, permafrost in our mountains is patchy, and not always easy to spot without drilling or digging. Some alpine areas have none. It is more pervasive northward, and it reaches lower elevations eastward, where there's less snow to insulate the soil.

Permafrost on a slope leads to "gelifluction," downslope creep of a partially thawed layer of soil an inch or two thick. This produces long-lasting characteristic gelifluction lobes and terraces, like saggy wrinkles of the ground surface. As a saturated soil surface starts to freeze up under colder air, the last 10 percent that's still slushy gets extruded up out of the ground before it, too, freezes. This "cryoturbation" makes life hard for plants, though some, like dryads, specialize in handling it.

"Frost riving" converts most mountaintop bedrock into loose rocks and boulders. This was traditionally explained by the simple fact that water expands by 9 percent as it freezes. The truth is a little more complex. The action happens not as the temperature drops through 32°F, but at several degrees colder. Thin films of water remain unfrozen well below 32°F in tiny cracks and pores in rock or soil. As some of this water turns to ice, the films migrate toward that new ice and join it, forming expanding lens-shaped slivers, typically parallel to the ground surface. These ice lenses do the work of breaking up rocks or soil, with help from the pressure that they exert and that the watery films transmit.

When the ice lenses work on soil or gravel, they push it around. When this "frost heaving" teams up with gravity, the net result is gradual earth movement downhill, the main

and varies by locale, with Gambel oak and big sagebrush being most common.

Broad, unforested dry basins or plains surround the Southern Rockies, making them especially well defined by dark greens on satellite imagery. The Continental Divide and its National Scenic Trail necessarily continue north in Wyoming, but only on the far side of a broad, nearly flat, semidesert gap. In addition to the entire Colorado part of our range, two dark-green fingers extend north into Wyoming and another two south into New Mexico: the Sangre de Cristos, and then across the Rio Grande Rift the Jemez Mountains, a sprawling, rather young volcano with its central Valles Caldera.

Pale, drab fire scars pock the dark greens. The Southern Rockies—especially in New Mexico—have been hit hard by wildfires of size and severity atypical of these forests' long-term fire regime. In 2022, New Mexico saw its largest fire, by far, in written history. In 2020, Colorado saw its *three* biggest fires in written history. In Colorado in 2002 and New Mexico in 2011, megafires left shockingly large "moonscape" areas. As of 2024, big chunks of these moonscapes show little promise of regrowing forest in the foreseeable future.

form of erosion on gently sloping tundra areas. It helps flatten the "summit flats" that are common from the Beartooths south to the San Juans. Here and there, small bedrock "tors" crop out, helping you to see how much bedrock has been broken and removed.

Frost heaving near the surface squeezes bigger stones out of place, nudging them upward. Flat sedimentary pieces may gradually form a mosaiclike pavement, but some long, thin ones get randomly stood on end like tombstones. Once the surface gets relatively stony, a feedback loop begins: during each freezing cycle, the 32°F freezing front extends deeper under stonier patches; the adjacent sandy patch then pushes its stray rocks toward that stony hollow as if it were "up." The stony parts get stonier. Given enough centuries, on slopes of less than 3 degrees, they form polygonal nets, or sometimes circles. (Some tundra plants form circles simply by growing outward while dying at the center.) On slopes between 6 and 15 degrees, gravity stretches these alignments into stone stripes running upslope-downslope. Steeper than 30 degrees is too steep to pattern.

All the above phenomena, called "patterned ground," adorn Arctic landscapes and many alpine ones. For years, explanations for them were long on intuition and short on data. Now, mathematical models can produce polygons, circles, and stripes by applying frost heaving and feedback loops to permutations of slope, rock texture, and depth-to-permafrost.

Stone stripes on tundra

Mount Timpanogos in Utah's Wasatch Range

THE UTAH ROCKIES

This region is a lot like the Southern Rockies. It has even more aspen, including Pando ("world's largest living thing," p. 80). Its extensive piñon-juniper woodlands add Utah juniper to the Rocky Mountain juniper that prevails in the Southern Rockies.

Utah's mountains are the wet parts of the second-driest US state. The Wasatch Range, especially, exploits orographic precipitation to the max: steep on its upwind flank, it lies downwind of a broad plain culminating in the vast Great Salt Lake, which supplies lake-effect snow. Thus, the Wasatch Range catches a rich bounty of snow that is famously dry, ideal for skiing—but its reliability and quality are likely to suffer over the coming decades. Utah's agriculture and population are utterly dependent on the water caught by the Wasatch Range, and they overexploit it. If Utah fails to sharply reduce its water use, the lake will soon dry up, with consequences going far beyond the mere loss of lake-effect snow.

Mountains here include four high ranges with distinct geologies. Running north-south just east of Utah's metropolitan area, the Wasatches are steepest (no relation to the Sawatches, in Colorado). They are rising today, on a Basin and Range fault, exposing many of the same sedimentary rock formations found elsewhere in the Rockies. A line of smallish granitic intrusions crossing the range near Park City delivered Utah's richest ore deposits more than 30 million years ago.

The Uinta Mountains are the tallest in Utah (13,528 feet) but display milder topography. They are a Laramide thrust arch—the only Laramide range that runs east-west. The thrusting, around 60 million years ago, raised billion-plus-year-old quartzite, shale, and slate to today's mountaintops. Uinta valleys filled up with Ice Age glaciers while the top of the range at its dry east end remained unglaciated, leaving vast tundra summit flats known locally as "bollies." In contrast, the western Uintas feature horns and cirques.

The little-known Tushar Mountains at the ecoregion's south end are actually slightly higher (12,174 feet) than the Wasatches (11,928 feet). Their rocks are volcanic, erupted from a supervolcano 25 million years ago.

Between the Tushars and the south end of the Wasatch Fault stand the Pahvant Range, the Wasatch Plateau, and other mountain groups with sedimentary bedrock, gentle summits, and a top elevation of 11,263 feet.

Don't Love Them to Death

There's one species whose numbers in these mountains keep going up—and that's us, followed by our dogs. While we may cringe at the prospect of ever-increasing crowds—around a third of all visits to US National Forests occur in the Rockies—wildlife and ecological communities like them even less.

Can we envision, and can we achieve, a future that accommodates more mountain recreation without incurring a lot more degradation of the very resources that drew us there in the first place?

Start envisioning by looking at Switzerland, a gorgeous mountain nation whose people love to hike. (The majority of Swiss people name hiking as their favorite sport.) From most European countries you can quickly reach Switzerland by train, and once there, you can easily take mass transit to mountain villages riddled with signed hiking trails. For multi-day treks the Swiss have 152 huts, with 9,200 bed spaces. (Worth thinking about, though I admit I'm not eager to give up sleeping out.)

Unfortunately, the Swiss model does not show that you can have high-density trail use without affecting wildlife. An Alps-like level of trail density would violate the US National Park Service's mission of preserving ecological integrity. If we want to increase the "human-carrying capacity" of our rugged mountains, it must be planned carefully and it must exclude the roadless parts of national parks.

Our Rocky Mountain ranges and our towns are much farther apart than those in Switzerland, and we have precious little mass transit currently, so we obviously aren't going to match Switzerland anytime soon. However, the Rockies' vastness also gives us an advantage over Switzerland: we have room to increase human visitation and still have little-visited areas and corridors for species that need them.

As we consider visiting these mountain regions—burning tons of carbon to get to our destinations—we should remember that the broadest single threat to our flora and fauna is rapid climate change. Thus, as my own explorations of the Rockies neared their culmination in writing this book, I had to face my obligation to look for—and share with you—ways to be less destructive.

The Mountaineers are committed to carbon-footprint reduction and endorse the following suggestions:
- Recreate responsibly: follow Leave No Trace principles, including respecting wildlife and disposing of waste properly.
- Steward the natural environment: protect nature through both small actions and organized stewardship activities.
- Reduce your carbon footprint: consider carpooling, flying less, and using lower-carbon transportation.
- Participate as a stakeholder: take part in land management and recreation planning decisions.
- Build an inclusive outdoors: help make the outdoors safe, accessible, and welcoming for all.

Bison in the Lamar River valley in Yellowstone National Park

These are well-worn parts of the Overthrust Belt. The thrusting was the Sevier orogeny (pronounced "severe")—named for the Sevier River, which originates here—beginning 125 million years ago and extending the length of the Rockies.

THE MIDDLE ROCKIES

The really big Middle Rockies features—distinct mountain ranges, scattered, some near, some very far, separated by big "holes," each range visibly brewing up its own weather on a summer afternoon—seem to make the sky bigger. All of the main eras of mountain-building in the Rockies are well represented here.

As in Colorado and Utah, Laramide arches and faults raised the highest Wyoming range, the Wind Rivers. This broad range has foothills to hide its glories from vehicle-bound tourists, but viewed from one of its peaks, it mixes spires, awe-inspiring blocks, and long, branching tundra-summit flats. A strand of glaciers near 13,804-foot Gannett Peak constitutes about 60 percent of all glacial area in the US Rockies. (*Not* in the Lower 48; that would be in Washington.) The Beartooths (12,807 feet) and the Bighorns (13,175 feet) are Laramide, as are several ranges in south-

west Montana, including the Spanish Peaks, Tobacco Root Mountains, Bridgers, and Ruby Range, as well as an eastern outlier, the Big Belt Mountains, where geologists first described the Belt Supergroup. These ranges expose granite and gneiss bedrock more than two billion years old—the oldest in the Rockies. To get to where it is, it rose more than seven vertical miles long ago on faults.

Basin and Range faulting is at work today on our fastest-rising ranges: the Teton, Madison, Lost River, and Lemhi Ranges. Whereas those last two, in Idaho, are largely sedimentary, the Teton and Madison Ranges rose on the sites of older Laramide uplifts and feature ancient granites and gneisses. Basin and Range faults drop the rock more than they raise it. That is, the Teton fault dropped one side (Jackson Hole) 16,000 feet while raising its other side (the 13,770-foot Tetons) 7,000 feet. Erosion of mountains on both sides filled Jackson Hole with sediment, and then a large Ice Age glacier flowed in from the Yellowstone ice cap, hauling off talus and gouging out lake beds. For the spectacular steepness of the mountain front, we can thank the pitch, speed, and freshness of the fault.

The much older Sevier orogeny shows itself in neatly parallel ridges and valleys composed of tilted sedimentary strata in the Salt River and Wyoming Ranges—Wyoming's and Idaho's part of the Overthrust Belt.

Held up by volcanic heat, the Yellowstone Plateau averages 7,500 feet, higher than any other broad plateau in our Rockies. Each major ice age gave it a huge icefield—at least 4,000 feet thick at Yellowstone Lake. The Yellowstone Hotspot, a present-day supervolcano, is coincidentally superimposed on Absaroka supervolcanoes (ab-**sor**-ka; p. 509) from 55 million years ago. The young and old volcanic soils support similar flora, though the older soils are more fertile. The Absarokas (12,435 feet) are forms carved by erosion from old andesite tuffs. Northeast of the Absarokas, the Crazy Mountains (11,209 feet) rose as a granitic intrusion during the same supervolcano era.

In climate as well, no Rockies ecoregion shows more diversity. Montana's and Idaho's highest ranges—the Beartooths (12,807 feet) and the Lost River Range (12,662 feet), respectively—represent its wet and dry extremes. Counterintuitively, the Beartooths have the more marine climate (more precipitation, more predominantly falling in winter) though they're farther from the sea. Moist winter air flowing from the west gets wrung out twice in western Oregon and then again by the Idaho Batholiths, leaving the Lost River Range very dry. Similarly, western Wyoming's mountains dry the prevailing westerlies before they reach the arid Bighorns (13,175 feet), the ecoregion's eastern edge.

The ecoregion is roughly congruent with a gap in the distribution of ponderosa pine. Limber pine commonly takes its place, then gives way to Douglas-fir upslope. Higher forests mix subalpine fir, Engelmann spruce, whitebark pine, and lodgepole pine, which can be abundant at almost any elevation. As a community matures, fir or spruce theoretically replace lodgepole pine, but lodgepole tends to invite recurring fires, which can perpetuate lodgepole indefinitely (p. 54).

Some southwest-facing slopes in the Lost Rivers and Lemhis are so dry that they have no timberline at all, just grassland and sagebrush or mountain-mahogany steppe from bottom to tundra. They look like Nevada: high fault-block ranges running north-northwest to south-southeast, separated by arid sagebrush flats and alluvial fans. To a geologist (and perhaps even to a botanist) they *are*

northern Nevada: Basin and Range mountains, which ran continuously from here to Nevada before the Yellowstone Hotspot blew up a swath to flatten the Snake River Plain.

The Tetons, the Salt River Range, and the Yellowstone Plateau get more snow than anywhere in Utah or Colorado. Prevailing westerlies that hit them haven't crossed serious mountains for hundreds of miles, so they've absorbed moisture and are primed to drop it. The Teton canyons are moist enough for blue spruce, beargrass, oak fern, woodrush, and thickets of fool's huckleberry. They get some spillover of west-slope snow in winter, and in summer they grow thunderstorms like weeds, while canyon walls enhance shade.

The Wind Rivers, somewhat drier, lack those moist-canyon plants. Their east-side limestones favor lodgepole pine; the west-side granites favor Douglas-fir, showing the influence of bedrock on vegetation. Aspen groves fringe the lower-timberline areas. Three Waters Mountain at the north end of the Wind Rivers drains to the North Pacific, the Gulf of California, and the Gulf of Mexico.

My favorite part of Yellowstone is the broad grassland valley of its northeastern quarter. Find a place away from the cars and get a feeling of the pre-settlement West: scan unfenced expanses looking for bison, pronghorns, elk, two kinds of bears, and wolves. This American Serengeti isn't perfectly "natural"—fear of humans is abnormally missing now, relative to at least eleven thousand years of hunting, and the island of protection may lure large populations into habitats higher than their optimal range—but a thousand years ago, you would have witnessed scenes a lot like this.

THE IDAHO BATHOLITHS

In much of central Idaho, the mountains refuse to line up. They defy individuation and naming as ranges. The entire jumble drained by the Salmon River is the Salmon River Mountains; the entire jumble drained by the Clearwater is the Clearwater Mountains. It makes sense to name these mountains after their rivers: the ridges and valleys were determined by stream erosion—by down-cutting forces, not by uplift or differences between rocks. Glaciers were generally scarce, and most of these mountains are erosional ridges, with long crests broad enough to carry a gravel road for miles. Many are craggy, though: extensive permafrost caused a lot of soil to slough down to the valleys when it thawed shallowly during Ice Age summers, leaving bedrock crags exposed. (More soil washed away during the early 1900s, trampled by eight million sheep hooves each summer—Ketchum was a sheep-shipping nexus.) Few ridges exceed 10,000 feet, but their breadth and the narrowness of the valleys produce an average elevation over 6,500 feet across the batholith, a larger area at this height than any area in Montana.

This may be the only place in the United States where the geological term *batholith* has entered the common parlance: locals call their region "the Batholith." Geologists,

Ecology of Economics

Alongside increased tourism, a parallel population explosion here is even more glaring: the real estate explosion. Americans from all over are buying mountain homes and ranches. Money, it seems, flows uphill toward scenery and clean air.

Amenity migration (i.e., moving to places chosen for their amenities like scenery and outdoor-recreation access) was seen here and studied as early as 1990, but it went into hyperdrive in 2020, when the pandemic showed that many workers can work far from their urban employment. Another accelerant had already been at work: Airbnb, which raised home values because short-term rental rates are much higher than long-term ones.

In addition, during these years income inequality rose sharply nationwide. Not only are most in-migrants well-off urbanites, but large Western ranches became a way for billionaires to flaunt their purchasing power. It's not necessarily just for show: as income gets more unequal, the top 1 percent have a lot more cash they need to invest, and they like to diversify beyond stocks and bonds. Land, being a finite resource, looks like a solid investment.

As of 2022, Teton County, Wyoming (i.e., Jackson Hole), has the highest per capita income in the country, beating second-place Pitkin County, Colorado (Aspen), by more than $100,000. The top twenty counties nationally include five more Rocky Mountain ski towns—Park City, Ketchum, Telluride, Steamboat Springs, and Vail. Their income numbers keep shooting upward. As intrepid journalist Jonathon Thompson writes, "This may be because the rich are getting richer, but also because more wealthy people are moving to places like Jackson and Aspen while the less wealthy are being forced out by rapidly increasing housing prices." While famous ski towns lead the way, lower-profile small towns are also booming.

The boom towns need their restaurant, hotel, ski-resort, housecleaning, and retail workers yet force them into poverty and even homelessness as rents become out of reach for wage-earners. Small businesses may cut back on their hours or their menus for lack of workers. Some shut down. A majority of service workers either camp out, perhaps in their cars—even in winter—or commute tens of miles. But long-commute towns are

meanwhile, split it into many batholiths. (A batholith is any mass of same-aged intrusive rock exposed at the surface over at least 100 square kilometers.) Ancient metamorphic rocks separate a younger northern and an older southern batholith; batholiths along the western margin are older still, formed where exotic terranes (see The Blue Mountains below for explanation) accreted to North America; and about 30 percent of the southern batholith turns out to be a few dozen younger intrusions. A strong pink tinge characterizes one of them, the Sawtooth pluton.

Standing out from the jumble of sinuous ridges are two spectacularly peaked, linear ranges that rose on north-south faults. The Bitterroots on the Montana-Idaho state line reach 10,157 feet, with granitic rocks in their south and metamorphics in the north. Idaho's Sawtooths reach 10,751 feet, get the batholith's heaviest snowpacks, and had sizable Ice Age glaciers, which left an unequaled legacy of alpine lakes. Redfish

also becoming unaffordable for service workers as they fill up with others who can't quite afford Jackson or Aspen. These include amenity-seeking in-migrants and also the elite towns' own mid-income class, now priced out. On top of being a hardship, all this unwanted commuting worsens traffic and carbon emissions.

Many booming towns have programs that provide below-market housing to workers, but no such town yet has met even half of its need. Acutely aware of their housing crises, local voters nevertheless often vote down potential remedies, preferring a fantasy of keeping their town looking the way it did thirty years ago.

Log-mansion sprawl and heavy traffic are obviously hard on wildlife. On the other hand, many billionaire ranch owners donate vast conservation easements. A few work to restore bison ecology. Those are upsides, ecologically, but meanwhile they exacerbate the housing shortage, and even for wildlife their boon is outweighed by billionaires' contributions to climate change, which grow every time they fly into or out of town. (Peer-reviewed estimates vary. To take a middling one, a US billionaire emits as much carbon annually as seventy-two average Americans, or five hundred average humans.)

This grotesque inequity is a throwback to the age of robber barons. It's hard to picture a solution, short of a return to the progressive tax rates of the mid-twentieth-century boom years. Locally, though some state laws foreclose some options, it should be possible at least to build worker apartments subsidized by fees (e.g., imposed on short-term rentals, on resort corporations, or on unoccupied houses) paid by those who generated the boom and who benefit from the underpaid workforce.

Possibly the most ironic sign in the world

Lake, the largest in the row of big moraine lakes at the Sawtooths' feet, is famed for its run of sockeye salmon (now on artificial life support). East of them are the higher White Cloud Peaks (11,815 feet), named for the cirruslike white of their highest granites. Rain-shadowed by the Sawtooths, they had tiny Ice Age glaciers. From the White Clouds north, over an area comprising around a quarter of the entire ecoregion, much rock—both volcanic and granitic—derives from the intense supervolcano era that once roiled the Rockies (p. 509).

Down at the ecoregion's southeast end stands its high point, 12,012-foot Hyndman Peak, part of the Pioneer Mountains' metamorphic core complex rather than of any batholith.

Climate in the western half of the ecoregion is marine influenced (most precipitation is in winter), but less so than in the two ecoregions to the north. Larch, yew, and western white pine are restricted to the fringes, and hemlocks are absent. Coarse soils derived largely from decomposed granite dry out all too well, leaving the region effectively more arid than its precipitation numbers suggest. More than half of the region's acreage burned between 1984 and 2020—more than in any other Rockies ecoregion. Where hotter, lower-elevation sites burned, a return to forest is in doubt.

THE BLUE MOUNTAINS

Wallowa high country fascinates botanists because of its close juxtaposition of limestone, granodiorite, and basalt substrates, each with distinct flora. The limestones, because they are far from other sizable limestone outcrops, have several endemic plants (species found nowhere else). Blue Mountains flora has coastal elements, such as Pacific yew trees, but overall is more like the Rockies than the Cascades.

Hell's Canyon also has endemics, because it's lower than anywhere else in our range: 6,000 awesome feet down from adjacent summits, and running north-south, it collects hot air on sunny days. Slopes in Hell's Canyon and nearby in the Salmon River Canyon display layer upon layer of Columbia River basalt lava, which flooded out from a swarm of fissures all over the region around 16 million years ago. This basalt caps the Seven Devils Mountains and about 40 percent of the Wallowas.

Beneath all the basalt lies a profound anomaly: exotic terranes, or pieces of the earth's crust that originated as volcanic island chains in the Pacific—at least one of them quite far across the ocean (p. 502). These sutured to North America as the oceanic plates they were riding on subducted under the continent's edge. Limestones and shales derive from the islands' offshore slopes. The Wallowa granodiorite rose and solidified when the terranes collided with North America; Columbia River basalt flowed much later.

Starting six million years ago, the Wallowa and Seven

Whitebark pines in the Sawtooths

Unnamed tarn in marble terrain, the Wallowas

Devils Mountains rose 6,000 feet on faults. They are still rising. The Wallowa Fault uplift produced extensive alpine vegetation and glaciated landforms, along with a top elevation of 9,838 feet. Outside of the Wallowas, this ecoregion has three small ranges exceeding 9,000 feet—the Seven Devils, Elkhorns, and Strawberrys.

THE COLUMBIA MOUNTAINS

Northern Idaho is our best area for growing trees—one tree in particular, the western white pine, which in 1911 was said to have the "highest commercial value of any species, wherever found." Its largest specimens ever measured grew here, as do the (much smaller) largest specimens alive today. In between, the species was devastated by white pine blister rust (p. 56), coming on the heels of 1910, the Rockies' worst fire season on record. But for many years, northern Idaho was yielding at least a third of the timber harvest value of the entire US Rockies. Our other high-value timber species also grow well here, in the marine-influenced climate.

Northern Idaho is also good at burning trees up, and has been so for millennia:

charcoal in lake beds indicates that fire has not increased in the region following white settlement. Winters are wet and relatively warm, growing a lot of vegetation, which becomes fuel during the dry summers. In summer, it's a lightning corridor. Burned areas that reburn within thirty years may turn into shrubfields and exclude conifers for centuries. Some high slopes are still recovering from 1910. Still, many valleys support verdant old groves.

In Montana, Belt sedimentary rocks rule this region. Westward, they give way increasingly to metamorphic and granitic bedrock, with Belt rocks cropping up here and there almost to the Columbia at Kettle Falls, Washington—the western edge of the North American continent, to a geologist. Beyond that, it's all exotic terranes. The Okanogan Valley bounds the ecoregion and the range of this book.

The region has relatively few lowland areas; mountains cover most of it, but none of the ranges are household names. The highest, Montana's Cabinet Mountains, reaches a mere 8,738 feet, but its stunningly rugged peaks demand to be taken seriously. (Under the Cabinet Wilderness lies an enormous copper deposit, which may get mined via deep shafts from outside the wilderness. If that happens, endangered bull trout and their pristine streams will suffer. On the other hand, climate change threatens far greater ecological harm, and the switch to clean energy demands desperate quantities of copper. We face many heartrending dilemmas like that one.) Rich mining districts abound in this ecoregion. The Coeur d'Alene River and Lake suffer from western mining's all-too-common toxic byproducts.

While alpine glaciers carved the Cabinets during the ice ages, broad ice sheets nearly overran the Selkirks in Idaho and the Purcells in Montana. Each of them still achieves an alpine timberline and an elevation just over 7,700 feet. If you follow them north into Canada, the Selkirks and Purcells grow into major ranges over 11,000 feet high, with huge glaciers.

Broadly, the region boasts great forested expanses with few people, offering critical habitat and corridors for megafauna of concern: wolverines, fishers, lynxes, wolves, grizzlies, and just possibly—hanging on by a thread—woodland caribou.

THE MONTANA OVERTHRUST BELT

The rocks that were folded and faulted here are sedimentary; none of the other six ecoregions have such a high proportion of sedimentary bedrock, and few places on Earth expose sedimentary formations of equal beauty. In the north, we're talking about Belt rocks 1.5 billion years old (p. 501), emplaced there by the Lewis Thrust Fault, overriding all younger rocks. The southern and eastern edges of this thrust sheet are roughly congruent with Glacier National Park, and that's no coincidence. Mountains in the Lewis sheet are higher and more dramatic because their strata are almost horizontal; farther south the strata tilt, enabling mountainsides to slide into rivers and get carried off.

South of the park, the tilted thrust faults expose Belt rocks in some of the many parallel ranges, and much younger Paleozoic sedimentary rocks in others. While layers of limestone, dolomite, shale, and sandstone predominate in rocks of both eras, the older Belt rocks tend to show at least some degree of metamorphic remineralization and hardening. The Mission Range on the ecoregion's western edge compares well with Glacier:

Lochsa River, a tributary of the Middle Fork of the Clearwater River in Idaho

Saint Mary Lake in Glacier National Park

it's got the Belt rocks and the vertical relief, but not the crowds. It rises on an active extensional fault.

The ecoregion's highest peak, 10,466-foot Mount Cleveland, stands just five miles from Canada. Triple Divide Peak, also in the park, drains to the Pacific, Hudson's Bay, and the Gulf of Mexico.

The Flathead watershed, west of the Continental Divide, has a marine climate somewhat resembling the Northwest Coast's. It is rich in western larch trees, and grows the world's largest specimens. A cousin, subalpine larch, steals the scene at timberline in the fall, turning golden before dropping its needles. Western white pine also appears, as well as both our species of hemlock. Western redcedar joins western hemlock in deep, lush, mossy valley forests. Glacier is the bull's-eye for wet airflow, which even spills over the divide to produce Alberta's one cranny of rainforest.

In contrast, the Montana area east of the Divide has the harshest, most continental climate in our region, and the shortest list of conifer species. Chinook winds scour those slopes in winter and spring, and kill a lot of trees. To us, Chinooks may be a relief from fierce cold, warming the air by as much as 54°F in four hours. But that's no blessing to a tree. The warm, dry wind sucks out the needles' moisture, which the tree cannot replace while its roots are frozen. The needles may all die at once, across entire swaths of trees ("red belts"), and some of those trees will die. Red belts logically ought to favor deciduous trees with no leaves to lose in winter, yet western larch cannot grow in this dry climate, and broadleaf trees aren't much more prevalent than in the west. Some lower timberlines in central Montana consist of limber pines and Douglas-firs that get increasingly bonsai'd eastward, until they are 3-foot bushes you could mistake for juniper.

2

Weather, Climate, and Fire

Weather, climate, and fire all shape this landscape over time.

WEATHER

Weather and climate are basically the same thing viewed on different time scales. Climate is what the weather has tended to do across a period of years; weather is what you can see at one moment in time, and in one place—say, all you can see from a mountaintop.

The Air Went over the Mountain

Hot air rises, right? But the higher you go in the mountains, the colder it is, right? What's going on? Weather reflects the instability of air caught between the conflicting forces of nature observed in those two truisms.

The atmosphere is too transparent for sunlight to heat it very much. Instead, it's the ground that the sun heats up every day, and in turn the ground heats the air in contact with it—the lowest air. As masses of low air heat up, they expand, which is to say they become less dense, or lighter, than the air above them. So they must rise.

As they rise, they become still less dense—not because of heat now, but because of less pressure: they've moved up to where a shorter column of atmosphere sits on top of them, compressing them less than before. The reduced pressure makes air thinner and

colder: the molecules are farther apart; they bounce off each other less often and slow down, which means they have less energy. Unlike a lake, whose water—an incompressible fluid with a distinct top boundary—can stratify, with warmer layers higher, the compressible atmosphere almost always really is colder the higher you get. (In theory, dry air cools 5.5°F with each 1,000 feet gained in altitude. Real-world lapse rates are usually much smaller, varying with moisture and other factors. And, somewhat confusingly, above seven miles up it gets warmer for a ways.)

The rising hot air and the sinking cold air can't make lasting headway against the laws of physics, but that doesn't mean they don't try. Their eternal struggle to turn things around is one way to produce wind. Their minor, temporary truces are temperature inversions: cold air settles under a layer of warmer air, and the air stills. Inversions are common in winter or at night, when the ground is no longer heating up. In contrast, strong daytime heating creates strong hot-air convection upward—conditions that, given enough moisture in the air, leads to thunderstorms.

In general, cold air gravitates toward low places. Valley bottoms have cool, moist microclimates due both to cold air drainage and to having far fewer hours of direct

Storm Warnings

Always go to the high country with enough insulation, shelter, and food to keep you alive, and enough navigation aids and skills to get you out again, should the weather turn bad. High-mountain showers are almost always cold showers; they can arrive as sleet or snow any month of the year.

Mare's-tails are cirrus clouds—the very high, thin, wispy family—arrayed in par-

Mare's-tail cirrus clouds

allel, most or all of them upturned at one end like sled runners. If blowing northward, they may presage a weather front by twelve to twenty-four hours. Broad sheets of cirrus whiteness, if northbound or thickening and lowering, may have the same meaning. However, scattered shreds of cirrus resembling pulled-out cotton puffs are common in good weather.

Lenticular (smooth, lens-shaped) clouds above or downwind of high peaks reveal an increase in wind speed or moisture in the air. These sometimes come and go without producing heavy weather, but more often they foretell rainier weather. Small, puffy clouds sitting all day around the heads of outstanding peaks aren't ominous unless they thicken steadily for hours. Ominous or not, bad weather is ahead if the peak they cap is your goal. Reconsider your plans: the view will be erased, and the wind will be strong and cold.

sunshine each day. The effect is strongest in valleys that run east to west, and weakest in south-draining valleys filled with midday sun. Alpine terrain, at the other extreme, receives copious sunlight and heats up intensely, but its thin air can't hold on to the heat; above tree line, the net daily rise and fall in temperature are much greater than in the lowlands.

The air contained in valleys expands in the daytime heat and contracts at night. The resulting "valley winds" and "slope winds" are Gaia's breath on your cheek. A valley wind is a main trunk flow aligned with the valley, whereas the slope wind is a thin sheet of air moving up or down the flanking slopes. Up in the day and down at night is the basic rule for both, and both winds are strongest in clear summer weather. The valley wind, being larger, lags behind the slope wind: in early morning, the upslope wind begins on valley flanks while the night's downval-

ley wind continues in the valley's center. Occasionally, the flow buffets in fierce pulses lasting a few seconds each, just after sunset, when downslope and downvalley winds join forces.

Our mountains lie within a very broad zone (the north temperate latitudes) of prevailing westerly winds: the average direction of high-altitude winds, clouds, and weather systems is from the west-southwest. Mountains' chief effects on prevailing winds are to keep them out of deep north-south valleys, and to strengthen them across mountaintops and in gaps in the range: air flow speeds up, just as water does, when constricted in a gorge. The high country does tend to be windier than the lowlands—though it can also be calm for hours on end.

Mountains Writing Rain
Mountain ranges with a wet side and a dry side are found worldwide. Mountain ranges

TOP: *Virga or phantom showers* BOTTOM: *Typical fair weather cumulus*

are rainmaking devices—not just barriers between moist marine air and drier continental air.

When air crosses mountains, it must rise and get thinner and cooler. Cool air can't hold as much moisture as warm air. Combine that law of physics with the one about rising air chilling, and mountains create clouds and rain—called orographic precipitation, from the Greek words for "mountain" and "write." The mountains write rain. Moving air meets mountains and is forced to rise, therefore cooling, eventually to the point where it cannot hold the water vapor it held easily before it rose.

Water vapor is the invisible gaseous state of water—individual water molecules evenly distributed in air. When the vapor turns back into a liquid, the molecules join in droplets too tiny to see until they get thick, as clouds, fog, or mist. Cloud droplets, about one-millionth the size of an average raindrop,

are too light to fall. Many stay suspended, warm up, and reevaporate after crossing the mountain crest.

Droplets can collide and coalesce until they're bulky enough to fall as rain, but that rarely happens over continental interiors. Over our mountains, they freeze first. They may become supercooled droplets—liquid droplets at temperatures well below freezing. These can't crystallize until they find tiny solid particles around which to do so (windblown dust, spores, bacteria, smoke, pollution, etc.). Eventually, given continued cooling, a significant number of them do, the supercooled droplets freeze, and then the droplets that bump into them can freeze to them, gradually adding so much weight that they fall as snow. Most raindrops are snowflakes that melted on the way down.

Droplets often remelt and refreeze before they reach Earth; in a thunderstorm, they can be blown upward to refreeze; downward to accrete more rainwater that then freezes on them, growing them; and up and down several more times to grow into hail. A single melting-and-refreezing cycle yields sleet. Layers of ice on a snowflake make graupel—like smaller, softer hail. Distant graupel showers look whitish. As they fall, they may enter warmer air and melt into rain, and then (commonly, in the Rockies in summer) while passing through dry, warm air they may evaporate back into vapor. These phantom showers, called virga, appear as dark, vertical streaks descending from clouds but not reaching the ground.

As a wrung-out airmass descends east of the Continental Divide, the same laws of physics may operate in reverse, creating a Chinook wind (p. 38).

Thermals and Cumulus

It can get wild, this interplay of air-mass movements, orographic temperature gradients, and convection cells of hot air rising off rocky terrain frying in the midday sun—all channeled by mountain topography. A

Mountains Writing Clouds

Air currents arching over a mountain range—just high enough to condense—and then descending create stationary clouds of several distinct types:

- A **cloudcap** envelopes a salient peak.
- **Lenticular clouds**—pure white slivers or crescents with the convex side up (their name means "lens shaped")—can form either directly above a salient peak or some distance downwind of it a little above peak level. Sometimes a few of them stack up over the peak or line up horizontally downwind. In the latter case, picture the airflow over the peak making a series of waves downwind, just as water in a riffle forms standing waves below a semisubmerged rock.
- **Rotor clouds** are puffy clouds in a row, downwind of and parallel to a range. They're pretty much the same thing as downwind lenticular clouds but with stronger wind, creating turbulence. Often you can see a forward-rolling motion, as the tops of the puffs ride on faster winds than the bottoms.
- A **banner cloud** is an eddy that hugs a ridge or a salient peak, just below crest level. Air is tumbling over and down, then eddying back up in the wind-blocked pocket. The upflowing portion chills and condenses into cloud, the same way upflowing air tends to do anywhere.
- **Fractocumulus and fractostratus clouds** are the little wisps that cling all over a mountainside in moist, fairly turbulent conditions.
- A **Chinook wall** (p. 38) of turbulent, heavy clouds aligns over the range crest. It looks threatening, but never descends to the plain nor brings any precipitation. It may extend waterfall clouds that pour over saddles and vanish into thin air.
- A **Chinook arch** is a vast, flat stratus layer over the plains, downwind, with blue sky below its edge.

Lenticular clouds

classic summer day in the Southern Rockies begins perfectly clear, gradually clouds up in the afternoon, climaxes in the form of local showers or even thunderstorms, and clears up again in the evening. This also happens farther north, but less often.

In these cycles, hot air rises off surfaces heating in the morning sun. This effect is strongest where vegetation is sparsest, due to either aridity or high elevation. Wide areas of sparse vegetation create scattershot patterns of rising warm air masses, or "thermals," like the slow-rising bubbles of a lava lamp.

All thermals on a given day in a given area have about the same water vapor content, and for that content level there is a temperature that will force the vapor molecules to coalesce into cloud droplets. The altitude

Lightning strike

where the thermals reach that temperature becomes the floor for a layer of puffy, white, flat-bottomed cumulus clouds, each the turbulent head on an otherwise invisible thermal. The change of state from gas to liquid releases heat, warming the thermal so that it may keep rising.

Where the cloud tops are crisply defined, like cauliflower, the droplets are liquid, even when they are below 32°F. If the rising cloud top gets colder still, it may become a cloud of tiny ice crystals. Crystalline clouds are usually filmy, white, and diffuse edged, and they can make a rainbow-colored sun halo or sundogs (a pair of weak "suns," left and right, mounted on the halo).

Thunder and Lightning

The cloud may keep rising until it hits a stable layer (the thermopause) that halts further rising and blows the uppermost ice crystals streakily out in front, forming an "anvil top." Now it's a cumulonimbus cloud, which can deliver a thunderstorm.

Under different conditions, thunderstorms that form along a range crest may rush down the canyons in the late afternoon; if their cold air slams into moist, warm air at the canyon's foot, it wedges in under the warm air and forces it rapidly upward, creating a new thunderstorm. Several storms can form at once, making a squall line along the downwind (usually east) foot of the range. Large-scale weather fronts also create squall lines as well as prolonged, powerful thunderstorms.

Squall lines may trail sheets of thin clouds that keep things murky and drizzly for hours. More typically, afternoon or evening thunderstorms dissipate into a clear, azure dusk.

Where hailstones or raindrops form a large mass within the cloud, they drag a lot of air with them—icy air from the top of the storm. This downdraft bursts outward as it hits the ground: you can feel (and often see) the blast of cold air arriving in advance of a downpour.

Thunderstorms build up electrical charges, for reasons that remain, well, cloudy. Positively and negatively charged layers form in the cloud, and then locally neutralize each other by means of overgrown sparks, which we call lightning. One type of bolt—cloud-to-cloud lightning—connects points within one cloud. Cloud-to-ground lightning begins as a descending negative leader, exploring randomly for a split second; when it gets close to a salient point on the ground, like a peak or a big tree, a positive streamer shoots up to meet it. They connect, and one or more return strokes, often branching, jet up into the cloud at about one-third the speed of light. Those make the jagged flashes we all love.

Air in their path heats instantly to around 30,000° (at that high temperature, does it really matter whether it is Fahrenheit or Celsius?) in an explosive expansion that we experience as loud noise. If you're close, you hear the full-spectrum *ker-rackkkk!* Farther away—up to 25 miles away—the high pitches fall off and you get a boom or a rumble. Farther than that, we hear nothing and call it heat lightning. Each five seconds that elapse between the flash and the boom, or crack, indicate roughly a mile of distance. At twenty or fewer seconds you should take precautions: lightning may hit your vicinity soon.

Both the storm's chill and its lightning are hazards. When we naturalists try to talk you out of your fear of big predators, we often say your odds of getting eaten are much less than of being struck by lightning. It's payback time: your odds of getting struck by lightning are way too high. You are reasonably safe in a forest, but if you don't have a forest, squat in a dry, low spot without a tree, or a few feet out from the base of a cliff. Little caves under cliff overhangs, unfortunately, are not good, as the current can use your body to span the gap. Dry moss and grass are good insulators; even snow is better than wet rock. Boot soles are insulators, but your hands and the seat of your pants are not, so don't sit. Spread your party out. If anyone develops a blue glow around them or their hair stands on end, they are building an electrical charge that precedes a lightning strike: drop everything metallic and *run* in diverse directions. Give lightning-strike victims immediate rescue breathing or CPR; you may save their life.

The Monsoon

Orographic precipitation of wet air from the Pacific (described above) comes here mainly in the winter, when the polar jet stream is overhead. In the summer, the jet shifts northward, and warmer, drier air masses move in.

While Pacific moisture controls the climate of much of the northerly half of the US Rockies, and does bring the snow for Utah and Colorado skiers, an entirely different weather pattern brings summer rains to the Southern Rockies. Wherever the year's precipitation chart shows a July or August peak within a fairly even (and relatively dry) year-round distribution, the North American Monsoon is in control.

A monsoon is a pattern of increased rain in the summer. The word *monsoon* may make you think of long deluges in India, where the term originated. The North American Monsoon is different. Rain can fall in buckets here, but most often this occurs briefly, locally—not day after day in any one place. "Widely scattered afternoon thundershowers" is the TV meteorologist's refrain. This happens every summer, with or without the monsoon; the monsoon increases the moisture content and thus the storminess.

The explanation of the standard monsoon is that summer sunshine heats the land intensely, creating a broad region of rising hot air, which then sucks in air from a nearby ocean, which does not heat up as much. While that explanation has been applied as well to the American Southwest, it's been challenged by a view of our monsoon as a version of mountain-driven precipitation. Mexico's Sierra Madre mountains play the pivotal role, in this latter view, deflecting the subtropical jet stream southward in summer (as opposed to the *polar* jet stream, which shifts north into Canada for the summer), and allowing warm, wet air masses to come up from the Gulf of California or sometimes the Gulf of Mexico.

Though its heart is in Mexico, our monsoon controls Arizona and New Mexico; southern Utah and Colorado somewhat less. Some people speak of the monsoon even in Montana, in the sense of a daily pattern of clear summer nights and mornings giving way to afternoon showers, exploiting moisture that arrives from the south. But only in the south does it add up to July being a lot wetter than May.

Climograms

To sum up, different rain and snow machines are at work in different proportions from place to place in the Rockies. Three things apply across the board: the mountains get much more precipitation than the lowlands;

Chinook forming a wall and raising dust on the Montana Front

a majority of mountain precipitation falls as snow; and west slopes of ranges are wetter than east slopes, even in ranges far from the Pacific Ocean.

A Pacific Northwest pattern dominates our northwestern ecoregions: the wettest months by far are in midwinter. In contrast, the monsoon pattern dominates in New Mexico and southernmost Utah and Colorado: the summer months are the wettest months, on average, but not by much, and in some years no strong monsoon arrives. In the remaining area, from central Colorado north to southern Montana, a majority of locations record April or May as their wettest months by a modest margin, often with a lesser peak in November. At mid to high elevations, April and November precipitation falls largely as snow.

As it turns out, winter-wet, May-wet, and July-wet climograms (bar graphs of precipitation by calendar month) are scattered across the range, rather than being a simple case of the Northwest versus the South. After all, westerly winds prevail throughout the Rockies. Each range that the Pacific air mass crosses wrings moisture out of it, leaving a drier air mass to flow eastward. The northwest coastal ranges get the first whack

and are the wettest places in the Lower 48. But that wrung-out air does steadily pick up new moisture, even from sagebrush plains. So, as a rule, each range in the Rockies gets precipitation to the degree that it is (a) distant from, and (b) higher than the next range to the west. The wettest stretches of the Continental Divide are at Yellowstone (downwind of the long Snake River Plain) and Glacier (downwind of modest ranges and broad valleys). Ranges as far inland as Wyoming's Medicine Bow Range may create so much orographic precipitation in winter that their higher parts are winter-wet even while valleys on either side of them are not.

Chinooks

The feet of mountains worldwide are subject to eerie spells of warm, dry wind. Each place has a name for them. Here, some Indigenous languages called them Snow Eaters; now we call them Chinooks. They are common in winter and spring, just east of the Rocky Mountain Front, at times when snow is dumping west of the front. In a Chinook, a normal orographic effect—westerlies heating up due to rapidly increasing pressure as they descend an east slope—is intensified by high pressure on the east side meeting low pressure on the west. Precipitation on the west slope pre-warms the air, thanks to another basic law of physics.

Chinook winds can reach 70 miles per hour, and they can stop and restart abruptly, sometimes with brief reincursions of cold air. Temperatures commonly rise 20°F in five minutes. (The all-time records are 103°F in twenty-four hours, from Loma, Montana; and 47°F in two minutes.) Snow gets gobbled up fast, with no visible runoff, because it either evaporates as fast as it melts or sublimates, bypassing the liquid state entirely. Frigid weather typically returns within a few days.

Many people feel the warmth as a pleasure, the wind as a thrill. Others experience migraines, malaise, or mental instability. Farmers dread Chinooks as thieves of their hard-earned precipitation, preventing snow from melting into the soil. Grazing animals throng where Chinooks expose grasses to eat. Trees, especially lodgepole pines, are threatened, life and limb, by red belt (p. 31).

CLIMATE

Any climatic subject you could chart on a regional map is macroclimatic; of equal concern to hikers and other creatures is the climate near the ground, the microclimate. There are also mesoclimatic processes, like slope winds, that operate on an in-between scale.

Mesoclimate

Just as the sun is hotter at noon than in the morning and evening, hotter at the equator than in the mid-latitudes, and hotter in summer than in winter, it heats south-facing slopes more than other slope aspects. Consequently, south slopes have hotter, drier plant community types than north slopes, and worse odds for conifer seedling survival as the climate warms.

In addition to that difference in sunlight angle, a ridge's windward and leeward sides get different precipitation. Although the orographic mechanism (and fog drip, p. 41,

where forested) produces the most precipitation directly above or a bit upwind of the ridgeline, the rain or snow blows downwind as it falls, so peak precipitation may fall just downwind of the ridge. Compounding that effect, snow blows over the ridge crest and settles in the wind lull, often building a cornice that may last and continue to release water well into the summer. Where the prevailing wind is from the WSW, the sun angle and precipitation effects both make the northeast side of a mountain wetter. In our mountains, you'll notice that cirques and the steepest faces (carved in the past by small alpine glaciers) favor the northeast side of the peaks.

The lowest slopes on the leeward side get the least precipitation, but have moist soils, partly because they receive subsurface drainage from higher lee slopes, which get the most rainfall but have soils too coarse to hold on to it. Over time, cool, moist soils are self-reinforcing: they grow shadier vegetation and often suppress fire better, so they retain more organic content, which does a better job of holding water.

The bottoms of steep-sided valleys also get reduced hours of sunlight, and may receive cold air drainage as well (remember, cold air sinks). The effect is strongest in valleys that run east to west, and weakest in south-draining valleys where midday sun hits the bottom head-on. Cold air settles in valleys and slight depressions protected from wind. Caves collect cold-air drainage so effectively that some hold ice year-round. At the other extreme, on high, non-forested peaks and ridges, the thin air can't hold the heat that the surfaces absorb, and tends to be chilly.

Valleys in winter can experience whiteout blizzard conditions without a cloud in the sky, when recent snow remains dry and loose and gets picked up on a windy day.

The warmest level in the mountains is a mid-elevation thermal belt subject to neither cold air drainage nor thin-air heat loss. If you want to sleep warmer, you may gain as

Avalanches

Avalanches come exploding down the same mountain slopes year after year. Some spots get an avalanche every one hundred or two hundred years, while others get several a month. Avalanche tracks (where avalanches recur) and basins (where they run out at the bottom) both produce brushy plant communities without intact tall trees—rich resources for wildlife.

Watch sand settling in an hourglass to visualize a powder avalanche—the resettling pattern of loose, fresh snow on a slope too steep for the snow to adhere. Less common, but more deadly, is the slab avalanche. This occurs where layers of snowpack from different weeks of the winter develop slippery internal boundaries, typically where a snow surface stood in the sun for days or weeks, gradually recrystallizing into larger, rounder grains. This surface got buried, then later recrystallized again as water vapor rose through it—a common sequence, especially in the cold, dry snow of Colorado. At some point, under a variety of weather conditions, a huge slab of subsequent snow up to several feet thick may slip on that recrystallized boundary and go plummeting downslope, quickly accumulating devastating force.

Skiers, snowshoers, and snowmobilers all set off avalanches readily. Each year, many die as a result. Do not venture onto winter snow steeper than a 25-degree slope without first studying avalanche safety.

Will avalanches get worse with climate change? Study results are mixed. It seems that warmer times will mean less snow and therefore smaller or fewer avalanches, but it's complicated. For example, meager snow early, followed by heavy snow in January, is a recipe for slab avalanches.

much as 15°F by leaving a stream bottom and camping on a slightly higher bench.

Microclimate

A microclimate may be much warmer or cooler than its surroundings for several reasons. First, the ground and lakes heat up in the sun, even on cloudy days, and heat the air next to them. High peaks are subject to intense radiation, including heat, thanks to reradiation from clouds, snow, or ice. Dark surfaces heat up to a much greater degree than pale ones. For example, dark, dry humus soil on a high south-facing slope has been measured at 175°F while the surrounding air was only 86°F. For an alpine lichen, seedling, or crawling invertebrate, 86°F summer afternoons may be a fact of life—or death.

Next, vegetation insulates. The tree canopy; the shrub, herb, and moss layers; and the snowpack are all blankets, keeping everything under them warmer in cold weather, and vice versa. The combination of earth-heat retention and snowpack insulation creates a winter-long 30–32°F environment for rodents that neither hibernate nor migrate seasonally. Deer and elk take "thermal cover" in forests during cold spells, but on summer days the forest is cooler than the clearings.

Vegetation and rough topography impede wind. This effect allows cold air collected by sinking (or air heated by warm ground) to stay put longer than it otherwise would.

A forest canopy can make its understory either drier or moister. Throughfall (drippage from a forest canopy) starts and ends later than individual showers in nearby clearings, and falls in bigger drops. Measured during a rain shower, it amounts to less rainfall reaching the soil; much water is absorbed by the canopy and the epiphytic plants on the trees, eventually evaporating again without ever reaching the ground. A light drizzle under a canopy may fail to wet the forest floor at all.

On the other hand, when fog sweeps through the forest canopy, moisture condenses on foliage, and some drips to the ground as throughfall even without any rain. Since low vegetation can catch only a fraction of the fog that a tall forest can, clearcutting a high watershed on the west side of a divide is estimated to immediately reduce its total precipitation by about half.

Rain, fog throughfall, wind, and sun all hit different parts of a tree differently, so each tree offers several microclimates for the small plants that grow on it.

The canopy's effects on evaporation are also mixed. Trees shade the forest floor from the drying sun, but they also suck up great volumes of soil moisture through their roots and transpire it into the air. Given how little rain falls in the summer here, that leaves many understories in our mountains parched, especially on gravelly, underdeveloped post–Ice Age soils. Small plants in competition with overstory trees may be handicapped even more in terms of water than light; many of our sparsest understory communities are found under somewhat open canopies. Non-green plants deal with this by borrowing water back from the tree roots, through fungal lifelines.

Forest cover also limits snowpack depth. Though most snow that settles in the canopy does reach the forest floor, some of it melts first, and most of it, falling in big clumps, is compacted on impact. Winter melting is greater in the forest, as the dark canopy absorbs solar radiation and reradiates some heat downward into the insulated forest microclimate.

In clearings, the bright snow reflects nearly all solar radiation that hits it, and doesn't heat up as much, at least in winter. But when warmer air masses arrive in spring, the canopy insulates the forest floor from this warmth, and many clearings melt out faster. Within meadows, individual trees hasten snowmelt because their heat-absorbing dark-body effect trumps their insulating effect.

Long-Term Climate Change

Over the past 3.5 million years (the Quaternary Period), Earth was much colder most of the time, cycling between ice ages averaging 50,000 years long and interglacial stages averaging 18,000, with abrupt hops up or steps down of many degrees that lasted several centuries. Times before the Quaternary were mostly warmer. Few had either large polar ice caps or alpine glaciers at temperate latitudes.

We don't know how stable the climate was at a century scale—even hundred-thousand-year-long ice ages can barely be detected—but we do see evidence of longer, ancient glaciations known as Snowball Earth periods. These lasted millions of years in the late Proterozoic Eon, freezing perhaps half of the earth's surface, or possibly all of it. Some ancient glaciations may have cycled on and off on a timescale of tens of millions of years. Think of those cycles as a youthful biosphere working out a greenhouse equilibrium: plants flourished until they removed too much CO_2, wrecking the greenhouse and icing the earth; volcanos continued to spout CO_2, unabated, until a strong greenhouse was restored, and so on. Greenhouse gases in the atmosphere—mainly CO_2 and methane— serve as a blanket on the earth, warming it by retaining solar radiation.

Scientists have attributed ice ages to an astounding variety of causes. Changes long-lasting enough to bring on a glacial era would almost have to trace back to either evolution or plate tectonics. Evolution

Dinwoody Glacier in the Wind River Range, Wyoming (shown in 2019), is the largest glacier remaining in the US Rockies

could have produced bursts of bigger, faster-growing plants, depleting CO_2. Plate tectonics opened and closed seaways between oceans many times, turning currents on and off; it reconfigured the continents nearer the equator at times, nearer the poles at others; and it produced epochs of greater and lesser mountainousness. Great mountain ranges reroute global air circulation, or they help deplete CO_2 by accelerating the weathering of rocks, which the ranges sequester as minerals. Any scenario could initiate a feedback loop via ice and snow reflecting solar radiation back into space.

As for smaller nudges to cause the glacial-interglacial cycles within the Quaternary, the mainstream hypothesis involves something called Milankovitch cycles: the intensity of solar radiation hitting Earth varies with at least five aspects of planetary motion (day/night, summer/winter, and three much slower processes). These cycles join forces at calculable intervals, reducing solar heating for long periods. Their compounded effect correlates fairly well with Pleistocene Ice Age ups and downs but seems subject to other influences: salinity-based ocean circulation; CO_2 sequestration and release from peat, swamp, forests, and so

on; methane release from methane hydrates on the seafloor; volcanic dust; extraterrestrial dust from impacts or from belts that the earth may pass through; dust as a feedback loop from glaciation; and the sun's intensity, which varies as described above.

The three warmest periods since the ice ages ended are the Holocene Climatic Optimum nine thousand to six thousand years ago (encompassing the dawn of agriculture and of civilization), the Medieval Warm Period from 700 to 1250 CE, and the present day. Warming will progress for several years even after we reach net-zero emissions, so by 2050 it will be warmer than it has been for at least four million years.

Recent Climate Change

The two principal "greenhouse gases," carbon dioxide and methane, keep the biosphere warm enough to survive. Their levels have been rising for more than one hundred years.

Human activities since the Industrial Revolution have released huge quantities of both CO_2 and methane. However, nature has ways of pumping out CO_2 and methane in quantities that dwarf ours. Another wild card, water vapor, is the most powerful greenhouse gas of all, and heats us, but as low clouds it can reflect heat back into space and cool us.

Global average temperature has risen for at least two hundred years, initially in a rebound from the Little Ice Age and later as a consequence of human activities. (For a timeline including these dates, see p. 506.)

Sea levels have been creeping up for several reasons. As a cause, melting of glacier ice has overtaken the simple expansion of seawater as it heats. A possible additional source is irrigation water as we pump it out of underground aquifers to eventually reach the sea. Ice Age fluctuations changed sea levels by hundreds of feet: if all the world's ice were to melt, it would drown the present homes of hundreds of millions of people.

Abrupt switching between glacially cold and interglacially mild phases lasting from 1,000 to 100,000 years has been the pattern for 2.5 million years. We don't know what triggers these shifts. A changing climate is more likely to flicker—hot/cold, floods/droughts—than to warm smoothly. Given gradual, predictable warming, agriculture might sustain decent yields by moving north, but it might not be able to adapt quickly enough to feed nine billion people if the climate flickers.

The nineteenth and twentieth centuries were something of a Goldilocks period relative to the preceding (post–Ice Age) ten thousand years, during which climate changes likely led to the collapse of civilizations and the abandonment of cities.

Humans carried on during the last ice age, in tiny numbers. There were probably local and even global human die-offs from time to time. Ecological communities thrived in many ice-free areas within a few centuries after abrupt climate shifts—not the same communities as before, of course. One hundred thousand years from now, there will be a diverse biosphere; there may well be humans, if we change our ways fast enough. Earth is resilient over the long run. Nature bats last.

Plant and Climate Feedback

The global increases in greenhouse gases and consequently in temperature would be far greater if it weren't for plants.

In any given year, plant photosynthesis consumes twenty times as much CO_2 as fossil fuels emit. Plants soon release about half of that CO_2, putting the other half into biomass. The biomass half will also be released if it burns or decomposes. The carbon in biomass is thus sequestered, in a sense, but only in the short term. Long-term sequestration of plant carbon only happens when carbon gets carried out to the anaerobic bottom of the sea and buried deep in sediment layers—like the shallow Paleozoic seas that produced today's fossil fuels. (Coal is the fossil carbon rock that comes to mind, but there's actually more carbon in limestone and dolomite.)

Northern bogs are valuable medium-term carbon sinks. Their acidity, coldness, and lack of oxygen nearly eliminate bacterial decomposition. Conversely, as they get warmer, bogs and permafrost areas will decompose and release tons of methane and CO_2, worsening global warming. Colder forests (i.e., northern or montane ones) are longer-term sinks than warmer ones, because cool soil temperatures slow down decomposition. But these sinks will turn into net carbon sources if boreal forests and peat deposits continue to burn up.

Even short-term biomass sinks are important. Global biomass is huge, and it is replenished about as quickly as it decomposes. For the last century or so, it has actually been growing, helping us. Though the causes of this growth are speculative, two are widely credited: parts of the temperate zone, notably eastern North America, reverted to forest after small farms were abandoned in favor of more concentrated agriculture elsewhere; and an atmosphere richer in CO_2 literally fertilizes plants, accelerating photosynthesis so that the plants consume CO_2 faster.

This benign feedback loop is confirmed, but it's limited and not expected to persist indefinitely; in fact, it may have recently halted globally. Trees often cannot benefit because their growth is hitting limits imposed by insufficient water or nitrogen. Increasing drought and fire mortality suggests that forests may already be maxed out in their ability to save us by sopping up our CO_2 emissions.

A study in our region does find tree growth rates increasing overall since 1850, mostly in boreal and high-elevation trees, the ones limited by cold weather—suggesting that the cause of the increase was not so much CO_2 as simple warming. In places where the immediate limit on tree growth is cold weather or a short growing season, climate change will benefit many trees. Conversely, in places where hot drought limits growth, climate change will slow tree growth. In the US Rockies, tree growth in this century may speed up in the north and slow in the south.

Industry-backed voices tell us that cutting "stagnant" old forests and replacing them with "vigorous" fifty-year rotations will help sink CO_2. Close study, however, shows the opposite: Old forests are the best carbon sinks. They are an irreplaceable resource.

While drought and fire may shrink the number of trees in the temperate and tropical zones, the boreal and Arctic zones may see vast new forests. This could help remove carbon from the atmosphere, but in terms of warming, any benefit is negated by the albedo effect: dark foliage absorbs solar heat, whereas bright snow on tundra bounces it back into space.

To be clear, planting trees in the West is vital; tree planting needs to increase. Currently we only produce enough seedlings to replant a small fraction of the area that burns. Holding deforestation at bay within burns is one of two good reasons to plant trees across the West. The second is to speed up the northward and upslope migration of genetic strains (and even species, in some cases) in response to warming. This migration

will occur naturally, but nowhere near fast enough to keep pace with warming. "Assisted migration" in forestry replanting, grounded in extensive study, has been Canadian policy for years, but it is barely done at all on US federal lands. Many ecologists here have cold feet, wary of creating runaway invasions. There is a sad history of intentionally imported species, like kudzu, becoming bad invasives, but no history, nor much likelihood, of this happening with trees planted on their native continent. We're speeding up something that nature will do inevitably anyway.

The wrong reason to plant trees here would be in the vain hope that denser forests would sequester more carbon. In reality, denser forests invite worse fires and worse beetle infestations, nullifying any carbon sink. Instead, *reducing* forest density is a guiding principle for forest management here. (As for mitigating our effect on the global greenhouse, the best ways are to drastically reduce emissions by switching from fossil fuels to clean energy sources, and to use less energy overall.)

Dense, young forests may need to be logged, replanted with a mix of species suited to a range of hotter climates, and then kept thin with prescribed fire and occasional selective cutting. Fire must remain a part of the prescription; western North American ecosystems evolved with fire. Many bird species, for example, require large burnt snags. In contrast to dense, young forests, our high-quality mature forests are few and far between, and fires will reduce them further. Over the short term, old forests should be maintained as libraries of species that may prove useful in a changing world.

The Rockies in a Warmer Future

Over the past century, year-round average temperatures in the Rockies have risen between 2 and 4°F, depending on location—slightly greater warming than the global average. Computer models of climate agree that global warming will accelerate. Modeled changes in precipitation, on the other hand, are less certain, varying from model to model and from place to place.

Since 2003, Gravity Recovery and Climate Experiment (GRACE) satellites have measured the weight of the earth directly beneath them, detecting month-by-month weight gain and loss, which is understood to reflect changes in the regional total mass of water in rivers, streams, lakes, snowpack, and especially ground-water. During a rainier-than-usual season, an area gets a little heavier. (Remarkably, this is further confirmed by GPS sensors detecting a few millimeters of decreased elevation, as the area sags isostatically when it gains weight.) The results from 2003 through 2014 align pretty well with trends predicted by a majority of climate models—mixed though they may be:

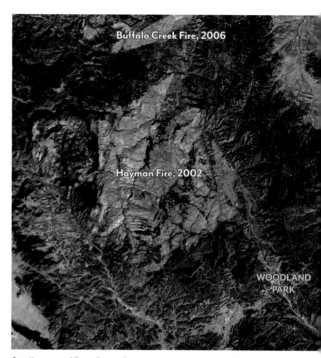

Satellite view of Front Range fire scars showing scant conifer recovery

- Mountains of Montana, northern Wyoming, and northern Idaho got somewhat wetter.
- Our Colorado and New Mexico range got slightly wetter, but the data there are weak.
- Areas of Utah and of southern Wyoming got either wetter or drier.
- Nevada and regions to our south, from California to west Texas, got drier.

Unfortunately, thanks to warming, most places that get measurably wetter get *effectively* drier, for four main reasons.

First, more precipitation falls as rain and less as snow, and the part that does fall as snow melts earlier. Either way, it flows downstream in winter and spring, rather than continuing to water soil and plants into summer, as it did in the past through snowmelt.

Second, much of the West sees increasing numbers of long rainless periods in summer, even in the places that see modestly increas-

ing precipitation thanks to their winters. These dry summers often turn into big fire years.

Third, hotter air in summer sucks moisture out of leaves, and out of the soil in the plants' root zones. Even where GRACE satellites detect increased groundwater, it may be too deep for the plants to use after the summer sun dries the near-surface soil.

Fourth, the increased rate of transpiration from leaves in warmer air means that for plants to avert drought stress, they need more water than before.

The change in the timing of streamflow—more in winter and spring, less in summer—will of course challenge fish and everything that lives in the rivers, though the exact effects aren't well known. And it will mean more floods and more water shortages for people.

Our remaining glaciers will continue to waste away, likely melting away altogether around mid-century. Newly deglaciated valleys may need centuries for soil to develop before they can support glacier lily meadows.

Predictions for subalpine and alpine flower meadows are mixed. Shrubby willows are already taking over from flowers on many alpine slopes in Colorado. A little lower, longer snow-free seasons enable conifers to invade subalpine meadows. On the other hand, some subalpine forests killed by fire or beetles are becoming flower meadows, at least for a while. In theory, some high slopes that are mostly rocks or snow now could see increased flower populations, but that tends to happen extremely slowly.

Invasive species may take advantage of fires opening up a lot of new territory, as well as humans causing enrichment of both CO_2 and nitrogen. Fires are the catalyst for long-lasting conversion of forest to brush or grassland. That happened during early Holocene times of warming, and it's happening today. In the parts of the West where ponderosa pine is common, almost half of the area burned in this century has zero or negligible conifer seedling populations. The current climate on these sites is survivable for pine *trees,* but not for pine *seedlings,* down in the top 2 inches of soil and the bottom 2 inches of air, which often exceed 140°F. Forest establishment at the dry margins of conifer growth has always happened only during rare wave years when a fine cone crop is immediately followed by two or three wetter-than-normal years. That lucky combo could regrow forest on a few burns that look hopeless today, but not on most of them.

The Southern and Utah Rockies already have a very limited ability to regrow conifers after a high-severity fire. They may come to look more like Arizona, with green forests at the highest elevations turning into ecological islands. Small-mammal populations get cut off from conspecifics, becoming prone to inbreeding and local extirpation; if they survive, they become genetically distinct groups.

The northerly parts of the Rockies are still able to regrow conifers after most fires, but this ability diminishes with continued warming. By 2050, they may face odds as bad as the Southern Rockies today. If you want children born today to enjoy forested Rocky Mountains all their lives, I can offer just two courses of action, both political: support forest management including managed fire to foster a higher proportion of future fire burning at low severity, and do all you can to mitigate the release of greenhouse gases into the atmosphere.

FIRE

Rocky Mountain forests are from fire born: fire is the characteristic ecological reset. The oldest trees in nearly every forest stand sprouted on ground cleared by fire.

Fire's influence on forests is widespread but especially dominant in the Rockies. Both to our east and our west, where tornadoes, coastal storms, derechos, and hurricanes blow, windthrow is a greater factor than it is here, whereas lightning is less of one, being infrequent on the West Coast and usually accompanied by rain in the East.

Rocky Mountain forests have probably always had enough lightning fires to account for the fire adaptations we see in our trees. They were additionally affected by Indigenous burning, especially in regions with permanent populations, like northern New Mexico. People set fires for many reasons: to improve visibility for hunting; to foster growth or browse for game; to drive game during a hunt, or enemies during war; to maintain huckleberry patches in subalpine areas; to stimulate new, straight willow shoots for basketry; to send signals; to clear a defensive line of sight around a village; and so on.

In New Mexico where the precipitation peaks in June and July, fierce fires burn mainly before the monsoon arrives, or in years when it fails to. In contrast, north-westerly areas get more precipitation, but mainly in winter; you can pretty much count on those forests getting bone-dry by September in at least one summer out of every ten, which is enough. And you can count on a few dry lightning storms.

In a fire-ruled ecosystem, each tree species has a fire strategy. Pines and larches have remarkable and varied strategies, from the fire-resistant thick bark of larch and ponderosa monarchs to the uncanny reseeding methods of lodgepole and whitebark pines. Western larch and ponderosa pine have adapted to frequent, low-intensity fires confined mostly to the understory. Individual trees may survive for three to seven centuries, their lower trunks bearing many scars that record a fire history.

A low-key fire in Yellowstone National Park

In contrast, when fire strikes stands of lodgepole pine, subalpine fir, or Engelmann spruce, most are killed right through their thin bark, or when fire climbs into the branches and torches the trees. These are stand-replacing fires (a.k.a. high severity—a measure of how much of the forest was killed; intensity, a different measure, involves heat and height). In typical forests of those species, all the trees are about the same age, dating from a few years after the last fire.

If you look at them on a broad-enough scale, most fires are a mixed-severity mosaic because fires are patchy. Even ferocious fires skip over some patches, leaving them unburned; at the other extreme, the tamest ground fires will still torch or kill a clump of trees here and there. (The strongest determinants of fire type and fire patches—even more than the species of trees that grow there—are weather and the quantity and moisture content of the fuels.) The Yellowstone fires of 1988 produced fine examples of such mosaics. This "pyrodiversity" is crucial to the forest's long-term resilience.

When you walk through a recent burn, you can infer the patches by looking at the needles. In some, the needles are still green and on the tree, while the low vegetation may have burned or been skipped over completely. In others, brown needles carpet the ground: that's from a somewhat hotter ground fire—hot enough to kill the trees but not high enough to consume the needles, which fell after things cooled. Sometimes a dead tree's bark is barely even scorched, as it doesn't take much heat to kill our fire-sensitive species. In other patches—after a crown fire, spreading from treetop to treetop—needles are gone, along with the fine twigs. And in some patches the trees are reduced to black snags, signaling a very intense crown fire. Living tree trunks have too much water in them to burn up, but a later fire of sufficient intensity may consume the snags.

We can divide Rocky Mountain history into four fire eras: Native American fire management; the Euroamerican settlement era; a century or so of fire deficit; and the current increase in fire severity. Climate is

Succession After Fire

As a plant community redevelops after a fire, we see patterns, called forest succession, of some species replacing others. Most pioneer plants either sprout from roots or charred stumps or grow from seeds adapted to withstand heat or to get transported abundantly. Their seedlings are quick to tap water, nutrients, and light. Their shade and transpiration create new microclimates. Their roots, in symbiosis with fungi and bacteria, work over the soil physically and chemically, depleting some nutrients and accumulating others. Many pioneers are fast-growing annuals that donate their entire corpses to the humus fund in the fall; perennials and shrubs contribute leaves. The seeds of more diverse and subtle competitors, trickling in on wind and fur and feces, soon find the environment more congenial than it was at first.

Increased shade may be the most obvious trend, but there are other crucial, less visible changes. An entire competitive/cooperative community of plants, animals, fungi, and microorganisms will develop.

Classical succession theory defined fire and other disturbances as aberrations from succession (seen as a path toward a "climax" community), but that view has given way to one that sees constant disturbance on one scale or another as the norm. Not only fires, windstorms, and loggers, but also slower, biotic events like insect pests, diseases, root rot, and dwarf mistletoe redirect successional trajectories.

In the Rockies, aspen and lodgepole pine are post-fire pioneers that may be replaced in succession by more shade-tolerant spruces and subalpine firs. At lower elevations, firs are more shade tolerant than ponderosa pine and can sometimes replace it—but in the natural order of things, frequent fires would stymie that trend indefinitely.

getting hotter and drier; the fuel structure is worse; and the policy of universal fire suppression is over, in theory. In practice we still do tons of firefighting (often unsuccessfully) and just a small amount of prescribed burning.

White settlement initially brought more fire: miners wanted to see the rocks, stockmen wanted more grass, and the Little Ice Age ended, a warming trend that contributed to fire. (The region also had a lot of fire nine thousand to five thousand years ago, when the earth was almost as warm as it is today.) But by 1880, railroads arrived, spelling the end of the frequent fire regime. Once railroads could deliver Western livestock to Eastern markets, sheep and cattle grazed the grassy understories down to nubbins that could no longer carry a low fire.

The conventional wisdom at the time held that "light burning" of the herb and shrub layers was beneficial. By 1900, the newly founded Forest Service challenged that wisdom with a new "modern" and "scientific" view, mocking light burning as "Paiute forestry." Gifford Pinchot, the first head of the service, saw incineration of marketable wood as a terrible waste—which it was his agency's job to prevent. He did not see the unintended consequences of fire suppression. He promoted his cause by smearing his chief opponent, the interior secretary. President Taft responded by firing Pinchot in January of 1910. That summer, northern Idaho erupted in megaconflagrations (1910 is still the record year for human deaths in North American forest fires). The pro-fire faction felt that the fires won the argument for them: the Forest Service can't put fires out, so why try? But the Forest Service went after public opinion, spinning the fires as a horrific tragedy that simply must never be allowed to recur.

Congress joined the fray. Within a year the interior secretary resigned, and pro-fire voices went almost unheard for several decades. Anti-fire PR ramped up, culminating in 1942 with the masterpiece of anti-fire propaganda, *Bambi*. Disney's cute fawn gained an ally in 1944: a shovel-wielding bear named Smokey.

At both the beginning and the end of the Smokey Bear period, effective change lagged several decades behind official policy. From 1910 to 1945, US policy was to put out all fires, but the ability to do so suffered from a lack of roads, funds, and technology. Efficacy improved in the 1930s, and still more after the war ended in 1945: with parachutes, planes, choppers, and funds, fires were put out pretty quickly. Fire suppression in Canada followed a similar arc. Smokey's era began to wind down in 1968, when national parks in the Western United States began letting wildfires burn under certain conditions. The Forest Service followed suit in 1974, at least on paper. Yet years later, the number of fires allowed to burn is still small outside of the national parks.

Smokey's management piled on top of the two previous changes that had already triggered the fire deficit—depopulation of the Indigenous people and overgrazing. We ended up with:

- Increased density, especially in ponderosa pine communities that would otherwise be thinned by frequent low fires
- Species shifts—a decrease in fire-adapted pines and larch; an increase in shade-tolerant grand fir, subalpine fir, Douglas-fir, and spruce; and net increases in noxious invasive weeds
- A drier forest, because thicker forests consume more water, leading to drought stress in trees and less water in streams
- Increases in diseases and pests, notably spruce budworm and pine beetles, which prefer denser, slower-growing forests; root and stem rots, which afflict grand fir and Douglas-fir; and dwarf mistletoe, whose slow advance across the land is broken up by each crown fire
- Decreases in wildlife diversity
- Decreases in soil fertility—at least in the Rockies, more frequent, smaller fires do the best job of recycling nutrients and creating the openings needed by nitrogen-fixing symbiotic plants
- Worse fires and less ability to control them—a run of especially fierce fire years has driven home the fact that fires will get out of control sooner or later, and they're going to be hotter and harder to contain in denser forests and forests half-killed by disease

Continued fire exclusion is not the no-fire path; it's the path to bigger and worse fires, with greater losses of human life and property.

Belief in fire had stayed alive in Indigenous people plus a few rebel scientists, and prevailed again in scientific circles by the 1970s. But the public, the politicians, and most of the timber industry have been slow to join in. Restoring fire-adapted structures is expensive. Once the low-severity fire regime has been missing for several decades, restoring it is a huge challenge. Techniques, funding, and public tolerance for prescribed burning present ongoing challenges, made all the more difficult by anthropogenic change, including invasive species, CO_2 and nitrogen enrichment, and climatic niches shifting northward and uphill. While we work on the politics, megafires keep wiping more forests off the map.

3

Conifers

In western North America, conifers dominate the forests. Globally, this is somewhat unusual for the temperate zone, though it's the norm in far northern forests. Since high elevations have cold climates, high mountain habitat zones often resemble boreal and Arctic ones.

"Conifer" is the common name for the phylum Coniferophyta. Many conifers, including all of those in the largest family, the Pines (p. 51), bear needlelike leaves and woody cones. The Yew family (p. 73) has needles but not anything you would call a cone. In the Cypress family (pp. 73–78), the cones can be either berrylike or conelike but small, and the leaves can either be needlelike (but quite short) or scalelike, tiny, and crowded. (There are additional conifer families, but not in the Rockies.)

Conifers produce resin, or pitch, a viscous blend of aromatic volatiles (terpenes) that evolved as a defense against herbivores and pathogenic fungi. Sticky pitch oozes into holes made by insects, trapping them or forcing them out, or into the wood near wounds in the bark, suppressing fungal attack. Resins provide the aromas of conifer needles and bark. (Don't confuse pitch with sap, the water-based, often sugary liquid that serves the circulatory systems of all plants, both broadleaf and conifer.) The pine family is pitchier than the cypress family, which evolved an additional class of volatiles that do an even better job of resisting insects and fungi.

All conifers are woody: they are trees or shrubs. They produce true seeds by sexual fertilization, but they lack true flowers. ("Flowering plants" denotes all seed plants other than conifers.) The young cones are flower counterparts. Small, short-lived male cones release pollen; young female cones receive it and slowly grow into the familiar woody cones.

Plant categories can be confusing. Flowering trees and shrubs are called broadleaf, even though a few, like heather, have needle-thin leaves, while some conifers, like the bunya-bunya, have fairly broad ones. To a forester or a lumberman, conifers are softwoods—even those few that are very hard, like yew. Evergreen and its opposite, deciduous, refer to whether the foliage remains alive through more than one growing season; people may think of them as synonymous with conifer and broadleaf, but in fact there are deciduous conifers, like larch, and a great many broadleaf evergreens.

CONIFERS WITH BUNCHED NEEDLES

The needles are bunched differently in the two genera of pines and larches: pines bear long evergreen needles in fascicles (bundles) bound together at the base by tiny membranous bracts. The number of needles per bundle (five, three, two, or one) is the easy place to begin pine identification; check a few bundles, since individual trees may be inconsistent. Five-needle pines are a subgenus loosely termed "white pines"; three-needle pines are sometimes called "yellow" or "red" pines. The Southwest has piñon pines with bracted fascicles of just one needle.

Larches (*Larix*), in contrast, bear soft deciduous needles, mostly in fat false whorls of fifteen to forty needles at the tips of peg-like spur twigs of about ¼ inch by ¼ inch. (Technically, the pegs and their whorls are twigs—very short ones, with compressed spirals of single needles, hence "false whorls.") However, on each year's fresh twigs the needles are single and spirally arranged.

PONDEROSA PINE

Pinus ponderosa (**pie**-nus: Roman name; pon-der-**oh**-sa: massive). Also western yellow pine; incl *P. scopulorum, P. brachyptera*. Needles 4–10", in bunches of 3 (or of both 2 and 3 southerly and in e MT), yellowish-green, clustered near branch tips; cones 3–5" × 2–3", closed and reddish until late in their second year, scales tipped with stout recurved barbs; young bark very dark brown, soon furrowing, maturing yellowish to light reddish-brown and very thick, breaking up into plates and scales shaped like jigsaw-puzzle pieces, and fragrant when warm; commonly 44" diam × 175' tall; tallest is 268' (sw OR); oldest is 929 years. Dry, low elevs. Pinaceae. The Montana state tree.

Our drive of the forenoon of [September 8, 1853] was still among the pine openings. The atmosphere was loaded with balm.
—Harvey Kimball Hines

I can almost say I never saw anything more beautiful . . . the forests so different from anything I have seen before. The country all through is burnt over, so often there is not the least underbrush, but the grass grows thick and beautiful. It is now ripe and yellow and in the spaces between the groves (which are large and many) looks like fields of grain ripened, ready for the harvest.
—Rebecca Ketcham

Oregon Trail emigrants like Hines and Ketcham loved the ponderosa forests of the Blue Mountains, and not just because they were easy to haul a covered wagon through. They're gorgeous, and ineffably aromatic in the summer sun—like warm caramel or vanilla, but with an edge. A little less saccharine, a little more toast.[1]

That classic parklike ponderosa stand is a product of frequent ground fires. Old ponderosas typically have fire scars, showing that fires came through at three-to-thirty-year intervals. Picture these as grass fires and brushfires, neither tall nor particularly hot. They weeded out most conifer saplings and some of the bigger trees. Here and there the flames leapt up and torched an old pine, but enough survived to provide all the "yellowbellies" the pioneers saw. Ponderosas are the most fire-resistant trees in their range, thanks to thick bark and high crowns. Since saplings are vulnerable, to reach full size it helps for them to grow well away from other saplings or from too much flammable litter under big trees.

Ponderosa pines don't just tolerate low-severity fires; they foment them. In contrast to puny needles that quickly decompose as duff, long pine needles dry out and persist as quick-flaring fine fuel, either on the ground or as "needledrape" on shrub twigs. Falling needles drape because they fall as three needles bundled at the base. Pine cones lie around just waiting to serve as kindling.

Those classic open stands of ponderosa are uncommon today, because that fire regime ended with white settlement and Indigenous depopulation. Next came sheep that overgrazed, eliminating grass fires. Later came high-grading—selective logging

1. The aroma is subjective. Many of us, in many regions, love the fabulous vanilla scent, but some plant guides persist in denying our noses, telling us that ponderosas never smell like vanilla, only like turpentine. Feh!

Stand of ponderosa pines

Ponderosa pine bark

All too often, logging companies have taken advantage of restoration thinning projects to log more big trees than restoration calls for. A vocal minority of ecologists see restoration as just an excuse to log, and as a cure worse than the disease. To back that up, they pose an alternative view of ponderosa-pine ecology in which crowded, mixed stands like today's—not ponderosa parks and low-severity fires—were the pre-settlement norm. However, the evidence supporting the standard view—starting with the sheer verifiable numbers of big, old ponderosa pines—is overwhelming.

Ponderosas less than a century old have dark gray bark and are called black pine. They don't compare with mature ponderosas in fire resistance or in timber value. Old ponderosas fetched top dollar. The few left today are key to ecological resilience.

While firs can, in the absence of fire, crowd ponderosas out from much of their range, our northerly regions have a ponderosa-only belt near lower timberline where no other tall tree survives. Ponderosas need 12 inches of annual precipitation. We aren't sure why this pine fails to invade the Great Plains, big parts of which do get at least 12 inches. There's a ponderosa gap between Idaho's Sawtooths and the east foot of the Beartooths and Crazies. Those ranges (and others to the east) have ponderosa pine, but within the gap, limber pine takes its place at lower timberline.

One way ponderosa pine deals with heat is by maintaining a high rate of transpiration through the needles to draw cool water up from the soil, cooling the seedlings by as much as 35°F. However, this only works when the roots can find moisture. In moist soil, a ponderosa seed can grow a 20-inch root within a few months. Its seedlings can thus get established during a hot summer, but not a hot, dry summer.

The seeds of all pines are tasty, fatty, and prized by birds and rodents, who bury countless seeds in small caches. They

of the biggest and most fire-resistant pines. Then fire suppression. Stands filled in with true firs and Douglas-firs. Crowded trees are more vulnerable to several kinds of lethal pests, and they carry flame to the ponderosa crowns. Today, forest fires in the range of ponderosa tend to be high-severity, stand-replacing fires. Foresters have figured all this out and begun programs to restore ponderosa forests, using prescribed fire and sometimes selective thinning. These show great promise, though there are kinks to work out and budgets are grossly inadequate to treat all the lands that need it.

Trees and Climate Change

Climate change is already affecting our forests, and these changes will only accelerate. Species will adapt to warming climate by shifting northward and upslope. During the ice ages, there were cycles of sudden warming that may have been at least as fast—Greenland is thought to have warmed 29°F in fifty years! Warming was slower down at latitudes that had trees. Species migrated northward, but the process was chaotic, taking a thousand years or more for species to recombine into lasting communities.

Predictions about where each of our tree species will grow in a future climate are astonishingly inconsistent. Many are studies of the climate envelope—the climate in which a species lives today—that look at the exact climate each tree is found in and plot where that climate may exist at various future dates under various climate models and scenarios. A few areas that get wetter might see conifer forests replace dry steppe vegetation. Some models foresee western larch, western hemlock, and redcedar thriving in the wettest parts of northwest Wyoming. But I expect shifts toward lusher forest to be rare.

Don't bet the farm on those predictions. There are too many wild cards, like unpredictable trends in insect pests, and the high uncertainty in modeling future precipitation. On top of that are the interactions between organisms. How fast does this tree migrate relative to its competitors? To its mycorrhizal partners (p. 263)? To potential new invasive species? For long-lived species, will individuals that already tower over the competition be able to live a normal life span after their climate envelope shifts away from them? The climate envelope tells you where the species currently competes successfully, but cultivated trees (protected from competition) thrive in a much wider range of climates, showing that competitors are the main factor keeping natural ranges small. In sum, future plant communities, like present ones, will result from hard-to-predict intersections where the trajectories of climate and of many interacting species meet.

Climate-envelope studies do offer some guidance on where to plant species northward. Many plantings may fail; some may succeed, multiply, and replenish the earth.

Aside from the trees invading some subalpine meadows, the predicted shifts within our range aren't clearly happening yet. However, there is somewhat more certainty regarding the general effects of warming:

- Faster growth with CO_2 enrichment (p. 44)
- Faster growth with longer snow-free seasons
- Slower growth and higher mortality with longer hot/dry seasons (drought)
- Pests extending their ranges
- More fire

As far as a plant is concerned, our region is getting drier even where annual precipitation increases (as explained on p. 45).

The worst pest of Douglas-fir and grand fir, the western spruce budworm, mounts outbreaks that tend to follow droughts. The overall worst recent insect epidemic—mountain pine beetles since 2000—swept into areas that had until now been too cold for this pest to mount epidemics, including higher elevations and much of British Columbia.

Where freezing and snow have previously been the limiting factors, most plants should benefit from warming; trees should grow a little more quickly. Over much larger portions of our range, drought and summer heat are the limiting factors. With climate change, drought and heat will more than offset earlier growth in spring. The growing season will be shorter on these sites, tree growth will slow, and drought-stressed trees will be more vulnerable to pests and fire.

intend to come back for them someday but inevitably overlook some caches, a few of which germinate. Ponderosas evolved spines on their cones to discourage seed eating, even though it appears that pines benefit from it: critters plant seeds where wind might never carry them. They also plant them deeper, in mineral soil that's often in litter-free spots, sparing the seedlings from drought and the eventual saplings from ground fire. If you see a clump of several pine seedlings within a square inch, it's a forgotten cache.

LODGEPOLE PINE

Pinus contorta. Needles in 2s, 1½–2½", yellow-green; cones 1½–2", egg-shaped, point of attachment well off-center, scales sharp-tipped; cones abundant, long-persistent on the branch, either closed or open; bark thin (less than 1"), reddish-brown to gray, scaly; commonly 20" diam × 100' tall; biggest (of our subspecies) is 41" diam × 135' tall. Ubiquitous exc s CO and NM; A–. Pinaceae.

Lodgepole pines are tricksters. They don't bother to compete with other conifers in size, longevity, shade tolerance, or fire resistance. They excel instead at rapid growth early in life, copiously produced and cleverly designed cones, and tolerance of any kind of soil. Prolific to a fault, they produce both pollen and seeds prodigiously year after year (a rarity among conifers). Their pollen drifts like an amber fog over midsummer's meadows. Lodgepoles release seeds at all times of year; they bear cones at five to twenty years of age, younger than other conifers; and their saplings grow fastest.

They may be even stronger competitors as atmospheric CO_2 increases. Outdoor experiments with high CO_2 in North Carolina found that fast-growing pine species grew even faster and began producing cones at even younger ages. Lodgepoles hold promise for northward afforestation as the climate warms, but that might require planting; they cannot keep pace with climate on their own.

Lodgepole pine sealed seed cones

Pollen cones and new twig growth

Their range will shrink at its trailing edge, in Colorado.

Lodgepoles unleash their signature punch after a fire: some of the cones on many lodgepole pines are sealed shut by a resin with a melting point of 113°F. The seeds inside, viable for decades, are protected through all but the hottest crown fires by the closed cone. While killing the pines, fire melts the cone-sealing resin, and the cone's scales open slowly, shedding seeds on a wide-open field. Lodgepoles can produce both cones of this type (called serotinous) and others that open and release seeds as they mature, in proportions varying by region and by age:

serotiny is common only after age thirty. Complete nonserotiny is common where poor conditions discourage other trees so well that lodgepole dominance can persist without fire—for example, in parts of the Idaho Batholith where the soil is coarse granite sand that scarcely retains water.

As in rabbits, prolificacy leads to overpopulation: a doghair stand—one stunted by its own extreme density. In this all-too-common circumstance, the speed demon lodgepole slows to a near halt. It looks dismal, but isn't bad in terms of species survival, so long as the doghair stand is able to mature and produce cones before the next fire. Some ecologists warn of lodgepole deforestation where fire cycles get shorter than fifteen years. A second Achilles' heel would be if fires burn so intensely that they consume all the cones, saplings, and seeds lurking in the duff. That intensity was seen where fire ripped a Colorado stand that had suffered 89 percent mortality from beetles four years before. Where aspen is already present in these situations, it may be able to take over and sustain a forest ecology of a different style.

Lodgepole pine is the most common tree in the Rockies from northern Colorado north, a local dominant tree in much of the spruce-fir elevation zone. It can grow anywhere from lower to upper timberline, but is rarely seen as alpine krummholz (prostrate shrubs).

Where Chinook winds blow, on the Rocky Mountain Front in Montana, lodgepoles are susceptible to red belt (p. 31), but they still do at least as well as any other tree.

The epidemic of mountain pine beetles (p. 462) between 1999 and 2013 killed around half of the mature lodgepoles over much of the species' range. The mortality was shocking in Wyoming and Colorado, and perhaps even more so in parts of central British Columbia that had formerly been too cold in winter for this beetle. By 2009, that was an area of dead trees the size of Wisconsin. Remarkably, lodgepole-pine dominance usually persists after a massive beetle kill: often the surviving saplings are lodgepole, and if not, there may be plentiful seeds dropping from the dead overstory.

Lodgepole's inner bark layer, or cambium, was an important Indigenous food. Too thin and dry to eat in winter, it plumps up in May and June when sap flows and pollen flies. Sweet and moist as well as nutritious, cambium was a treat at that time of year when fresh berries were only a memory. Though best fresh, it could also be dried and stored. Women stripped the bark using bear shoulder blades, deer ulnas, or juniper branches, and then removed the cambium (from either the wood or the bark, depending on the season) with a scraper made from another bone or a sheep's horn. To avoid killing the tree, they scraped a big patch from just one side. Ponderosa pine, western larch, and western hemlock all provided good cambium, but the abundant lodgepole provided the most. As its name records, it was widely used for teepee and lodge frames.

TWO-LEAF PIÑON PINE

Pinus edulis (**ed**-you-lis: edible). Needles mostly in pairs, ¾–1⅝" long, rather stout and stiff, blue-green with pale stomatal stripes; cones 2", broad, almost spherical when open, often with crusted exuded pitch blobs; cone scales thicken toward tip; bark red-brown, scaly, furrowed; broadly conical to rounded small trees or shrubs. A champion in NM was 68" diam × 69' tall; one reached 973 years in UT. Often with juniper, at lowest elevs that support trees; southerly. Pinaceae. The New Mexico state tree.

Lower timberline is the ecological edge where forest gives way to grassland or steppe because the soil is a little too dry for forest. In the southern half of our range, lower timberlines are the province of piñon-juniper woodlands—"P-J" for short. (In the northern half, it may be ponderosa or limber pine.) Four species of piñon pine (peen-**yawn**) and at least six species of juniper hyphenate with each other. The piñons divide the Southwest into territories, so

Two-leaf piñon pine

umbrella pine, *P. pinea.* Piñon nuts are every bit as tasty, but on the smaller side.

Over the past century, P-J (or junipers alone) advanced across great expanses of sagebrush steppe. Ranchers see this as an existential threat to grasses, and thus to ranching, and have destroyed a great many junipers and piñons to enhance grazing terrain. Ferreting out a few centuries of history, ecologists find there were multiple sweeping invasions, and equally sweeping diebacks. This is no surprise, given that the P-J ecosystem requires just a tiny bit more moisture than the sagebrush ecosystem, and could respond to small changes in climate.

A major dieback arrived in 2003, as the Southwest's megadrought got underway. In some P-J stands, essentially all the piñons died while the junipers were okay. Blame this at least partly on the piñon ips, a bark beetle that lay waiting for drought to weaken its hosts.

most piñon-juniper communities have one or two juniper species and just one piñon. In the Southern Rockies, it's generally *P. edulis.* In the Wasatch Range, it is likely to be single-leaf piñon pine, *P. monophylla,* with most needles borne singly; otherwise that tree grows farther west. Two other piñons grow to our south.

Piñons all have large, oil-rich seeds, a crucial food resource for Puebloan peoples as well as many animals. Pinyon jays, Mexican jays, Steller's jays, scrub jays, and Clark's nutcrackers all go for the pine nuts and serve to disseminate them. This is much like the symbiosis between whitebark pines and Clark's nutcrackers, except that piñon pine isn't so dependent on a single bird species, and its cones don't require a bird to open them. Pinyon jays average fifty-six seeds at a time, carried in a pouch below the tongue, and carry them several miles. The most effective disseminators among the five jays, they are in sharp decline, posing a threat to piñons. In a part of New Mexico with numerous gas wells, studies found that the constant din of gas compressors drives scrub jays away, reducing piñon seedlings by 75 percent.

Most grocery-store pine nuts come from Asian species; the huge Italian ones are from

WESTERN WHITE PINE

Pinus monticola (mon-**tic**-a-la: mtn dweller). Needles in fives, 2–4" long, blue-green with white bloom on inner surfaces only, blunt-tipped; cones 6–10" × 2–4", thin-scaled and flimsy for their size, often curved, borne by a short stalk from upper branch tips; young bark greenish gray, maturing to gray with a cinnamon interior, cracking in squares; commonly 3' diam × 120' tall; biggest living tree is 6'9" diam; tallest is 232'. n ID, w MT, WA, OR.

Western white pine once dominated vast forests, and the lumber trade in northern Idaho, and was named the state tree. It is sadly diminished today. Many that you now see in natural forests are young, and sick. The Idaho Giant, 210 feet tall and 80 inches in diameter, succumbed to bark beetles in 1997. Plenty of larger trees, up to 101 inches in diameter, succumbed before 1955.

Commercial success led to this evil fate, via an introduced fungus: white pine blister rust. America's logging industry—after feasting on eastern white pine, *P. strobus,* until that

Western white pine

breeding. But blister rust can develop its own counter-resistance, and natural selection may spread that too. In nature, species and their enemies coevolve over long periods, with selection ultimately discarding genetic strains that fail to develop mutual survivability. The main reason we have so many catastrophic pests (and weeds) in modern times is that all our trade and travel continually make new bad matches between pests and hosts. When it comes to living organisms, free trade is a terrible thing.

WHITEBARK PINE

Pinus albicaulis (al-bic-**aw**-lis: white bark). Needles in 5s, 1⅝–3" long, yellow-green, in tufts at branch tips; cones 1¾–3" long, egg-shaped, purplish, dense, long persistent on the tree, never opening unless forced open; cone scales thicken toward tip; pollen cones red; bark thin, scaly, superficially whitish or grayish; commonly 20" diam × 65' tall; tallest is 90'; greatest diam 110"; oldest is 1,270 years. Alp/subalp; MT, ID, WA, OR, and w WY. Pinaceae.

With their broad crowns and tufted, paler foliage, whitebark pines are easy to tell from the other high-country conifers—even in death.

species was depleted—was thrilled to find a bigger white pine species in the Northern Rockies. (The hottest demand, oddly, was for wooden matches in the 1920s, thanks to a low content of crackling resins.) Western white pine production peaked then. As the pines were logged, demand for replanting stock grew so fast that foreign nurseries entered the market. A 1910 shipment of French seedlings to Vancouver brought blister rust, *Cronartium ribicola,* a European disease that was not a big problem in Europe.

Since the rust fungus requires a currant or gooseberry plant as an alternate host, *Ribes*-extermination programs (p. 105) went on for decades. They proved futile. Western white pines died off almost as inexorably as American elms and chestnuts—each a victim of a different European fungus.

Some white pines today grow big enough to produce seeds, sustaining a small, scattered population. Natural selection should increase the number of rust-resistant trees, and foresters assist that process through

If you see a subalpine ghost forest with whitened, forked, crooked dead tree trunks towering over young spruces and firs, that was once a fine whitebark-pine grove. These ghost trees may stand for decades, giving an exaggerated sense of recent death, yet it's true that we're looking at a catastrophic decline of whitebark pines. As of 2023, they are listed as a threatened species in the United States. The culprits are pine beetles (p. 462) and the introduced white pine blister rust disease, abetted by fire suppression and climate change.

Since cold suppresses both pine beetles and blister rust, warming exacerbates their threats. Beetles ravaged whitebark pines in a warm phase around 1930, then were killed off in whitebark habitat by the cold winter of 1933. After 2000, the warming climate

Stand of whitebark pine

Whitebark pine

cached in suitable soil and then forgotten. This enables whitebarks to rapidly recolonize large burns where wind-disseminated trees can only crawl back, generation by generation, from the green periphery.

Nutcrackers cache as many as fifteen pine nuts together. When several nuts germinate, they may grow as a clump, or they may fuse together at the base and appear to be one multi-stemmed tree. In the alpine zone, whitebark pine grows as krummholz; at its lowest elevations, it may grow straight and single-stemmed, resembling lodgepole pine.

Nutcrackers came to North America from Asia only two million years ago, perhaps bringing whitebark pine's ancestors with them. While whitebarks were coevolving with nutcrackers, their European relatives, Swiss stone pines and Eurasian nutcrackers, evolved a very similar mutualism. Key elements in both are, on the trees, big, fat-rich seeds and cones that don't open unless damaged; and, on the birds, beaks and muscles to force cones open, and food-caching behavior. Whitebarks and stone pines have traits with no apparent adaptive value other than to accommodate birds. Their cones, for example, grow on vertical branches near the top of the tree, easy for birds to see and work on.

Nutcrackers are not as dependent on whitebarks, however. Unlikely to return to an area where whitebarks become too scarce, they will turn to ponderosa or Douglas-fir seeds. Humans will need to lend a hand in the replanting effort. Forest Service nurseries across the West are hard at work identifying and reproducing the whitebarks (and other white pines) with rust-resistant genetics. About 50 percent of these nursery offspring inherit the resistant genetics—a much better rate than wild populations. Thanks to the nutcracker, restoration plans can focus on planting a lot of whitebarks in strategically located core areas scattered across the range, requiring fewer seedlings than an across-the-board plan. Then the birds can replant in between.

meant that few, if any, US whitebark stands were too high or cold for pine beetles. A hard freeze early enough in fall can still wipe them out, though, as happened in 2009 in parts of Montana.

Whitebark pine nuts travel on "adopted" wings, flying as far as 20 miles in the throat pouches of Clark's nutcrackers, who then cache them to retrieve later. The cones remain stuck to the branch, and their heavy, wingless seeds wouldn't go far in the wind even if the cone did open. Nearly all whitebark seedlings originate from the tiny fraction that are

Red squirrels and both black and grizzly bears eat tons of pine nuts. Bears get them by robbing squirrel middens, or sometimes by breaking off entire branches. Grouse find dense whitebark crowns cozy in winter.

LIMBER PINE

Pinus flexilis (**flex**-il-iss: flexible). Needles in fives, 1½–3" long, yellow-green, in tufts at branch tips; cones 2–6" long, pointed, green ripening to russet brown, falling whole after a year or more on the tree; cone scales thin toward the tip; pollen cones yellow-green, may mature reddish; young bark thin, scaly, grayish, eventually darkening; ± crooked trees 12–50' tall; often multi-stemmed due to nutcracker caching (see above). Champion, in UT, is 96" in diam, 61' tall; oldest perhaps 1,670 years. Diverse, dry rocky sites; A–. Pinaceae.

Near Fort Collins, Colorado, limber pines grow at alpine timberline (elevation 11,155

Limber pine

feet), at lower timberline (5,250 feet) on the Great Plains nearby, and at sites scattered in between. No other conifer here displays such elevational range. Geographically, they range from north of Calgary to south of Palm

Corkscrew Trees

Notice the old, barkless dead tree trunks with sharply spiraled wood grain. In any one species, most individuals lack this spiral, but most of the twisty ones all twist in the same direction—to the right upward. Spiral grain appears mainly in harsh mountain environments as trees get old: it's only in the outer portion of their wood, which grew in advanced age.

The one scientific paper devoted to explaining corkscrew trees points out that a large majority of asymmetrical tree crowns twist in the direction the prevailing wind pushes them. Our broad prevailing wind direction is west-southwest, and our trees tend to grow a little more foliage on their south sides, where the sun hits them. (The opposite applies in the Southern Hemisphere's temperate zone, where most conifers twist to the left.) Our limber pine is an exception, twisting left. Perhaps the key is simply the broader point that spiraling makes the trunk stronger, an advantage in windy places.

Another hypothesis holds that spiral grain distributes sap (water and nutrients) from each root to branches on all sides,

Dead limber pine twists to the left

rather than only to the branches directly above that root. This would be an advantage where age and environmental hard knocks have damaged the vessels on one or more sides of the trunk.

Springs—also impressive, though surpassed by Douglas-fir and ponderosa pine.

Given such climatic range, ecologists have a hard time projecting where limber pines will thrive in a warmer world. They could be major winners, but that outcome is threatened by white pine blister rust, mountain pine beetles, dwarf mistletoe, and drought itself. Blister rust was late to invade Colorado, and the limber pines there show a greater prevalence of genetic resistance to it than most five-needle pines, so scientists have written up detailed, proactive plans hoping to preserve them. Alberta, on the other hand, lists them as an endangered species, and they've suffered severe mortality in Glacier National Park.

Limber and whitebark pines look much alike; limber, however, has much longer cones and they open at maturity. (Greenish pollen cones, vs. red on whitebark, works pretty well in the Rockies.) For young trees with no cones, even specialists find it almost impossible to tell the two trees apart; when doing research, they'll test DNA. The pine nuts are as big as those of whitebark, and as attractive to bears; their importance in bear diets hasn't been quantified. They play similar ecological roles in the alpine, including the mutualism with Clark's nutcrackers (p. 402). Near Fort Collins, limber pine shares lower timberline with ponderosa pine, but at lower timberlines in Montana, Wyoming, and eastern Idaho, it seems to thrive mainly in ponderosa's absence. On windswept, soil-poor ridges, it extends fingers into midmontane and subalpine forests. In sum, limber pine grows in an amazing range of climates, but can outcompete other conifers only on very dry, sandy soils. Ecologists would say that its fundamental niche is far broader than its realized niche.

From the San Juan Mountains south through New Mexico, limber pine hybridizes and intergrades with the Chihuahua or southwestern white pine, *P. strobiformis*. The trees there tend to be taller, slenderer, and more vertical than typical limber pines, and are generally though of as southwestern white. Recent genetic studies found limber pine genes predominant in all the Colorado hybrid trees analyzed, and pure *strobiformis* only in Mexico. The authors professed hope that this genetic mixing, which enhances limber pine's summer drought tolerance, could help the trees adapt to warming.

ROCKY MOUNTAIN BRISTLECONE PINE

Pinus aristata (air-iss-**tay**-ta: with bristles). Needles in fives, 1–1½" long, dark blue-green with a thin line of white bloom, sharp-pointed, older ones flecked with pitch; young branches like bottle brushes with many years of accumulated needles; cones 2–4½", initially purplish; cone scales thicken toward tip, which bears a prickle; bark gray, aging red-brown; small, contorted tree. Greatest girth 44"; greatest age by ring counts is 2,570 years. Near upper and sometimes lower timberlines; CO, NM. Pinaceae.

The Rocky Mountain bristlecone is a remarkable tree that few people see. Longevity and tolerance of cold are its exceptional traits: it lives at the highest elevations that can support trees in Colorado and New Mexico.

Until 1970, the species *P. aristata* also included the Great Basin bristlecones of California and Nevada (now separated as *P. longaeva*), which boast the world's oldest plants, exceeding five thousand years. Prior to 1970, scientists studying the species mostly looked at California trees, leaving the Colorado type poorly known. With its very small, mostly high-elevation range, it's clearly vulnerable to climate change. Exactly how vulnerable is less clear; a few stands grow at lower elevations in Colorado, next to Douglas-firs and ponderosa pines, and it does well for decades in cultivation at still-warmer urban sites. We infer that it can survive in a fairly wide climatic range, but it competes so poorly with other conifers that we only find it on sites that stress them, at the margins of where tree growth is possible. Perhaps in a future climate it will exploit unforeseen opportunities. One major roadblock: though today's populations are almost all healthy, it's susceptible to white pine blister rust (pp. 56–57), and likely to suffer rust mortality in the future. Bristlecone saplings are growing at one Colorado site 100 feet higher than any other conifers, demonstrating tree line advancing in response to warming.

Wings on the seeds suggest that wind is their main disperser, but birds also serve that role at times, as they do for whitebark and several other pines.

Rocky Mountain bristlecone pine

One longevity strategy it shares with *P. longaeva* is called strip bark growth. During exceptional droughts, a sector of the tree can die—a branch, a root, and the vertical swath of inner bark that conducts sap between them can all die, together, and be walled off, leaving the rest of the tree unharmed. That strip of bark will fall away eventually, exposing dead wood, but the tree survives as long as there's even a single strip of living bark.

Western larch bark

WESTERN LARCH

Larix occidentalis (**lair**-ix: Roman name; oxy-den-**tay**-lis: western). Also tamarack. Most needles deciduous, soft, pale green, 1–1¾", 15–30 in "false whorls" on short peglike spurs; however, on this year's twigs needles are single and spirally arranged, and on seedlings they remain through winter; cones 1–1⅝", reddish until dry, bristling with pointy bracts longer than the scales; young bark thin and gray, maturing yellowish to cinnamon or purplish brown, 3–6" thick, furrowed and flaking in curvy shapes ± like ponderosa pine bark; commonly 52" diam × 170' tall; largest is 87"; tallest 192'. ID, nw MT, WA, OR. Pinaceae.

A larch is something many people mistakenly think of as a contradiction—a deciduous conifer. The deciduous needles always set it off visually, even from a distance: intensely chartreuse in spring, then a subtler but still distinctive grass-green through summer, smashingly yellow for a few weeks in October, and conspicuous by their absence for a five- or six-month winter. You can tell a larch in winter from a maple or cottonwood by its coniferous form (single, straight trunk, and symmetrical branching) and from a dead evergreen by its warty texture (pegs on its twigs). The trunks are often skinnier than nearby conifers of similar height. (Other deciduous conifers you may have seen are not native here—bald-cypresses and dawn redwoods, both in the cypress family.)

An Arduous Beeline

David Lyall was perhaps the last of the rugged Scots prominent in early exploration of the West. As surgeon-naturalist on the British contingent (Canada being a British colony) of the Northwest Boundary Survey of 1857–62, he followed a beeline across unknown, precipitous terrain—a heroic task on its own—yet he found energy and enthusiasm for science along the way. He published *Botany of Northwest America* in 1863.

Relative to other Rocky Mountain conifers, larch is fast-growing, long-lived, shade-intolerant, fire-resistant, and water-demanding. Since evergreen competitors photosynthesize through much of winter, a larch has to make up for lost time with high photosynthetic efficiency. This requires full sunlight and ample groundwater through the dry months. Deciduousness helps larches recover from defoliating insects or fires; a larch is going to produce a whole new crop of needles every year anyway. It can afford to have a few grouse munching on its irresistibly tender needles.

Montana's Clearwater/Swan Valley is larch heaven. The champion lives there. Floating larch needles carpet the downwind shores of Seeley Lake in winter, and wave action rolls them up into spheres, typically golfball sized but ranging up to more than a foot in diameter.

Western larch has few lethal pests or diseases, resists fires, and grows fast once it gets past the first few decades. It looks like an excellent candidate for planting both within and north of its present range. It's a superior timber product—strong and beautifully reddish—while also promoting tourism: those forests are gorgeous! Favoring larch too much, though, could risk depriving large mammals of the thermal cover (dense, insulating canopy) they need in winter.

Western larch rivals ponderosa pine as an exemplar of trees that thrive under a regime of frequent low-severity fires, and suffer when deprived of those fires. Shade-tolerant trees will tend to replace larch. Several wildfires have burned parts of the Bob Marshall Wilderness, with surprisingly good effects—namely, an increased proportion of larches among the survivors, forming a more open forest.

Salish and Kootenai peoples used to draw off larch sap to make a syrup.

SUBALPINE LARCH

Larix lyallii (lye-**ah**-lee-eye: for David Lyall, see above). Also woolly larch. Needles deciduous, soft, pale green, 1–1⅝", 30–40 per bunch on short peglike spurs, exc that needles are single and spirally arranged on this year's twigs, and are ± evergreen on lowest branches of saplings; cones 1½–1¾", bristling with pointy bracts much longer than the scales; this year's twigs densely, minutely woolly; bark gray, up to 1" thick; tree broad-crowned, much-branched, and/or multi-stemmed; rarely a low shrub; biggest tree is 7' diam, tallest is 126' (both in WA); oldest is 1,011 years (MT). Near tree line; ID, nw MT, WA. Pinaceae.

Though evergreen conifers inhabit colder climates than broadleaf trees, on average, the most cold-loving of all trees are deciduous conifers, the larches. They are the most northerly and the most alpine genus of trees all around the Northern Hemisphere.

Being deciduous offers some protection in winter from both cold desiccation and storm breakage. Thus we sometimes find larches growing tall where the other conifers can only grow as krummholz. Sometimes an early frost freezes the needles in place through the

Subalpine larch

Subalpine larch

and rawer substrates (e.g., talus and recent glacial moraines). They keep no low limbs, a habit with one big disadvantage (they can't layer) and one big advantage (ground fires rarely threaten them).

CONIFERS WITH SINGLE NEEDLES

Other than pines and larches, our conifers bear needles singly.

DOUGLAS-FIR

Pseudotsuga menziesii (soo-doe-**tsoo**-ga: false hemlock; men-**zee**-see-eye: for Archibald Menzies, p. 111). Needles ½–1½", varying from nearly flat-lying to almost uniformly radiating around the twig, generally with white stomatal stripes on the underside only, blunt-pointed (neither sharp to the touch nor notch-tipped nor broadly rounded); cones 2½–4" × 1½", with a paper-thin 3-pointed bract sticking out beneath each woody scale; soft young cones sometimes crimson or yellow briefly in spring; young bark gray, thin, smooth with resin blisters; mature bark dark brown, grooved, with tan and red-dish-brown layers visible in cross-section; winter buds ¼" long, pointed, not sticky; trunk usually straight. Mid-elev forest; A. Pinaceae.

Common virtually throughout our range, Douglas-fir grows from 55° N. in northern British Columbia, down to 19° N. in central Mexico—the greatest north-south range of any American conifer.

It is also the world's tallest species. Most authorities rank it second, behind coast redwood, but that's because early logging targeted the finest Douglas-fir stands, taking no prisoners and few measurements. Two firs measured by professional foresters were 400 and 393 feet tall—both taller than any redwood on record. One with slightly less reliable stats may have stood 415 feet. The 400-footer was 13 feet, 8 inches in diameter, rivaling redwoods in bulk as well as height.

But all those were the coastal variety of the species, growing in the benign coastal climate, where firs can keep adding

winter; they drop when they thaw in spring, and are soon replaced. While the tree's base is still deep in snow, its upper branches leaf out, providing spring's first greens—a treat for grouse that survived winter on a diet of tough old fir needles. Larch needles taste like tender, young grass, with an initial spicy-resinous burst. They're a visual treat too, contrasting dramatically with other needles twice a year—bright grass-green in June, and yellow in late September to October.

On congenial sites, subalpine larches reach impressive sizes, and on tough sites they can live a thousand years. Compared to other trees here, they tolerate greater cold

Douglas-fir

firs, and spruces. After some severe burns, Scouler willows quickly resprout to take over, and later, Douglas-firs come up in the willows' shade. In the parts of our range that lack ponderosa pine, Douglas-fir is likely to be the tall tree near lower timberline, perhaps alongside smaller trees like limber pine, aspen, or mountain-mahogany.

Douglas-fir may yield in succession to redcedar or the hemlocks on moist sites where they can grow. It can generally hold its own against lodgepole pine or grand fir, and may slowly replace ponderosa pine (especially if there are no fires) where they grow together.

After certain hot-weather conditions in interior British Columbia and northeast Washington, sweet sap sometimes crystallizes all over Douglas-fir needles. This "Douglas-fir sugar" was once legendary; but Indigenous uses of Douglas-fir were less central than the uses of, say, redcedar or paper birch. They chewed the sap and burned the thick bark or the cones for heat. Douglas-fir wood was economically unimportant until white men came with steel tools.

The seedlings are a winter staple for deer and hares, while the seeds are eaten by small birds and rodents. Bears strip the bark to eat its succulent cambium; this wounds the tree, often fatally, by making an opening for invasions of insect larvae or rot fungi.

This tree is named for David Douglas, who exported its seeds for English gardens in 1826; it was first collected for British science before Douglas was born, by Archibald Menzies in 1791, and again by Lewis and Clark in 1805. Douglas prepped for exploring the Northwest by visiting Menzies, then an old man in London. Douglas called the tree a pine; later taxonomists tried out "yew-leafed-fir," "spruce," and finally "false-hemlock," while sticking with fir for the common name. It is none of the above. Like our hemlocks and cedars (more misapplied European names), *Pseudotsuga* is a Pacific Rim genus, with three species in Japan and China and one in a tiny mountainous range

considerable wood in their fifth, sixth, and seventh century of life, and may live to 1,300. Our variety, Rocky Mountain Douglas-fir, grows minimally after age 200, and rarely reaches 400. The tallest may be a 209-footer in Clearwater County, Idaho. Western larch and western white pine can beat that.

For much of the twentieth century, Douglas-fir was the top lumber species in the United States. The tree that commerce so desires is, again, the big coastal Douglas-fir. The smaller Rocky Mountain variety is often less valuable than western larch or white pine or ponderosa pine. Worse, in the Rockies, it is so prone to attack by bark beetles, tussock moths, spruce budworm, and mistletoe that commercial foresters prefer other species in most locales.

By maturity it develops thick, corky bark and loses its lower limbs, making it fairly fire-resistant, exceeded here only by western larch and ponderosa pine. Being more shade-tolerant than those two, Douglas-fir tends to increase through succession in all forests that have it but lack hemlocks, true

A Patron Saint for Backpackers

If I were Pope, I would canonize David Douglas. Time and again he set off into the wilderness, usually with Indigenous guides or Hudson's Bay Company trappers, but also often alone. He packed a cast-iron kettle, a wool blanket, a lot of tea and sugar, tradable goods such as tobacco and vermilion dye, his rifle and ammunition, and pen, ink, and reams of paper for wrapping plants, seeds, and skins—no shelter usually, no dry change of clothes, no waterproofing but oilcloth for the papers and tins for the tea and gunpowder. Often without food in his pack, he might hunt and eat duck, venison, woodrat, or salmon, or dig up wapato roots; other days he consoled himself with tea, and berries if he was lucky. Once while boiling partridge for dinner, he fell asleep exhausted and, waking at dawn to a burnt-through kettle, counted himself clever to boil up a cup of tea in his tinderbox lid.

Douglas figured he walked and canoed 6,037 miles of Washington and Oregon in 1825–26. He was in either the Blue Mountains or the Selkirks when he collected the type specimen of ponderosa pine. In 1827, he crossed the Canadian Rockies to Hudson's Bay to catch a ship back to England. For all that, the Royal Horticultural Society paid him their standard collector's salary of one hundred pounds a year, plus sixty-six pounds for expenses. His mission was to ship them seeds or cuttings to grow lucrative exotics for English gentlemen's gardens.

During his third voyage to America, he sailed to Hawaii where, at the age of thirty-four, he came to a gruesome end. Out walking alone with Billy, his faithful Scotty dog, he somehow ended up in a pit trap for feral bulls, where he was gored and trampled. Did he fall, or was he pushed? Rumors that he was pushed persist to this day.

As a boy in Scotland, Douglas was too rebellious (hyperactive?) for school. His stonemason father pulled him out at age eleven to apprentice in gardening. As his interest grew, he gleaned a botanical education wherever he could, eventually auditing lectures by William Jackson Hooker (p. 121). Hooker took young Douglas on field trips in the Highlands, was impressed with his fanatical drive and enthusiasm, and sent him off to London and fame. Despite his spotty education, Douglas wrote eloquently in his journals. In contrast to fellow Scot Thomas Drummond (p. 136), whose laconic response to the Canadian Rockies was that they "gratified him extremely," Douglas wrote of:

> mountains towering above each other, rugged beyond all description; the dazzling reflection from the snow, the heavenly arena of the solid glacier, and the rainbow-like tints of its shattered fragments . . . the majestic but terrible avalanche hurtling down from the southerly exposed rocks producing a crash, and groans through the distant valleys, only equalled by an earthquake. Such gives us a sense of the stupendous and wondrous works of the Almighty.

in Southern California. Douglas-fir itself is a top forestry species in many countries. One planted in New Zealand in 1859 may be the world's tallest cultivated-tree specimen, at 229 feet.

SUBALPINE FIR

Abies lasiocarpa (**ay**-bih-eez: Roman name; lazy-o-**car**-pa: shaggy fruit). Also *A. bifolia*. Needles ¾–1½", bluish green with a broad white stomatal stripe above and 2 fine stripes beneath, often

Subalpine fir

curving to crowd the upper side of the twig, tips variable; cones purplish gray to black, barrel- to cigar-shaped, 2½–4 × 1¼", borne erect on upper branches, dropping their seeds and scales singly while the core remains on the branch; bark thin, gray, smooth on younger trees, without superficial resin blisters; upper branches short, horizontal, low branches right at ground level, long; or a prostrate shrub; biggest of the RM variety is 31.5" diam, 122' tall, in sw CO. Abundant in the subalpine; rarely down to low elevs; A. Pinaceae.

The peculiar narrow spires of subalpine firs are the visual archetype of a timberline tree. As in other true firs, the limbs are stiffly horizontal and somewhat brittle, so on this one they need to be short to hold up to subalpine snow and wind. Long lower limbs escape those stresses by spending the winter buried in the snow.

Hugging the ground puts them where they need to be to "layer," or reproduce by sprouting new roots and stems from branches in contact with soil. Subalpine fir is good at layering and at pushing both tree line and scrub line upward. At scrub line it grows in krummholz (prostrate) form, and spreads almost exclusively by layering. Occasionally

it produces an asymmetrical, half-dead-looking little tree with voluminous krummholz skirts. Such trees were confined for years to the shape of the snowpack; any foliage above the snowpack was killed during winter by a combination of wind desiccation, frost rupturing, and abrasion by driven snow. This is the krummholz way of life. Then, perhaps, there was deeper snow for a few winters, providing growing room for half a dozen little vertical shoots. The next time a normal-snowfall winter came, one of the shoots survived with some needles on its downwind side—the side relatively protected from desiccation and abrasion. Years later, the little tree is likely still "flagged," its surviving limbs positioned exclusively downwind and above the snow abrasion zone (the first 10 inches above protective snow).

Though a little bit of lasting snow may be a conifer's friend up on windswept alpine ridges, deep and long-lasting snow is a hindrance to tree establishment in subalpine parkland. Conifers need a longer growing season than herbs and shrubs.

On steep meadows, tree seedlings are often wiped out by snow creep. Some limbs, after being encased in snow the better part of the year, spend the remainder matted with a weird black fungus called snow mold, *Herpotrichia nigra* (meaning "black creeping hair," but less hairlike than horsehair lichen). Luckily, snow mold isn't as deadly as it looks. Other hindrances include soil too sodden, arid, or shallow, or sedge turf too dense.

In Glacier National Park, historical photographs show that subalpine firs have been expanding their clumps, filling in the spaces below tree line but not pushing it upward. This may reflect fire exclusion more than global warming.

The competition between spruce and fir looks almost neck-and-neck, but is often deflected by wild cards. The fir produces far more seedlings that hang around in the understory, as you may notice. If they are released by, say, a blowdown, firs have a head start, but the spruce grows faster for a while, and eventually co-dominates because it lives much longer.

If the disturbance is a fire, both species have to seed in from outside the burn, which may take a while; resprouting aspens or post-mortem-seeding lodgepoles may take over first. If enough decades then go by without another fire, spruces and firs can return because their seedlings tolerate shade.

Since 2000, mortality agents that target firs or spruces have come to the fore. A spruce beetle epidemic peaked between 2013 and 2016, giving firs a brief advantage locally. Firs, for their part, are subject to Subalpine Fir Decline, a group of issues whose relative lethality remains unresolved. Beginning in 1957, the balsam woolly adelgid, from Europe, killed many firs in the northwest part of our range; you can recognize this aphid's victims by their extremely swollen branch tips. Colorado barely has them; instead it has epidemic western balsam bark beetle, but that only inflicts 12 percent of the subalpine fir deaths in a typical study: firs are dying without any obvious pest, hence the vague term "decline." The likely culprits are heat and drought. After mortality events strike, subalpine fir seedlings do not survive really hot mornings, so we expect the species to disappear from many east- and south-facing slopes.

WHITE FIR

Abies concolor (**con**-col-or: same color). Needles ¾–2¼", quite broad, usually round-tipped, usually upcurved, whitish thanks to stomatal stripes both above and below (hence the name

White fir

concolor; cones dense, heavy, ± cylindrical, 3–5 × 1½", greenish, borne erect on upper branches, dropping their seeds and scales singly while the core remains; pollen cones often red; bark gray, resin-blistered in youth, becoming deeply furrowed in age; branches horizontal; biggest is 78" diam. Mid-elev forest; CO, NM, UT, se ID. Pinaceae.

White fir is a fine-looking and fragrant tree. Deer and elk feed on its new spring foliage, squirrels on its seeds. It is relatively fast-growing and short-lived, rarely reaching age three hundred. A tree in the San Juans reached 162 feet tall before dying by fire in 2019. Two specimens of California white fir (arguably a separate species) served as the US Capitol Christmas tree.

The northern reach of white fir is ambiguous. Foresters in Oregon and western Idaho talked about their white firs for generations, but Oregon Flora, based in the Department of Botany and Plant Pathology at Oregon State University, determined that all of those trees are white/grand hybrids (A. grandicolor, facetiously) and that Oregon has no true A. concolor. Newer maps show white fir only as far north as extreme southeast Idaho.

Being fire-susceptible, white and grand firs have been major beneficiaries of Smokey's firefighting efforts over the past century. This has not gone well: most of the notoriously sick, half-dead, and conflagration-prone stands of eastern Oregon and parts of Idaho are overly dense hybrid-fir stands that filled in where ponderosa pines were logged out. Leading mortality agents are the fir engraver beetle and several kinds of rot. Studies of mixed conifer forests in Oregon's Blue Mountains found that in pre-settlement times they had few firs, and mostly pines. When doing restoration thinning of stands with pines and firs, foresters often try to restore those proportions, seeing firs as ladder fuels that endanger the more resilient pines (as they do giant sequoias, in California).

GRAND FIR

Abies grandis (**gran**-dis: big). Also lowland white fir. Needles ¾–2", quite broad and thin, spreading in a flat plane from the twig, notch-tipped, dark green above, 2 white stomatal stripes beneath (needles of topmost branches

Tree Line and Timberline

Timberline is a broad elevation belt between two other lines, each irregular: tree line, the upper boundary of erect tree growth, and scrub line, the upper limit for conifers to grow in the prostrate, shrubby form called krummholz ("crookedwood").

Krummholz and treeless tundra are alpine, whereas subalpine parkland is the tree-grove-and-meadow mosaic below krummholz. Downslope, it merges with subalpine forest, which you can think of as all the forest dominated by spruce and subalpine fir.

Cold, wind, and the length of the snow-free season are all factors in how well trees will—or will not—grow at high elevations. Winter wind can suck all the moisture out of needles, killing them, when the soil is frozen, which prevents the roots from replenishing the lost moisture. Burial in the snowpack offers protection, allowing conifers to grow as krummholz where they cannot grow as trees. The outlines of the krummholz in summer show the contours of the winter snowpack.

Though lasting snow is a conifer's friend up on windswept alpine ridges, deep, long-lasting snow can hinder tree establishment in subalpine parkland. Needles need enough time to grow and then harden to protect themselves against freezing. Once hardened, they can usually handle winter temperatures here. (Snow also hinders tree establishment via downslope snow creep, which scours seedlings away.)

Once seedlings are several feet tall, they can start photosynthesizing long before the snow is gone, but making it past the seedling stage in the open requires a run of longer-than-average snow-free seasons. Such a run around 1950 produced invasions of trees across subalpine meadows ranging from Glacier National Park to the Sangre de Cristos.

Tree invasions contrast with the more typical, slow tree-clump expansion. It's easier for seedlings to get their start next to an existing tree for two reasons. First, most subalpine trees can layer, or grow a new stem where branches in contact with the earth take root; the parent limb feeds the new shoot intravenously, a big advantage over growing from seed. Second, during spring thaw, tree foliage, being dark, absorbs more of the sun's heat than open snow does, melting itself a little tree well in the last few feet of snow. The well is a microsite with a growing season several weeks longer than the surrounding meadow—just what seedlings need. Hence trees in subalpine parkland typically grow in slowly expanding clumps, often elongated downslope into a teardrop shape. Sometimes the old trees in

often neither flat-spreading nor esp dark); cones dense, heavy, ± cylindrical, 2½–4 × 1½", greenish, borne erect on upper branches, dropping their seeds and scales singly while the core remains; bark gray, resin-blistered in youth, becoming ± ridged and flaky with age; branches horizontal. w ID, nw MT, WA, OR. Pinaceae.

The foliage on a grand fir sapling catches your eye, the tidy, flat array of long, broad needles showing off the glossy green color.

Flat leaf arrays are primarily an adaptation to deep, shady forest, where most rays of light beam from directly overhead. The flat array keeps a maximum amount of leaf surface facing up, to catch the rays efficiently. Grand firs are only slightly less shade-tolerant than western hemlocks, and require less rainfall, especially on valley floors where drought is not an issue. Climate models suggest habitat for grand fir could increase in the

Timberline in Colorado

the center die and only shrubs replace them, producing a hollow clump—a timber atoll.

Charcoal in meadow soil profiles reveals that many subalpine meadow areas grew some trees at least once since the ice ages. The case can be made that most meadow and heather communities below tree line owe their existence to past fires from which recovery, at these cold elevations, is slow.

Timberline is a dynamic equilibrium made visible. It responds very, very slowly to slight changes in climate, and then gets kicked around by faster disturbances. It should move upward in response to warming climate, but how much it will do so in reality is unknown. So far, upward trends are seen in a few places, but overall they are not as strong a pulse today as during mid-twentieth-century decades. Upward movement is restrained by slow soil development, by the pests and diseases that plague our timberline trees, and, increasingly, by fires. The ecologist Alina Cansler predicts that tree line and scrub line will inevitably rise, while forest line may descend patchily, following fires and pest outbreaks. That could expand subalpine meadows. In some areas, with hotter droughts, we may see brush replace subalpine meadows, grassland replace tundra, or plants grow in combinations never seen before.

This discussion has been about alpine timberlines, as opposed to lower timberlines—the low-elevation transition between forest and steppe or grassland. Lower timberlines are mainly a function of soil moisture and, again, fire history. At both upper and lower timberlines, there's less drought stress on north-facing slopes, allowing trees to persist there while not growing on nearby south slopes at the same elevation.

We have a lot of territory moist enough for the survival of mature conifers, but not conifer seedlings. If a severe fire takes out the trees there, forest probably won't come back. In some of those burns that look almost hopeless today, there's a chance of forest coming back someday if shrubs grow to provide shade, and then seeds find their way into that shade, and if there's a series of several wetter-than-average years.

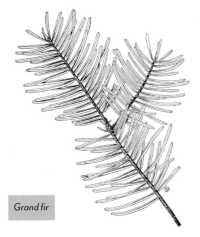

Grand fir

SPRUCE

Picea spp. (**pis**-ia: Roman name, from "pitch," for some conifer). Needles blue-green, stiff, sharp, 4-sided, with stomatal stripes on all sides, typically pointing outward (bottlebrush-like) around the twig, bad-smelling when crushed; they grow from short peglike bases that are conspicuous on the twig after needles fall; cones hang from branches; branchlets often tipped with conelike galls; bark thin, scaly. Pinaceae.

Engelmann Spruce

P. engelmannii (eng-gell-**mah**-nee-eye: for Georg Engelmann, see below). Needles ¾–1¼"; young twigs minutely fuzzy (through 10× lens); cones 1½–2¼", scales very thin, flexible, irregularly toothed or wavy along outer edge; bark thin, scaly, ages to cinnamon; crown dense, narrow; or prostrate and shrubby at timberline; tallest living tree is 223'; greatest girth 87"; oldest is 911 years. A subalpine forest dominant throughout our region.

Colorado Blue Spruce

P. pungens (**pun**-jenz: sharp-pointed). Needles ¾–1¼"; young twigs hairless; cones 2¼–4", scales thin, irregularly toothed or wavy along outer edge; bark gray, furrowing in age. Mainly near streams at lower mid-elevs; tallest tree (the tallest in CO of any species) is 180'; greatest girth 59"; CO, NM, UT, WY, and extreme e ID. The CO state tree, and the UT state tree until 2014.

future, perhaps even "invading" east of the Continental Divide.

Grand firs in eastern Oregon and nearby parts of Idaho are hybrids with white fir, whose genetic influence shows up as longer needles with more stomatal bloom in the upper groove, more upward curve from the twig, and less notch at the tip.

Given a natural regime of frequent, low-intensity fires, pines, larches, and Douglas-firs would typically dominate forests where we see grand fir today. Older grand firs may survive low to moderate fires thanks to reasonably thick bark, but the species is vulnerable overall because of low branching, shallow roots, and susceptibility to rot, especially after fire wounds its bark.

An Herbarium for the West

Georg Engelmann never explored the Rockies; he spent his career in St. Louis receiving, describing, and naming specimens from all the major Rocky Mountain collectors of the day (1833–84). He founded the Missouri Botanical Garden, where his collection forms the heart of the herbarium—today the world's sixth largest, with more than eighty thousand specimens, each the basis of one named species or other taxon.

When we drink wine, we have him and his entomologist colleague to toast. After the Phylloxera bug catastrophically devastated vineyards in the 1870s, their solution was to graft wine-grape vines onto roots from American wild grape species that resist Phylloxera. Most wine today comes from such grafts.

White Spruce

P. glauca (**glaw**-ca: pale). Needles ½–⅞", blue-green with pale grayish waxy coat, most commonly crowding upward and forward from the twig; young twigs hairless; cones 1–1½", scales stiff, broadly round at tip; n MT, rare in ID.

SPRUCES ARE LOVELY, with denser, drapier, darker, bluer foliage than our other conifers. Horticultural blue spruces were bred from the bluest *P. pungens* in Colorado. In the wild, blueness varies too much to help you tell blue spruce from Engelmann; cone size, aging-bark color, and elevation are more useful metrics, but some specimens cannot be absolutely assigned to either species. All three of these species hybridize and intergrade. Many spruces of Idaho and western Montana are white-Engelmann hybrids.

Spruces are the second-most northerly conifer genus, after larches. In the Rockies and north to Alaska, forests just below timberline are dominated by Engelmann spruce, subalpine fir, and lodgepole pine. Spruce does a little bit better than fir in easterly parts of our range where the climate is strictly continental (dry winters with extremes of cold), and it does better on limestone soils. Subalpine fir is more successful where the maritime (winter-wet) influence is felt.

Grouse select dense spruce crowns that offer warm, dry roosts. Commonly reaching to the ground, spruce crowns catch fire easily. Engelmann spruces can grow slowly but steadily for three to four hundred years, but rarely achieve that potential because they are fire-prone. Wet locations may protect some of them. They like soils with aerated moisture—streamsides, not marshes.

Many spruce branch tips bear curious conelike appendages—galls, or "houses" for adelgid larvae (p. 456). Gall tissue is secreted by a plant in response to chemical stimulation, usually by a female insect laying eggs. A spruce adelgid's gall envelops and terminates new growth at a branch tip, but rarely harms the tree. The dead needles and the gall

Engelmann spruce

Engelmann spruce bark

Colorado blue spruce

turn tan; together, they look like a 1-to-2-inch cone with needle-tipped, melted-together "scales" each hooding an opening into a larval chamber. The gall may hang from the branch for years, long after the larvae mature and move on; other insects may move in.

Engelmann spruce lumber has found a market in Japan, as it resembles certain Japanese woods that are scarce now.

Spruce pitch is chewable, fragrant, and sweetish, but sure sticks to your teeth.

The Rockies are seeing rapidly increasing mortality of Engelmann spruces, from fires and from the spruce beetle (p. 462). Epidemics of that bark beetle may kill 90 to 99 percent of the mature spruce in an area, while leaving other areas almost unscathed. The San Juan Mountains have taken a vicious hit. Wildfires that followed 90 percent spruce mortality within a few years produced huge, high-severity patches in which basically all trees were dead and seeds in the soil were consumed. Aspen is taking over those patches; conifers may return eventually. Granted, high severity is normal for fire in spruce-fir forests, and it's normal for aspen to resprout and flourish after fire. But zero—no conifers surviving the fire, and no conifer regeneration, in large patches across the San Juans—is a shockingly low number.

Mortality due to spruce beetles at typical levels—say, 70 percent of the mature trees and few of the saplings—leads to increased proportions of subalpine fir and enhanced resistance to subsequent spruce-beetle epidemics, at least for several decades.

MOUNTAIN HEMLOCK

Tsuga mertensiana (**tsoo**-ga: Japanese name; mer-ten-zee-**ay**-na: for Karl Mertens, p. 120). Needles ½–¾", bluish-green with white stomatal stripes on both top and bottom, ± ridged, thus ± 3- or 4-sided, radiating from all sides of twig, or upward- and forward-crowding on exposed timberline sites; cones 1–2½", light (but coarser than spruce cones), often purplish, borne on upper branch tips; bark furrowed and cracked; mature crown rather broad; also grows as prostrate shrub at highest elevs. Subalpine; WA to extreme nw MT. Pinaceae.

The compact, gnarled shoulders of mountain hemlocks shrug off the heaviest snow loads in the world, in the Northwest's coastal mountains. At every age, this species' form is brutally determined by snow. Their crowns grow ragged from limbs breaking. The seedlings and saplings are gently buried by the fall snows, then flattened when the snowpack, accumulating weight, begins to creep downslope. Tramping across the subalpine snowpack on a hot June afternoon, you can almost hear the tension underfoot of young trees straining to free themselves and begin their brief growing season. The stress of your foot on the surface may upset an unseen equilibrium, snapping a hemlock top a few feet into the air. After the trees grow big enough to take a vertical stance year-round, they may keep a sharp bend at the base ("pistol-butt") as a mark of their seasons of prostration. Even in maturity they may get tilted again where soil creeps downslope.

After the ice ages, mountain hemlocks managed to leap across 125 miles of uninhabitably dry terrain to reach northern Idaho from the Cascades. How did they do that? Studies have concluded that their winged, lightweight seeds can occasionally travel that far in the wind. They set up camp in northern Idaho 4,100 years ago, but

Mountain hemlock

waited another 3,300 years for further cooling before fully populating that region, and never made the additional leap to Glacier National Park, which offers good habitat. Competition from other conifers held them back. (Western hemlock made the same leap, and did make it to Glacier.) This finding supports a bit of optimism about the tree's ability to migrate long distances. Generally, subalpine trees including this one have relatively poor prospects, as trees need to migrate uphill in response to warming, and there's precious little territory available uphill from where they live now.

Western hemlock

WESTERN HEMLOCK

Tsuga heterophylla (het-er-o-**fill**-a: varied leaves). Needles of mixed lengths, ¼–¾", round-tipped, flat, slightly grooved on top, with white stomatal stripes underneath, spreading in ± flat sprays; cones ¾–1" long, thin-scaled, pendent from branch tips; mature bark up to 1" thick, platy, checked (almost as much horizontal as vertical cracking); inner bark streaked dark red/purple; branch tips and treetop leader drooping. n ID, nw MT, WA, OR. Pinaceae.

My image of western hemlock is of a sapling's limbs, their lissome curves stippled a soft green made incandescent in the understory dimness by a stray swath of sunlight.

Western hemlocks are prolific: notice the profusion of little hemlock cones on the forest floor, or on the tree. Each year, a mature hemlock drops more than one viable seed per square inch of ground under it. Only an infinitesimal fraction will make it to tree size.

On the Oregon coast, very young, pure hemlock stands produce organic tissue ("biomass") at the fastest rate yet measured in the world. Hemlocks achieve their efficiency partly by sheer leafiness—a 6-inch trunk can support over 10,000 square feet of leaf surface area, almost twice as much as Douglas-fir. While the greater leaf area catches more light, it also loses more moisture; shade tolerance seems to be a tradeoff against drought tolerance. (The productivity of forests within the historical range of sea-going salmon owes a lot to those fish; see p. 445.)

Mature hemlocks, with thin bark and shallow roots, are frequent victims of fire, wind, and heart rot. Nearly all hemlocks that reach two hundred years old in the Rockies are developing heart rot and will become hollow. *Echinodontium tinctorium,* Indian paint fungus, is the usual culprit. It produces huge, hard conks, once prized and traded throughout the West for the best red face paint. Conks were ground to a powder and mixed with animal fat. Hemlock pitch also served as a dark base for face paint. Tannin-rich hemlock bark was used to tan skins; to dye and preserve wood (sometimes mashed with salmon eggs for a yellower dye); to make nets invisible to fish; and as a styptic to halt bleeding.

The word "hemlock" traces back to 700 CE as the English word for the deadly parsleys used in Socrates's execution. The English apparently smelled a resemblance to hemlock in the crushed foliage of a New England conifer that was new to them; they called it hemlock spruce—later shortened to hemlock.

WESTERN YEW

Taxus brevifolia (**tax**-us: Greek name; brevif-**oh**-li-a: short leaf). Also Pacific yew. Needles ½–¾", grass-green on top, paler and concave

Western yew

beneath, spreading flat from the twig, broad and thin, drawing abruptly to a fine point but too soft to feel prickly; new twigs green; ♂ and ♀ organs on separate plants; juicy, red, cup-shaped ¼" fruit holds 1 seed; bark thin, peeling in purple-brown scales to reveal red to purplish, smooth inner bark; branches sparse; trunk often crooked or sometimes a sprawling shrub. Champion is 58" diam; tallest is 84'. Moist, for-ested valleys; n ID, nw MT, WA, OR. Taxaceae.

Our yew is a conifer without cones. It bears its seeds singly (and only on female trees) cupped within succulent red seed coats loosely termed "berries" but techni-cally "arils." Seeds of Eurasian yews contain alkaloids capable of inducing cardiac arrest, but there are no records of serious poisoning from Pacific yew, which is far less toxic. Birds love yew berries, and moose and elk browse the foliage.

Woodworkers class all conifers as "soft-woods," but yew is among the hardest of woods. It can be worked with power tools or carved to make extraordinarily durable and beautiful utensils. The sapwood is cream, the heartwood orange to rose. Indigenous people made it into spoons, bowls, hair combs, drum frames, fishnet frames, canoe paddles, dig-ging sticks, splitting wedges, war clubs, deer-trap springs, arrows, and bows. Ceremonially, it imparted strength. Worldwide, yew species were the wood of choice for bows. The Greek name for yew, *taxos,* spawned both "toxin" and *toxon,* meaning "bow."

The beautiful, smooth underbark can be almost cherry red. In 1987, Western science suddenly wanted yew bark. An order was filled for 120,000 pounds of bark from which to extract a tiny amount of paclitaxel, the chemotherapy drug sensation of the 1990s. For a few years, many people feared that bark stripping might wipe the species out in its native habitat. Fortunately, the drug indus-try soon found they could more economi-cally synthesize the drug from cultivated European yew needles and so stopped buying yew bark. Paclitaxel remains an example of lifesaving drugs discovered growing under our noses.

Paclitaxel is produced not by the yew itself but by endophytic fungi living within yew needles. Future paclitaxel production will likely be from cultured fungi or bacte-ria. Some fungi may have evolved paclitaxel to defend themselves from other fungi and from *Phytophthora,* a mysterious genus that includes notorious killers like sudden oak death and Port Orford cedar root rot. Western yew is the only species known to carry both of these pathogens, yet it has not been decimated by them anywhere—perhaps thanks to paclitaxel.

Yew has completely taken over some valley floors in Idaho and Montana, defying a com-mon assumption about forest succession—that taller plants shade out and replace shorter ones. In this case, the medium-small tree creates both heavy shade and allelopathic chemistry in the soil; few saplings but its own can survive. But climax yew stands would likely be rare without fire suppression.

CONIFER TREES WITH TINY, SCALELIKE LEAVES

This group is the cypress family, Cup-ressaceae. Needles are arranged in opposite pairs or in whorls of three. More strikingly, most of them are compressed and scalelike, not needlelike. One shrub species, common juniper, instead has closely packed, spinelike needles up to ¾ inch long. Short, sharp needles like that also occur in our other species, but only in their first few seedling years or on a few low branches later in life.

WESTERN REDCEDAR

Thuja plicata (**thoo**-ya: Greek name for some tree; plic-**ay**-ta: pleated). Leaves tiny, yellowish-green, in opposite pairs, tightly encasing the twig, flattened, the twig (incl leaves) being 4–8 times wider than thick; foliage dies after 3–4 years, turning orange-brown but persisting for months before falling; cones ½", consisting of 3 opposite pairs of seed-bearing scales, plus a narrow, sterile pair at tip and 0–2 tiny, sterile pairs at base; bark reddish, thin (up to 1"), peeling in fibrous vertical strips; leader drooping; lower trunk flaring, often fluted and buttressed. Moist lowland soils; n ID, nw MT. Cupressaceae.

Western redcedar branch tip

Cedars are a breed apart, easily recognized by their droopy sprays of foliage and vertical, fibrous bark. The bark is too acidic to encourage lichens, fungi, or moss. Though slow-growing, our cedars resist windthrow, rot, and insect attack, and can live more than a thousand years. Western redcedars, the family's largest trees, may develop buttressed waistlines 30 to 60 feet around.

Redcedar groves are a thing apart—quieter, deeper. They occur on moist valley floors (where soil moisture is 12 percent through August). Cedars tolerate drier climates than hemlocks do, as long as their roots are wet.

Throughout their range, redcedars provided invaluable materials. The inner bark, shredded with deer bone and then woven, made warm clothing. Unwoven, it was soft enough for cradle lining or menstrual pads; torn in strips and plaited, it became roofing, floor mats, hats, blankets, dishes, or ropes. Buds, twigs, seeds, leaves, and bark each had medicinal uses. Cedar charms sanctified or warded off spirits of the recently deceased. Cedar-bough switches were skin scrubbers for both routine and ceremonial bathing.

Easy to work with stone tools and fire, cedar made up in durability and aesthetics what it lacked in strength. Cedar heartwood is warm red, weathering to silver-gray; it smells wonderful, resists rot, and splits very straight if it comes from an old, slow-grown tree. Split cedar shakes made great roofs in pioneer days, but should be avoided today due to their flammability.

The true cedars, genus *Cedrus* in the pine family, are native to the Himalayas and the Mediterranean, and are much planted elsewhere. Being fragrant may have been the sole reason for naming our cedars after them. *Thuja* is a Pacific Rim genus, with other species in Northeast Asia. Eastern North America has a *Thuja* referred to as a "white-cedar," while junipers in the region are referred to as "redcedar." All these cedars have pungent fragrance due to their rot- and insect-repellent chemistry.

Rocky Mountain juniper

Utah juniper

JUNIPER

Juniperus spp. (ju-**nip**-er-us: Roman name). Also *Sabina* spp. Mature leaves tiny, scalelike, yellowish-green, tightly encasing the twig, not flattened; leaves on seedlings, saplings, and lowest limbs of young trees needlelike, ¼", prickly; cones berrylike, blue to blue-black, rather dry, resinous, 1-to-3-seeded, ¼"; bark red-brown, fibrous, shreddy; dense, small, conical trees, limbs nearly to ground, or sprawling shrubs. Near lower timberline. Cupressaceae.

Rocky Mountain Juniper

J. scopulorum (scop-you-**lor**-um: broomy). Leaves opposite, making the stems 4-angled; berries have heavy bluish bloom, typically 2 seeds; ♀ berries and inconspicuous ♂ flowers on same plant. A.

Utah Juniper

J. osteosperma (os-tee-o-**sperm**-a: bony seed). Leaves opposite; berries ½", often tan or brown, typically with 1 seed; ♀ berries and inconspicuous ♂ flowers usually on same plant; often multi-stemmed. Severe sites, often with limber pine; UT, e ID, WY, w CO, barely into NM.

One-seed Juniper

J. monosperma. Leaves opposite; berries ½", blue and somewhat juicy when fresh in summer, typically with 1 seed; ♀ berries and inconspicuous ♂ flowers on separate plants; often multi-stemmed. Severe sites, NM, CO, se UT.

Western Juniper

J. occidentalis (oxy-den-**tay**-lis: western). Leaves in whorls of 3, each whorl rotated 60° from the next. Lowlands of OR, w ID.

JUNIPER TREES ARE dry-country species with little defense against fire; their strategy is to grow where fire can't reach them—that is, where the vegetation is too sparse to spread fire. Until this century, that left most junipers on rocky sites with little soil, but with fire suppression they have spread. After reaching a large size, a juniper tree may survive low fires because its own shade and litter create a small, grass-free firebreak. In New Mexico, you'll find frequent-fire juniper savannas because their grasses, such as blue grama, are too fine to raise a tall, hot flame.

In the Wasatch Range stands a Rocky Mountain juniper 80 inches in diameter and 39 feet tall. I previously reported a western juniper twice that size, and one 2,675 years old, but then taxonomists moved those California trees into a separate species, *J. grandis.*

Whereas the pine family employs resins as a defense against insect pests, the cypress family (our "cedars") is far less resinous, instead relying on richly aromatic tropolones that combat both rot and insects. This has worked well for them, as you can tell from the number of oldest, biggest, and tallest tree species that are in this family. The family includes Chile's *alerce,* the new kid

One-seed juniper

One-seed juniper bark

Creeping juniper

in the oldest-plant competition, and was recently enlarged to include redwoods and giant sequoias.

Indigenous peoples boiled juniper leaves to steam sickness out of a house, or to prepare a bath for a sick person. Back east, juniper trees are made into moth-repellent "cedar" chests and closets.

Junipers grow on the edge—the driest margins of tree-sustaining climate—and climate change is starting to push them off that edge. They have a remarkable ability to shut down their metabolisms during drought, forgoing photosynthesis and growth and resuming when rain returns. Still, there are limits imposed by carbon starvation during photosynthesis shutdowns or by air gaps forming within the tree's sap-carrying vessels. As the Southwest's megadrought intensified in early 2022, dieback of Rocky Mountain juniper was seen in several areas. Though pests and pathogens are usually present, this dieback is attributed to the hot drought directly. Expect some piñon-juniper woodlands to revert to grassland and steppe in the warmer years to come.

Utah juniper tolerates more heat, and shows little or no decline so far.

CONIFERS THAT GROW EXCLUSIVELY AS LOW SHRUBS

Up at timberline, several conifer tree species adapt by growing as low shrubs; in contrast, the two juniper species below grow at various elevations but never achieve tree stature.

CREEPING JUNIPER

Juniperus horizontalis. Also *Sabina horizontalis.* Mature leaves tiny, scalelike but sharp-tipped, waxy-green, tightly encasing the twig, oppo-

Common juniper

COMMON JUNIPER

J. communis (com-**you**-nis: common). Leaves all ¼–¾", sharp, closely packed along the twig in whorls of 3, from a ± distinct joint at each leaf base; berries blue-black with bloom, round and quite fleshy, ¼–⅝" across, 1-to-3-seeded, resinous but sweetish; bark red-brown, thin, shreddy; low, mat-forming shrubs. Rocky sites from steppe to tundra; A. Cupressaceae.

The world's most widespread conifer species, common juniper is humble, but well suited to cold windswept ridges and slopes where even tall conifer species grow as low, creeping krummholz.

Junipers are anomalous among conifers in enclosing their seeds in fleshy, edible fruits. (Yews cup—but do not enclose—their poisonous seeds in "berries.") Properly speaking, a berry is a fruit, and a fruit is a thickened ovary wall, so neither junipers nor yews have true berries; juniper berries are technically cones, with few fleshy, fused scales. They have a sweetish, resiny flavor of suspiciously medicinal intensity. Used with restraint, they season teas, stuffings, gin, or your water bottle. People with inquisitive palates will try them straight off the bush. Indigenous people here said that a few juniper berries would stave off hunger all day, thanks perhaps to that resiny persistence on the palate. Birds love them, and disseminate the indigestible seeds.

site (making the stem 4-angled); juvenile leaves needlelike, ¼", prickly; cones berrylike, blue-purple, often with waxy bloom, rather dry, resinous, 1-to-6-seeded, ¼"; ♀ berries and inconspicuous ♂ flowers on separate plants; bark red-brown, fibrous, shreddy; low, mat-forming shrubs. Foothills, mainly e of CD; WY, MT. Cupressaceae.

"Currently holds the record as the smallest conifer, since I have found cone-bearing individuals only 4 cm tall."
—Chris Earle, conifers.org

4

Flowering Trees and Shrubs

The distinction between trees and shrubs is a loose one, having little correlation with plant classification. Anything that has a single, woody, upright main stem at least 4 inches thick or 26 feet tall at maturity I call a tree; any woody plant falling short of those dimensions I call a "shrub." Several species in this chapter grow as either trees or shrubs depending on environment.

DECIDUOUS

With Alternate Leaves and Catkins

Catkins are clusters of flowers that lack showy petals for attracting insects; wind pollinates them, not animals. In our species, male and female catkins are dissimilar. In the final three species in this group, the flower clusters are not catkins technically, though they are still wind-pollinated clusters with neither petals nor petal-like sepals.

Our two mountain-mahogany species fudge the line between deciduous and evergreen; I've placed one in each category.

QUAKING ASPEN

Populus tremuloides (**pop**-you-lus: Roman for "the people," hence poplar for unknown reasons; trem-you-**loy**-deez: like Eurasian aspen).

Also trembling aspen. ♀ and ♂ catkins on separate trees; ¼" conical seedpods, in long strings, splitting in 2 to release tiny seeds; leaves 1–2½", broadly heart-shaped to round, pointed, bumpy-edged to fine-toothed, on leafstalks 1–2½" long and flattened sideways; bark greenish to white or pale orange, smooth; or dark and rough on old trees or where it's been chewed by browsers; 10" diam × 40' tall. A. Salicaceae.

Quaking aspen is the widest-ranging American tree, growing from Alaska to New England and down through the Rockies and Sierra Madre to Guanajuato. Here it provides much of our fall color—a brilliant pale yellow—and countless benefits to wildlife. Almost two hundred species of birds and mammals use it, from elk and beaver to grouse and pika.

Aspen leaves quake or flutter in light breezes because their flat leafstalks (try rolling one between your fingers) flex easily in one direction only. This fluttering suggested the name *tremula* for the Eurasian aspen, and thence *tremuloides* for its American cousin. Quaking has advantages in photosynthetic efficiency: it allows more light to reach the lower tree canopy, evens out the light hitting all parts of the tree, and may absorb CO_2 more quickly. It also minimizes drag, to keep branches from snapping or trees from uprooting. Leaves of each branchlet clump

Quaking aspen

Quaking aspen female catkins

green layer from sun-scald; that is, together the orange and white layers let just enough light through, but not too much. (This has yet to be scientifically tested.) Some people powder their skin with white aspen cells, as a sunscreen. This strikes me as plausible, and worth trying if you're out on a trail and forgot your sunblock.

Minimized bark offers minimal defense from damage, fungi, or fire, yet aspens are outstandingly fire-adapted. Aspen groves halt some fires simply because their foliage and their understories are moist compared to those of, say, doghair lodgepole pine. Hotter fires, under worse conditions, sweep aspen groves and kill all the trunks; however, the roots, protected by four or five inches of soil, survive all but the most intense fires, and soon send up suckers. This resembles the famous fire strategy of lodgepole pines: die, then immediately reproduce like crazy and shoot up before the competition. But aspens respond even more quickly, since roots store far more energy than seeds.

Fire also triggers germination of aspen seeds buried in the soil; recent burns are the main place where we see aspen seedlings. Most aspen trunks originate not as seedlings but as root suckers. Whole groves are clones, or genetic individuals, all of one sex, with a single, huge root system. A clone's extent is conspicuous in autumn: the timing and hue of fall color are identical throughout a clone, while differing from one clone to the next. Aspen clones can be vast and live a very long time. (A clone near Fish Lake, Utah, dubbed Pando, had 47,000 stems at its peak, covered 106 acres, was around 10,000 years old from the day its original seed germinated, and weighed about 6,600 tons—making it arguably the world's largest living thing [but see p. 268]).

Russian botanist Nikolai Lashchinsky told me that the *P. tremula* clones in his area "never die." Aspen roots there—after three hundred years of invisibility—sent up abundant suckers after grazing cattle were taken away.

together in high winds, forming a single weathervane.

Aspen leaves often end up decorated by leaf miners (p. 472). Swollen points along aspen twigs are galls of the poplar twig-gall fly. When they initiate on young main stems they are a few feet tall, but over the decades they may grow into large swellings encircling the aspen trunk.

Aspens have green trunks that photosynthesize. *What?* you protest. *They're white!* Well, some aspens, especially young ones, actually look green, but most mask their green layer under orange bark cells with a bleached-out, powdery outer layer that continually sloughs off. This is close to the thinnest bark a tree could have, and that's the point—to minimize cork (bark) cells so they don't block incoming light needed for photosynthesis. Paradoxically, the white layer may also, by reflecting light, protect the sensitive

Deer, elk, and cattle so heavily browse young aspens that wide areas commonly go decades with hardly any new stems reaching tree size. Populations languish. People fret. Scientists struggle to tease out data revealing whether this is a normal cycle, or something we can blame on lack of fires or on the absence of wolves leading to heavy deer and elk browsing. Perhaps more cohorts thrived back when Indigenous people set fires and hunted elk. At Yellowstone, studies disagree over how well the reintroduced wolves are solving the browsing problem.

Black cottonwood trunk with lungwort

A newer threat is tellingly called "sudden aspen decline," or SAD. Areas of dieback since 2000 show aspen range shrinking at its drier edges due to climate change. In central Arizona, populations of a widespread sap-sucking insect called oystershell scale have exploded and delivered the coup de grâce—a new development that will be watched. Over its whole range, though, aspen's prognosis is not sad. While aspens lose range at warm, dry, low elevations, they are gaining ground at their cold and Arctic limits, including higher elevations in the Rockies, while growing 50 percent faster than before—fertilized by increased CO_2 in the air—in temperate range where rain increases, such as Wisconsin. In the Rockies and in boreal Canada, they commonly gain ground at the expense of conifers following wildfires and bark beetle outbreaks.

Black cottonwood

COTTONWOOD

Populus spp. ♀ and ♂ catkins on separate trees; round seedpods, in long strings, split to release many tiny seeds with cottony fluff; leaf buds sticky with resin; bark initially smooth, pale gray, breaking into deep furrows toward tree base. Low to mid-elev streamsides. Salicaceae.

Black Cottonwood

P. trichocarpa (try-co-**car**-pa: hairy fruit). Also *P. balsamifera.* Most pods split 3 ways; leaves have a long, pointed tip, and a broad, round to heart-shaped base; glossy dark green above, light gray beneath, bright yellow in fall, 3–6" long (longer and ± diamond-shaped on saplings); leaf buds and their fallen scales honey-scented in spring; leafstalk more than ⅓ as long as leaf blade; vase-shaped tree 30–130' tall × 6' diam. WA to UT and nw WY. Similar Balsam poplar, *P. balsamifera,* with capsule splitting 2 ways, grows (not abundantly) in CO, WY, and MT.

Narrowleaf Cottonwood

P. angustifolia (an-gus-tif-**oh**-lia: narrow leaf). Pods split in 2; leaves most often narrowly lanceolate,

Narrowleaf cottonwood

Plains cottonwood

ing petroleum. A plantation of "hybrid poplars" is all one clone, with cottonwood genes predominating, and they look eerily uniform in their vast grids, thanks to identical genes and highly mechanized irrigation. Some can reach 60 feet by age six.

Alders, willows, and cottonwoods share an ability to colonize fresh gravel bars, a nitrogen-poor substrate. Alders can do that because they fix nitrogen. The mystery of willows' and cottonwoods' similar success may have been solved recently. The microbiologist Sharon Doty found that willow and cottonwood also fix nitrogen, in a new way. It was long thought that plants can only host nitrogen-fixing bacteria in root nodules, which are found in only a tiny fraction of all plant species (not including cottonwoods). Then, in the 1990s, sugarcane was identified as a non-nodulated nitrogen-fixer. Doty's work greatly expands the field. She has successfully fertilized crop plants by inoculating them with bacteria from cottonwood. Like other nitrogen-fixing shrubs, young cottonwoods are choice browse for elk and deer, especially in winter when the trees' protein content is highest.

tapered to the base, 2–4" (longer and much wider on saplings), slightly darker above than beneath; leafstalk longer than a willow's, but shorter than a black cottonwood's; buds mildly fragrant; slender tree, to 65' tall. All but WA and OR.

Plains Cottonwood

P. deltoides (del-**toy**-deez: triangular). Also eastern cottonwood. Pods split 3 or 4 ways; leaves triangular, with coarse, often rounded teeth, 2–4" long and wide, almost same color on both sides; leafstalks flattened, so leaves flutter like aspen; spreading tree, commonly 65' × 4' diam. More common eastward; A–. Similar Rio Grande or Fremont cottonwood of UT, NM, and w CO is either a species, *P. fremontii*, or a subspecies, *P. deltoides* ssp. *wislizeni*.

COTTONWOODS ABOUND ALONG rivers and streams. Their high, stout crotches provide choice nest sites for river-fishing ospreys, bald eagles, and herons. In spring, they ravish with their honeylike scent, which will stick to your fingers if you pick up the sticky bud scales littering the ground.

Black cottonwoods are easily our largest broadleaf trees and our fastest-growing trees of any kind, leading to ideas that they are ideal for plantations. Early plantations trying to produce cottonwood lumber were an abject failure. A hundred years later, the idea is back, this time to produce pulp for paper, chips for fiberboard, biomass (fuel) for power generation, or feedstock for plastics, replac-

WILLOW

Salix spp. (**say**-lix: Roman name). ♂ and ♀ catkins on separate plants; ♂ catkins are the fuzzy or fluffy ones, called "pussy willows" if they appear before the leaves (not true of these spp., exc *scouleriana* and sometimes *pseudomonticola*); winter buds are capped by a bud scale, which they push off in 1 piece; shrub or rarely a small tree, often thicket-forming. Streambanks, wet places, open alp/subalp slopes. Salicaceae.

Umpteen species and hybrids of *Salix* line Rocky Mountain streams and shores. Another three are mat-forming alpine subshrubs about 2 inches high. Most are notoriously hard to identify; the above descriptions enable a guess, at best.

An anomaly among willows, the Scouler willow can grow in shade or on slopes far from water. It flourishes after severe fires, either invading with wind-carried seeds or resprouting from deep roots; the seeds typically grow into trees, the sprouts into multistemmed shrubs. It can also grow on gravel bars, but rarely exceeds 4 feet there.

Around timberline in Colorado, grayleaf and short-fruit willow dominate vast thickets that are increasing, aided by both climate change and nitrogen deposition. Scientific papers express alarm at this trend, which comes at the expense of tundra wildflowers. Meanwhile, other scientific papers raise the alarm over precipitous *declines* in willows in Colorado; these would be the riparian willows, those that benefit from groundwater near streams. Their decline is blamed on many interconnected changes: changes in streamflow, populations of beavers, populations of hoofed browsers, and depth and duration of snow, as well as the absence of predators of the hoofed browsers and an epidemic of a fungal canker that infects the holes sapsuckers drill.

Willow shoots are choice browse. They're the staple food of moose. Beavers eat some willows, but greatly increase the net number of willows as they expand wetlands: the cut stems beavers emplace in dams often take root and produce new willows.

Indigenous people twisted willow bark into twine for fishnets, baskets, and tumplines (forehead straps for loads carried on the back). Poles were cut from willows to frame sweat lodges or to support fishing platforms—which became all the more secure when they took root and grew where implanted in the riverbed. The bark and roots were used in many ways to relieve pain and inflammation, reduce fever, or stop bleeding, including menstruation. European herbalists discovered similar uses, leading to the synthesis in 1875 of a less acidic form, acetylsalicylic acid (a.k.a. aspirin). Though aspirin is used as a blood thinner, another willow bark component, salicin, works as a topical astringent to stanch blood flow.

Scouler willow

Scouler Willow
S. scouleriana (scoo-ler-ee-**ay**-na: for John Scouler. Also fire willow. Flowering early, occasionally before snowmelt; leaves 2–4", broadly obovate with a pointed tip, whitish- or reddish-velvety underneath; newer stems velvety; bark often skunky-smelling when crushed; shrub or small trees 3–40' tall. Diverse habitats; A.

Whiplash Willow
S. lasiandra var. *caudata* (la-zy-**ann**-dra: woolly stamens; caw-**day**-ta: tailed). Also *S. lucida* ssp. *caudata*. ♀ catkins 1–5" long, on short twigs with a few leaves; leaves 2–7", fine-toothed, with a long drawn-out tip (the "whiplash" or "tail"),

Whiplash willow

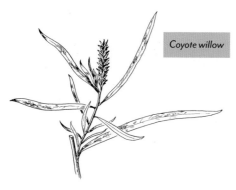
Coyote willow

about equally green on both sides; leafstalk has tiny, dark, spherical glands just below the leaf base; shrub or multi-stemmed small tree, 3–35' tall. Abundant at low to (rarely) subalpine elevs; A.

Coyote Willow

S. exigua (ex-**ig**-you-a: short). Also sandbar willow. Leaves 10× longer than wide, sparsely toothed, same greenish gray on both sides; mostly 5–10' tall. Higher parts of river bars and banks; A.

Geyer Willow

S. geyeriana (guy-er-ee-**ay**-na: for Karl Geyer, p.142). Catkins tan, short, ¼–1"; leaves lanceolate, 6–8× longer than wide, paler beneath but often ± hairy on both sides in midsummer; smaller stems bluish-waxy-coated; 8–18' tall. A–.

Barclay Willow

S. barclayi (**bar**-clay-eye: for a Mr. Barclay, who collected it at Kodiak in 1858). ♀ catkins are on short side-branchlets with a few leaves; leaves

elliptic to obovate, toothed, green at every age, finely hairy even at maturity, at least on midrib; older stems shiny dark red; 4–12' tall. Thicket-forming at alp/subalp elevs. Northerly.

Long-beak Willow

S. bebbiana (beb-ee-**ay**-na: for Michael Bebb). ♀ catkins ¾–3", appearing loose due to long pedicels; each capsule has a long, drawn-out tip; leaves dark green above, waxy-pale beneath with prominent netted veins; new twigs purplish, fuzzy; older ones white-streaked gray; large shrub or small tree 7–16' tall. A.

Serviceberry Willow

S. pseudomonticola. Leaves elliptic to obovate, toothed, reddish, and finely hairy when young, maturing smooth and green; mostly 3–12' tall, but can grow as an alpine mat. Northerly.

Rock Willow

S. vestita (ves-**tie**-ta: with a coat). Catkins on 3-leaved side branchlets; leaves oval, heavy, dark green above, with veins ± deeply incised in a net pattern; 6–48" tall. n ID, n MT.

Grayleaf Willow

S. glauca (**glaw**-ca: whitish-coated). ♀ catkins ¾–2½" long; young leaves and branchlets whitish-woolly; older leaves glossy green above, gray beneath, elliptic to oblanceolate, 1–3"; mostly 2–5' tall. Rampant on alp/subalp slopes in CO; A.

Short-fruit Willow

S. brachycarpa (bracky-**car**-pa: short fruit). Similar to grayleaf willow, and they hybrid-

Long-beak willow catkins

Serviceberry willow

Grayleaf willow

Rocky Mountain willow

Snow willow

Cascade willow

ize, so ID is not always possible, but the ♀ catkins (the ones like strings of seeds) tend to be shorter, just ¼–¾" long. A.

DWARF WILLOW

Salix spp. ♂ and ♀ catkins on separate plants; twigs yellowish to purplish; mat-forming shrub usually less than 4" tall. Alpine.

Rocky Mountain Willow

S. petrophila (pet-**rah**-fil-a: rock-loving). Also *S. arctica* var. *petraea*. Catkins maturing after leaves, 20-to-80-flowered, ¾–2½" long; leaves ¾–2" long, ¼–¾" wide, round or point-tipped, often grayish with hair on both sides. A.

Snow Willow

S. nivalis (niv-**ay**-lis: of snow). Also *S. reticulata*. Catkins maturing before leaves, 4-to-17-flowered, ¼–⅝" long; leaves oval, round-tipped, ⅜–1½", green above, grayish but hairless beneath, conspicuously net-veined, leathery. A.

Cascade Willow

S. cascadensis. Catkins maturing after leaves, 15-to-35-flowered, ⅜–1" long; leaves elliptic, point-tipped, ¼–½" long, barely ¼" wide, about equally green above and below, leathery. CO, WY, UT, MT.

TRUE WILLOWS REDUCED to carpet stature abound in alpine tundra, as they do in the Arctic. Some of these species grow taller elsewhere. An Arctic willow in Greenland had 236 annual rings.

GAMBEL OAK

Quercus gambelii (**quirk**-us: Roman name; **gam**-bel-ee-eye: for William Gambel). ♂ catkins release copious yellow pollen; tiny ♀ catkins, on same tree, consist of a 3-styled pistil in a cup-shaped involucre that hardens into the cap of a ½–1" acorn; leaves 2–5", deeply pinnately blunt-lobed, rather glossy above, often velvety beneath, coloring yellow to dark red in fall; bark gray, furrowed; medium to tall shrub, or occasionally a tree up to 18" diam × 35' tall. Foothills up to mid-elevs; southerly. Fagaceae.

The montane brush fields of Colorado and New Mexico feature an abundance of this oak in its shrubby form. It provides a lot of acorns, a vital food resource for many

Gambel oak catkins and leaves

animals, yet most of the young oaks grow from root sprouts rather than acorns.

Expect to see more Gambel oak in the future. The trend toward large, severe wildfires favors plants that resprout after their aboveground parts burn. As the Southern Rockies lose forested area, a frequent replacement is montane brush dominated by Gambel oak. (Chaparral in California and Arizona plays a similar role, but differs in being mostly evergreen shrubs.) The Jemez Mountains have large brush patches that replaced forest 115 years ago and show no sign of succeeding to conifers. Montane brush is highly flammable, and when it reburns, resprouting perpetuates the conversion from forest to brush.

On the other hand, in some settings oaks are nurse plants, helping conifer seedlings get through a dry summer, both by shading them and by drawing moisture from deep soil. In the moonscape part of the Las Conchas Fire scar, planting ponderosa pines experimentally under Gambel oaks more than doubled their chance of surviving three years. Still, growing an actual new forest might require keeping fire out for many decades—a tall order in a warming climate.

ALDER

Alnus spp. (**al**-nus: Roman name). ♂ and ♀ catkins on the same tree; ♀ catkins woody, like miniature spruce cones, ½–1" long; leaves 1½–4", oval, pointed, toothed or doubly toothed, ± wavy-edged, turn from green to brown late in fall; bark ± smooth, gray or reddish. Betulaceae.

Mountain Alder

A. incana (in-**cay**-na: whitish). Also thinleaf or river alder; *A. tenuifolia*. Catkins appear in summer, and lengthen and open the next spring before leaves appear; ♀ catkins on short stalks; leaf undersides whitish; tall shrub or tree often 20–33'. Bogs, moist forest, abundant on streamsides; A.

Green Alder

A. alnobetula ssp. *sinuata* (al-no-**bet**-you-la: alder-birch; sin-ew-**ay**-ta: wavy-margined). Also slide or Sitka alder; *A. viridis, A. sinuata*. Catkins appear in spring, simultaneous with leaves; ♀ catkin stalks often longer than catkins; leaf undersides often sticky, rarely whitish; sprawling shrub 5–16' tall. Higher-elev streamsides, seepy slopes, avalanche basins and tracks; northerly.

ALDERS HOST BACTERIA in nodules on their roots that convert atmospheric nitrogen for plant use, fixing some in the soil and some directly into the plant. Alder leaf litter is

Mountain alder with red gall on catkins

Green alder catkins

Water birch   Resin birch   Paper birch

plentiful, nitrogen-rich, and quick to decompose and cede its nitrogen to the soil.

A green alder has the genetic information for growing upright, as a tree, but it also knows how to flex and bow and sprawl where its environment demands. We know it best where it's at its worst—the downhill-sprawling "slide alder" of avalanche tracks. What the roaring avalanche finds accommodating, the sweating bushwhacker finds maddeningly intransigent, a tangle of springy, unstable stems always slipping us downslope like flies in the hairy throat of a pitcher-plant. If, like the avalanche, we could stick to downhill travel, we'd have little problem with slide alders.

Almost every Native group that lived near these alders used their inner bark for red and orange dyes—one application being on fishing nets, to make them invisible underwater. Catkin tea was said to improve children's appetites. Alder seeds, buds, and catkins provide critical winter food for small birds that stay year-round, such as pine siskins and chickadees.

BIRCH

Betula (**bet**-you-la: Roman name). ♀ catkins (on same plant as ♂) erect; their core stems persist after dropping bracts and winged seeds; twigs usually downy; bark has raised, horizontal dark streaks ("lenticels"). Betulaceae.

Water Birch

B. occidentalis (oxy-den-**tay**-lis: western). Leaves pointed-oval, 1–2½", with teeth of 2 sizes, turn yellow in fall; young branches covered with tiny, wartlike resin dots; bark smooth, dark red to bronze; shrub or tree up to 35' tall. Streambanks and shores; A.

Resin Birch

B. glandulosa (gland-you-**low**-sa: with sticky glands). Also swamp birch. Leaves oval to round, ⅜–1", sawtoothed, heavy, shiny, leathery, often sticky, turn orange to russet in fall; young branches covered with sticky, wartlike glands; bark dark red or gray, not peeling; shrub 3–10' tall. Streambanks, willow thickets, bogs, seeps; A.

Paper Birch

B. papyrifera (pap-er-**if**-er-a: paper-bearing). Leaves pointed-oval, 1½–4", with teeth of 2 sizes, turn yellow in fall; leaf undersides have tiny hair tufts in the forks of veins; bark bronze on young trees, becoming ± chalky white, peeling in papery sheets; tree up to 100' tall × 3' diam. Sunny moist slopes, bogs; northerly, plus just a few in CO and s WY. All these birches hybridize rather freely: ne OR and c ID have hybrids with light brown bark.

AN EXCELLENT WOOD for arrow shafts, birch was used in the Rockies ten thousand years ago to make meter-long darts that hunters hurled with an atlatl. Interior British Columbia peoples used paper birch bark to make canoes and beautiful baskets; the Dakelh even made foldable watertight canoes. By carefully peeling only the outer layer, they could harvest it without killing the tree. (Don't try this if you aren't trained in the tradition.) Peeled bark made a reliable tinder for starting fires, or a long-lasting torch when rolled up. Eastern peoples inscribed rot-resistant birch bark with pictographic scrolls

Greasewood leaves and seed pods

Spiny hopsage

Fourwing saltbush

inner bark was used for a brown dye. Birch bark tea found use as a contraceptive and abortifacent.

GREASEWOOD

Sarcobatus vermiculatus (sar-co-**bay**-tus: fleshy bramble; ver-mic-you-**lay**-tus: wormlike). ♂ flowers in 1"-high spikes range from pale green to dark pink; each ♂ flower a rhomboidal shield hiding stamens; at base of spike a few ♀ flowers, 2 styles in a tiny cup, enlarge into a flat, crinkled seedpod, which may ripen pink; leaves wormlike, ¾–1½"; twigs pale gray; spiny, bushy shrub 3–8' tall. On heavy alkaline soils of dry flats, often with saltbush, which it resembles; A. Sarcobataceae.

> *extreemly troublesome;…when I mention it hereafter I shall call it the fleshey leafed thorn.*
> —Meriwether Lewis, May 11, 1805

Livestock seem to like the salty taste of greasewood leaves and find some nourishment in them, but if they eat too much in a short time they can be poisoned by excess oxalate crystals. Die-offs have felled flocks of sheep.

HOPSAGE

Grayia spinosa (**gray**-ia: for Asa Gray, p. 217). Also *Atriplex grayi*. Usually with ♂ and ♀ flowers on separate plants; both kinds of flowers tiny, greenish, clustered; seed enclosed in an oval or round papery bract ⅜–⅝" long, initially greenish, maturing pink (showy) in midsummer; leaves around 1", thickish, ± elliptic, tips often salt-dusted; young twigs reddish, side branches spine-tipped; bushy shrub 1–5' tall. Juniper, sagebrush; w CO and NM to WA and sw MT. Amaranthaceae.

FOURWING SALTBUSH

Atriplex canescens (**at**-ri-plex: Roman name; cay-**nes**-enz; turning gray). Usually with ♂ and ♀ flowers on separate plants; ♂ flowers like small round buds, yellowish, in small spikes in

recording events, birch trunk have endured for generations.

Water birch bark was also used for canoes and baskets, when big-enough trees were found, but that was rather rare. Its

leaf axils; ♀ crowded in ½–2" spikes, lacking petals or sepals, but each has a bract that, as the seed matures, grows 4 lengthwise wings up to 1" long and wide; bracts initially whitish green, drying to tan and conspicuous through winter; leaves ± narrow, entire, to 2" long, coated with scurfy gray stuff; bushy shrub 1–6' tall. Juniper, sagebrush, desert, often in saline soil; A–. Amaranthaceae.

Puebloan people preferred saltbush ashes for nixtamalizing maize (dehulling kernels in an alkaline soak before turning them into tortillas or hominy). Navajo ate saltbush seeds after grinding them into flour.

Quite a few plant species have separate male and female plants. Saltbush goes much further: from any given year to the next, around 20 percent of individuals change sex, typically becoming male either following unfavorable (dry or frigid) months or after exhausting their fecundity in a very good preceding year.

Rocky Mountain maple

Pair of winged seeds

With Opposite Leaves

Opposite leaves are a quick and easy first step in ID, provided you don't mistake the leaflets of pinnately compound leaves for opposite leaves.

ROCKY MOUNTAIN MAPLE

Acer glabrum (**glab**-rum: smooth). Also Douglas maple. Flowers small, green, clustered; 5 petals and 5 similar sepals; seeds with 1" long, straight-backed wings, in pairs at ± right angles; leaves opposite, 3- (or 5-) lobed, toothed either sharply or bluntly, yellow to red in fall, 2–4½", on equally long leafstalks; twigs ± red; bark gray to purplish; shrub or tree up to 40'. In var. *glabrum* the twigs are gray and leaves are smaller, deeply lobed to almost compound. Ravines, floodplains, burns, clearcuts, and slopes, below 10,000 feet; A. Aceraceae.

Brilliant red blotches that appear on maple leaves in midsummer may be induced by either fungal infection or velvet gall mites. In either case, the mechanism is like that of fall color: vessels carrying photosynthetic sugars out of the leaves get blocked, and the trapped sugars are converted into red pigments called anthocyanins. In winter, the twigs provide vital browse for deer, elk, and sheep.

The hard, strong wood bends well after soaking and heating, making it a favorite material for snowshoe frames, lodge frames, fishnet handles, drum hoops, digging sticks, spoons, arrows, and more. Maple bark supplied fiber for tying and weaving.

BOX ELDER

Acer negundo (neg-**oon**-doe: Sanskrit name of a supposedly similar plant). Also Manitoba maple; *Negundo aceroides*. ♂ and ♀ flowers on separate plants; no petals; ♂ flowers make dangling, silky pink tassels 2" long; ♀ flowers make somewhat shorter, less showy tassels; seeds with 1" long, curving wings, in pairs, persist on tree well into winter; leaves compound; 3 or 5 leaflets resembling poison ivy (pointed, with a

Box elder female   Box elder male   Red osier dogwood

few coarse teeth); side leaflets have short leaf-stalks; yellow in fall. Trees to 35' tall, 3' diam, or sometimes thicket-forming shrub. Streamsides and washes in dry lowlands; 48 states and 9 provinces.

Box elders are fast-growing, short-lived, rather weak trees. Their winged seeds, like those of other maples, spin like helicopters as they fall; they are fatally toxic to horses if eaten in quantity—a connection first discovered in 2012, after the syndrome had been a mystery for decades.

The Cheyenne boiled maple syrup from box elder sap. "Box elder" is a double misnomer based on supposed resemblances to boxwood and elderberry plants.

RED OSIER DOGWOOD

Cornus sericea (ser-**iss**-ia: silky). Also creek dogwood; *C. alba, C. stolonifera.* 4 white petals ⅛"; 4 stamens; 4 tiny sepals; flowers in flat-topped clusters; berries dull pale bluish or greenish, ¼", single-seeded, unpalatable; leaves opposite, 2–5", elliptical, pointed, wavy-edged, the veins curving around to merge along the margin, coloring richly in fall; new twigs deep red or purple; shrub 6–16' tall. Wet places; A. Cornaceae.

This dogwood's flowers lack the big white bracts we think of as dogwood flowers. You can see a resemblance, though, in the tiny true flowers, as well as in the leaves' outline, venation, and fall coloring. The young stems (osiers) also turn dark red through winter.

Osier is an old French word meaning a long, new shoot, originally of willow, suitable for wicker. "Dogwood" derives from the Scandinavian *dag*, for "skewer." Shuswap and Okanagan peoples also liked this species for skewers, and for racks for roasting and drying salmon—for the nice, salty flavor it infused. They relished the dogwood-smoke flavor in dried berries and smoked fish. The bark was dried for smoking. The adventurer Prince Maximilian of Wied noted in 1833 that Native people "inhaled the smoke, a custom that is, doubtless, the cause of many lung diseases. The mixture the Indians of this part of the country smoke, is called kini-kenick, and consists of the inner green bark of the red willow, dried, and powdered, and mixed with the tobacco of the traders."

You may wonder why these berries, like many others, evolved to be dry, mealy, and flavor-challenged. Berries exist to be eaten and then pooped out by wildlife, distributing seeds away from the parent plant. What would attract an animal to berries offering precious little sugar or protein?

Black twinberry

Mountain snowberry

Utah honeysuckle

The attraction is that they are equally unattractive to bacteria. They become a critical food resource in winter, when all the sweet, juicy berries have decomposed or been eaten. In the case of red osier dogwood, the berries' fat content makes them an important resource for bears, while the stems are a top winter food for moose and elk.

This very cold-tolerant shrub is common in Arctic tundra, surviving winter by forcing so much water out of its cells that freezing can't damage them.

BLACK TWINBERRY

Lonicera involucrata (lo-**niss**-er-a: for Adam Lonitzer; in-vo-lu-**cray**-ta: with involucres). Also bearberry honeysuckle; *Distegia involucrata.* Corolla yellow, hairy, ⅜–¾" long, tubular with 5 short lobes; calyx inconspicuous; flowers in pairs on a short stalk from leaf axils, flanked by 2 pairs of heavy, hairy bracts that mature deep to carmine red around glossy black ¼" berries; leaves elliptic, pointed, 2–5", smooth

above, often hairy beneath; shrub 2–8' tall. Streambanks, seeps; A. Caprifoliaceae.

In times past, black twinberries were called "crow food" (Crow being the only spirit crazy and black enough to relish such bitter black fruit), "grizzly berries" (same reason), or "inkberry" because you could paint dolls' faces or dye graying hair with the juice.

Black twinberry plants twin to the nth degree: opposite leaf axils bear opposite stalks, each bearing a pair of flowers between two pairs of hairy bracts, two of them two-lobed. The bracts turn deep magenta and reflex downward over time, bracketing paired purple-black berries. This display usually hides in a damp thicket.

UTAH HONEYSUCKLE

Lonicera utahensis. Flowers without conspicuous bracts or calyx; corolla white to pale yellow, ⅜–¾" long, tubular, with 5 flaring lobes, hairy inside but not out; berries red to pink, naked; leaves oblong, rounded at tip, 1–3", smooth above, sometimes hairy beneath; shrub 3–6' tall. Moist, ± wooded slopes; UT to WA.

SNOWBERRY

Symphoricarpos spp. (sim-for-i-**car**-pus: gathered fruit). Also waxberry. Flowers pink or white, ¼–⅜"; berries white, paired or clustered, pulpy, 2-seeded; leaves and twigs opposite, most leaves oval to elliptic. Caprifoliaceae.

Mountain Snowberry

S. rotundifolius (ro-tund-if-**oh**-lius: round leaf). Also roundleaf snowberry; *S. oreophilus.* Flower a tubular, bell shape, about twice as long as it is wide, petals fused at least ½ their length;

Common snowberry   *Blue elderberry*   *Red elderberry*

berries ¼–⅜"; leaves ½–1½", ± crowded, firm, veiny, slightly rolled under at edges, rarely lobed; robust shrub 2–4½' tall, spreading by root suckers. Open forest and dry meadows; A.

Common Snowberry

S. albus (**al**-bus: white). Flower full of long hairs, broadly bell-shaped, scarcely longer than wide, petals fused at least ½ their length; berries ½"; leaves 1–2" or much larger, often lobed; shrub 2–10' tall. Thickets, open forest; A.

Western Snowberry

S. occidentalis (oxy-den-**tay**-lis: western). Flowers broadly bell-shaped, about equally long and wide; 5-lobed flaring corolla fused less than ½ its length; style protrudes from corolla; twigs opposite, reddish, maturing gray and flaking; 1–3" leaves may have a few blunt teeth; shrub 1–3' tall, spreading by rhizomes to form dense thickets. Streamsides at low to mid-elevs; A–.

"POPCORN" CAPTURES THIS berry's essence better than "snow" or "wax," though at the risk of falsely tempting hungry hikers. The lightweight berries, though eaten and disseminated by birds and rodents, are mildly poisonous to humans, causing vomiting, dizziness, and mild sedation in children.

Fast-growing, non-flowering shoots of *S. albus* have variable leaves, sporting all numbers and sizes of odd-shaped lobes.

ELDERBERRY

Sambucus spp. (sam-**bew**-cus: Greek name derived from use as a flute). Flowers creamwhite, tiny, in dense clusters; flower parts in 5s; pinnately compound (rarely twice-compound), 5–12" long; leaflets narrowly elliptical, pointed, fine-toothed, ± asymmetrical at base; berries ¼", in large clusters, 3-to-5-seeded; stems pithfilled, weak; shrub or shrubby trees 3–20' tall. Streamsides, thickets, moist forest, or clearings; A. Viburnaceae.

Blue Elderberry

S. cerulea (se-**rue**-lia: sky blue). Also *S. nigra, S. mexicana*. Flowers and (later) berries in ± flat-topped clusters without a single central stem; berries blue (blue-black with a waxy bloom); leaflets usually 7, 9, or 11; leaf undersides and twigs whitened.

Red Elderberry

S. racemosa var. *racemosa* (ras-em-**oh**-sa: bearing racemes). Also

S. racemosa var. *microbotrys, S. microbotrys*. Flowers/berries in ± conical racemes with a main central stem; berries red, bitter; leaflets usually 5 or 7, not whitened.

Rocky Mountain elderberry

Mooseberry

Rocky Mountain Elderberry

S. racemosa var. *melanocarpa* (me-lon-o-**car**-pa: black fruit). Also *S. racemosa* var. *pubens, S. melanocarpa, S. microbotrys* var. *melanocarpa.* Like red elderberry, but berries black to very dark red; shrub mostly 3–6' tall.

WHERE RED AND BLUE elderberry grow together, the red elderberry blooms, sets fruit, and ripens a good month ahead of the blue. Some botanists consider our blue elderberry conspecific with the Old World black elderberry, *S. nigra,* the source of most elderberry wine, and they are indeed better eating than our red or black ones. Native people ate berries of both species, fresh or more often steamed, and stored until lean times. The Okanagan devised a way to refrigerate them: they waited to harvest blue elderberries until just before the first snows, then spread the bunches on a thick layer of ponderosa-pine needles. Over this they laid another layer of needles, soon to be covered by snow. The berries stayed moist and just above freezing. They were easy to locate once their juice stained the snow surface.

Elder flowers, bark, leaves, twigs, and roots—and also berries—have long been seen as both toxic and medicinal. Foliage and bark contain cyanide, a defense against browsing. Indigenous people used the bark to induce perspiration, lactation, or vomiting, or alternatively to reduce swelling, infection, or diarrhea—an odd mix of prescriptions. They hollowed the soft pith out of elder stems to make peashooters for children, pipestems, whistles to lure elk, and drinking straws for girls subject to ritual restrictions during puberty. Rich in antioxidants, the berries have health benefits confirmed in many studies. They can be mildly toxic when underripe, and are safer to eat after cooking.

MOOSEBERRY

Viburnum edule (vie-**burn**-um: Roman name; **ed**-you-lee: edible). Also squashberry, bush-cranberry. Flowers white, in small clusters, ¼", with 5 petals and 5 calyx lobes; berries red-orange, juicy, tart, aromatic, with 1 flat seed; leaves opposite, 1–4" long and ± equally wide, sharply toothed, the bigger ones 3-lobed, turning crimson in fall; shrub 3–7' tall. Low to mid-elev moist to boggy forest; all exc UT and NM. Viburnaceae.

These sour berries were popular with Indigenous people, especially because they can often be picked and eaten, frozen, in midwinter.

CANADIAN BUFFALOBERRY

Shepherdia canadensis (shep-**er**-dia: for John Shepherd). Also soopolallie, soapberry. Flowers tiny, yellowish, 4-lobed, ♀ and ♂ on separate plants; berries ¼", yellow to red, bitter, tipped with 4 flat-spreading calyx lobes; leaves opposite, elliptical, ½–2¼", green above; undersides have brown, scabby dots and a silvery, scurfy coating; shrub 3–10' tall. Shrubfields and open forests below timberline; A. Elaeagnaceae.

From Montana to the Bering Sea, "Indian ice cream" was made by whisking bitter buffaloberries and water into a rich froth. It's an acquired taste. Folks were, however, happy to add sugar once it became available.

Canadian buffaloberry

White virgin's bower

Blue virgin's bower

Buffaloberry seems almost indifferent to habitat: a Forest Service report on Montana forests divides them into one hundred types, and found buffaloberry in seventy-two, but no one of them had buffaloberry covering more than 13 percent of the ground. Hosting nitrogen-fixing bacteria on its roots undoubtedly helps.

CLEMATIS

Clematis spp. (**clem**-a-tis: Greek name for some vine). 4 petal-like sepals; no petals; many stamens; seeds with long, plumy tails, in a dense ball; leaves compound. Ranunculaceae.

White Virgin's Bower

C. ligusticifolia (lig-oos-tis-if-**oh**-lia: leaves like privet). Sepals white, ½", spreading flat; ♀ plants have numerous ½" pistils; usually 5 or 7 leaflets; vine climbing 20' or more on trees or fences, becoming covered with fluffballs (seed heads). Valley floors; A.

Blue Virgin's Bower

C. occidentalis (oxy-den-**tay**-lis: western). Also *Atragene* or *C. grosseserrata*. Sepals pale blue-purple, 1¼–2¼" long, wrinkly and rather floppy, may open wide; flowers on 2–6" stalks scattered singly along the vine; 3 lanceolate leaflets; low climbing or trailing woody vines to 8' long. Open forest; A.

Rock clematis

Rock Clematis

C. columbiana. Also *Atragene columbiana, C. tenuiloba, C. pseudoalpina*. Flowers blue-purple, 1¼–2¼" long, usually bell-shaped; 7 or 9+ leaflets, often deeply lobed; either a 4" subshrub mat (the variety among alp/subalp rocks) or a low, trailing vine (lower-elev open forest); A.

LIKE MANY OTHER members of its family, virgin's bower contains toxic alkaloids and has been applied externally. The Okanagan used it as a rinse to prevent gray hair, but The Nlaka'pamux used it for eczema. The Blackfoot

Botanizing Alpine Tundra

Edwin James became, in 1820, the first botanist to visit the alpine zone in Western North America, and the first Euroamerican to ascend a 10,000-foot-plus peak there, Pikes Peak (14,115 feet). His tundra firsts include discovering and naming the alpine bluebells. Major Stephen Long, the expedition's commander, named the mountain James had climbed "James' Peak," mistakenly believing that they had already passed Pikes Peak farther to the north. That northern peak ended up with the name Longs Peak, and poor James ended up without a peak in his name.

(Captain Zebulon Pike had set off to climb his peak in 1806, but turned around some 15 miles short, reporting that "no human being could have ascended to its pinical [sic]." Since hordes ascend Pikes Peak today, the statement is often ridiculed. Unfair! He was claiming that no others could have done better under the circumstances, namely with "only light overalls on, and no stockings," on snow, at –4°F, with no food except what game they might hope to shoot and no sign of any game since leaving distant river bottoms.)

Long's expedition was trumpeted as the Yellowstone Expedition, but budget-cutters in Washington trashed that objective and redirected Long when he reached present-day Omaha. (He traveled that far by steamboat in 1819, just twelve years after Lewis and Clark's return; the West was developing fast!) A second botanist, **Thomas Say**, was assigned to cover "Zoology, &c.," including "the diseases, remedies, &c. known amongst the Indians." Say remains better known to birders than botanists. Late in the trip, a group of soldiers deserted after stealing the saddlebag containing all of Say's notes, which they doubtless jettisoned while searching in the bag for anything useful.

used it as a charm against ghosts and to heal "ghost bullet" wounds.

CLIFFBUSH

Jamesia americana (**james**-ia: for Edwin James, see above). Also waxflower. Flowers ½–1" wide, with 5 white or pinkish petals, fragrant, in clusters of 7–35; 5 sepals; fruit a dry capsule; 2" leaves oval, toothed, veins recessed, white-woolly beneath; bark red-brown, shreddy; shrub 2–6' tall. Wet rocks; southerly; in CO mainly e of CD. Hydrangeaceae.

CLIFF FENDLERBUSH

Fendlera rupicola (**fend**-ler-a: for Augustus Fendler, p. 246; roo-**pic**-a-la: rock dweller). Also *F. falcata*. Flowers 1–1¾" across, with 4 broad, spoon-shaped petals red-tinged

Cliffbush

Cliff fendlerbush

Littleleaf mock-orange

white, from deep-pink buds; often numerous and showy; 4 sepals, 8 stamens; leaves narrowly elliptic, untoothed, ½–1¾"; shrub 3–8' tall. Rocky slopes, cliffs; w CO, NM, se UT. Hydrangeaceae.

Navajo made arrow shafts, weaving forks, and planting sticks from the wood, and used the plant ceremonially.

MOCK-ORANGE

Philadelphus spp. (fil-a-**del**-fus: for Ptolemy Philadelphus, king of Egypt). Flowers 1–1½", white, in showy clusters; 4 oval petals, 4 sepals, many yellow-tipped stamens; leaves oval to lanceolate, with 3 main veins from the base; shrub 3–10' tall. Hydrangeaceae.

Lewis's Mock-orange

P. lewisii (lew-**iss**-ee-eye: for Meriwether Lewis, p. 241). Also syringa; *P. trichotheca, P. inodorus.* Flowers fragrant; leaves 1½–3"; shrub 3–10'. Streamsides, open woods; WA to MT.

Littleleaf Mock-orange

P. microphyllus (micro-**fill**-us: small leaf). Flowers sometimes conical; 4 white petals, many yellow-tipped stamens; leaves 1–1½"; shrub 2–5' tall.

Lewis's mock-orange

Rocky slopes, open woods, brushfields; w CO, NM, UT, sw WY.

AS IF A MOCK-ORANGE thicket in the July sun weren't heavenly enough already, you're likely to find it aflutter with swallowtails as well. Idahoans made an excellent choice of state flower, but I'm puzzled that they call it *syringa,* the Latin name for garden lilacs (no relation).

With Alternate Leaves, 5 Petals, 10+ Stamens

Some, but not all, Rose family shrubs have ten or more stamens.

SERVICEBERRY

Amelanchier alnifolia (am-el-**an**-she-er: archaic French name; al-nif-**oh**-lia: alder leaf). Also saskatoon, shadbush. Inflorescence of 3–15 flowers is often a jumble of petals; berries ½", several-seeded, red, ripening to purplish black, with bloom; leaves 1–2", oval, toothed at tip but not at the base; petals white, ½–1" long, narrow; low, spreading shrub or tree up to 30' tall, usually less than 10'. Sunny slopes; A. Rosaceae. Similar Utah serviceberry, *A. utahensis,* typically has smaller flowers.

Meriwether Lewis said "sarvisberry," as many Easterners still do. The word derives

Serviceberry   *Utah serviceberry*   *Chokecherry*

not from "serving" but from *Sorbus,* the European service tree, which is closely related to our mountain-ash (p. 98).

The berry is sweet and good, resembling duller, seedier blueberries. They rivaled chokecherries as the chief berry for pemmican. Birds, bears, and elk love them, and also love the shrub's branches—sometimes too well, preventing the plants from bearing fruit.

CHOKECHERRY

Prunus virginiana (**prune**-us: Roman for "plum"; vir-gin-ee-**ay**-na: of Virginia). Also *Padus virginiana.* Flowers ⅜", white, with 20–30 bright-yellow stamens, showy, in dense racemes 2–6" long; cherries oval, ¼", sweetish but astringent, crimson to black; leaves 2–4", oval, pointed, fine-toothed; tall shrub or small tree up to 30' tall. Thickets and edges of clearings; A. Rosaceae.

Chokecherries are named for their powerful puckering effect. Fortunately, their sweetness preserves better than their astringency; Indigenous people pounded them into pemmican, while settlers boiled them into jelly and jam. Pemmican was said to contain pounded chokecherry pits; we now know that *Prunus* pits all contain amygdalin, which breaks down into cyanide, and have been known to kill small children. (The genus includes cherries, prunes, and plums.) Ethnographer Nancy Turner speculates that traditional drying may reduce or eliminate the toxins. Chokecherry leaves are choice browse for deer and elk, but in large quantities have killed cattle.

Chokecherry is a common host for tent caterpillars (p. 472). These moths have a population outbreak every ten years or so, consuming billions of leaves down to the midvein and tenting the remains under cobwebby shelters. This stunts the plants' growth for the year, but rarely does any lasting damage.

Infusions of chokecherry leaves, twigs, or bark were popular medicines in times past. Both Anglo- and Native Americans prescribed them for diarrhea, intestinal worms, and other stomach ills. The Okanagan used the infusions to prevent postpartum stretch marks. The Blackfoot and others soothed sore throats and coughs with cherry juice, either fresh or reconstituted from dried cherries—the original cherry cough syrup. Meriwether Lewis, camped on the Upper Missouri in 1805, cured his own case of severe stomach cramps with two pints of "a strong black decoction of an astringent bitter taste," brewed from chokecherry twigs. He already knew the plant, since it grows all the way to the East Coast (whence the name *virginianus*). While crossing the Bitterroots, the party held starvation at bay with chokecherries.

Western mountain-ash

Alderleaf mountain-mahogany

MOUNTAIN-ASH

Sorbus spp. (**sor**-bus: Roman name). Flowers white, fragrant, ½", in dense, flat-topped 1¼–4" clusters; berries red to orange, ⅜", several-seeded, mealy, bitter; leaves pinnately compound, 7–13 leaflets 1¼–2½", oblong to elliptical, fine-toothed at the tip but not at the base (compare serviceberry, p. 96, and wildroses, p. 100), bright yellow and orange in fall; shrub (rarely small tree) 3–14' tall. Moist, ± open forest. Rosaceae.

Western Mountain-ash

S. scopulina (scop-you-**lie**-na: of rocks). Leaflet tips pointed. A.

Sitka Mountain-ash

S. sitchensis (sit-**ken**-sis: of Sitka, Alaska). Leaflet tips round in outline. OR to nw MT.

PERSISTING ON THE BUSH through winter, mountain-ash berries are important forage for birds and small subalpine mammals. Judging by reports of birds flying "under the influence," the berries have enough sugar to ferment on the bush, though an acrid taste and mealy texture hide the sweetness. There are few reports of Indigenous people using them. The sweeter berries of the European mountain-ash or rowan tree, *S. aucuparia,* have been known to produce a wine.

ALDERLEAF
MOUNTAIN-MAHOGANY

Cercocarpus montanus (sir-co-**car**-pus: tailed fruit). Flowers yellow, 1–3 per leaf axil, ½" across, with numerous stamens their showiest parts; 5-lobed calyx has fine wool; seeds have 2–3" feathery tails; leaves oval, toothed, blades ¾–1½" long, finely woolly at least underneath;

Red hawthorn

branches ± reddish; shrub 3–8' tall with mostly vertical branches. Dry, rocky slopes at all elevs; WY to NM. Rosaceae.

Some call this an evergreen, as the leaves last through winter in some areas. Curl-leaf mountain-mahogany (p. 115) is more reliably evergreen.

HAWTHORN

Crataegus spp. (cra-**tee**-gus: strong). Flowers ½–¾" across, white, in small clusters; 10 pink/white stamens; berries ⅜", edible but dry and insipid; leaves 1–3", obovate, lanceolate, or wedge-shaped, fine-toothed all around and also coarse-toothed at the outer end; branches armed with stout ½–2" thorns; tall shrub or small tree up to 18' tall. Streamsides up to timberline. Rosaceae.

Red Hawthorn

C. chrysocarpa (cris-a-**car**-pa: golden fruit). Also *C. piperi, C. columbiana.* Berries scarlet. A–.

River Hawthorn

C. rivularis (riv-you-**lair**-iss: of rivers). Berries purplish-black when ripe; leaves narrower than ½ their length. WY and e ID to NM.

Black hawthorn

Black Hawthorn
C. douglasii (da-**glass**-ee-eye: for David Douglas, p. 65). Berries purplish-black when ripe; leaves wider than ½ their length. WA to UT and WY.

ELK RELISH HAWTHORN leaves, and manage to browse them, thorns be damned.

THIMBLEBERRY
Rubus parviflorus (**roo**-bus: Roman name; par-vi-**flor**-us: small- or few-flowered). Also *R. nut-kana, Rubacer parviflorum.* Petals white, ½–1", nearly round, crinkly; berries red, like thin, fine-grained raspberries; leaves 3–8", at least as broad as long, palmately 5-lobed (maplelike), fine-toothed, soft, and fuzzy; stems erect, 4–7' tall, ± woody but weak, thornless. Moist forest and thickets; A. Rosaceae.

Some find thimbleberries delicious, others insipid, which I attribute to colorblindness of the tongue. I concede that they're gritty, sometimes dry, acidic to the point of canker-ing, and exasperatingly sparse on the bush, yet I find in a good thimbleberry one of the more exquisite berry flavors on Earth.

WILD RASPBERRY
Rubus idaeus (eye-**dee**-us: of Mount Ida, on Crete). Flowers white, ¾", in small clusters; berries bright red, juicy, obviously raspberries; leaves 3- (or 5-) compound; leaflets elliptical, pointed, toothed, sometimes with a few lobes; thorny canes, 3–6' tall, with yellowish, peeling bark, live for 2 years. Canyon bottoms, forest, or moist talus (incl alpine); A.

The American red raspberry is now included in the European species from which horticultural strains were bred.

BOULDER RASPBERRY
Rubus deliciosus. Also *Oreobatus deliciosus.* Petals white, 1", nearly round; berries dark red, obviously raspberries, often rather dry; leaves simple, up to 2" wide, kidney-shaped, with fine teeth upon coarser teeth; stems up to 5' tall but often flopping instead, with peeling bark. Streambanks, canyons, foothills to mid-elevs; southerly.

Thimbleberry

Wild raspberry

Boulder raspberry

Subalpine spiraea

Birchleaf spiraea

Hardhack

SPIRAEA

Spiraea spp. (spy-**ree**-a: Roman name). Also meadowsweets. Flowers pink to white, tiny, fuzzy with 25–50 protruding stamens; seedpods tiny, 2-to-several-seeded; leaves 1–3", oval, toothed on outer ½ only. Rosaceae.

Subalpine Spiraea

S. splendens. Also *S. densiflora.* Flowers pink in rounded heads about 2" across; leaves downy underneath; shrub, often prostrate or leaning. Moist meadows, thickets, forest; northerly.

Birchleaf Spiraea

S. lucida (shining). Also *S. betulifolia.* Flowers white to very slightly pink, in rounded heads 2–3" across; leaves smooth; usually erect shrub. Open forest; northerly.

Hardhack

S. douglasii (da-**glass**-ee-eye: for David Douglas, p. 65). Flowers pink, in tapered, cylindrical heads often 4" high; leggy shrubs 2–7' tall. Wet soils; WA, OR, ID, nw MT, and Routt County, CO.

SHRUBBY CINQUEFOIL

Dasiphora fruticosa (days-if-**or**-a: wool-bearing; froo-tic-**oh**-sa: shrubby). Also *Pentaphylloides floribunda, Potentilla fruticosa.* Petals yellow, ½" long; sepals apparently 10 (5 smaller bracts alternate with 5 true sepals); compound leaves of 3, 5, or 7 leaflets, hairy, not toothed; seedpods long-haired; dense, rounded to matted shrub 6–24" tall. Diverse sites, often rocky, but not dry; A. Rosaceae.

Shrubby cinquefoil or "yellow rose" grows on rocky hills all across the country and in Eurasia, and is cultivated in cities. It's the dominant shrub on some Colorado sites.

WILDROSE

Rosa spp. Flowers pink; fruit orange, turning red or purple, many-seeded, dry and sour; leaves pinnately compound; leaflets 5 to 9, oval to elliptical, toothed exc at base, 1–2½"; ± thorny shrub. Moist forests and openings. Rosaceae.

Woods Rose

R. woodsii (**woods**-ee-eye: for Joseph Woods). Flowers 1–2", in small clusters; thorns usually few; paired thorns below axils are heaviest; plants 1–8' tall. A.

Shrubby cinquefoil

Woods rose

Prickly rose

Mallow ninebark

Nootka rose

Baldhip rose

Prickly Rose

R. acicularis (ay-sic-you-**lair**-iss: needling). Also *R. sayi*. Flowers 1½–2½"; fruits usually bluish or purplish; plants 1–5' tall, bristling all over with fine, straight thorns. CO, WY, MT, ID.

Nootka Rose

R. nutkana (noot-**kah**-na: of Nootka, B.C.). Flowers 2¼–3½"; fruits also large, single; thorns mainly in pairs below leaf and branch axils; plants 2–6' tall. A.

Baldhip Rose

R. gymnocarpa (jim-no-**car**-pa: naked fruit). Flowers ¾–1", single; fruits ⅜", round, unique

among these spp. in not retaining the 5 sepals; plants 1½–4' tall, bristling all over with fine thorns. MT, ID, WA, OR.

ONE WONDERS IF this pale, unassuming, five-petalled bloom could ever have carried the mythic and symbolic weight of the rose that troubadours sang of. Had horticultural wizards been at work multiplying rose petals and colors even before the Middle Ages? Yes, they had. "Hundred-petaled" roses were grown by 400 BCE.

MALLOW NINEBARK

Physocarpus malvaceus (fie-zo-**car**-pus: bladder fruit; mal-**vay**-see-us: mallow—). Also seven-bark. Flowers small, white (stamens often pink), in dense hemispheric heads 1¼–2" across; 2 pistils per flower become two 1-seeded seedpods per cluster; leaves 1–3", palmately veined and 3- (rarely 5-) lobed, coarsely toothed; bark peeling, reddish- or yellowish-brown; shrub 3–8' tall. Dry open forest; WA to n-most CO. Rosaceae.

Ninebark bark shreds and peels in layers—but rarely, if ever, so many as nine, or even seven.

OCEAN-SPRAY

Holodiscus discolor (ho-lo-**dis**-cus: entire disk; **dis**-col-or: variegated). Also ironwood, arrow-wood, creambush, rock-spiraea; *H. dumosus*. Flowers tiny, white, cream, or pinkish, profuse

Ocean-spray

With Alternate Leaves, 5 Petals, 5 Stamens

Five stamens typify our Currant family, and fold in a few shrubs from several other families.

BITTERBRUSH
Purshia tridentata (**pur**-shi-a: for Friedich Pursh, see below; try-den-**tay**-ta: 3-toothed). Also antelopebrush. 5 petals yellow, flat-spreading, ⅜"; calyx funnel-shaped, fuzzy, 5-lobed, capsules 1-seeded, long-pointed, not plumed; leaves aromatic, ¾" long, very narrow, 3-pointed and -veined, grayish, white-woolly underneath, edges rolled under; stiff, bushy shrub 2–6' or up to 10' tall; leaves sometimes persist through winter. Steppe, open forest; A. Rosaceae.

in 4–7" conical clusters; 5 petals; seeds single in tiny, dry pods; clusters of pods persist in place through winter; leaves 1–2½", roughly oval, with coarse teeth and often fine teeth upon those; shrub 4–12' tall. Cliffs, forest openings; A. Rosaceae.

Blooming with masses of tiny cream-white flowers in parallel sprays, ocean-spray resembles an ocean wave breaking. "Arrowwood" and "ironwood" refer to Indigenous uses for the straight branches after fire-hardening— arrows, fishing spears, digging sticks to harvest clams and roots, roasting tongs, drum hoops, and hoops for cradles.

Bitterbrush thrives on sites marginal for forest growth, like the lower timberline zone. It enjoys an advantage in being able to fix nitrogen, fertilizing itself and its plant community. Of course, plants can't fix nitrogen themselves: bacteria in nodules on their roots fix it. They can fix even more if the same roots also host mycorrhizal partners (p. 264).

Up North, His Good Luck Went South

Friedrich Pursh spent several years in the eastern United States where he fell into the opportunity to describe and publish his era's prize set of unpublished plant specimens: those from the Lewis and Clark Expedition. Lewis paid Pursh sixty dollars to begin the job. Before it was done, Lewis died and Pursh, falling out with his American employers, absconded to London with one quarter of the collection. In 1814, he published his *Flora Americae Septentrionale*, the first-ever attempt at a coast-to-coast American flora. The specimens eventually found their way back to these shores, but Pursh remained—unfairly, perhaps—an object of intense resentment among American botanists. He did a creditable job of naming and describing. He devoted twelve years to a second magnum opus, a flora of Canada, only to see it lost in a fire. Drunk and broke, he died and was buried in an unmarked grave in Montreal.

Bitterbrush — Swamp gooseberry — Red gooseberry

Though bitter to us, "antelopebrush" is rated just about tops by all Rocky Mountain browsers. Protein content is a major concern to browsers, and bitterbrush is protein-rich thanks to the augmented nitrogen. Years of heavy browsing (hedging) often give the shrubs a low, mounded form.

Bitterbrush and big sagebrush share the name *tridentata* and the leaf shape it refers to, as well as a fuzzy, gray leaf surface and many of the same sites. Their flower structures, however, are utterly different, and the two are unrelated; the leaves evolved convergently in adapting to similar habitat. (Sagebrush leaves do not roll under at the edges, and are equally woolly on top and bottom.)

CLIFFROSE

Purshia stansburyana (for Howard Stansbury, p. 217). Also *P. stansburiana*. Flowers abundant, white to cream; 5 petals nearly round, ¼–½"; numerous stamens; pollinated styles grow up to 2½" long and feathery on the seeds; sticky-dotted, leathery, 5-fingered leaves, ⅝", often

Cliffrose

remain green through winter; flowers sweet-scented, foliage more acrid; twisty shrub 2–9' tall. Piñon-juniper, open forest, brushland; NM, CO, UT, and disjunct in nc ID.

CURRANT AND GOOSEBERRY

Ribes spp. (**rye**-beez: Arabic for "rhubarb"). Five sepals, united about ½ their length; 5 petals smaller and less colorful than sepals, attached just inside calyx mouth alternately with the 5 stamens; leaves palmately lobed; rather weak-stemmed shrub. Grossulariaceae.

Swamp Gooseberry

R. lacustre (la-**cus**-tree: of lakes). Also black prickly currant. Flowers saucer-shaped, pinkish, 10–15 in pendent clusters; berries hairy, purplish black; leaves hairless, ½–2½"; stems with tiny prickles plus larger spines at stem nodes; straggly or sprawling shrub. Forested wet spots; A.

Red Gooseberry

R. montigenum (mon-**tidge**-en-um: montane). Also alpine prickly currant. Flowers saucer-shaped, pinkish, 4–8 in pendent clusters; berries hairy, red-purple; leaves sticky-hairy, ½–1¼"; numerous large prickles at stem nodes, no fine prickles; spreading shrub 8–24" tall. Rocky sites, esp alp/subalp; A.

Canadian Gooseberry

R. oxyacanthoides (oxy-a-can-**thoy**-deez: sharp thorn—). Also *R. irriguum, R. setosum, R. cognatum.* Flowers white to greenish or pinkish, 1–4 in

Wax currant

Golden currant

Sticky currant

pendent clusters, bell-shaped; berries purplish-black to reddish, ½", without bristly hairs; leaves ½–1½" wide, with coarse teeth upon 3–5 lobes, often fuzzy or sticky-hairy, but variable; 1–3 large prickles per node; smaller prickles between nodes; 1–5' tall. From n-most CO n.

Whitestem Gooseberry

R. inerme (in-**ur**-me: unarmed, an inaccurate name). Flowers maturing white, pendent, bristly on the outside; calyx lobes bent back; straplike petals are shorter than stamens; smooth berries ripen purple or gray-black; leaves 1–2½" wide, with coarse teeth upon 3–5 lobes; 0–3 prickles per node; 3–8' shrub with pale gray stems. Moist sites at mid elevs; A.

Wax Currant

R. cereum (**see**-ree-um: wax). Flowers cream to strong pink, sticky-hairy, tubular, ¾" long, ending in short, spreading calyx lobes; berries red; leaves grayish, ± round, ½–1¼" across, toothed, gummy, musky when crushed; no prickles. Sunny slopes, to timberline; A.

Golden Currant

R. aureum (**aw**-ree-um: golden). Flowers yellow, ⅝" across, 5–18 in loose clusters; 5 flat-

spreading calyx lobes; berries orange to dark red; leaves fanlike, ¾–2" wide, with blunt teeth upon 3 big, round lobes; stems hairless, no prickles, 3–10' tall. Brushland, streambanks; A.

Sticky Currant

R. viscosissimum (vis-co-**see**-sim-um: stickiest). Flowers greenish or pinkish white, bell-shaped, 3–10 in loose clusters; berries black; leaves, young stems, berries, and flower bases downy and sticky; leaves musky, 1–3½", with 3 (or 5) shallow rounded lobes; no prickles. Forest, aspen groves; A–.

Northern Black Currant

R. hudsonianum. Flowers white, 6–15, almost saucer-shaped, in a slender ± erect raceme; berries black, often with waxy bloom; entire plant ± sticky-hairy, rank-smelling; leaves 1½– 4½", with 3–5 pointed lobes; no prickles. Streambanks, forests, meadow; UT and w WY to WA.

Trailing Black Currant

R. laxiflorum (lax-if-**lor**-um: loose flowers). Flowers reddish, small, 6–8 in sparse racemes often hiding under leaves; berries ripen black, bristly; leaves 1½–5" long, with 5 or 7 pointed lobes; thornless shrub usually low and sprawling. Forest openings; southerly, plus a few in w ID.

Wolf's Currant

R. wolfii (**wolf**-ee-eye: for John Wolf). Flowers white, in rounded erect racemes; berries ripen black, bristly; leaves 1–2¾" wide, with 5 broad lobes; thornless shrub up to 16' tall but often sprawling close to the ground. Aspen and spruce forest; southerly, plus a few in OR and w ID.

Northern black currant

Wolf's currant

GOOSEBERRIES HAVE NOTHING to do with geese; the word derives from the French *groseille* ("currant"), as does the family name, Grossulariaceae. Gooseberry plants differ from currants in having prickles. *Ribes* fruits are edible and nutritious, and most were widely eaten both fresh and dried by Indigenous people; today, phytonutrient fanatics seek them out. (Our species have not been analyzed, but European black currants score very high.) Some currants take your mouth to the movies, running sweet moments sandwiched between sour and bitter episodes; others are only insipidly bitter.

Gooseberry thorns inflict painful allergic reactions in some people. Nonetheless, Indigenous people used big ones for tattooing or removing splinters, or as fishhooks or needles.

Ribes species are prey to white pine blister rust, a fungus introduced to the West in 1910 that devastates five-needle pines (p. 56). The fungus alternates between *Pinus* and *Ribes* hosts, so Plan A for saving pines was the "War on *Ribes*." Unhappily for *Ribes,* this coincided with Plan A for fighting the Great Depression—to create of millions of subsistence-wage public service jobs. Soon, thirteen thousand men were on the Works Progress Administration and Civilian Conservation Corps payrolls killing shrubs, and when those men went off to fight World War II, the war on *Ribes* conscripted delinquent teens and POWs; later it turned to the herbicides 2,4-D and 2,4,5-T. Some million currant and gooseberry plants died between 1930 and 1946.

The currants won. They far outnumbered us. Recently, scientists learned that the fungus doesn't need currants as alternate hosts after all; widespread species of paintbrush and lousewort serve just as well. Another pointless ecotragedy brought to you by Hubris Productions.

Currants share their fire adaptation with *Ceanothus:* buried seeds remain viable for decades, germinating after shade is removed.

SUMAC

Rhus spp. (rooce: Greek name). Flowers clustered, ♂ and ♀ usually on separate plants; 5 greenish-yellow petals, 5 sepals; berries fuzzy; leaves compound, turning brilliant scarlet in fall. Anacardiaceae.

Fragrant Sumac

R. aromatica. Also skunkbush; *R. trilobata.* Flowers ⅛" long, in pendent clusters appearing before leaves in spring; berries red with yellow, ¼", in long clusters; 3 coarsely lobed leaflets 1–3" long; bushy shrub 4–10' tall. Dry canyons, woodlands; A–.

Smooth Sumac

R. glabra (**glab**-ra: smooth). Flowers ⅛" long, forming dense, erect, conical clusters, as do the hairy dark-red ⅛" berries; 7–29 fine-toothed lanceolate leaflets 2–5" long; 4–10' leggy shrub sprouts from roots to form thickets. Foothills; some parts of each state.

Fragrant sumac

Smooth sumac

SUMACS ARE FALL-COLOR superstars. Their berries have a nice lemonade tartness, making them a popular seasoning in Middle Eastern cuisine. Try them in your water bottle. Southwestern Native people wove the stems and bark into baskets.

WESTERN POISON IVY

Toxicodendron radicans (tox-ic-oh-**den**-dron: poison tree; **rad**-ic-enz: with rooting stems). Also *T. rydbergii, Rhus radicans.* Flowers ¼", greenish white; berries white, up to ¼", in dense, erect clusters, single-seeded, ± striped longitudinally; leaves compound, leaflets 3 (rarely 5), 2–5", narrowly pointed, with a few variable lobes; shrub up to 3' tall. Fairly dry low-elev clearings, outcrops; A. Anacardiaceae.

Western poison ivy

- Leaflets in threes almost always
- Some leaflets crimson by late summer, and the rest by fall
- New foliage reddish and glossy in spring
- Translucent white berries (for a mnemonic aid, think of blisters) in bunches, present by late summer and persisting into winter

Unlike nettle stingers—an elaborate and effective defense against browsing—the poison in poison ivy, oak, and sumac has little survival value to the plant. Call it an accident of biochemistry, or one of the commonest allergies in *Homo sapiens.* Other animals are rarely affected; some even gather the nectar, eat the berries, or browse the leaves.

In humans, susceptibility can be acquired but rarely shed. Many people sure of their immunity have tried to show it off, only to get a rude shock a couple of days later. Apparently this never happened to Leo Hitchcock, author of *Flora of the Pacific Northwest,* who showed off by casually picking specimens with his bare hands on our class field trips. Lore has it you can cultivate immunity by eating tiny leaves over the period of their development in spring. Don't try it.

Symptoms, if any, appear twelve to seventy-two hours after contact. If you itch within minutes of laying eyes on poison ivy, that's from other irritants, like anxiety. If you wash all exposed parts within ten minutes with soap and hot water (or, in the field, with Tecnu) and quarantine your exposed clothes and pets until you can launder them, your odds of escaping are excellent. The allergen, urushiol, may be anywhere on the plant, definitely including bare twigs in winter. A few people are alarmingly sensitive. Burning destroys the toxin, but smoke may carry unburned particles. There have been fatalities following inhalation of poison ivy smoke—drownings in a sea of blister fluid in the lungs.

For most of us, our best defense is simply knowing when we're headed into areas where poison ivy grows, and then knowing the characteristics, spotting the plants, and circumventing them.

DEVIL'S CLUB

Oplopanax horridus (op-lo-**pan**-ux: heavily armed cure-all; **hor**-rid-us: horrid). Flowers ¼", whitish, in a single erect spike up to 10" tall; berries bright, glossy red, 2–3-seeded, up to ¼"; leaves 6–15" across, palmately 7-to-9-lobed,

Devil's club

Redstem

fine-toothed, all borne ± flat near top of stem; leafstalks and undersides of main veins densely spiny; stems 3–12' tall, ½–1½" thick, punky, crooked, usually unbranched, entirely covered with yellowish-tan prickles. Seeps and small creeks, often under cedar; n ID, nw MT, WA. Araliaceae.

Devil's club prefers cold, shaded, sopping, "gloomy" spots; a devil's club thicket is thorniness incarnate, knobby, twisted, tangled, untapering stalks rising out of wet, black earth. In summer, these hide devilishly under an attractive umbrella of huge leaves. Worse, the spines inject a mild irritant. The scarlet berries, eventual centerpiece to the broad table of leaves, aren't recommended either.

The name *Oplopanax* is itself an oxymoron. *Oplo* implies weaponry, while *panax* is a cure, as in panacea. The "cure" part may refer to its relative ginseng (genus *Panax*, one of the most cross-culturally recognized of all herbal panaceas). Yet devil's club under its thorns grows a thin bark that stakes the plant's own claim as a panacea. Some Native peoples used it to alleviate colds, arthritis, bronchitis, diabetes, tuberculosis, body odor (a concern of hunters when near game), excessive milk, and amenorrhea. To cure headaches, the Cheyenne and Crow smoked devil's club roots (obtained through trade, along with tobacco they mixed with it). The Tlingit make it into an antibiotic salve today; my informant swears by it.

REDSTEM

Ceanothus sanguineus (see-an-**oath**-us: thistle, a puzzling name; san-**gwin**-ius: blood-colored). Also buckbrush. Flowers tiny, white, in dense, fluffy 2–5" clusters; seedpods of three 1-seeded cells; leaves 1½–4", oval, fine-toothed, sometimes ± sticky above, hairy along the veins beneath, with 2 veins from base nearly as heavy as midvein but not reaching the tip; stems purplish; shrub 3–10' tall. Sunny slopes; w MT, n ID, OR, WA. Rhamnaceae.

Though redstem was once called Oregon-tea, any tea made from its leaves would be ersatz (or an inferior imitation): the name alludes to New Jersey–tea, an eastern *Ceanothus* that was actually brewed by patriotic colonists wanting to declare their independence from tea taxed by King George.

Redstem is a post-fire pioneer, sprouting either from root sprouts or more often from seeds that germinate in spring following "scarification" by fire heat the previous summer. A nitrogen-fixing plant, it nourishes both four-legged browsers and nearby plants.

With Alternate Leaves, 4 or 5 Petals/Sepals, 8 or 10 Stamens

This diverse group starts with species that can grow as either trees or shrubs. Then it moves to the Heath family, Ericaceae, which looms large in regional importance though not in number of species; Ericaceae land in four different sections of this guide: this

Cascara

one, the broadleaf evergreen shrubs, the non-green herbs, and the five-petaled herbs.

CASCARA

Frangula purshiana (pur-she-**ay**-na: for Friedrich Pursh, p. 102). Also *Rhamnus purshiana.* Flowers tiny, greenish, 4 or 5 white petals; berries ¼", 3-seeded, red ripening to purple-black, toxic; leaves 2–6", round-tipped, minutely toothed, with 10–12 pairs of recessed side veins strikingly parallel to each other; shrub or tree up to 33' tall. Low-elev streamsides, moist forests; n and w ID, nw MT, OR, WA. Similar alderleaf buckthorn, *Rhamnus alnifolia,* in the same range, has even smaller, green flowers and rarely exceeds 5' tall. Rhamnaceae.

Most parts of the cascara plant are powerfully laxative. For some decades, there was a lucrative trade in cascara bark for this purpose—and many girdled trees died—but fortunately the "cascara sagrada" trade has dwindled.

SALTCEDAR

Tamarix chinensis × *Tamarix ramosissima* (**tam**-a-rix: either their Arabic name or a Spanish river; chy-**nen**-sis: of China; ram-o-**see**-sim-a: extremely branched). Also tamarisk. Flowers pink, tiny, in long, slender, much-branched panicles; 5 petals, 5 stamens; tree has innumerable, very fine branchlets coated with tiny grayish leaves rather like juniper leaves, only deciduous and

often salt-crusted; shrub or tree up to 16' tall. Banks of streams and reservoirs in arid country; A. Tamaricaceae.

This lacy Asian import is ornamental enough to have been brought to our shores centuries ago. It then took over thousands of miles of shoreline throughout the West, often displacing willows. Land managers fight it with every weapon at their disposal. Their first glimmer of success comes from a beetle introduced to eat it. Meanwhile, hydrologists have reduced their alarming early estimates of how much water it sucks from watersheds and transpires into thin air: the numbers are still bad, but not catastrophic.

Though distinct in their Asian homelands, these two species hybridize so thoroughly in North America that it seems no one can identify any specimen as being definitively one or the other.

SILVERBERRY

Elaeagnus commutata (ee-lee-**ag**-nus: Greek name for some willow; com-you-**tay**-ta: traded). Also wolf-willow. Flowers have a strong aroma—love it or hate it; yellow inside, silvery outside, bell-shaped, 1–3 per leaf axil, ¼–½" long, persisting on the tip of the berry; 4 yellow sepals, no petals, 4 stamens; berries silvery, mealy, ⅜"; leaves elliptic, 1–3", heavily silver-scurfy on both sides; branches dark red,

Saltcedar

Silverberry

brown-scurfy when young; thicket-forming shrub 3–12' tall. Gravelly soils, either arid or streamside; northerly. Elaeagnaceae.

Silverberry is a top food plant for moose, but low on the list for other wildlife. It excels at revegetating mine tailings and tar sands, and contributes to soil by fixing nitrogen. The bark was used in basketry, twine, and whisks for making "Indian ice cream" out of soapberries, a relative. The stripy seeds were sometimes used as beads, and sometimes eaten, either with the dry berry or after stripping that off.

A close relative, Russian olive, *E. angustifolia,* was originally imported as an ornamental tree but is now notoriously invasive in the arid West.

HUCKLEBERRY AND BLUEBERRY

Vaccinium spp. (vac-**sin**-ium: Roman name). Flowers pinkish, small, globular, with 5 (rarely 4) very short, bent-back corolla lobes, and similar calyx lobes on the tip of the berry; leaves elliptical. Ericaceae.

Dwarf Blueberry

V. cespitosum (see-spit-**oh**-sum: in clumps). Berries light blue with heavy, waxy bloom, without conspicuous calyx lobes; flowers much longer than wide; twigs red to greenish, angled; leaves often finely bumpy (not regularly sharp-toothed), prominently net-veined beneath; 2–12" tall. Meadow and open forest, all elevs, esp alp/subalp; A.

Low Huckleberry

V. myrtillus (mer-**til**-us: myrtle). Also *V. oreophilum.* Berries black or bluish (to reddish in hybrids with grouseberry); twigs green, sharply angled, downy; leaves ½–1¼", fine-toothed; ± broomy shrub 4–14" tall, resembling (and often associating with) grouseberry. Subalpine fir/lodgepole; A.

Black Huckleberry

V. membranaceum (mem-bra-**nay**-see-um: thin). Also thinleaf or tall huckleberry; *V. globulare.* Berries black to dark red, juicy, delicious; leaves 1–2½", thin, pointed, minutely toothed; 2–6' tall. Low subalpine; WA to UT and WY.

Dwarf blueberry

Low huckleberry

Black huckleberry

Oval-leaf Huckleberry

V. ovalifolium. Berries blue with bloom or purplish black without bloom, generally sour and seedy; leaves ¾–2½", smooth-edged; shrub 1½–5' tall. Subalpine forests; northerly.

Bog Blueberry

V. uliginosum (you-li-gin-**oh**-sum: wet). Also *V. occidentale.* Flowers/berries often in clusters of 2–4, from leaf axils; berries blue, with 5 calyx

Bog blueberry

Grouseberry

lobes still showing; leaves without teeth. Alp/
subalp bogs, shores, pine forests; WA to UT
and WY.

Grouseberry

V. scoparium (sco-**pair**-ium: broom—). Berries
bright red, ⅛"; leaves ¼–½"; 4–14" tall; twigs
green, with angled edges, numerous, broomy.
Subalpine, in sun or shade; A.

THIS DIVERSE AND well-distributed genus
has always been of intense interest to
bears, birds, some Native people, and hikers. Toward summer's end, some bear scats
are little more than barely cemented heaps
of huckleberry leaves; after the "cement"
decomposes, you may wonder how these leafy
heaps came to be so neatly molded. The bears
lack the patience to pluck the berries singly.

Who can blame them, with only a month or
two left to fatten up for hibernation?

Blueberries are good for us! Their blue
pigments are flavonoids that act in our bodies as antioxidants. Rats fed the rat equivalent of one cup of blueberries a day reversed
the effects of aging on memory: older rats
beat young rats on memory tests. But as
with all too many promising nutrition studies, others with underwhelming results have
countered them.

Northwest Indigenous people used carefully timed fire to maintain extensive berry
patches. That legacy dwindled under Smokey
Bear's misbegotten anti-fire administration.
One Forest Service report assigned twice as
high a cash value per acre per year to berries
as to the trees likely to replace them on subalpine sites.

"Huckleberry" versus "blueberry" is a
murky issue. In the East, a related genus
(*Gaylussacia*) with seedy black berries has
first claim to the name "huckleberry." Some
authors use the old British names "whortle-
berry" and "bilberry," but I have trouble
saying those with a straight face. Another
author calls them all blueberries, even the
"red blueberries."

The dwarf blueberries of the subalpine
parkland are best in the West, to my taste.
The black huckleberry is the mainstay of
the West's huckleberry industry, combining
greater availability with good flavor and texture. It grows and bears fruit most lavishly
in sunny recent burns, while abounding less
fruitfully in montane forests.

FOOL'S HUCKLEBERRY

Rhododendron menziesii (roe-doe-**den**-dron:
rose tree, a doubly misleading name; men-**zee**-
zee-eye: for Archibald Menzies, p. 111. Also mock
azalea, rustyleaf; *Menziesia ferruginea*. Flowers
¼", pale rusty orange, jar-shaped, pendent on
sticky-hairy pedicels, calyx and corolla shallowly
4-lobed, stamens 8; seed capsules 4-celled, dry;
leaves 1½–2½", elliptical, seemingly whorled
near branch tips, often hairy, coloring deeply in

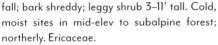
Fool's huckleberry

White rhododendron

fall; bark shreddy; leggy shrub 3–11' tall. Cold, moist sites in mid-elev to subalpine forest; northerly. Ericaceae.

This plant has no berries to tempt anyone, no matter how foolish, so I'd call it a "fool's fool's huckleberry." True, its summer foliage might be carelessly mistaken for the other tall heath-shrubs, black huckleberry and white rhododendron. It often grows with them, but in some Montana ranges it is a thicket dominant—and damned tough to bushwhack through all on its own. Sometimes the tall heaths encircle an expanding tree clump or fill in among scattered trees invading subalpine meadowland.

Heath-family species typically have five petals and five sepals, with fours on a few aberrant individuals. Fool's huckleberry is one species where fours are the norm.

WHITE RHODODENDRON

Rhododendron albiflorum (al-bif-lor-um: white flower). Also *Azaleastrum albiflorum*. Flowers white, ¾", broadly bell-shaped, petals fused no more than ½ their length, 1–4 in clusters just below the whorl-like leaf clusters at branch tips; leaves elliptical, 2–3½", glossy, bumpy, slightly reddish-hairy; capsules 5-celled; bark shredding; leggy shrub 3–6' tall. Subalpine forest and shrubfields; OR to w MT, plus rare outliers in n CO. Ericaceae.

It's hard to see these blossoms—so modest by garden "rhodie" standards—as rhododendrons. But few who reach timberline by foot

Ashore, Briefly

Archibald Menzies was the first European scientist ashore in the Northwest south of Alaska. On the strength of a brief and rather inconsequential first visit to Vancouver Island in 1787, he was appointed surgeon-naturalist on *HMS Discovery* under Captain George Vancouver in 1791–95. The ship's mission allowed Menzies little time for botanizing ashore, and his live collections all died on board. However, his plant descriptions and the dried specimens that reached England fired up interest in the Northwest, eventually leading to voyages by David Douglas and others. Menzies was an old man in 1824 when young Douglas visited him for a briefing on Northwest plants.

would slight their beauty. The plant mixes with fool's huckleberry and black huckleberry, but can grow higher up than either.

With Alternate Leaves and Composite or Irregular Flowers

The huge Sunflower family, Asteraceae, with its composite flower heads (p. 155) comprises many herbs plus a few shrubs, below. The same is true of the irregular-flowered Fabaceae, represented in this chapter by one shrub.

BIG SAGEBRUSH

Artemisia tridentata (ar-tem-**ee**-zia: Greek name for wormwoods, honoring either a goddess or a queen; try-den-**tay**-ta: 3-toothed). Also *Seriphidium tridentatum*. Composite flower heads drab yellowish, tiny, in loose spikes, blooming late Aug and later; leaves spicy-aromatic, grayish-woolly, wedge-shaped, ± shallowly 3-lobed at the tip, avg ½" wide; bark shreddy; avg 2–6' tall on steppe; more often 1½–3' in mtns. All elevs exc alpine; A. Asteraceae. Similar silver sage, *A. cana*, has narrow, linear leaves, a few of them with a pair of side lobes midway; 1–2' tall.

The prairie is barren and inhospitable looking to the last degree. The twisted, aromatic wormwood covers and extracts the strength from the burnt and arid soil.
—J. K. Townsend in Wyoming, 1834

Big sagebrush

John Kirk Townsend (p. 312) grudgingly conceded that sagebrush "perfumed the air, and at first was rather agreeable." Euroamerican pioneers reacted to it with distaste or suspicion. Subsequent overfamiliarity led to feelings ranging from ironic affection to contempt. Cattle won't eat it, so it increased under heavy grazing.

Today we have come to see the sagebrush community as one in need of saving. Fire, cheatgrass, and development are converting one to three million acres a year, out of 150 million sagey acres. More shocking still, global warming is projected to render most of the US sagebrush empire inhospitable to sagebrush by 2090.

Sagebrush's fiefdom is the steppes below lower timberline. The steppe variety is defenseless against fire; it dominates where aridity keeps plants from growing close enough together to carry a fire—or at least it did until cheatgrass invaded. Our mountains mostly have the smaller *A. tridentata* ssp. *vaseyana,* which does stump-sprout after a fire.

Sagebrush likes moderately arid conditions, with at least 7 inches of annual precipitation. Fuzzy-surfaced gray leaves help by reflecting light and by stilling the air at the leaf's surface, but don't retain water as effectively as waxy-surfaced leaves, which in turn are less effective than leafless photosynthetic stems (cacti, Mormon-tea). Big sage does have a few moisture tricks up its sleeve: it produces two generations of leaves per year—big, three-lobed ones grow in spring when moisture is available, and tend to wither and drop in midsummer, whereas leaves produced in late summer and persisting into winter may lack lobes and are much smaller, so they don't allow as much water to transpire. Meanwhile, below ground, sagebrush sends a few very deep, fine roots to get some moisture during the drier seasons, while most of its roots form a wide, dense network in the top 15 inches of soil— the optimum pattern for grabbing water during and after brief thundershowers.

Domestic sheep and pronghorns eat some sagebrush but prefer other foods. Pronghorns in many parts of their range depend on sagebrush browse to get them through the harsher winters. The leaves are nutritious—high in both protein and fat—but fiercely defended by unpalatable volatile terpenoids (that signature fragrance).

Rubber rabbitbrush

Singlehead goldenbush

Deer are able to belch the volatiles out after segregating them while chewing cud; few animals can do that. When a sagebrush is nibbled on, airborne concentrations of at least one volatile, methyl jasmonate, increase as much as tenfold. Nearby plants—including unrelated neighbors—step up their own production of defensive chemicals in response. Some scientists in this field like to call that "communication between plants," while others scoff: "If I smell smoke and run from a burning house, was that communication between the fire and me?"

Zane Grey's "purple sage" was actually good old sagebrush. The Southwest has real purple sage, a *Salvia* species related to culinary sage. Sagebrush, in contrast, is related to culinary tarragon and absinthe (p. 171). Astringent aromas of both *Artemisia* and *Salvia* species led to their widespread use as medicinal and spiritual purifiers, often via smoke.

RUBBER RABBITBRUSH

Ericameria nauseosa (er-i-ca-**me**-ria: heath—). Also chamisa; *Chrysothamnus nauseosus*. Composite flower heads bright yellow, forming massed showy, rounded clusters (blooming Aug–Oct and sometimes persisting, dried and pale, into the next summer); each head taller than wide, consisting of about 5 disk flowers (no rays); leaves linear or threadlike, ± limp; branches numerous, flexible; plant gray-green overall due to fine, feltlike wool on leaves and stems; aromatic broomy shrub 1–4' tall. Low to mid-elev grassland-steppe and pine forest; A. Asteraceae. Similar green rabbitbrush, *Chrysothamnus viscidiflorus*, is dark green overall, with little or no wool, and brittle twigs.

Rubber rabbitbrush—a stump-sprouter that livestock avoid—tends to increase after fire or overgrazing. Rabbits as well as pronghorns browse it. It exudes a latex that engineers were able to turn into good rubber. The Navajo used it as an emetic or cathartic; others chewed it to relieve thirst. In the Northern Rockies it cured diarrhea or coughs, taken either as a tea or a smoke of whole branches.

SINGLEHEAD GOLDENBUSH

Ericameria suffruticosa (suf-fruit-ic-**oh**-sa: semishrubby). Composite flower heads bright yellow; disk florets ⅜" long, 5-lobed, yellow styles protruding; rays few, paler, ⅝"; leaves narrow, ½–1½", crowded; leaves and stems rough and somewhat sticky; dense clumps 4–16" high, can be mistaken for an herb. Rocky places; OR, ID, sw MT, w WY.

SPINELESS HORSEBRUSH

Tetradymia canescens (tet-ra-**dye**-mia: 4 together; cay-**nes**-enz: graying). Also gray horsebrush. Flower heads vase-shaped,

Spineless horsebrush

Trapper's tea

New Mexico locust

clustered, each holding 4 individual flowers with 5 cream to yellow corolla lobes; in seed, 4 silvery bracts frame a mass of long, silvery bristles; leaves linear, ¼–1½", silky or woolly; gray-woolly rounded shrub 2–3' tall in sprawling masses from rhizomes. Sagebrush, piñon-juniper; A. Asteraceae. Similar catclaw horsebrush, *T. spinosa*, has curved spines.

NEW MEXICO LOCUST

Robinia neomexicana (ro-**bin**-ia: for Jean Robin). Massive clusters of ¾" purple, fragrant, pealike flowers; calyx bristly-hairy; fruit a 3" pendent bean pod; leaves pinnately compound, of 15–21 finely hairy leaflets 1–1½"; thicket-forming shrub with stout 2" thorns, or rarely a tree up to 25' tall. Severe burns, streamsides, ravines; southerly. Fabaceae.

Though its showy, fragrant flowers earn it a place planted in yards and roadsides (where

it also grows wild), if you encounter the locust in a brushfield, it's the thorns that will grab you. With Gambel oak, it took over broad areas of former ponderosa forest after the Las Conchas Fire in the Jemez Mountains. Both shrubs resprout from the roots and are very flammable; once they have taken over in a warming climate, it may be hard for conifers to return.

Black locust, *R. pseudoacacia,* is a fine, tall tree native to the eastern United States and widely planted by settlers in the West.

BROADLEAF EVERGREENS

Our broadleaf evergreens (a few of which occasionally reach tree size) have heavy, rigid leaves that can dry out partially in summer without wilting. This reduces transpiration, conserving water. They can also get a bit of photosynthesizing done during warm breaks in late fall and early spring. Our cacti also land in this section, since they are equally green in all seasons.

Shrubs Taller Than 1 Foot

TRAPPER'S TEA

Rhododendron columbianum (roe-doe-**den**-dron: rose tree, a doubly misleading name). Also *R. neoglandulosum, Ledum glandulosum.* Flowers white, ½", in showy, rounded clusters; 5 petals,

5–10 stamens; leaves ¾–1½", oval, thick, evergreen, resinous, white or green beneath, the edges ± rolled under; twigs sticky-hairy; shrub 1½–5' tall. Subalpine meadows, bogs, frost pockets in open forest; all exc CO and NM. Ericaceae.

Leaves of a more northerly relative, Labrador tea (*R. groenlandicum*), were popular as a stimulant tea. Trapper's tea, with stronger alkaloids, is not recommended.

Curl-leaf mountain-mahogany

CURL-LEAF MOUNTAIN-MAHOGANY

Cercocarpus ledifolius (sir-co-**car**-pus: tailed fruit; lee-dif-**oh**-lius: leaves like Labrador tea). Flowers yellowish, 1–3 per leaf axil, ½" across, with numerous stamens their showiest parts; 5-lobed calyx has fine wool; seeds have 2–3" feathery tails; leaves aromatic, winter-persistent but sometimes drought-deciduous, lanceolate to elliptic with strongly downrolled edges, untoothed, ½–1½" long, finely woolly at least underneath; branches ± reddish; gnarly-branched shrub 3–10' or small tree to 40' tall. Dry, rocky slopes at all elevs; OR to w-most CO. Rosaceae.

Mountain-mahogany is so named because its wood is exceptionally heavy and hard. As firewood it is nearly smokeless, hence prized by rustlers and others wishing to escape notice. The plant is most conspicuous around the steppe-forest margin, where it forms pure stands on some slopes in the southernmost part of our range. Browsing mammals love it, as they do most of our nitrogen-fixing shrubs, and may reduce it to a low, hedgelike form.

SNOWBRUSH

Snowbrush

Ceanothus velutinus (see-an-**oath**-us: thistle, a misleading name; ve-**lu**-tin-us: velvety). Also tobacco-brush, cinnamon-bush, sticky-laurel. Tiny white flowers in dense, fluffy, ± conical 2–5" clusters; petals, sepals, and stamens 5; seedpods of 3 separating, 1-seeded cells; leaves 1½–3½", oval, fine-toothed, often tightly curling, heavily spicy-aromatic, shiny and sticky above, pale-fuzzy beneath, with ± 3 equally heavy main veins from base nearly to tip, robust shrub 2–6' tall. Dry, sunny clearings; A–. Rhamnaceae.

Snowbrush invades burns, including slash-burned clearcuts. Its seeds are activated by the heat of a fire and the increased soil warmth of a new clearing. The young shoots grow slowly, but fertilized by nitrogen-fixing bacteria they host in root nodules, can crowd out annuals like fireweed within a few seasons. Snowbrush can grow densely enough to prevent conifer establishment. That has been seen as a problem in the Cascades, but in central Idaho, Douglas-fir seedlings are said to actually prefer snowbrush cover. Certainly, in the long term, its nitrogen should aid conifer growth. (Dozens of herbs in the legume family also fix nitrogen, as do six other shrubs common in our region: bitterbrush, redstem, snowberry, mountain-mahogany, and the alders. Snowbrush is the only one whose nitrogen legacy has been calculated and found to amount to a large fraction of total usable nitrogen in its soil—perhaps simply because so much of it grows in one place.)

Between fires, snowbrush reproduces mainly by stem sprouts and root suckers. The seeds fall to the earth and usually lie dormant—and viable as long as two hundred years—waiting to be reawakened by a fire. The reappearance of a full-blown snow-brush community centuries after snow-brush was shaded out of the forest will seem miraculous.

Though not as delicious to browsers as its deciduous kin, redstem, snowbrush becomes critically important to deer and elk in the winter when other browse plants are leafless.

The name "tobacco-brush" stems from the leaves' fragrance in the afternoon sun, and may have inspired false rumors of use in smoking mixtures.

OREGON-BOXWOOD

Paxistima myrsinites (pa-**kis**-tim-a: thick stigma; mir-sin-**eye**-teez: myrrhlike). Also mountain-boxwood, myrtle-boxwood, mountain lover.

Flowers ⅛", clustered in leaf axils, dark-red petals and whitish stamens and sepals each 4 (but flowers are few and far between some years); capsules splitting in 2; leaves opposite, ½–1¼", elliptical, shallowly toothed, glossy, dark above; twigs reddish, 4-angled; dense shrub 10–40" tall. Forested slopes; A. Celastraceae.

Opposite evergreen leaves (unique here) distinguish this plant from the heath shrubs. (Actually, though the leaves pass for opposite, botanists who look at them under a microscope declare them alternate.) Sprays of the foliage are gathered as florist's greens.

YUCCA

Yucca spp. (**yuk**-a: Carib name for a very different plant, mistakenly applied). Giant rosettes of swordlike, stiff, evergreen leaves, occasionally bearing tall stalks of huge, showy, pendent cream-colored flowers; leaf margins bear sparse, coarse fibers; leaves grow either from the ground or from a short, woody stem. Asparagaceae. The New Mexico state flower.

Datil Yucca
Y. baccata (bac-**cay**-ta: with berries). Also banana yucca. Flowers 2–5", initially pink-tinged, from deep-pink buds; fruit 2–11", pendent, fleshy, green, with big black seeds; leaves 1–3' × 1–2½"; stem, if present, prostrate on ground. Piñon-juniper, canyons; NM, sw CO.

Soapweed Yucca
Y. glauca (**glaw**-ca: whitish). Flowers 2"; fruit erect, 2–3½" long, dries and splits into segments

Oregon-boxwood

Datil yucca

Soapweed yucca

Creeping Oregon-grape

revealing shiny black seeds; leaves 24–36" × 1"; stem up to 2' tall. Dry lowlands mainly e of CD; MT, WY, CO, NM.

YUCCA MOTHS COME at night and work hard to pollinate yucca, but they also sabotage the results by laying their eggs in the flower ovaries. The resulting fruits show a borehole, with many seeds missing inside, eaten by the caterpillars.

The roots can resprout after the aboveground plant burns up in a fire. Joshua trees are a tall species of *Yucca*, while agaves—the source of tequila and mezcal— are close relatives. Indigenous people used yucca leaves for fiber, the roots for soap or shampoo, and the seedpods and flowers for food—but only after preparation.

Shrubs Less Than 1 Foot Tall

The distinction between small subshrubs and herbs can get fuzzy. Herbs wither to the ground in winter. Cacti and yucca definitely do not—so they're shrubs, even without parts we would call "wood." But if you're looking at small plants in summer, woody parts can be hard to detect. If you see a low plant with true evergreen leaves—they're thicker and stiffer than deciduous leaves, and often glossy—held up off the ground by stems, figure that those stems also remain year-round; they must be woody (i.e., shrubs). A very few plants, like rattlesnake-plantain, have evergreen leaves borne right at ground level; the aboveground stalk is just for the flowers, and withers in fall—not a shrub. For some other borderline cases, I just made judgment calls: I kept all the penstemons together as herbs.

CREEPING OREGON-GRAPE

Mahonia repens (ma-**hoe**-nia: for Bernard M'Mahon; **ree**-penz: creeping). Also *Berberis repens, B. aquifolia* var. *repens*. Flowers yellow, in a terminal group of 3–7" spikes amid a cluster of sharp ½–2" bud scales; petals/sepals in 5 concentric whorls of 3, outer whorl(s) ± green; berries ⅜", grapelike, purple with a heavy blue bloom; leaves compound, crowded at top of stem, 10–16" long; leaflets 11–21, spiny-margined (hollylike), pointed-oval, palmately veined; stems unbranched, 4–18"; inner bark yellow. Open pine forest, slopes; A. Berberidaceae.

Oregon-grape leaves are evergreen, but a few may burst into crimson at any time of year. The stamens snap inward at the lightest touch to shake their pollen onto a bee.

The berries (not grapes) have an exquisite sourness not balanced by much sweetness. They're juicy, so you might try one or two for a little mouth excitement. Native Americans considered them starvation fare, but quickly learned to love them once they obtained sugar. Oregon-grape jelly and wine have been traditional since pioneer days.

Brittle prickly pear

Plains prickly pear

Claret cup cactus

The roots are another story. Virtually every American plant in this family was used medicinally, especially for stomach ills and dysmenorrhea. Today, one ingredient (berberine) is a recognized antibiotic. Another was found to help overcome bacterial resistance to berberine and other antibiotics—showing promise to combat antibiotic-resistant strains of staphylococcus, a serious threat in hospitals.

The roots are buried deeply enough to resprout after moderately hot fires.

PRICKLY PEAR

Opuntia spp. (o-**pun**-ti-a: Roman name for some plant). Flowers showy; many tepals and stamens; stamens bend inward when touched; stems fleshy, in scarcely flattened segments, with spines in clusters of 3 to 7; mostly prostrate, mat-forming cactus. Dry, rocky soils at low to mid-elevs. Cactaceae.

Brittle Prickly Pear

O. fragilis (fra-**jil**-iss: brittle). Flowers 2" across, yellow, sometimes with red at tepal bases; fruits tan to red, oval, 1" long, usually dry; stem segments break apart easily; spines ½–1¼". A.

Plains Prickly Pear

O. polyacantha (pol-ly-a-**can**-tha: many spines). Flowers 2–3" across, yellow or hot pink; fruits becoming dry, brownish, 1–1½" long; stem segments usually don't break apart easily; spines 1–2¼". From c ID e and s.

> *Against the thorns of this plant I found that mockasons are but a slight defence.*
> —John Bradbury, 1811

PRICKLY PEAR FLESH is palatable once you get past the spines. The Salish learned to burn off the spines and squeeze out the flesh to eat it, and ranchers learned to use grass fires to burn off the spines so that four-legged browsers would eat it, thus reducing cactus infestation. The sweet edible "pears" are fruits of a third species, *O. phaeacantha*, found mainly at lower elevations.

CLARET CUP CACTUS

Echinocereus triglochidiatus (eck-eye-no-**see**-ri-us: spiny cactus; try-glo-kiddy-**ay**-tus: 3 stiff hairs). Also kingcup cactus, hedgehog

Alpine laurel | Pipsissewa | Kinnickinnick

cactus; *E. mojavensis.* Flowers very showy, 2–4" long, spiny at base; many scarlet tepals and pink or purple stamens; fruits yellow-green to red, juicy, ¾–1⅜"; stems fleshy, ribbed-cylindrical, branching or not; dozens or hundreds of branches forming broad mounds; 1–4" spines in clusters of 5 to 12, usually all roughly equal. Rocky places, juniper, open woodland; UT, CO, NM.

The Colorado state cactus does not actually grow in that state if you treat *E. mojavensis* and *E. coccineus* as full species, as some taxonomists do. *Flora of Colorado* keeps them all in *E. triglochidiatus* because visible distinctions between them are unreliable.

ALPINE LAUREL

Kalmia microphylla (**kahl**-mia: for Pehr Kalm; mic-ro-**fill**-a: small leaf). Corolla pink, bowl-shaped, ½"; sepals tiny, green; flowers 3–8 in terminal clusters; capsules 5-celled, with long style; leaves opposite, ½–1" long, narrow, often with rolled-under edges; spreading subshrubs to 6". Marshy, subalpine soils; A–. Ericaceae.

Alpine laurel's profuse pink blossoms brighten up high bogs and soggy alpine slopes right after snowmelt. Look closely: ten little bumps on the odd-shaped buds hold the ten anthers (stamen tips). When the flower opens, the stamens are spring-loaded to throw their pollen on the first insect to alight.

This plant and its relatives are toxic. Laurel is a common name for *Kalmia* of all sizes, though they aren't related to the Laurel tree (*Laurus*) that wreathed champions in ancient Greece. Pioneers also called rhododendrons "mountain laurel."

PIPSISSEWA

Chimaphila umbellata (kim-**af**-il-a: winter-loving; um-bel-**ay**-ta: bearing flowers in umbels). Also prince's-pine. Stamens and pink-to-white petals flat-spreading, pistil fat, hublike; flowers ½" across, nodding; leaves 1–3", very dark, narrowly elliptical, saw-toothed, ± whorled on lower ½ of stem; 4–12" tall. Shady forest; A. Ericaceae.

Pipsissewa's success under heavy shade suggests dependence on mycorrhizae, which often leads to spotty distribution. Its leaves, *Foliachimaphilae,* used to sit on apothecary shelves as a remedy for bladderstones. They are harvested still (overharvested in some locales) more to flavor the world's most famous soft drink than for herbal remedies.

KINNICKINNICK

Arctostaphylos uva-ursi (arc-tos-**taf**-il-os: bear grapes; **oo**-va-**ur**-sigh: grape of bears). Also bearberry; *A. adenotricha.* Flowers pinkish-white, jar-shaped, with 5 bent-back lobes; berries bright red, ¼", dry and mealy, flat-tasting; leaves ½–1¼", round-tipped, widest past mid-length; thin, gray bark flakes off, revealing smooth, red inner bark; prostrate shrub usually rising 2–6" tall. Rocky sites; A–. Ericaceae.

Kinnickinnick is an Eastern intertribal trading word meaning "smoking herbs."

Alpine wintergreen

White mountain-heather

Hudson's Bay Company traders brought the word west and applied it to this plant Native peoples taught them to smoke, often mixed with tobacco. The berries seem to please bears and heather voles, but among Native peoples they were starvation fare or adulterants for sweeter berries. This is a valuable pioneer on volcanic or glacial soils, and a fine ground cover in gardens.

ALPINE WINTERGREEN

Gaultheria humifusa (galth-**ee**-ria: for Jean-François Gaultier; hue-mif-**you**-sa: trailing). Flowers white to pinkish, bell-shaped, about ⅛" long, from leaf axils; berries smooth, red, up to ¼", delicious; leaves thick, dark green, oval, ± pointed, ½–¾"; spreading shrub 1–6" tall. Subalpine forest and meadow edges; A. A similar plant with fine hair on calyx and berries, mainly in nw MT, is *G. ovatifolia*. Ericaceae.

Aromatic wintergreen oil comes from *Gaultheria* leaves, mainly the checkerberry of eastern North America. Wintergreen gum and candy today derive the key oil, methyl salicylate, either synthetically or from black-birch twigs. It has long been popular both as a flavoring and as a medicine whose best use is as a topical counterirritant—that is, a balm that soothes deep aches by heating up the surface flesh. This versatile volatile is released by many plants when aphids attack them, by tropical orchids to attract bees, and by butterfly males as an anti-aphrodisiac they leave on females, post-coitally, to decrease the female's chance of mating again. The spicy little berries have some of it in them, so don't overdose.

MOUNTAIN-HEATHER

Cassiope and *Phyllodoce* spp. (ca-**sigh**-a-pee and fil-**od**-os-ee: characters in Greek myth). Ericaceae. Alp/subalp.

White Mountain-heather
C. mertensiana (mer-ten-zee-**ay**-na: for Karl H. Mertens, see below). Also moss heather. Corolla bell-shaped, white; flowers pendent

Father and Son Botanists

Karl Heinrich Mertens accompanied the Russian **Count Fyodor Lütke** on his globe-circling voyage of 1826–29, collecting well over a thousand plants, many of them new to science. In London in 1829, Mertens's tales so impressed David Douglas that Douglas became obsessed with completing his second trip to the West by sailing from the Russian colony at Sitka, southeast Alaska, to Siberia. He planned to walk the length of Siberia on his journey home, collecting plants as he went.

Mertens died the following year, at age thirty-four. His plant discoveries at Sitka include species named *mertensiana*, *mertensianus*, or *mertensii*, and *sitchensis*. They do not include *Mertensia*, the bluebell genus that was named earlier in honor of his father, botany professor **Franz Karl Mertens**.

from axils near branch tips; capsules ± erect; leaves ⅛" long, densely packed along the stem in 4 ranks, thus square in cross-section; spreading, mat-forming shrub 2–12" tall. OR to MT.

Four-angled Mountain-heather

C. tetragona (teh-**trag**-a-na: 4-sided). Like the preceding, but with a pronounced groove down the center of each rank of leaves. nw MT.

Pink Mountain-heather

P. empetriformis (em-pee-trif-**or**-mis: crowberry-shaped). Corolla pink, bell-shaped; flowers 5–15 in apparent terminal clusters, erect in bud, pendent in bloom, then erect again as dry capsules; leaves needlelike, ¼–½"; dense matted shrub 4–10" or up to 15" tall. Northerly.

Yellow Mountain-heather

P. glanduliflora (gland-you-lif-**lor**-a: sticky, glandular flower). As above, exc corolla cream yellow to off-white, narrow-necked jar-shaped. Northerly.

Pink mountain-heather

Yellow mountain-heather

HEATHER-DOMINATED COMMUNITIES in the Rockies are found on cold, wet sites in the alpine and subalpine zones. These often result from deep, late-lying snow, such as the drifts that form on a ridge's lee side. Along its lower edge, the pink heather may grade into a black alpine sedge bed that has even later-lying snow, perhaps in the hollow of an alpine gelifluction terrace. The same sequence occurs in open subalpine forests where airflow and shade patterns around the trees create snowdrifts.

Scottish heather in gardens is *Calluna vulgaris*. Vast "heath" communities—species of *Cassiope, Phyllodoce, Erica,* and *Calluna*—took over much of Scotland hundreds of years ago following deforestation and heavy sheep grazing. A warming climate in the Rockies may provoke heather-forest transitions in either direction: longer snow-free seasons invite conifers to take over, but then more frequent fire can kick it back to meadow or heather.

Cassiope foliage is almost like clubmoss; *Phyllodoce* foliage is more like common juniper.

DRYAD

Dryas spp. (**dry**-us: a kind of nymph, in myth). Also mountain-avens. Petals 8 (rarely 9 or 10), ⅜–½"; calyx 8-to-10-lobed; stamens and pistils numerous; as seeds mature, the styles grow long and plumy, and twist together into a point when immature or wet, then open into a fluffy tuft; leaves evergreen, leathery green above, white-woolly beneath, oblong with scalloped, rolled-under edges, ± prostrate on ground; plants mat-forming, spreading by prostrate, woody stems; 1-flowered stalks 1–8" tall. Mid-elev gravel bars to alpine tundra, usually on calcareous terrain. Rosaceae.

White dryad

White Dryad
D. hookeriana (hook-er-ee-**an**-a: for William Hooker, see below). Also *D. octopetala, D. punctata.* Petals white. A–.

Yellow Dryad
D. drummondii (dra-**mon**-dee-eye: for Thomas Drummond, p. 136). Petals yellow. n MT, OR, WA.

Yellow dryad

MATS DOMINATED BY dryads form "spotted tundra," a classic component of both Arctic and alpine tundras all the way around the Northern Hemisphere. Dryads pioneer on rocky, unstable soils that are often bare of snow in winter. Their evergreen leaves are ready to photosynthesize, however briefly, whenever the temperature ventures above freezing. Their ground-hugging shape helps them avoid drying winds, and freely rooting prostrate stems hold fast in mobile soil. The roots host nitrogen-fixing bacteria, an invaluable feature of many pioneer species that isn't found in many small alpine plants. The bowl-shaped solar-tracking flowers focus and hold solar heat, speeding the flowers' maturation and making a warm spot attractive to pollinators that must bask to get their mornings going. *D. hookeriana* augments solar heat with its own respirative heat, raising air temperature in the bowl by 27°F in one study, and a visiting fly's body temperature by 15°F.

The Godfather of British Plant Hunters

Sir **William Jackson Hooker**, a leading scientist of the nineteenth century, developed Kew Gardens in England into a fantastic collection of plants sent from all over the world during the era of plant hunters. Earlier, while teaching botany in Glasgow, he noticed the astonishing zeal of a teenaged gardener there, and took this David Douglas as his protégé on field trips, and eventually sent him off to explore North America. Hooker catalogued and named hundreds of new plants sent back by Douglas, Scouler, Menzies, Drummond, and others. Hooker's onion and Hooker's thistle are named for Sir William, but Hooker's fairy bells were named later for his son Sir **Joseph Hooker**, a plant hunter and colleague of Charles Darwin.

The Father of Binomials

Carl Linnaeus was the father of systematic biology, or taxonomy. Before his day, naturalists improvised multi-word Latin descriptions for any plant or animal under discussion. Linnaeus made everyone's life easier with his idea that the name should be two words, a binomial, that all scientists would agree upon for each type (species) of organism, with the first word (genus) being a broader category often embracing several species. During his lifetime (1707–78), he published more than seven thousand species, including most of our genera that occur in Europe and quite a few from the New World.

He took many genus names from the classical Greek and Roman natural history authors Theophrastus, Pliny, and Dioscorides. He didn't always know exactly what plant those authors had in mind, so when I translate a genus name as "the Greek name," I mean it was a name that ancient Greeks used for some plant, not necessarily the plant Linnaeus gave that name to.

Linnaeus's system for classifying plants was "sexual" in that he based it on the numbers and arrangement of female and male flower parts. In promoting his system, he scandalized the British by trying out every anthropomorphic sexual metaphor he could think of—"husbands" "promiscuously" sharing the "bedroom" with "wives and concubines."

His father coined the name "Linnaeus" by putting a Latin ending on the Swedish word for linden, a tree growing on the family farm. Swedish peasants up until then had not had surnames that passed from generation to generation. Carl was born Carl Linnaeus, Latinized it to Carolus Linnaeus for his publications, and Swedified it to Carl von Linné when the king elevated him to the nobility.

Generally considered calciphiles, dryads are found on sedimentary bedrock in Montana, but on some edges of their range they grow on igneous substrates—on basalt dikes in the Wyoming Absarokas, for example. Rather than the limey chemistry, dryads may be seeking high water content, which leads to frost heaving, a problem few competitors handle as well as dryads.

Another typical habitat subject to soil-movement stress is the upstream ends of gravel bars in mountain rivers. Here, it's raging spring floodwaters that sweep over the top of the dryad mat while shifting the gravel particles around its roots. After the water recedes, having taken most of the smaller soil particles with it, the coarse gravel that remains drains quickly, making it the most drought-stricken part of the river bar.

A popular rock-garden plant was bred from European varieties of our white dryad, but our natives don't transplant happily.

MAT ROCK-SPIRAEA

Petrophytum caespitosum (pet-ro-**fie**-tum: rock plant; see-spit-**oh**-sum: in clumps). Flowers minute, many, in a dense ⅜–2" spike, bristling with numerous stamens; 5 petals initially white but

Mat rock-spiraea

Twinflower

withering brownish and persisting; leaves ⅛–⅝" long, crammed against the ground in low mats that can be extremely dense and hard; flower stalks (but not the leaves) may reach 8" tall. Limestone rocks at all elevs; A–. Rosaceae.

After the white flowers turn brown, rock-spiraea flower spikes might be mistaken for alpine willow catkins, but are found in dry rock crevices, not on moist soil like willows.

TWINFLOWER

Linnaea borealis (lin-**ee**-a: for Carl Linnaeus, p. 123; bor-ee-**ay**-lis: northern). Flowers pink to white, 2 per stalk, conical, pendent, ½" long, 5-lobed, stamens 4; capsules 1-seeded; leaves opposite, very shiny, dark, ¼–1"; spicy- or anise-fragrant esp in warm sun; flowering stalks 3–5", reddish, with 2–6 leaves on lower ½ only; from long, leafy runners. Moist forest; A. Caprifoliaceae.

Carl Linnaeus, who chose the scientific names for thousands of plants, didn't name any for himself, but is said to have asked a colleague to name this one after him. He then wrote, "*Linnaea* was named by the celebrated Gronovius and is a plant of Lapland, lowly, insignificant, flowering but for a brief space, after Linnaeus who resembles it." False modesty! Though tiny and simple, the twinflower grows throughout the cooler third of the Northern Hemisphere and is widely admired. Linnaeus liked to hold a sprig of it when posing for portraits.

5

Flowering Herbs

Defined as seed plants without woody stems, the flowering herbs include most plants we call "wildflowers," as well as grasslike plants—which do flower, technically, though their flowers fall well short of corsage-ready. Some wildflowers, even small ones, are actually shrubs. It can be tricky to distinguish herbs from shrubs. Slightly woody little "sub-shrubs" appear in this chapter unless they have heavy evergreen leaves at least an inch off the ground, proving that there's a woody stem bearing them. This latter type includes heathers and twinflower (pp. 120–124).

- Sedge stems are triangular in cross-section, with V-shaped leaves in three ranks along the three edges ("Sedges have edges").
- Grass stems are round and hollow, with a swollen node at the base of each leaf.
- Rush stems are round and pith-filled; their leaves too are often tubular, especially near the tip; the pistils and seedpods are three-celled.

"Sedges have edges" is true throughout genus *Carex* (most of our sedges), but some sedges (e.g., cottongrass, p. 134) have rounded stems with a hint of triangularity near the top.

Positive identification of grasslike plants requires a microscope and a whole vocabulary of grass parts. For this book, I chose a modest number of species, out of hundreds growing here, and have described them minimally. If habitat, description, and illustration fit what you're looking at, then you have an educated guess of what it is.

Beargrass (p. 147) is a lily you might mistake for a sedge when its flowers are absent; its leaves are V-shaped in section, dry, pale, abrasive, and robust, in thick clumps. Blue-eyed-grass (p. 149) is a slender iris that blooms early and briefly.

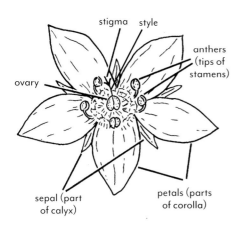

stigma   style

anthers (tips of stamens)

ovary

sepal (part of calyx)

petals (parts of corolla)

GRASSLIKE HERBS

Grasses, sedges, and rushes are the three huge families of grasslike plants. To tell them apart, roll a stem between thumb and forefinger:

Grasslike, with Round, Hollow Stems (Grasses+)

Though classically personified as humble, grasses are in reality taking over the world.

The Poaceae are the most successful plant family, if judged by their rate and breadth of genetic diversification in recent geologic time. The ascendency of grasses is reflected in—and has been magnified by—the development of agriculture, which has always focused on grains (grass seeds) and on grazing (grass-eating) mammals.

Grasses are flowering plants with simplified, undecorated, single-seeded dry flowers (florets), each with three short-lived stamens and a usually two-styled pistil. The florets are flanked by several scales or bracts, arranged alternately (in contrast to the whorls of six tepals on a rush flower). The bracts may end in stiff hairs (awns); look for awns when you identify grasses. One or more florets and their bracts along a single axis make a spikelet. Several spikelets attach to the main stalk either directly, making a spike, or by small stalks (often again branched), making a panicle.

Bearded Grasses (with Awns at Least 1 Inch Long)

FOXTAIL BARLEY

Hordeum jubatum (**hor**-dee-um: barley, a misleading name; jew-**bay**-tum: bearded). Also *Critesion jubatum*. Spikelets 1-flowered, each functional spikelet tightly flanked by 2 slenderer sterile ones; awns ½–3" long, purple to bronze to frosty green; bristly spikes arch in youth. Rampant on roadsides; A.

Though native and gorgeous, foxtail barley grows like a weed on disturbed ground, and antagonizes cattle and their owners by piercing soft tissue when grazed.

BOTTLEBRUSH SQUIRRELTAIL

Elymus elymoides (**el**-im-us: Greek name; el-im-**oy**-deez: resembling *Elymus*—an absurd name now that this species is in that genus). Also *Sitanion hystrix*, a beautiful name I am loath to part with. Spikelets of 2–6 florets in a 1½–6" spike that looks strikingly brushlike thanks to numerous awns ¾–4" long; plants 4–20" tall. Dry, rocky soil, pine forest to alpine; A.

COLUMBIAN NEEDLEGRASS

Eriocoma nelsonii (area-**co**-ma: woolly hair; nel-**so**-nee-eye: for Aven Nelson). Also *Achnatherum nelsonii*, *Stipa occidentalis*, *S. columbiana*, *S. nelsonii*, *A. occidentale*. 1-floret spikelets held tightly erect against the main stem, in a 2–12" spike; awns ¾–2¼" long, twice-bent, spreading at various angles; leaves often inrolled; plants 8–40" tall. A.

NEEDLE-AND-THREAD

Hesperostipa comata (hes-per-o-**sty**-pa: western grass; co-**may**-ta: hairy). Also *Stipa comata*. Spikelike

Foxtail barley

Bottlebrush squirreltail

Needle-and-thread

Bluebunch wheatgrass

Giant wildrye

3–10" panicle bristles with 3–8" awns that are ± twisted and curled and/or crooked; leaves narrow, rough, inrolled; stems clumped, 12–28". Rocky soil at low to mid-elevs; increases with overgrazing; A.

The corkscrew shape of the awn spins the falling seed, helping to drill it into the soil.

Grasses with Narrower Flower Spikes

BLUEBUNCH WHEATGRASS

Pseudoroegneria spicata (soo-doe-reg-**nair**-i-a: after another genus; spic-**ay**-ta: spiked). Also *Elymus spicatus, Agropyron spicatum.* Large (6–8 floret) spikelets form a 3–6" slightly zigzagging spike; bent awns ¼–¾" long or (esp in OR, ID) absent; plants 24–40" tall, in big clumps. Grasslands, all elevs but primarily low; A.

In general, moderate grazing—even by cattle—stimulates the health and productivity of many grasses. The trouble with cattle includes their inclination to hang around by streams and ponds, damaging riparian zones, and to herd up and not move around much until shooed, leading to patchy overgrazing. Badly overgrazed and compacted soils (a common condition 70 to 120 years ago) are prone to invasion by non-native plants. This is especially true in intermountain grassland whose grasses—chiefly bluebunch wheatgrass and Idaho fescue—evolved with few bison and with little grazing by other hoofed animals. Trampling both breaks and damages the bunchgrasses and the cryptobiotic crusts (p. 288) in the soil. Many scientists believe that livestock grazing, while potentially benign on sod-forming grasslands, is inherently harmful to intermountain bunchgrasses.

BLUE WILDRYE

Elymus glaucus (**el**-im-us: Greek name; **glaw**-cus: bluish-pale). 3–5-floret spikelets in a 2–8", often-nodding spike; awns vary from 1" to absent; leaves flat, ⅛–½" wide; plants 16–48" tall. All elevs; often in gravelly soils, incl river bars; A.

GIANT WILDRYE

Leymus cinereus (**lay**-mus: anagram of *Elymus;* sin-**ee**-ree-us: ash gray). Also *Elymus piperi.* Massive 4–10" spike with several spikelets clustered at each node; awns short or lacking; leaves and stems robust, often harsh with fine hairs; 3–10' tall, in huge clumps. Plains, ravines, streambanks at low elevs; A. Eaten as a grain by Shoshone and others; stout stems used as toy arrows, fire pokers.

CRESTED WHEATGRASS

Agropyron cristatum (ag-ro-**pie**-run: wild wheat; cris-**tay**-tum: crested). Also rescuegrass; *A. desertorum.* Short-awned spikelets in 2 tidy ranks, forming flattened, erect 1¼–4" spikes; leaves flat; 12–40" tall, in clumps. Asian; widely invasive; much planted for "revegetation"; A.

One bunchgrass known to be easy to establish on disturbed ground is crested wheatgrass, but it is itself a non-native invasive weed. Cattle love crested wheat, so vested interests often support planting it instead of more ecologically desirable plants like bluebunch wheatgrass and sagebrush.

Blue grama

Buffalograss female florets

Alpine timothy

BLUE GRAMA

Bouteloua gracilis (boo-tel-**oo**-a: after the Boutelou Brothers, botanists in Madrid; **grass**-il-iss: slender). Also *Chondrosum gracile*. Bristly, often purplish spikelets tightly aligned on 1 side of stem, in 1–3 spikes that often arc gracefully when dry; leaves often hairy, curly; plants 8–22" tall. Dry foothills, piñon-juniper; ID to NM.

BUFFALOGRASS

B. dactyloides (dac-til-**oy**-deez: fingerlike). Also *Buchloë dactyloides*. ♂ and ♀ florets on separate plants; ♂ spikes 1-sided (flaglike), straight, 1–4 per stalk; ♀ clusters sheathed by special leaves, in low leaf axils; leaves short and curled, with long, straight hairs; plants 2–8". Dry foothills mainly e of CD; MT to NM.

Grasses coevolved with grazers. Here, that would be bison, bighorn sheep, many small mammals, and—to lesser degrees—elk, pronghorn, and moose. More recently, the rise of agriculture, which focused on grass seeds (grain) and grass-eating mammals (grazers), magnified the ascendency of grasses.

Millions of cattle and sheep grazed the West before the science of grazing ecology got up and running. Before their arrival, blue grama and buffalograss co-dominated the shortgrass-prairie ecosystem—the high, dry, western belt of the Great Plains, butting up against the Rockies on the west and merging eastward into the slightly wetter, lusher tallgrass prairie. Tallgrass produced far more biomass, but these two scruffy, durable shortgrasses were better able to nourish tens of millions of half-ton grazers (bison) thanks to anomalously high protein ratios. Though described below as growing 8 to 18 inches high, blue grama was more often 1 to 4 inches when subjected to both bison and typical High Plains aridity: a long-term study found that reintroducing bison increases the amount of buffalo grass and with it, overall plant diversity and drought resistance. Despite appearances, it thrived under the abuse. The trampling herds were harder on competing grasses than on blue grama, and they repaid grama with a lot of nitrogen in their urine and manure. Blue grama recycled the nitrogen back into protein, which nourished the bison, and so on.

ALPINE TIMOTHY

Phleum alpinum (**flee**-um: Greek name). Also *P. commutatum*. Short-awned, often purplish spikelets form a single, dense, neatly cylindrical ½–1¾" spike; leaves flat, edges feel raspy; plants 4–20" tall. Moist alpine to mid-elev meadows; A.

Downy oatgrass

Purple reedgrass | Northern sweetgrass

DOWNY OATGRASS

Trisetum spicatum (try-**see**-tum: 3-awned; spic-**ay**-tum: spiked). Spikelets of 2–3 florets forming a 1–3" spike, often purplish; awns ¼", bent outward; plants 4–16" tall, often fuzzy all over. Common from alpine ridges to spruce-fir forest; A.

PINEGRASS

Calamagrostis rubescens (cal-a-ma-**grah**-stiss: reed grass; roo-**bes**-enz: reddish). Flowering stems abundant in recent burns, otherwise few or absent, the plant spreading mainly by rhizomes; 3–6" spikelike panicle of 1-floret spikelets; awns barely protruding, bent at midlength; leaves narrow, usually flat, rough to touch; plants 16–40" tall. Pine forest, often with elk sedge; A–.

PURPLE REEDGRASS

C. purpurascens (pur-pur-**ass**-enz: purplish). Like pinegrass, exc awn is only slightly bent, leaves often rolled inward; panicle usually purplish; plant 12–30" tall. Higher elevs; dominates some tundra turfs on calcareous soil; A.

SPIKE FESCUE

Leucopoa kingii (lew-co-**po**-a: white bluegrass; **king**-ee-eye: after Clarence King, p. 217). Also *Festuca kingii, Hesperochloa kingii*. Florets 3–5 in awnless spikelets; panicle dense, almost spikelike, the branches ± erect; leaves ⅛–¼"

wide, bluish; stems 1–2', in dense clumps with many dry bases from previous years. Dry, rocky soil; mid-elev forest to tundra; A–.

Grasses with ± Awnless Florets in Open Panicles

NORTHERN SWEETGRASS

Anthoxanthum hirtum (anth-a-**zanth**-um: yellow flower; **her**-tum: hairy). Also vanilla sweetgrass; *A. nitens, Hierochloë odorata*. Yellowish 1-floret spikelets in a delicate 2–5" panicle; leaves vanilla-scented; stems 12–20" with purplish bases. Widely scattered in mtn meadows; A.

Indigenous people craved this fragrance, using the plant as a sachet, an incense burned ceremonially, a hair rinse, a strand in hair braids, and so forth. Anglos use one aromatic, toxic ingredient, coumarin, in perfumes and pharmaceuticals.

REDTOP

Agrostis stolonifera and *A. gigantea* (ag-**ros**-tiss: Roman name; sto-lon-**if**-er-a: with runners). Also *A. alba*. Purplish 1-floret spikelets in a delicate 2–10" panicle; leaves flat, midvein prominent; plants 8–40" tall. European; invasive on streamsides and moist, disturbed areas; A.

ALPINE BLUEGRASS

Poa alpina (**po**-a: Roman name). Florets 3–7 in broad, flattened spikelets; outer scales purple when young; leaves rather short, flat for much of their length; plants 4–16". Alp/subalp; A.

Alpine bluegrass

Muttongrass

Purple hairgrass

P. pratensis (Kentucky bluegrass, actually from Europe) invades widely, up to subalpine. Abundant native *Poa* species include *P. wheeleri* and *P. secunda* from grassland-steppe to alpine.

MUTTONGRASS

Poa fendleriana (for Augustus Fendler, p. 246). Florets 4–6 per spikelet, often red-stained, developing a whitish cast, in a robust panicle initially thick, later opening out; stems and leaves rough; leaf tips folded and upcurved like a canoe; bunchgrass 8–20" tall. Dry soils, sagebrush up to alpine; A.

Grasses with Short-Awned Florets in Open Panicles

BLUEJOINT REEDGRASS

Calamagrostis canadensis (cal-a-ma-**grah**-stiss: reed grass). Spikelets of 1 floret in a 3–8" panicle with branches either flat-spreading or erect; purplish where alpine; each floret encloses 1 awn among many finer, straight hairs; leaves usually flat, droopy, rough; plants 1–4' tall. A dominant grass of wet meadows, floodplains; A.

TUFTED HAIRGRASS

Deschampsia cespitosa (desh-**amp**-si-a: for Jean-Louis-August Loiseleur-Deslongchamps; see-spit-**oh**-sa: growing in clumps). Also salt-and-pepper grass. Panicle 4–10"; spikelets pale and/or purplish-black (often bicolored); 2 florets barely stick out between 2 slightly longer scales (glumes); awns protrude slightly; leaves creased; plants 8–36" tall. Higher meadows and open forest; abundant on moist, snow-protected alpine sites; A.

PURPLE HAIRGRASS

Vahlodea atropurpurea (va-**load**-i-a: for Martin H. Vahl; at-ro-pur-**pew**-ri-a: black-purple). Also *Deschampsia atropurpurea.* Purplish spikelets of 2 florets sandwiched between and nearly hidden by 2 longer scales (glumes); awns completely hidden, less than ⅛", bent inward; panicle 2–4"; leaves flat; plants 6–24" tall. Subalpine; A.

IDAHO FESCUE

Festuca idahoensis (fest-**you**-ca: Roman name). Spikelets of 4–7 florets forming a 3–6" spikelike panicle, with ⅛" awns; leaves narrow, wiry, dark green; plants 12–32" tall. All elevs, grassland or open forest; A.

SHEEP FESCUE

F. saximontana and *F. brachyphylla* (sax-i-mon-**tay**-na: RM; bracky-**fill**-a: short leaf). Also *F. ovina.* Spikelets of 3–4 florets in a narrow 1–3" panicle; awns less than ⅛"; leaves inrolled, in a dense basal tuft; plants 4–16" tall. Alp/subalp; both A.

Idaho fescue

Cheatgrass

Cattail

MOUNTAIN BROME

Bromus carinatus (**bro**-mus: Roman name; kar-en-**ay**-tus: keeled). Also *B. marginatus, B. sitchensis, Ceratochloa carinata.* Robust, flattened spikelets of several florets; panicle 4–10", branches ± erect; leaves ± flat, soft; stems 1–4', usually hairy. Mid-elev meadow and forest; A.

CHEATGRASS

B. tectorum (tec-**tor**-um: of roofs, once a common habitat in Europe). Also downy brome, downy chess; *Anisantha tectorum.* Delicate-looking 8–26" grass in small tufts, finely hairy all over; lower panicle branches often droop; awns ½", about as long as the floret; an annual, it comes up in winter or spring, dries out by midsummer. European; invasive in dry rangeland and elsewhere; A.

Cheatgrass, a non-native, has taken over an estimated 100 million acres. It expands thanks to a nefarious cycle involving sagebrush and fire. In sagebrush grasslands, most native grasses are perennial bunchgrasses; they grow in tight clumps. With scant rain, clumps and sagebrush alike grow rather widely spaced, their roots reaching out to seize moisture from the nearly bare ground in between. Fires, where they can't jump from clump to shrub to clump, spread spottily, sparing plenty of individuals.

Cheatgrass tufts grow new each year from seeds. Compared to perennial grasses, they grow closer together, being smaller and faster-growing, and they build up strong root systems in fall and winter, giving them the jump on bunchgrasses, which grow their roots in spring. By midsummer cheatgrass sets seeds and dries out completely, becoming totally unappealing to grazers but highly attractive to fire. Cheatgrass spreads fire so well that Great Basin brushfires now tend to be all-consuming. With plenty of cheatgrass in the basin to provide seeds, the post-fire plant cover will have even more cheatgrass, and the next fire will come sooner, before sagebrush has a chance to reestablish.

Moreover, this cheatgrass-fire cycle reduces soil nitrogen. Believe it or not, the intermountain sagebrush ecosystem is threatened. Most range managers now agree on the need for a massive campaign to restore perennial bunchgrasses and sagebrush. It won't be easy to come up with either the money or the techniques.

CATTAIL

Typha latifolia (**tie**-fa: Greek name; lat-i-**foh**-lia: broad leaf). Flowers minute, chaffy, in a dense, round, smooth spike of 2 distinct portions, the upper ♂ thicker when in flower but withering as the lower ♀ thickens and turns dark brown in fruit; stalks 3–10'; leaves ½ as tall by ¼–¾", smooth; from rhizomes in shallow water; A. Typhaceae. (It is not a grass.)

Migrating waterfowl—and the hunters thereof—avidly seek out cattail marshes. The

Holm's Rocky Mountain sedge

Water sedge

Salish and others weave the stalks into thick, spongy mats for mattresses, kneeling pads in canoes, packsacks, baskets, rain capes, and temporary roofs in summer. Some Native groups eat cattails. The rhizomes and inner, basal stalk portions are pretty good baked, raw, or ground as flour.

Grasslike, with Triangular Stems (Sedges)

Carex spp. (**cair**-ex: Roman name). Cyperaceae.

Clearly resembling grasses, but with triangular stems, the sedge family includes bulrushes and tules, and the more tropical genus *Cyperus,* which includes the papyrus that Egyptians used for the first paper. Here, most sedge species are in genus *Carex.* Some like saturated or even submerged soil, but many prefer cold, dry meadows. The common denominator of sedge habitats is stress—that is, they benefit when the grasses they compete with are in any way handicapped. Sedges favor high altitudes and latitudes; most require full sun. Elk sedge and Ross's sedge are forest sedges that root-sprout to increase after fire.

Few sedges are grazed much, but elk sedge is an exception that provides forage important to elk and bears, and water sedge is valuable forage for waterfowl.

When a *Carex* sedge blooms, stamens in conspicuous disarray adorn parts of the flowering spikes. Stamen-bearing units are male flowers. A female flower is tipped with either two or three headless, threadlike styles, and a dark scale covers much of its outer surface. Stamens are typically much shaggier than styles, and yellower when fresh, though later they wither and may be browner than styles. Sedge identification relies on the number of styles and the distribution of stamens. For an oversized illustration, picture the grass plant we know as corn: the male spikes (tassels) form a terminal cluster atop the plant, while numerous styles (silks) show that the lateral spikes (ears) are all female. Both spike types are unisexual. Bisexual spikes, in a clever twist of jargon, are either *androgynous* or *gynecandrous,* depending on whether males or females are on top. In the sequence that follows, the first two species have two styles per female floret; the remainder, with three styles, are arranged with low plants first, then taller ones.

HOLM'S ROCKY MOUNTAIN SEDGE

C. scopulorum (scop-you-**lor**-um: RM). Flower spikes ½–1¼", several, the terminal one usually ♂; ♀ flowers dramatically black (or purplish black) with tan or light-green edges; plants 4–16" tall. Wet alp/subalp meadows; A.

WATER SEDGE

C. aquatilis (a-**qua**-til-iss: of water). Spikes up to 2", terminal one ♂ or androgynous, lateral ones ♀, erect; plants 16–40" tall, rhizomatous. In water or wet soil; common; A.

DUNHEAD SEDGE

C. phaeocephala (fee-o-**sef**-a-la: dun head). Flower spikes pale, few, small, closely clustered;

spikes with ♀ flowers above ♂, or sometimes all ♀; plants 2–12" tall. Alpine tundra and rock fields; A.

ROSS'S SEDGE

C. rossii (**ross**-ee-eye: for polar explorer John Ross). Flowers fewer than 10 per stem, scarcely ¼" long, framed between 2 bracts that extend well past flowers; often some stems very short, hiding spikes down among the leaves; plants 4–12" tall, in tufts. Dry forests; A.

SMOOTH-FRUIT SEDGE

C. heteroneura (het-er-o-**new**-ra: differing veins). Also *C. atrata, C. epapillosa*. Spikes 3–6, ⅓–¾", ♀ flowers dark brown against pale greenish; ♂ on lower part of top spike; 6–16" tall in clumps. Alp/subalp meadows; A.

BLACKROOT SEDGE

C. elynoides (el-in-**oy**-deez: looks like a spikesedge, genus *Kobresia*). Spike single, slender, cylindrical, tan to dark brown, ♀ flowers fewer than 10, enlarging and spreading out at maturity; leaves wiry, resembling the 2–6" stems. Extensive dense turf on dry tundra ridges where soil has developed; favors limestone; A.

CURLY SEDGE

C. rupestris (roo-**pes**-tris: of rocks). Spikes single, slender, cylindrical, brown to purplish; leaves flat at base, up to ⅛" wide, curly at tip; plants 2–6" single or in small tufts, not dense turf. Dry, wind-blasted alpine slopes, esp over limestone; may form cushions scattered across harsh alpine gravels, or may form alpine turf with grasses; A–.

BLACK ALPINE SEDGE

C. nigricans (**nye**-grik-anz: blackish). Flower spike single, quite dark, ♂ flowers above ♀, which typically droop after maturing; leaves 4–9 per stalk; plants 2–10", forming thick turf in small hollows. Alp/subalp; more abundant northward; A.

Dense, turfy beds of black alpine sedge underlie the latest-melting patches of snow. Once these beds dry out, they offer a perfect spot for basking, tumbling, or sleeping. What's good for you is, in this instance, tolerated by the flora too, as these sedges are relatively resilient—but not immune—to impact: camp on sedge beds only when away from trails, pick beds unmarked by previous campers, and move on after one or two nights.

To deal with its short growing season, black alpine sedge is a speed demon among grasslike plants, setting seed as soon as

Dunhead sedge

Smooth-fruit sedge

Black alpine sedge

Elk sedge

Mertens sedge

Raynolds sedge

Showy sedge

MERTENS SEDGE

C. mertensii (mer-**ten**-zee-eye: for Karl Mertens, p. 120). Flower spikes dense, cylindrical, ¾–1½", several; ♂ flowers (darker) take up the lower ½ of the terminal spike, and at most, the lowest few scales of the other spikes; stem edges sharp, rough; leaves flat, bigger ones all at least a few inches up the stem; plants 1–3' tall. Moist openings up to near timberline; WA to w MT; a pretty sedge, its spikes nodding gracefully on arcing stems; northerly.

RAYNOLDS SEDGE

C. raynoldsii (ray-**nuld**-zee-eye: for William Raynolds, p. 217). Flower-spikes ½–1¼", typically 4, the top one ♂ and much more compact than the others; ♀ flowers ± puffy, dramatically black and green; plants 8–28" tall. Meadows, mainly below forest line; A–.

thirteen days after its release from snow. (Typical nearby plants take forty-two to fifty-six days.) Even that is rarely quick enough, so it spreads mainly by rhizomes, producing a turfy, rather than clumpy, growth habit.

ELK SEDGE

C. geyeri (**guy**-er-eye: for Karl Geyer, p. 142). Just 1 spike, topped with thin filaments from close-packed ♂ spikelets; 3 thicker styles ⅜–⅝" long from each of the 1–3 ♀ flowers splayed at bottom; leaves about as tall as the stems, 6–20", in clumps; a huge, fibrous root system hides below its modest stems. Dry forests; clumps that aren't flowering at all may dominate understory; A.

SHOWY SEDGE

C. spectabilis (spec-**tab**-il-iss: showy) and Payson's sedge, *C. paysonis* (pay-**so**-niss: for Edwin Payson). Spikes ⅜–1¼", 2–5+, the terminal one ♂, lowest one may droop; lateral ones may have a few ♂ at top; blackish scales with a pale-green midrib; plants 6–22" tall. Subalpine meadows, WA to Wind Rivers and Uintas. The seeds are snackable. Very similar *C. paysonis* and *C. podocarpa* cover much the same range.

COTTONGRASS

Eriophorum angustifolium (airy-**ah**-fur-um: wool-bearing; angus-tif-**oh**-lium: narrow leaf). Also Alaska-cotton; *E. polystachion*. Spikelets 2–5, becoming white, cottony tufts ¾–1¾" long (in seed); leaves triangular near tips; stems

Cottongrass

triangular to nearly round, 8–36" tall. High bogs; A. Cyperaceae.

Up in Alaska, the sight of mile after mile of cottongrass blowing in a breeze is hard to forget. Here, it is common in subalpine bogs.

Grasslike, with Round, Pith-Filled Stem (Rushes)

Our Rush family plants (Juncaceae) are in two genera: *Luzula,* which looks like grass with wide leaves, and *Juncus,* which contrasts with its seemingly leafless look and its darker bluish-green shade.

WOODRUSH

Luzula spp. (**luz**-you-la: light—). Flowers tiny, dry, green to brown, with 6 tepals and 2 sepal-like bracts; seed capsule has 3 cells with 1 seed each. Juncaceae.

Like the glacier lily, the subalpine woodrush can melt its own hole to bloom through a few inches of dwindling snowpack. Its ancient name, *gramen luzulae,* or "grass of light," observed the grace of an otherwise inconspicuous plant when bearing dewdrops in the morning light. More shade-tolerant than most grasslike plants, the woodrush can indeed grow in the woods, particularly near timberline, but it's more abundant in open meadows and on moraine gravels. The Rush family's three-celled seed capsule contrasts with one-celled, one-seeded fruits of grasses and sedges.

Small-flowered Woodrush

L. parviflora (par-vif-**lor**-a: small-flowered). Inflorescence a loose, arching array of single flowers; leaves grasslike, wide, flat, finely hair-fringed, from sheathing bases without swollen nodes; plants 6–20" tall. Abundant in alp/subalp forest and meadows; scattered in mid-elev forest; A. Similar species incl northerly *L. hitchcockii.*

Spike Woodrush

L. spicata (spic-**ay**-ta: spiked). Inflorescence a single, usually nodding, bristly spike ½–1¼" long; plant 2–16" tall. Alpine; A.

RUSH

Juncus spp. (**junk**-us: Roman name). Flowers of 6 dry green to brown tepals and 2 outer bracts; seed capsule has 3 cells, each with many seeds; leaf blades (if any) tubular, resembling the dark-green, tubular, pith-filled stems. Juncaceae.

Rushes are tough, reedy, deep-green, round-stemmed, round-leaved plants with chaffy tufts, often nearly black, for flowers. Most are small. (Big "bulrushes" or "tules" are not rushes, but sedges.) Rushes commonly grow in wet places, including bogs and marshes, but in the high country you often see them in dry meadows and gravels. Since

Small-flowered woodrush

Spike woodrush

Parry's rush

Baltic rush

Mertens rush

rushes become utterly unpalatable by maturity, Baltic rush, for one, can become weedy in heavily grazed meadows.

The Okanagan allegedly used Mertens rush in witchcraft. Large rushes, like Baltic rush, were woven into baskets and sleeping mats.

Drummond's Rush

J. drummondii (dra-**mon**-dee-eye: for Thomas Drummond, see below). Flowers 1–3, green, seemingly borne on the side of the stem because a leaf borne just below them appears to be the stem continuing; capsule blunt; tepals ¼"; stems 6–14" in dense clumps. Alp/subalp; A.

Parry's Rush

J. parryi (**pair**-ee-eye: for Charles Parry, p. 212). Flowers 1–4, narrow, green, seemingly borne halfway up the stem because the 1–5" upper bract looks like more stem; capsule sharp-tipped; tepals ¼"; stems 4–12".

Wintering Alone in the Rockies

Thomas Drummond and Sir John Richardson (p. 447) came to Canada with Sir John Franklin's second expedition. Drummond left the party and spent winter alone in an improvised brush hut in the Alberta foothills, relying on game for food. He went two months without seeing a soul, and many months with only dormant, snow-covered plants to study. The doughty Scot summed up his relations with grizzlies thus: "The best way of getting rid of the bears, when attacked by them, was to rattle my vasculum or specimen-box, when they immediately decamp." He used even less ink to sum up the Rockies near Jasper: "They gratified me extremely."

His daily routine along the Saskatchewan River gives an idea of how hard naturalists worked. "When the boats stopped for breakfast, I immediately went on shore . . . proceeding along the banks of the river, and making excursions into the interior, taking care, however, to join the boats, if possible, at their encampment for the night. After supper, I commenced laying down the day's plants, changed and dried the papers of those collected previously; which occupation generally occupied me till daybreak, when the boats started. I then went on board and slept till the breakfast hour."

He had met David Douglas (p. 65) in Scotland. The Hudson's Bay Express (an annual transcontinental canoe voyage) reunited them on the Saskatchewan in 1827, and they took the same ship back to England. Douglas, apprehensive about sharing his findings with a competitor, was stunned when Drummond freely shared his own.

Before his unexplained death in Cuba in 1835, Drummond spent two years on the first extensive botanical exploration of Texas. He dreamed of settling in Texas and starting a family, if only he could save up enough money for five acres and two cows.

Streambanks to dry meadows, mainly alp/subalp; A.

Baltic Rush

J. balticus. Flowers in a seemingly lateral spray; stem 1–4', apparently leafless (upper bract looks like a continuation of the stem, and lower leaves are reduced to sheaths) stems rising singly from the rhizome. Forms broad patches (not clumps) in marshes at all elevs; A.

Mertens Rush

J. mertensianus (mer-ten-zee-**ay**-nus: for Karl Mertens, p. 120). Flowers dark brown, tiny (⅛" long), many, in a compact, round terminal cluster; leaves 1–4, the uppermost one angled off just below the inflorescence; stems 4–12"; in dense clumps; alp/subalp; A.

Skunk-cabbage

SHOWY MONOCOTS

Monocots and dicots are named for their seed leaves, or cotyledons—the first green part to sprout from a newly germinated seed—being single on monocots and a pair on dicots. Since those disappear early on, we recognize monocots by less definitive traits, mainly these two:

- Parallel-veined leaves
- Flower parts in threes (three petals and three sepals, but when these are nearly identical, we call them six "tepals")

The monocots that follow are distinguished from the grasslike monocots by (most of them) broader leaves and (all of them) moist flower parts evolved for visual attractiveness—in a word, showiness. Very few dicots reliably have their flower parts in threes. Only three such genera—buckwheats (pp. 237–238), wild ginger (p. 241), and dock (p. 268)—appear in this book, and they don't have parallel leaf veins.

Showy Monocots with 4 or Many Stamens

Since monocots typically have flower parts in threes, having four stamens makes these two families oddballs.

SKUNK-CABBAGE

Lysichiton americanus (lye-zih-**kite**-on: loose tunic). Flowers of 4 lobes, 4 stamens, greenish yellow, ⅛" across, many, in a dense spike 2½–5" tall, partly enclosed or hooded by a yellow, parallel-veined "spathe"; leaves all basal, ± net-veined, oval, eventually up to 3' × 1' or even bigger; from an enlarged, fleshy vertical root. Wet ground; northerly. Araceae.

Many plants evolved sweet fragrances that attract sugar-loving pollinators, like bees. Others evolved putrid smells that attract flies and beetles. Our skunk-cabbage gets skunkier when its leaves are damaged. Its main pollinators, rove beetles, congregate in skunk-cabbage flowers to eat pollen and to mate. The flowers are at first functionally female, then a few days later become male (producing pollen, but no nectar).

The Calla-lily family, Araceae, has a voluptuous characteristic inflorescence consisting of a spadix (a fleshy spike of crowded flowers) and spathe (a large bract enfolding it). In the case of our skunk-cabbage, spathe and spadix thrust up from wet ground in early spring. Leaves come later, and keep growing all summer to reach sizes unmatched north of the banana groves; they were widely used as "Indian wax paper" to wrap camas bulbs, berries, and other foods for steam-pit baking

Wapato

Rough-fruit fairy bells

Hooker's fairy bells

♀ flowers, less showy and usually borne lower on the stem; flowers (both sexes) in whorls of 3; leaf blades from narrowly 3-pointed to roundly arrow-shaped, 1–5" long × 1–3" broad, on long leafstalks from the base, which is usually submerged in water; A. Alismataceae.

In fall, the slender wapato rhizomes produce egg-sized tubers that taste a little like roasted chestnuts. Traditionally retrieved from the mud with the toes (by humans) or with the bill (by ducks), wapato roots were widely traded among the first peoples, and by those on the Lewis and Clark expedition: "it has an agreeable taste and answers very well in place of bread."

Showy Monocots with 6 Stamens (Lilies)

All the following genera were in one large Lily family from Linnaeus's time until fairly recently. Despite recognizing several families today, botanists still think of them all as "lilies *sensu latu*," or "in a broad sense."

FAIRY BELLS

Prosartes spp. (pro-**sar**-teez: attached). Also *Disporum* spp. Flowers ¼–½", white, bell-shaped, pendent from branchtips, usually in pairs; berries yellow to red, egg-shaped, ¼–½", ± edible (juicy, sweetish but insipid); leaves 2–5", wavy-edged; stems 12–30", much branched. Liliaceae.

Rough-fruit Fairy Bells

P. trachycarpa (tray-key-**car**-pa: rough fruit). Berry finely pebbly, holds 6–12 seeds; leaves smooth, at least above. Forests and aspen groves; A–.

Hooker's Fairy Bells

P. hookeri (**hook**-er-eye: for Sir Joseph Hooker, p. 122). Berry smooth, holds 4–6 seeds; leaves finely hairy. Forests; northerly.

ONE CURE FOR SNOW BLINDNESS, according to the Blackfoot, would be to place a fairy bells seed under each eyelid and hold it there overnight.

or for storage. The leaf bases and the roots are marginally edible after prolonged cooking or storage breaks down the intensely irritating, "hot" oxalate crystals. Elk and bears eat them, with no complaints recorded. Food or not, skunk-cabbage was regarded as strong medicine—to induce labor, for example. Later, some white man patented and sold it under the name "Skookum."

WAPATO

Sagittaria cuneata (sadge-it-**air**-ia: arrowlike; cue-nee-**ay**-ta: wedge-shaped). Also arum-leaf arrowhead. Petals 3, white, spreading, nearly round, ⅜–¾"; 3 sepals green, pointed; many stamens; pistils on separate, ball-shaped

Twisted-stalk   Starry false-Solomon's-seal   Plumed false-Solomon's-seal

TWISTED-STALK

Streptopus amplexifolius (**strep**-ta-pus: twisted foot; am-plex-if-**oh**-lius: clasping leaf). Also *S. fassettii*. Flowers bell-shaped, ½" long, dull white, 1 (sometimes 2) beneath each leaf axil; berries red, ¼–½", juicy, sweetish but insipid; flower stalklets sharply kinked; leaves 2–5", tapered, elliptical; stems 12–40", much branched. Moist forest; A. Liliaceae.

Twisted-stalk discretely hides its flowers under its leaves—quite a trick, since the crotch *above* each leaf bears the flower. Look closely to see the flower stalk where, fused to the stem, it runs up it to the next leaf base. (The top leaf can't have a flower under it.) The berries were called "snakeberries" by the Lillooet, "witchberries" by the Haida, and "Grizzly Bear's favorite food" by the Kootenai, who considered them poisonous. Other nations (including the Okanagan, according to one source) ate them with gusto.

FALSE-SOLOMON'S-SEAL

Maianthemum spp. (my-**anth**-em-um: May flower). Also *Smilacina* spp. Flowers white, fragrant, many, in 1 terminal inflorescence; berries ¼", round; leaves heavily veined, pointed, oval to narrowly elliptical, 2–7"; stems arching, unbranched, often zigzagging at leaf nodes. Asparagaceae.

Starry False-Solomon's-seal

M. stellatum (stel-**ay**-tum: starry). Also star-flowered lily-of-the-valley. Tepals ¼", flat-spreading; flowers 6–18; berries longitudinally striped, ripening dark red to blackish; stems 8–24". A–.

Plumed False-Solomon's-seal

M. racemosum (ras-em-**oh**-sum: bearing flowers in racemes, a misleading name). Also *M. amplexicaule*. Stamens ⅛", longer than the minute tepals; flowers in a ± fluffy, conical panicle; berries speckled at first, ripening red; stems 1–3'; A.

FALSE CLUES RUNNING wild: this genus bears scant resemblance to Solomon's seal (genus *Polygonatum*). *Racemosa* isn't racemose, but *stellata* is. I find false-Solomon's-seal berries sometimes delicious (not purgative, as some authors report) but often insipid. Both species grow in deep woods but also in clearings and open slopes, and their form varies accordingly. In deep shade, *stellata* spreads its leaves flat from a bent-over stem; in sun, it holds its stem upright, grows narrower leaves, angles them upward, and folds them sharply at the midvein.

TOFIELDIA

Triantha occidentalis (try-**anth**-a: 3 flower). Also false asphodel; *Tofieldia glutinosa*. Tepals white, ⅛–¼", persisting while the 3-styled pistil grows out past them into a fat, reddish, 3-celled capsule; flowers/fruit in a dense cluster atop an 8–20", sticky-hairy stem; seed capsules red; 1–3 grasslike basal leaves 2–6", and sometimes 1–2 smaller stem leaves. Wet places, mainly subalpine; WA to Yellowstone. Tofieldiaceae.

Look closely at tofieldia stems and you may see tiny beetles and flies immobilized among sticky hairs. A recent study found that the plants digest and utilize animal

Tofieldia

Pacific trillium

Glacier lily

proteins from these little victims. In other words, tofieldia is a carnivorous plant, a fact that had escaped notice because it has no parts obviously devoted to predation. The sundew, a well-known carnivore, uses similar but larger bulbous-tipped reddish glands, and also uses the same phosphatase enzymes for digestion. Typical carnivores make a point of locating their insect traps well away from their flowers, to avoid killing pollinators. Tofieldia does not, but accomplishes the same purpose: its trapping glands are too small to catch likely pollinators.

Analysis of the plant family tree finds *Triantha* on a branch far from the lilies—where it had long been grouped—not just in a different family, but a different order: Tofieldiales.

PACIFIC TRILLIUM

Trillium ovatum (**tril**-ium: triple; oh-**vay**-tum: oval). Also wake-robin. 3 petals 1–3", white, aging through pink to maroon; 3 sepals shorter, narrower, green; flowers single, on a 1–3" stem from the whorl of 3 leaves, 3–7", often equally broad, net-veined exc for the 5 or so ± parallel main veins; plants 6–16", rhizomatous; rarely with 4 or 5 leaves and/or petals and sepals, instead of 3 each. Low forest; A–. Melanthiaceae.

For most flowers that shift color as they age, as trilliums do, the advantage is in using the aging blooms to maximize floral display to attract pollinators from afar, but then to redirect them to whiter blooms that have more viable pollen. Later on, the trillium packs its seeds in a gummy oil tasty to ants, the usual disseminators.

GLACIER LILY

Erythronium grandiflorum (air-ith-**roe**-nium: red—, the flower color of some species; gran-dif-**lor**-um: large-flowered). Tepals golden yellow, 1–1½", spreading to reflexed either in an arc or from the base; flowers ± nodding, single or sometimes 2–6 on a 4–12" stalk; capsule erect, 1" tall, 3-celled and -sided; 2 leaves 4–8", basal, wavy-edged; from a scallionlike bulb. Abundant around timberline, scattered at lower elevs; A. Liliaceae.

Glacier lilies can generate enough heat to melt their way up and bloom through the last few inches of snow (see p. 222). Fluttering with illusory fragility in subalpine breezes, they offer themselves as vehicles for those anthropomorphic virtues we love to foist on mountain wildflowers—innocence, bravery, simplicity, perseverance, patient suffering, and so forth. The seedpods are food for hoofed browsers, and the bulbs are an important "garden vegetable" (p. 342) for grizzlies. Glacier lilies survive fires well.

MARIPOSA LILY

Calochortus spp. (cal-o-**cor**-tus: beautiful grass). Petals creamy white (or pink/purple) drying yellowish, ⅝–1½" long, almost as broad; sepals pointed, much smaller; flowers 1–5; seedpods have persistent 3-branched styles; the only large leaves basal, flat; stem 4–20", from a small, deeply buried bulb. Liliaceae.

Gunnison's Mariposa Lily

C. gunnisonii (gun-ih-**so**-nee-eye: for John Gunnison, p. 217). Each petal has a purplish basal spot, a purplish transverse band about ⅓ of the way up, and branched or knobby-tipped hairs in between; sepals also have a purplish basal spot and band; seedpods narrow, erect; leaves narrow, inrolled, several. NM to s MT.

White Mariposa Lily

C. eurycarpus yu-ree-**car**-pus: broad fruit). Petals have a large purple blotch near center, and just a few hairs below that; seedpods erect, with 3 broad wings; basal leaf up to 1" wide. OR to WY.

Sego Lily

C. nuttallii (nut-**all**-ee-eye: for Thomas Nuttall, p. 191). Tepals pink or white with yellow base, fine magenta crescents; seedpods erect; leaves narrow, inrolled. Arid lowlands; A–. The UT state flower.

Pointed-tip Cat's Ears

C. apiculatus (ay-pic-you-**lay**-tus: with a small point). Petals have a small, dark dot near the base, an abruptly narrowed basal neck, and dense white (and yellow) hair over at least their lower ⅓; seedpods nodding, 3-winged; basal leaf up to ⅝" wide. Northerly.

Elegant Cat's Ears

C. elegans. Long, white hairs cover at least the lower ¾ of petals; petals and sepals often have a ± crescent-shaped purple spot; seedpods nodding, with 3 broad wings; leaf flat, much taller than stem. Northerly.

"CAT'S EARS" IS the most pictorial of many flattering names given to these sensuous blossoms, all the more admired for being prohibitively hard to cultivate. Most species need bone-dry soil before the bulb goes into healthy winter dormancy. There was an era when gardeners and nurseries attempted to cultivate them on a grand scale; many species were nearly loved to death in the process. The leading *Calochortus* taxonomist was also a leading collector: he actually boasted of his "record" pace of digging up four thousand bulbs per day! Plainly there was nothing like today's awareness of species endangerment.

Beetles frequently pollinate *Calochortus*, apparently drawn by such features as its hair-filled, off-white bowls, which collect and hold heat; large glands on the petals, which

Gunnison's mariposa lily

White mariposa lily

Sego lily

Pointed-tip cat's ears

Elegant cat's ears

Not a Happy Camper

Karl Andreas Geyer came west with Sir William Drummond Stewart's 1843 expedition, the first pleasure trip to the Rockies. The party camped on Persian carpets under crimson canopies. Geyer left the party at the Big Sandy in Wyoming, carrying on westward, at first alone but soon with Jesuit missionaries—a giant step downscale: "the hospitality the Jesuits showed to me was scant and beggarly." From the Jesuits' Flathead Valley mission to the Coeur d'Alene area was "one of the most terrible journeys I have ever made, especially in the midst of winter, crossing 76 times streams. Some we had to swim . . . Owing to the difficulties . . . I could not pay proper attention to the vegetation. But this much I do know, that I saw [a coast redwood]."

An apparently competent botanist, Geyer was, on the other hand, short on charm and perhaps ethics. He reneged on a contract to deliver his collections to the eminent botanist Georg Engelmann (p. 70); overstayed his welcome at the Chemokane Mission (the missionary's wife wrote, "[We] are determined to be rid of him"); and left Chief Factor John McLoughlin, of the Hudson's Bay Company, scrambling to find out whose credit line Geyer had been charging supplies on.

On his return to Germany after eleven years away, Geyer looked, to his friends, "at least 20 years older." His several thousand specimens were offered for sale to wealthy collectors at ten dollars per hundred. This was the income field botanists relied on.

Alexander Gordon, another botanist who started out in Stewart's party, had the temerity to write to Sir William Hooker (p. 122) politely arguing for higher prices for his rarer specimens, but had to settle for the usual two pounds per hundred—only to be thwarted from selling any specimens at all. He and **Joseph Burke**, a third botanist in the party, failed to make their marks in botanical history because of shipping damage to their collections: Burke's rotted while taking fifteen months at sea, and Gordon's went down with their ship.

---

beetles feed from; and prominent stamens and pistils.

A Cheyenne about to race a horse would feed her a mariposa lily bulb to help her win.

WILD ONION

Allium spp. (**al**-ium: Roman name). Flowers ¼–⅜" long, several, on stalklets all from 1 spot between pointed, onionskin–like segments of the spathe that encase the inflorescence in bud; stems often bunching; from small onions with the trademark aroma. Amaryllidaceae.

Nodding Onion

A. cernuum (sir-**new**-um: nodding). Inflorescence broad, nodding (often erect in fruit); tepals pink or white, oval, much shorter than stamens, all ± alike but in separate inner and outer whorls; stem 8–20". In sun, all elevs; A.

Short-styled Onion

A. brevistylum (brev-is-**tie**-lum: short style). Tepals pink, all ± alike, 2× longer than stamens;

Nodding onion

Short-styled onion

Hooker's onion

Netskin onion

leaves 3+, flat; stem 8–24". Wet meadows and streambanks at all elevs; from CO n.

Geyer's Onion

A. geyeri (**guy**-er-eye: for Karl Geyer, p. 142). Tepals pink or white, all ± alike, not quite hiding stamens; leaves 3+, channeled; stem 4–20"; in our common variety, *tenerum,* flowers are few and sparse, many being replaced by little bulbils, which reproduce clonally. Wet meadows; A.

Hooker's Onion

A. acuminatum (a-cue-min-**ay**-tum: pointed). Tepals pointed, purple or pink to (occasionally) white, the outer whorl ± spreading, bell-shaped, inner whorl smaller, narrow; leaves 2–5, withering before flowers open, much shorter than the 4–12" tubular stem. Dry foothills; A.

Netskin Onion

A. textile (**tex**-til-ee: woven, referring to the bulb coat). Tepals white (rarely pink) with reddish midrib, all ± alike, ¼" long, longer than pistil and stamens; leaves 2, basal, inrolled, often taller than the 2–14" stems; several bulbs wrapped together in a gray-brown, net-textured coat. Dry sites; A.

Wild Chive

A. schoenoprasum (skee-no-**pray**-zum: reed leek). Tepals pink or white, all ± alike, longer than stamens; flowers closely clustered, blooming in sequence from center outward; leaves 2, tubular, their bases sheathing the stem for several inches. Gravelly streambanks, wet meadows; A.

AN ONION IS easy to recognize (and sometimes hard to miss) because it smells like an onion. If it doesn't, don't try a taste test; it might be the toxic death camas. If you want it for seasoning, just clip a few leaves. Digging up the puny bulb isn't worth the trouble and kills the plant. Along with the above common species, we have several rare ones in need of protection.

BEAD LILY

Clintonia uniflora (clin-**toe**-nia: for NY governor DeWitt Clinton; you-nif-**lor**-a: 1 flower). Also queen's-cup, bride's-bonnet. Flowers single, white, ± face-up, broadly bell-shaped

Wild chive

Bead lily

Alp lily

Wood lily

Yellow bell

Leopard lily

WOOD LILY

Lilium philadelphicum (**lil**-ium: Roman name; fil-la-**del**-fic-um: probably for the city). Flower 2–3" long, erect atop 1–2' stem, red-orange, purple-spotted near base, bell-shaped, usually single; capsules fleshy, 1"; leaves 2–4", narrow elliptical, the upper ones in 1 or 2 whorls; from a many-cloved bulb. Clearings, mainly near or e of CD. Liliaceae.

Stave off the temptation to pick or dig up this attention-grabbing flower, and just leave it for others to enjoy au naturel. Rampant picking has badly reduced its numbers.

YELLOW BELL

Fritillaria pudica (frit-il-**air**-ia: checkered; **pew**-di-ca: blushing). Flowers yellow, pendent, narrowly bell-shaped, ⅜" long, usually single; 6 tepals all alike; leaves ± grasslike, 2 or several; stem succulent, 4–12". Blooms early; grasslands, ponderosa woods, from nw CO n. Liliaceae.

LEOPARD LILY

Fritillaria atropurpurea (frit-il-**air**-ia: checkered; at-ro-pur-**pew**-ri-a: black-purple). Tepals ½–¾", inward-curving, brownish purple mottled with yellowish green; flowers pendent, 1–4; capsule ½", 6-winged; leaves 2–5", narrow, both whorled and single; stem 5–36", from a bulb of a few large garliclike cloves with many tiny ricelike bulblets. Grassy slopes below tree line; A–. Liliaceae.

These elegant flowers sell themselves short, with camouflage coloring and a smell that draws flies. The bulbs and ricelike bulblets were a major food in the old days; they're bitter, and too rare to dig up now. The similar chocolate lily, *F. affinis,* with larger, darker

to nearly flat-spreading; 6 tepals ¾–1"; berry intensely blue, ⅜", many-seeded, inedible; 2 or 3 leaves 3–6" × 1–2", heavy, smooth, shiny, basal, sheathing the 2–4" stalk. Shady forests; northerly. Liliaceae.

The beady blue berry—more striking than the formal white blossom—was used sometimes as a stain, but never as a food.

ALP LILY

Lloydia serotina (**loy**-dia: for Edward Lloyd; sir-**ot**-in-a: delayed). Also *Gagea serotina*. Flowers 1 or 2, face-up, conical to saucer-shaped; 6 tepals barely ½", white with dark veins and often a purplish basal tinge; seedpod egg-shaped, ¼"; leaves very narrow, few, sheathing the 2–6" stalk; bulb sheathed in dried leaves of previous years. Alpine, esp on limestone-derived soil; A. Liliaceae.

flowers, grows from the Northwest Coast to northern Idaho. When the Haida were introduced to rice, they named it "fritillary-teeth."

SAND LILY

Leucocrinum montanum (lew-co-**cry**-num: white lily). Also starlily. White flowers spread 2" wide from several white floral tubes emerging from center of a rosette of strap-shaped leaves up to 8" long; no stems; withering after blooming in May–June. Dry lower elevs; A–. Asparagaceae.

The base, and ovary, of each sand lily flower is a couple of inches belowground, and the shiny black seeds are still there when they mature. Ants may disseminate the seeds, which can take eleven months to happen after the next year's flowers push the old seeds out. This odd strategy has not been studied conclusively. The lily's distribution is also odd: it's common in parts of the High Plains and adjacent foothills, and widespread in the Great Basin, but in between it's missing from most west-draining parts of the Rockies.

DOUGLAS'S TRITELEIA

Triteleia grandiflora (try-tel-**eye**-a: 3 ends; grand-i-**flor**-a: big flower). Also gophernuts; *Brodiaea douglasii.* Flowers ½" long, sky-blue, several, on stalklets all from 1 spot between narrow bracts; stamens attached at unequal heights on the 6 tepals' inner faces; leaves 1 or 2, flat, grasslike. Low forests or sagebrush from w-most CO and WY n. Asparagaceae.

Triteleias and the similar brodiaeas resemble onions, taste rather like onions, and were a major food resource. For gophers, they still are.

CAMAS

Camassia quamash (ca-**mass**-i-a **qua**-mosh: 2 versions of a Chinook Jargon name for camas, from an originally French word). Tepals blue-violet, ¾–1½", narrow, the lowermost one usually noticeably apart from the other 5; inflorescence roughly conical; capsule ½–1", splitting 3 ways; leaves narrow, basal/sheathing, shorter than the 8–24" stem. Seasonally moist meadows; A–. Asparagaceae.

Camas bulbs were the prized vegetable food of most Indigenous groups from the coast to western Montana, while camas cakes were second only to dried salmon in trade volume. In fact, so valuable were the bulbs that the plowing up of camas prairies for settler's pastures in 1877 provoked the Nez Perce War.

Native Americans may have purposefully spread camas to new areas. A family would mark out, "own," and maintain a camas patch year-round for generations, weeding and burning it. Compared to the early nineteenth-century reports of vast seas of blue, camas prairies have since lost much of their camas, probably for want of caretakers. Year-round tending made it possible to weed out death camas when it's easiest to recognize—while in bloom; then it would be safer to dig camas

Sand lily   Douglas's triteleia   Camas

the following spring before flowering, when the bulbs are best. Nevertheless, people did die from inadvertently eating death camas. Camas bulbs are not recommended to hikers.

In Quamash cuisine, according to David Douglas:

> A hole is scraped in the ground, in which are placed a number of flat stones on which the fire is placed and kept burning until sufficiently warm, when it is taken away. The cakes, which are formed by cutting or bruising the bricks and then compressing into small bricks, are placed on the stones and covered with leaves, moss, or dry grass, with a layer of earth on the outside, and left until baked or roasted, which generally takes a night. They are moist when newly taken off the stones, and are hung up to dry. Then they are placed on shelves or boxes for winter use. When warm they taste much like a baked pear. It is not improbable that a very palatable beverage might be made from them. Lewis observes that when eaten in a large quantity they occasion bowel complaints... Assuredly they produce flatulence: when in the Indian hut I was almost blown out by strength of wind.

Blame the flatulence on inulin, an indigestible sugar that takes the place of starch in camas bulbs, Jerusalem artichokes, and a few other vegetables.

MEADOW DEATH CAMAS

Toxicoscordion venenosum (tox-ic-oh-**scor**-de-un: toxic garlic; ven-en-**oh**-sum: poisonous). Also *T. gramineum, Zigadenus venenosus*. Flowers white, saucer-shaped, in a tall raceme (like beargrass but with fewer, larger flowers); tepals less than ¼", with a yellow or green oval spot near base; capsules ½–¾", splitting, 3-celled, 3-styled; most leaves basal, narrow, sheathing, plus often 2 or more on the 8–30" stem. Grassy places, primarily lower; A. Melanthiaceae.

Back during camas-digging days, death camas may have killed more people in the West than any other plant ever will. Today it maintains its reputation with an occasional sheep death. No one would confuse the small white flowers of death camas with the big blue ones of camas, but mistakes occur when populations of the two are intermixed and both have gone to seed

Meadow death camas

Mountain death camas

or withered—and camas bulbs are ripe for eating.

In contrast to most plants that use chemistry to defend themselves from herbivores but proffer harmless, nutritious nectar and pollen, this plant's nectar and pollen are lethal to most insects. That would have been terminally maladaptive for the species if it weren't for one species of bee—the sole pollinator of Toxicoscordion—co-evolving tolerance to the toxin, which now protects the bee from brood parasites. Look for dead bugs surrounding death camas.

Bronze bells

Beargrass

Similar foothill death camas, *T. paniculatum,* has several flowers on its lower side-branches, making the inflorescence a panicle.

MOUNTAIN DEATH CAMAS

Anticlea elegans (antic-**lee**-a: Odysseus's mother; el-eg-enz: elegant). Also *Zigadenus elegans.* Tepals all alike, ⅜", greenish cream with a heart-shaped greenish spot near the base; flowers in a tall, loose spike; leaves, stem often with whitish coating. Subalpine forest to alpine tundra; A. Melanthiaceae.

Few if any humans have been killed by *Z. elegans,* which may be less toxic than the preceding, and prefers elevations where edible camas is uncommon.

BRONZE BELLS

Anticlea occidentalis (oxy-den-**tay**-lis: western). Also *Stenanthium occidentale.* Flowers several, pendent, narrowly bell-shaped, ⅜" long, varying from burgundy red to greenish cream; 6 tepals all alike; 2–3 leaves basal, up to 1" wide; stem 8–15". Moist to wet alp/subalp sites; northerly. Melanthiaceae.

BEARGRASS

Xerophyllum tenax (zero-**fill**-um: dry leaf; **ten**-ax: holding fast). Stamens longer than tepals; flowers white, fragrant, saucer-shaped, ½" across, many; inflorescence at first nippled, bulbous, 3–4" across, elongating up to 20", the lowest flowers setting seed before the highest bloom; capsules 3-celled, dry; leaves narrow, tough, dry, V-grooved, with minutely barbed edges; basal leaves largest, 8–30", in a large, dense clump; stalk to 60", covered with small leaves. Locally abundant in subalpine meadows and open forests; WA to Tetons. Melanthiaceae.

In optimal meadows, beargrass blooms abundantly most years. Under partial forest cover, it may bloom rarely, but it still reproduces year after year by producing new offset clumps from the rhizomes. And then one year, hundreds bloom at once. Wildfires can also pass over, burning the vegetative clumps but leaving the soil cool enough for the roots to survive and resprout.

Beargrass flower heads and seeds are good, relatively rich bear food, but don't rank among grizzlies' favorite foods. Spring's tender leaf bases feed deer, elk, and gophers. Neatly clipped leaf bases you see here and there are more likely the work of a "brush-picker" gathering foliage for the florist trade.

California corn lily

By summer the leaves are wiry and strong, making them an important material for baskets and hats. David Douglas crowed, "Pursh is correct as to their making watertight baskets of its leaves. Last night my Indian friend Cockqua . . . brought me three of the hats made on the English fashion, which I ordered when there in July; the fourth, which will have some initials wrought in it, is not finished, but will be sent by the other ship. I think them a good specimen of the ingenuity of the Natives and particularly also being made by a little girl, 12 years old . . . I paid one blanket (value 7 shillings) for them."

Douglas's imaginative biographer, William Norwood, read between those lines to argue that Douglas, in his praise for a little girl's craft, was disguising a romantic entwinement with a "Chinook princess."

CORN LILY

Veratrum spp. (ver-**ay**-trum: true black). Also false-hellebore. Flowers saucer-shaped, numerous; styles 3, persistent on the 1"-long, 3-celled capsules, but lacking from the (staminate) lower flowers; leaves mostly 5–12", coarsely grooved along veins, oval, pointed; 3–7' tall. Wet meadows, mainly subalpine. Melanthiaceae.

Green Corn Lily

V. viride (**veer**-id-ee: green). Also *V. eschscholtzii*. Flowers pale green, ½–¾" across, in a loose panicle with drooping branches. Northerly.

California Corn Lily

V. californicum. Also *V. tenuipetalum*. Flowers dull white, or only slightly greenish, ¾–1½", in a dense panicle with ascending branches. A.

HEAVY BEDS OF snow lying on steep meadows tend to creep downslope through the winter, scouring vegetation from the surface. Woody seedlings are frustrated year after year, while herbs with fat storage roots and fast spring growth are favored, perpetuating the meadow. On many wet slopes, blunt corn lily shoots are the first plants to thrust upward as the snow recedes. Looking almost Venusian—startlingly clean and perfect—when they first unclasp from the stalk, they may soon become ragged, with help from foraging elk and insects.

Be warned that signs of browsing do not mean the plant is safe to eat: ewes that ate *V. californicum* gave birth to one-eyed lambs. Scientists blame an alkaloid they named cyclopamine after the mythical one-eyed monster. Long listed as promoting tumors, cyclopamine now turns out to *kill* some cancer cells very nicely, particularly medulloblastoma, a brain cancer in children that has no approved treatment to date. Better yet, no harm has been apparent in either the ewes themselves or lab mice dosed with cyclopamine. But don't munch on the plant: both roots and young shoots are toxic, and used to be ground up for a crop insecticide (by farmers) or for a sinus-clearing snuff (by Native peoples). Every group that knew the plant used some part of it—sometimes burned as a fumigant, sometimes worn around the neck to ward off evil.

Showy Monocots with 3 Stamens (Irises)

The Rockies have just two genera in the Iris family, both lovely but utterly different in that Sisyrinchium has six identical tepals, whereas Iris has a unique arrangement of

Rocky Mountain iris

Blue-eyed-grass

three petals, three sepals, and three petal-like styles (pistil branches).

ROCKY MOUNTAIN IRIS

Iris missouriensis (**eye**-ris: rainbow; miz-oo-ree-**en**-sis: of the Missouri River). Flowers pale blue, 2 (rarely up to 4); consisting of 3 spreading sepals, 3 erect petals, and 3 smaller petal-like styles ± resting on the sepals and hiding the stamens; stem 10–20"; leaves grasslike, basal; capsule 3-celled, splitting. Moist-in-spring sites up to subalpine; A. Iridaceae.

Iris was the Greek goddess who flashed across the sky, bearing messages: the rainbow. Some *Iris* species in the West were tough enough to use in cordage, but this one found diverse medicinal uses. The Shoshone, for example, put it on tooth cavities, venereal sores, earaches, burns, and rheumatism.

BLUE-EYED-GRASS

Sisyrinchium montanum and *S. idahoense* (sis-er-**ink**-ium: Greek name, derived obscurely from "pig snouts"). Both formerly *S. angustifolium*. 6 tepals blue with yellow base, ¼–¾"; stamens fused into one 3-anthered column; flowers 1 or a few; leaves grasslike, basal, shorter than the 2-edged stem, plus 2 bracts from base of flower stalk: the longer bract looks like a stem continuing above the flowers; 6–14" tall. Dry, grassy sites briefly moist in spring; A. Iridaceae.

These delicate perennials complete their active season in wet soil in a few weeks of spring, then wither and die back for the rest of the year, going dormant through the summer (p. 222).

Showy Monocots with Irregular Flowers (Orchids)

Orchid flower structure is a snap to recognize: One petal, lower-most and thrust forward, is always utterly unlike the others and usually much larger. It's called the "lip" and serves as a platform for pollinators. Above the lip is a combined stamen and pistil called the "column."

Orchid flowers are among the most elaborate insect lures on Earth. While the flowers evolved outlandishly, the roots mostly atrophied: a majority of orchid genera are rootless tropical lianas. Our orchids have meager vestigial root systems that tap into preexisting networks of fungal hyphae.

BOG ORCHID

Platanthera spp. (plat-**anth**-er-a: broad anther). Also rein orchids; *Limnorchis* spp., *Habenaria* spp. Flowers in a tall spike; lip with a long, down-curved spur to the rear, 2 sepals horizontal, the other sepal and 2 petals erect, hooding the column. Orchidaceae.

White Bog Orchid

P. dilatata (dil-la-**tay**-ta: widened). Also bog candles, *P. leucostachys*. Flowers white, spicy-fragrant, in a dense 4–12" spike; stem 8–40", with many clasping leaves, lower ones up to 10" × 2",

White bog orchid

Round-leaf bog orchid

Slender bog orchid

much smaller upward. Wet ground, often sub-alpine; A.

Round-leaf Bog Orchid

P. orbiculata (or-bic-you-**lay**-ta: circular leaf). Flowers white to greenish or yellowish, in a loose spike; stem 8–24", leafless exc for a few tiny bracts; basal leaves typically 2, oval to nearly round, 2–6". Shady forest; northerly.

One-leaf Bog Orchid

P. obtusata (ob-too-**say**-ta: blunt leaf). Also *Lysiella obtusata*. Flowers yellow-green to almost white, in a loose spike; stem 2–8", leafless; single basal leaf blunt elliptic, 1–4". Moist spots in forest; A.

Slender Bog Orchid

P. stricta (**stric**-ta: drawn tightly together). Also *H. saccata*. Flowers green, sometimes purple-tinged, in a loose spike; lip narrow, much longer than the round, ± scrotum-shaped spur; leaves oval to elliptic, sheathing the stem; 8–20". Wet ground from Tetons n.

THE DISTINGUISHING FEATURE of bog orchids is a narrow, nectar-filled pouch or "spur" projecting rearward from the lip. The spur is a key element in the grand pattern of orchid evolution, which allies each variety and species of orchid with one, or at most a very few, species of insect. The right insect is not only powerfully attracted, but also physically unable to extract nectar from two successive blooms without picking up pollen from the first and leaving a good dose of it on the stigma of the second. Bog orchids' devices include proboscis-entangling hairs to engage the insect for a little while; adhesive discs that stick to the insect's forehead while instantly triggering the stamen sac to split open; and little stalks that each hold a cluster of pollen to the adhesive disc (now on the insect's head), at first in an erect position that keeps the pollen away from the stigma of that same flower, but then (when the insect flies on) drying out and deflating into the right position to push the pollen onto the gluey stigma of the next flower of the same species, where the insect repeats her routine exactly. All this just to minimize wasting pollen in the wrong places. Efficiency!

The inch-long apparent stalk between the flower and the stem is actually the flower's ovary: a 180-degree twist in it shows that the lip evolved from what was originally the uppermost petal.

RATTLESNAKE-PLANTAIN

Goodyera oblongifolia (**good**-yer-a: for John Goodyer; ob-long-if-**oh**-lia: oblong leaf). Flowers greenish white, many, in a 1-sided spike up to 5"; lip shorter than and hooded by the fused, ¼"-long upper petals, all connected

Rattlesnake-plantain

Broadlip twayblade

Northern twayblade

to the stalk by a twisted ovary; leaves 1½–3", in a basal rosette, thick, evergreen, very dark glossy green, mottled white along the veins; stem unbranched, 10–16". Shady forest; A–. Orchidaceae.

The intensity of the snakeskin pattern on the leaves varies, even among side-by-side plants. Sometimes only the midvein is white, but usually there is enough white pattern to tell these leaves from those of white-veined pyrola (p. 213). This is one orchid with enough leaf area to make it nearly independent of its mycorrhizal partners by the time it matures.

LADIES'-TRESSES

Spiranthes romanzoffiana (spy-**ranth**-eez: coil flower; rom-an-zof-ee-**ay**-na: for Nikolai Rumiantsev, p. 200). Flowers ± white, ½" long, seemingly tubular, in (usually 3) ranks in a 2–6", dense, coiled-looking spike (perhaps suggesting a 4-strand braid); leaves sheathing, mostly basal; stem up to 24". Wet meadows and streambanks, up to timberline; A. Orchidaceae.

Ladies'-tresses

TWAYBLADE

Neottia spp. (nee-**ott**-ia: bird's nest). Also big-ears; *Listera* spp. Flowers pale greenish, small (about ½"), several, in an open, short-stalked spike; broad lip petal, upper 2 petals slender, resembling the 3 sepals; seed head like a tiny bowl of eggs; leaves 2, broad, 1½–2½", apparently opposite, clasping the stem at mid-height, tips pointed; stem unbranched, 4–12". Forest shade. Orchidaceae.

Heartleaf Twayblade

N. cordata (cor-**day**-ta: heart-shaped). Lip petal splits into 2 long, pointed lobes, like an inverted Y. A.

Broadlip Twayblade

N. convallarioides (con-va-larry-**oy**-deez: like lily-of-the-valley). Lip petal narrow at base, widening to 2 broad, short, round lobes, like a narrow inverted heart. A–.

Northern Twayblade

N. borealis (bor-ee-**ay**-lis: northern). Lip petal broad at base, ending in 2 broad, short, round lobes. A.

Northwestern Twayblade

N. banksiana (banksy-**an**-a: for Sir Joseph Banks). Lip petal has 2 tiny but long and pointed side-lobes at its base, and almost no notch at the tip. WA to Tetons.

Mountain lady's-slipper   Yellow lady's-slipper   Calypso orchid

THOUGH DIMINUTIVE AND dull, a twayblade is every inch an orchid, as testified to by its glues and mechanisms for sticking pollen onto visiting insects—often mosquitoes. (And who better than an *orchid* to *testify*? These two words derive from *orchis* and *testis*, the Greek and Latin words, respectively, for testes.) The twayblades are so closely related to the European non-green plant *Neottia* that taxonomists moved them into its genus after a century in *Listera*.

LADY'S-SLIPPER

Cypripedium spp. (sip-rip-**ee**-dium: Venus's slipper). Also moccasin flower, Venus-slipper. 1 to 3 flowers; lip 1" long, very bulbous, with purplish veins and a yellow staminate structure at its base; upper sepal and lateral petals brownish-purple, about 2" long, slender, often twisted; stem 6–24", with several broad, clasping-based, 2–6" leaves, a smaller bract beneath each flower. Wet soil in bogs, forests; rare. Orchidaceae.
Mountain Lady's-slipper
C. montanum. Lip white. WA to Bighorns.
Yellow Lady's-slipper
C. parviflorum (par-vif-**lor**-um: small flower). Also *C. calceolus*. Lip yellow. A.

CALYPSO ORCHID

Calypso bulbosa (ca-**lip**-so: "Hidden," a sea nymph; bulb-**oh**-sa: bulbous). Also fairy slipper, deer's head orchid. Flower single, pink; lip slipper-shaped, almost white, magenta-spotted above, magenta-streaked beneath; other petals and sepals all much alike, narrow, ¾"; leaf single, basal, growing in fall, withering by early summer; stem 3–7". Moist, mature forest; uncommon; A. Orchidaceae.

A close look reveals these little orchids to be just as voluptuously overdesigned as their corsage cousins, which evolved bigger to seduce the oversized tropical cousins of our insects. Get down onto their level to see and smell them, using a hand lens if you like, but don't pick them. Their bulblike corm is so shallowly planted that it's almost impossible to pick them without ripping the corm's lifelines.

A calypso is dependent on its fungal and plant hosts (p. 263); its single leaf withers and is gone early in the growing season. Like other orchids, it produces huge numbers of minute seeds (3,770,000 seeds were found in one tropical orchid's pod) with virtually no built-in food supply, and an abysmal germination rate. They germinate only if the right species of fungi are already growing there and can supply nutrients. The tiny black specks in vanilla beans and vanilla ice creams are orchid seeds.

CORALROOT

Corallorhiza spp. (coral-o-**rye**-za: coral root). Entire plant (exc lip) dull pinkish-brown, or rarely pale yellow (albino), but no parts green; flowers ¾–1¼" long (½ of that being the tubular ovary), 6–30 in a loose spike; leaves reduced to inconspicuous sheaths on the 6–20" stem. Forest. Orchidaceae.

Spotted coralroot | Striped coralroot | Western coralroot

Spotted Coralroot

C. maculata (mac-you-**lay**-ta: spotted). Lip usually white, with many magenta spots (sometimes also on petals). A.

Striped Coralroot

C. striata (stry-**ay**-ta: striped). All petals and sepals brownish- to purplish-striped. A.

Western Coralroot

C. mertensiana (mer-ten-zee-**ay**-na: for Karl Mertens, p. 120). Lip redder than plant, often with 1 or 2 spots or blotches. WA to w WY.

CORALROOTS USUALLY GROW in forest stands with few herbs or shrubs. They blend in with the duff and sticks until a shaft of sunlight hits, suddenly incandescing their eerie, translucent flesh. Their rhizomes do resemble coral—curly, short, knobby, and entirely enveloped in soft fungal tissue. As in other orchid genera, a seed's embryo develops only if penetrated, nourished, and hormonally stimulated by a minute strand ("hypha") from a fungus in the soil. Fungal hormones suppress root-hair growth and stimulate the

orchid to produce mycorrhizae instead (p. 263). In the case of coralroots, no root hairs ever form.

DICOTS

The remainder of this chapter is all dicots, a huge category that isn't perfectly user-friendly; it's safe to say all of its members in this chapter either don't have parallel-veined leaves or don't have flower parts in threes.

Dicots Without Green Parts

Many plants in unrelated lineages took the radical evolutionary step of giving up on photosynthesis. In the orchid family (above) and the heath family (following), it's done by living off the underground fungal web (pp. 154 and 263), whereas broomrape does it by directly parasitizing other plants.

PINEDROPS

Pterospora andromedea (tair-**os**-por-a: winged seed; an-drom-ed-**ee**-a: a name from Greek myth). Entire plant gummy/sticky, monochromatically brownish red exc for the amber, 5-lobed, jar-shaped corollas; flowers many, on downcurved stalks; capsules ± pumpkin-shaped; leaves brown, small, and sparse on the 12–48" stem. Forest; A. Ericaceae.

This year's glowing amber pinedrops stalks shoot up alongside last year's dry, brown stalks, still standing. They're our tallest non-green plants: I measured one at 5.5 feet. As the name implies, they may be found under (and mycorrhizally linked to) ponderosa pines, but also under Douglas-fir or many kinds of trees all across the continent.

Pinedrops

Freeloaders on the Fungal Web

Some of our most intriguing herbs lack chlorophyll, and obtain all their nutrients from fungi. Since chlorophyll is the green pigment in plants, these plants are not green. They get carbohydrates from living green plants via mycorrhizal fungi (p. 263), which pass them along to the non-green plants below ground.

The word *saprophyte* ("decay plant") was used for over a hundred years for these plants—and for decay fungi in the same breath—by scientists who saw them all as plants that live by decaying things. Since fungi are not plants, and plants cannot decay things (they lack the enzymes for it), saprophytes don't exist. Fungi that rot things are now called saprobes or saprotrophs. The words used by scientists for the *plants* formerly known as saprophytes range from long (*mycoheterotrophs* or *epiparasites*) to short (*cheaters*).

Mycorrhizal symbiosis lets fungi and plants utilize each other's strengths: the fungus's efficient uptake of water, phosphorus, and nitrogen, and the plant's ability to make carbohydrates out of air, water, and sunlight. Some mycorrhizal plants, it seems, gradually contributed less and less to this exchange, and got away with freeloading as long as there were other plants feeding the same fungus. As the cheaters evolved, any organs they no longer needed atrophied. They ended up with vestigial leaves, little or no chlorophyll, and no real roots—nothing but a stalk of flowers reproducing, while using the trees above for leaves and the fungi below for roots.

Proving whether this symbiosis is mutualistic or parasitic has resisted most efforts to date. Non-green plants do produce chemicals that some researchers call "vitaminlike" because they stimulate growth of both the fungal and green hosts. But that doesn't prove they're contributing: parasites such as mistletoe commonly stimulate growth in their hosts—sapping their hosts, not energizing them.

This plant is called Albany beechdrops in the East; it was near Albany, New York, that David Douglas, a year before coming to the Northwest, sought beechdrops out and was excited to find it, accurately observing that in "soil so dry that every other vegetation refused to grow . . . it seems to be a sort of parasite like *Monotropa* or *Orobanche*." Enthusiastically, but less accurately, he went on to write, "I have no doubt but it will cultivate."

INDIAN PIPE

Monotropa uniflora (ma-**not**-ra-pa: flowers turned 1 way, a meaningless name for a 1-flowered plant; you-nif-**lor**-a: 1 flower). Also ghost pipe, corpse-plant. Entire plant fleshy, white or pink-tinged, drying black; flower single, narrowly bell-shaped, ½–¾", pendent, but erect in fruit—a soft, round capsule; usually 5 petals; leaves translucent, small; stems densely clustered, 2–10". Dense forest; northerly. Ericaceaae.

A strange plant, but a familiar one all across the continent. Fungi in the Russula family are its usual connection to nutrients, and Douglas-fir is often at the far end of the pipeline. Don't pick Indian pipes—they'll just turn black and ugly within hours.

PINESAP

Monotropa hypopitys (hye-**pop**-it-eez: under pine). Also *Hypopitys monotropa*. Entire plant fleshy, pink-tinged yellow or red to straw-colored, drying black; flowers several, bell-shaped, ⅜–¾", mostly 4-lobed (rarely 5), initially downturned all in 1 direction from each stem, turning erect in fruit; seed capsules round, soft; leaves translu-

Indian pipe *Pinesap* *Naked broomrape*

cent, small; stems clustered, 2–10". Shady forest; A. Ericaceaae.

Digging up a pinesap (don't!) would reveal a soft mycorrhizal rootball only a couple of inches deep. Species of *Boletus* are common partners of this species, and a pine or other conifer is almost always hooked up to the same bolete.

Genetic studies show that pinesap and Indian pipe are not each other's closest relatives. It is unclear to me why authorities have not yet responded by reverting to the name *Hypopitys monotropa*, which has been the accepted name at various times past.

BROOMRAPE

Aphyllon spp. (ay-**fill**-ahn: no leaves). Entire plant non-green, fleshy, bristling with sticky hairs; flowers tubular, ⅝–1½" long, with 5 corolla lobes; leaves reduced to scaly bracts, often buried. Orobanchaceae.

Clustered Broomrape

A. fasciculatum (fas-ic-you-**lay**-tum: clustered). Also *Orobanche fasciculata*. Plant incl corollas mostly pale yellow and/or dull purple; corolla more narrow-tubular, with smaller lobes, than the following; flowers single on 3–10 slender stalks branching from a short, hard stem. Near sagebrush and kin, its host plants; A.

Naked Broomrape

A. purpureum (pur-**poo**-rium: purple). Also *Orobanche uniflora*. Corollas blue-purple, dull pink, yellow, or almost white, with lobes spreading wide; flowers single on 1–4 slender stalks. Found near its host plants—stonecrops, saxifrages, asters; A.

Clustered broomrape

BROOMRAPE LIVES AS a direct parasite on plant roots, in contrast to most of our non-green plants, which obtain their photosynthates indirectly, via fungi.

Until recently, the family Orobanchaceae was small, consisting of parasites. Wildflower aficionados may be dismayed to see their beloved paintbrushes reclassified as Orobanchaceae. However, broomrapes and paintbrushes were always seen as closely related, and many paintbrushes are partly parasitic. The parasitic mode evolves so easily that it does not warrant family-level distinction.

Dicots with Composite Flower Heads

Picture a daisy. Seeming petals radiate from a central disk. On close examination, the base of each petal-like ray sheaths a small pistil. The ray and pistil together constitute a ray flower, not a petal; the petal-like length is an entire corolla of petals, fused then split down one side and flattened. The central disk turns out to be a lot of little flowers too—disk flowers; each has a (usually two-branched) style poking out of a minute tube of five fused stamens, within a larger tube, which is the (usually five-lobed) corolla. What we saw at first as one flower is a composite flower head, the inflorescence of the family Asteraceae, or composites. Where a calyx of sepals might cup a flower in another family, an involucre of green bracts forms a cup under a composite flower head. (Technically, sepals may also be present in the form of a pappus—a brush of hairs, scales, or plumes that remain attached and grow as the seed forms, usually to provide mobility via wind or fur.)

The tight fit of the stamen tube around the style is a mechanism that prevents self-pollination. As the pistil grows, it plunges all the pollen out of the tube; once its tip has grown free of the tube, it can split, exposing stigmas on the inner faces of its two branches.

The composites in this chapter are separated into three types—like daisies, or like dandelions, or like thistles—depending on the presence or absence of ray flowers and disk flowers. The dandelion type has no disk flowers, just long, colorful, strap-shaped,

bisexual ray flowers with stamen tubes as well as pistils. Other composites have only disk flowers, or else *appear* to have only disk flowers because their inner and outer flower types are equally unshowy, Either way, I'll call those thistlelike.

Composite flowers excel at producing abundant, well-nourished seeds, and have thrived and diversified in recent geologic time. With more than twenty thousand species worldwide (and hundreds in the Rockies), composites are even more diverse than grasses.

Composites with Disk and Ray Flowers (Daisylike)

If you picture a daisy, it has ray flowers radiating from a cushionlike central disk of disk flowers.

DAISY AND FLEABANE

Erigeron spp. (er-**idge**-er-un: soon-aged). Rays relatively narrow and numerous; in many spp. they can be white, purple, or anywhere in between; disk yellow; involucre has just 1 row of equal bracts. Asteraceae.

Cutleaf Daisy

E. compositus (com-**poz**-it-us: compound). Head single, rays 20–60, white (may be pinkish or bluish); leaves basal, compound with slender leaflets in 3s. Many individuals are rayless, setting seed without pollination. Sandy or rocky soils, mid-elevs and up; A.

Showy Fleabane

E. speciosus (spee-see-**oh**-sus: showy). Also aspen fleabane. Heads several; rays 60–150, blue-purple (rarely pink or white), very narrow; leaves all the way up stem, narrow elliptical, edges smooth. Moist sites in sun; A.

Subalpine Daisy

E. glacialis. Also *E. peregrinus.* Heads usually single; rays 30–80, violet (rarely pink or white), ⅜–1"; involucre woolly or sticky from top

Cutleaf daisy

Showy fleabane

Subalpine daisy

One-flower daisy

Low daisy

to bottom; leaves basal (few and tiny on alpine plants) and a few small ones on stem. Moist sunny sites, mainly alp/subalp; A.

One-flower Daisy

E. grandiflorus (grand-if-**lor**-us: big flower). Also *E. simplex*. Head single, rays 50–120, lavender to white, ½" long; leaves narrowly oblong, mainly basal; stem sticky-hairy; stem, involucre, and often the leaves woolly; 1–8" tall. Alp/subalp; A–.

Woolly Daisy

E. lanatus (lan-**ay**-tus: woolly). Head single; rays 30–80, white to pink, ⅜"; leaves narrowly oblong, with 2 side teeth on some; involucre ± purple, ¼–⅜" tall; leaves all basal; plants 1–4", with long, crinkly hairs all over. Alpine; MT, WY, CO.

Low Daisy

E. humilis (**hue**-mil-iss: humble, i.e., low). Head single; rays 50–150, white to blue-purple, ¼"; involucre deep purple, ⅜–¾" tall; basal leaves spoon-shaped or narrowly oblong; smaller leaves on stem; plants 1–8" tall, bristling with white hair. Seeps and wet meadows at mid to alpine elevs; ID, MT, WY.

DAISIES GENERALLY BLOOM during the spring flowering rush, earlier than most composites, suggesting the Latin name. Insecticidal properties implied by "fleabane" are medieval superstition, or arise from confusion with *Pyrethrum* daisies. The name "daisy" (originally for the English daisy, *Bellis perennis*) traces back to "day's eye" in Old English. "Daisy" and "fleabane" are essentially interchangeable.

ASTER

Spp. formerly incl in genus *Aster*. Rays 10–40 exc where noted, ¼–¾"; disk yellow; involucre has unequal bracts in 3 rows. Late summer and fall. Asteraceae.

Leafy Aster

Symphyotrichum foliaceum (sim-fee-**ah**-trick-um: joined hairs; fo-lee-**ay**-shum: leafy, ref to the bracts). Also *Almutaster foliaceus*. Rays 16–60, violet or blue to white; several heads; non-sticky involucre bracts dramatically bigger, looser, and more leaflike than on most similar flowers, the lowest row the biggest by far; 8–24" tall. Moist, ± sunny sites, all elevs; A.

Engelmann's Aster

Doellingeria engelmannii (der-ling-**ee**-ria: for Ignaz Döllinger; en-gle-**mah**-nee-eye: for Georg Engelmann, p. 70). Also *Eucephalus engelman-*

Leafy aster

Engelmann's aster

Arctic aster

Showy aster

Gray aster

Hayden's aster

Arctic Aster

Eurybia sibirica (yu-**rib**-ia: wide/ few, ref to rays). Rays 12–50, violet: a few heads; disk yellow to burnt orange; involucre purplish; leaves narrow, stiff, toothed or not, gray-fuzzy; stems 2–12", not sticky. High elevs; northerly.

Showy Aster

E. conspicua. Rays violet to blue; several heads; involucre sticky; leaves sharp-toothed; 1–3' tall. Moist, ± sunny sites, up to mid-elevs; northerly.

Gray Aster

E. glauca (**glaw**-ca: whitish). Also *Herrickia glauca, Eucephalus glauca, Aster glaucodes.* Rays 10–16, lavender; disk yellow or dark red; many heads; involucre tall but compact; leaves grayish, fine-toothed; stems 1–3'. Rocky sites at all elevs; southerly.

Hayden's Aster

Oreostemma alpigenum var. *haydenii* (oreo-**stem**-a: mountain garland; al-**pidge**-en-um: alpine; hay-**den**-ee-eye: for Ferdinand Hayden, p. 217). Rays purple; head single; leaves basal, linear, pointed, 2–4"; stems 1–3, ± sprawling, 2–6". Alp/ subalp; northerly.

ASTERS GENERALLY BLOOM later than daisies and have fewer, wider rays.

TOWNSENDIA

Townsendia spp. (town-**zen**-dia: for David [*not* J. K.] Townsend). Striking 2–3" flower heads on low plants—often resting on the rosette of leaves; disk yellow, rays various; deep involucre with several rows of bracts; leaves mainly basal, spoon-shaped, entire. Dry meadows and gravels, all elevs. Asteraceae.

nii. Rays usually 8 or 13, white or pale pink; involucre reddish; stems 2–5', ribbed, mostly unbranched, leafy all the way up. Moist, ± open forest; A–.

Wyoming Townsendia
T. montana. Also *T. alpigena.*
Rays 13–21; deep violet is a
common color; stem ½–2".
OR to c MT and nw CO.

Parry's Townsendia
T. parryi (**pair**-ee-eye: for
Charles Parry, p. 212). Rays
21–64; pale blue to lavender
most common; stem 3–8".
OR to WY.

Stemless Townsendia
T. exscapa (ex-**cay**-pa: stemless). Rays 11–40, white, often
pink-stained; stem up to 1¼".
Low to mid-elevs; MT to NM.

Wyoming townsendia

Stemless townsendia

THE BLACKFOOT GAVE their
horses townsendia tea as a
medicine.

ARNICA

Arnica spp. (**ar**-nic-a: lamb-
skin, referring to leaf texture).
Heads entirely yellow, single
to few, blooming in early to
mid summer; 4–8 opposite leaves, plus some
leaves nearby in nonflowering whorls from the
rhizome; stems 4–24". Asteraceae.

Heartleaf arnica

Hairy arnica

Rydberg's Arnica
A. rydbergii (rid-**berg**-ee-eye: for Per Axel
Rydberg, CO botanist). Rays 7–10, ½"; leaves
± smooth-edged, narrow-elliptical, with 3–5 ±
parallel main veins; stem 4–12". Alp/subalp; A–.

Heartleaf and Broadleaf Arnica
A. cordifolia and *A. latifolia* (cor-dif-**oh**-lia:
heart leaf; lat-if-**oh**-lia: broad leaf). Rays 8–15,
½–1¼"; leaves ± toothed, broadly oval, or often
heart-shaped on *cordifolia.* Open forest and
meadows; together, possibly the most ubiquitous flower in our range; A.

Hairy Arnica
A. mollis (**mol**-iss: soft). Rays 12–20; disk flowers
large, 5-lobed; involucre sticky-hairy from base
to top; leaves lanceolate to elliptic; green parts
usually hairy and sticky. Mid-elev to subalpine
meadows; A.

HERBALISTS USE ARNICA
roots as a plaster for sore or
injured muscles or joints.
The Okanogan-Colville
used it—mixed with heart
and tongue of a robin—as
a love potion. *A. cordifolia* readily resprouts from
deep roots after a fire.
Arnicas commonly set
seed without pollination,
resulting in broad patches
that are clones.

Rydberg's arnica

GROUNDSEL

Senecio spp. (sen-**ee**-cio: old man). Groundsel,
butterweed, and ragwort are ± interchangeable
names. Rays sparse, disorderly, and few (4–10
or rarely none); disk yellow; involucre deep,
bracts all in 1 row; several heads. Asteraceae.

Arrowleaf groundsel

Lambstongue groundsel

Dwarf mountain groundsel

Arrowleaf Groundsel

S. triangularis. Rays ¼–½", yellow; leaves narrow-triangular, 3–7", all on 1–5' stem. In saturated soil; A.

Lambstongue Groundsel

S. integerrimus (in-te-**jer**-im-us: very smooth-edged, which is not always true). Rays ¼–½", yellow to white, or lacking; leaves on long stalks, ± elliptical, mainly basal, smaller upward; 8–24" tall. In sun; A.

Dwarf Mountain Groundsel

S. fremontii (free-**mont**-ee-eye: for John C. Frémont, p. 297). Rays ¼–⅜", yellow; leaves alternate, somewhat succulent, oblanceolate, ⅜–1½" long, usually coarsely toothed, hairless, often deep maroon when first coming up; clumps of ± sprawling stems 4–14" tall. Alp/subalp rocky or wet sites; A. In CO and NM it's var. *blitoides* (a.k.a. *S. blitoides*), with sharp, hollylike leaves.

Big-head Groundsel

S. megacephalus (meg-a-**sef**-a-lus: big head). Also Nuttall's ragwort. Rays ½–1", yellow; leaves 3–7", with long gray hairs, narrow, mainly near base; 8–20" tall. Alp/subalp; MT, ID, nw WY.

Colorado ragwort

Colorado Ragwort

S. soldanella (sold-a-**nel**-a: little coin). Rays ½", yellow; disk wider; flower single; leaves, stem, and involucre purplish to entirely purple; leaves oval; 2–8" tall. Alpine; CO and NM.

BUTTERWEED

Packera spp. (for Alberta botanist John G. Packer). Also *Senecio* spp. Rays 7–14, ½–¾"; disk and rays yellow; deep involucre with 1 row of bracts. Asteraceae.

Streambank Butterweed

P. pseudaurea (sue-**dor**-ia: false gold). Several heads; basal leaf blades elongated-heart-shaped, fine-toothed, on long stalks; stem leaves narrower, few, with long lobes on basal ½, teeth on outer ½; stem 2–30". Moist places in forest; A–.

Streambank butterweed

Rocky Mountain Butterweed

P. streptanthifolia (strep-tanth-if-**oh**-lia: leaves like twistflower). Like the preceding, exc basal leaf blades taper to the stem end; A.

Woolly Butterweed

P. cana (**cay**-na: gray). Leaves and stems white-hairy; basal leaves usually entire, stem leaves few and small, often toothed; 5–16" tall. Sunny, dry sites at all elevs; A.

Alpine Meadow Butterweed

P. subnuda (sub-**new**-da: almost hairless). Also *S. cymbalarioides*. Leaves and stems scarcely hairy; basal leaves often scalloped, with a ± round blade on a long leafstalk; stem leaves few and small; 2–12" tall. Alp/subalp; WY, MT, ID, OR.

SEVERAL SPECIES IN this genus poison livestock, and several, including *P. pseudaurea,* have been used medicinally.

Woolly butterweed

Alpine meadow butterweed

Dusky butterweed

Blanketflower

DUSKY BUTTERWEED

Tephroseris lindstroemii (tef-rah-**see**-riss: ash-colored chicory; lind-**strö**-me-eye: for Axel Lindström). Also twice-hairy or fuscate groundsel; *Senecio fuscatus, S. lindstroemii.* Alpine dwarf with large heads of a distinctive, rich saffron-orange color; 1 or a few heads; rays 7–12, ½–¾"; involucre a deep cup of bracts all in 1 row; green parts have thick, cobwebby hair. Beartooths and Absarokas (and also boreal). Asteraceae.

BLANKETFLOWER

Gaillardia aristata (guy-**lar**-dia: for Gaillard de Marentonneau; air-iss-**tay**-ta: with bristles). Also brown-eyed Susan. 1–4 large (2–3") heads per stem; rays golden-yellow, ending in 3 long lobes; disk at first orange and nearly flat, expanding to a bulbous, bristly red-brown button; stems and leaves hairy. Low-elev grassland, forests; A. Asteraceae. Similar black-eyed Susan—*Rudbeckia hirta,* of

the lower Great Plains—invades some Rocky Mountain roadsides.

HAIRY GOLDEN-ASTER

Heterotheca villosa (het-er-o-**thee**-ca: varied cups; vil-**oh**-sa: soft-hairy). Also *Chrysopsis villosa.* Heads single, golden yellow, 1" across; rays 10–25; all green parts (variably) coarsely hairy; leaves all on stem, elliptical; stems 4–20", numerous. Can persist in bloom from

Hairy golden-aster

Four-nerve daisy

Alpine sunflower

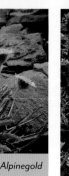

Alpinegold

Woolly sunflower

ALPINE SUNFLOWER

Hymenoxys grandiflora (hi-men-**ox**-iss: sharp membranes). Also old-man-of-the-mountains; *Rydbergia grandiflora, Tetraneuris grandiflora.* Huge yellow heads 2–3", facing e or toward sun; disk often 1" or wider, rays 15–34, 3-toothed; leaves of several narrow lobes, both basal and on stem; plant woolly, 2–12". Alpine rock crevices and meadows, to highest elevs; c MT s. Asteraceae.

ALPINEGOLD

Hulsea algida (**hul**-see-a: for Gilbert Hulse; **al**-jid-a: cold). Head yellow, single, facing up, ½–1¼" across, rays 26–60, slender, disk broad; involucre, stem, and leaves sticky-hairy; leaves mainly basal, grayish, with short, pointed lobes or coarse teeth; 6–16" tall. Alp/subalp; OR, ID, s MT, nw WY. Asteraceae.

midsummer into fall; rocky sites at all elevs, disturbed ground; A. Asteraceae.

SHOWY GOLDENEYE

Heliomeris multiflora (helio-**me**-riss: sun part). Also *Viguiera multiflora.* Heads yellow, several, ¾–2"; rays 10–14; disk hemispheric; involucral bracts long and narrow; leaves mostly opposite, narrow, pointed, rough, 1–3"; stems 1–4'. Late summer into fall; steppe, gravel bars, dry clearings, roadsides; sw MT s. Asteraceae.

FOUR-NERVE DAISY

Tetraneuris acaulis (tet-ra-**new**-ris: 4 veins; ay-**caw**-lis: stemless, i.e. flower heads on stalks, not leafy stems). Also goldflower; *Hymenoxys acaulis, Actinea acaulis.* Heads entirely yellow, single, ¾–2", facing up; rays 9–15, ⅜–¾", 3-lobed; leaves all basal, ± elliptic, without teeth or lobes; 1–15" tall. All elevs; A. Asteraceae.

WOOLLY SUNFLOWER

Eriophyllum lanatum (area-**fill**-um: woolly leaf; lan-**ay**-tum: woolly). Also Oregon sunshine. Rays about 8, ⅓" long; heads all yellow, single; leaves, stems, and involucres all thickly white-woolly; leaves on stem, usually linear; stems 5–24", many, weak, in thick clumps or sprawling mats. Dry sites at all elevs; WA e to nw WY. Asteraceae.

LITTLE SUNFLOWER

Helianthella spp. (he-lee-anth-**el**-a: little sunflower). Big (1¾–4") yellow flower heads mainly face s or se; leaves ± opposite, long and narrow, entire; 1–5' tall. Asteraceae.

One-flower Little Sunflower

H. uniflora. Flower head single; leaves have 3 main veins. Clearings; WA to w CO and nw NM.

Five-veined Little Sunflower

H. quinquenervis (kwin-kwe-**ner**-vis: 5 veins). Usually 3 flower heads; leaves with 5 main veins, rough. Aspen groves, clearings, from ID and c MT s.

THESE PLANTS ARE often covered with ants that eat nectar oozing around the involucre. While swarming the plant for nectar, the ants chase away flies attempting to lay eggs on the flower, where the fly larvae, later in the season, would eat the sunflower seeds. The plants' reproductive success correlates with ant presence, showing that the symbiosis is mutual. Ant-plant mutualisms are common in the tropics, but less so here.

The common sunflower, *Helianthus annuus,* found in some mountain valleys, is a coarser, rougher plant with dark-brown flower disks.

Five-veined little sunflower

Arrowleaf balsamroot

Hoary balsamroot

Yellow mule's ears

BALSAMROOT

Balsamorhiza spp. (bal-sam-o-**rye**-za: balsam root). Heads all yellow, numerous, 1 per stem; leaves basal, plus a small bract or 2 on stem. Sunny, often s-facing dry slopes. Asteraceae.

Arrowleaf Balsamroot

B. sagittata (sadge-it-**ay**-ta: arrow-shaped). Heads 2½–4" across; basal leaves triangular, 12" × 6", on long stalks; plant covered with soft, feltlike silvery wool, esp under leaves when young; plants 10–30", in massive clumps. A–.

Hoary Balsamroot

B. incana (in-**cay**-na: whitish). Heads 2–3" across; basal leaves pinnate, 4–14" long on shorter petioles, covered with thick, long, silky white wool. WA to all of WY.

ARROWLEAF BALSAMROOTS PUT on spectacular massed displays. We have several similar balsamroot species, and they hybridize. Deer fatten on the huge flower heads, and the Salish, Nez Perce, and others ate the young shoots, the seeds, and the fragrant, slightly woody taproot. With a root that survives most fires, balsamroot may flourish in burns.

MULE'S-EARS

Wyethia spp. (wye-**eth**-ia: for Nathaniel Wyeth, p. 192). Heads 2½–6" across, 1 to several; leaves have thick white veins; basal leaves up to 20" × 6", tapering at both ends, stem leaves similar but 1–4" long; plant 10–24", in thick clumps. Sunny slopes up to timberline. Asteraceae.

White Mule's-ears

W. helianthoides (he-lee-anth-**oy**-deez: sunflowerlike). Heads white to cream-colored; leaves fuzzy. Moist meadows; northerly.

Yellow Mule's-ears

W. amplexicaulis (am-plex-i-**caw**-lis: clasping the stem). Heads bright yellow; leaves dark, heavy, hairless, ultraglossy with an aromatic varnish; stem leaves clasp stem. A–.

Pygmy goldenweed

Lyall's goldenweed

Curlyhead goldenweed

Western canada goldenrod

Missouri goldenrod

MULE'S-EARS AND BALSAMROOT are close cousins, and were eaten in similar ways—for example, a mush made from roasted, ground seeds.

GOLDENWEED

Tonestus spp. (ton-**es**-tus: anagram of *Stenotus,* a relative). Also serpentweed, iron plant; *Haplopappus* spp. 1 flower head per stem, yellow; 15–42 rays; involucre has both leafy and membranous bracts; leaves narrow but stalkless, basal and then smaller up the stem. Alpine; WA to c CO. Asteraceae.

Pygmy Goldenweed
T. pygmaeus. Plant hairy but not sticky all over; 1–5". Southerly.

Lyall's Goldenweed
T. lyallii (lie-**ah**-lee-eye: for David Lyall, p. 62). Plant sticky, 1–6". Northerly, plus rare in c CO.

CURLYHEAD GOLDENWEED

Pyrrocoma crocea (peer-a-**co**-ma: orange hair; **cro**-see-a: orange). Also *Haplopappus croceus.* 1 flower head per stem; 25–70 narrow rays up to

1½" long, distinctively yellow-orange, crinkling when they dry; leaves long and narrow, clasping the 8–30" stem. Showy in mid-elev to subalpine meadows, roadsides; southerly. Asteraceae.

GOLDENROD

Solidago spp. (so-lid-**ay**-go: healing, based on medieval beliefs). Heads yellow, numerous, in 1 terminal cluster; rays 5–18, small. Asteraceae.

Western Canada Goldenrod
Solidago lepida (**lep**-id-a: scaly). Also *S. canadensis.* Heads yellow, 100+, in a tall, fluffy, often pyramidal cluster; leaves alternate, narrow elliptic, largest near mid-stem; stem grayish-hairy, more so upward; 1½–5' tall. Moist meadows and clearings; A.

Missouri Goldenrod
S. missouriensis. Inflorescence ± round-topped; stem smooth, 1–3' tall, largest leaves in a basal cluster. Moist meadows, prairies, and clearings; A.

Rocky Mountain Goldenrod
S. multiradiata (multi-ray-dee-**ay**-ta: many-rayed). Rays 12–18, ¼"; leaf edges have fine

teeth and a fringe of hairs; upper stems finely hairy, 2–14". Rocky sites, often alp/subalp; A.

Sticky Goldenrod

S. simplex. Also *S. glutinosa.* Rays 8–15, ¼"; flowers often in a tall, narrow column; leaves hairless, basally crowded, tapering to leafstalks; stem smooth, 2–24". All elevs; A.

Rocky Mountain goldenrod

Yarrow

THE WIDESPREAD AND familiar Canada goldenrod—a scapegoat for a lot of hay fever—was split several ways by *Flora of North America,* with most Rocky Mountain populations falling into *S. lepida.* Missouri goldenrod is fairly distinct. Identification is tough among the several dwarf goldenrods, which often hybridize.

YARROW

Achillea millefolium (ak-il-**ee**-a: for Achilles; mil-ef-**oh**-lium: thousand leaf). Also *A. lanulosa.* Rays 3–5, white (rarely pink or lavender), ⅛" long and wide; disk yellow; heads many, in a flat to convex inflorescence; leaves narrow, very finely dissected, fernlike, aromatic; to 3' tall. In sun, all elevs; in at least 49 US states. Asteraceae.

Achilles, the Greek hero in the Trojan War, was taught by Chiron the centaur to dress battle wounds with yarrow. Indigenous Americans used yarrow poultices, drank yarrow tea for myriad ailments, and steamed the homes of sick people with the pungent smell—rather like rosemary and sage—of yarrow leaves. In China, the yarrow-stalk oracle was systematized by Confucius and his followers in the *I Ching.* Europeans still use yarrow both medicinally and as a flavoring in bitters and beer. The herbalist Gregory Tilford claims it repels insects, at least briefly, when rubbed on the skin, but he warns of irritating sensitive skin.

Some divide this species several ways on the basis of degrees of leaf dissection; but it

Prairie coneflower

was found that transplants, over a few years, came to nearly match yarrows around them, altering their leaf shape to fit their new environment.

PRAIRIE CONEFLOWER

Ratibida columnifera (ra-**tib**-id-a: a mystery name; col-um-**nif**-er-a: column-bearing). Also Mexican hat. Rays 3–7, yellow (or occasionally purple), drooping, ½–1½"; disk cylindrical or conical, much taller than wide, its flowers blooming progressively upward in a visible ring; leaves pinnate; 1–4' tall. Grassland or steppe of valleys and foothills; A–. Asteraceae.

The Turkish-born Franco-American-German child prodigy Constantine Rafinesque (1783–1840) taught evolution a half-century before Darwin. He authored countless names of both plants and animals—too many for his own good. His later

Cutleaf coneflower

Western coneflower

Dusty maiden

Also *R. ampla.* Rays 6–16, yellow, droopy; disk barrel-shaped, yellowish to grayish; leaves compound and lobed, 4–10" wide; stem 3–6'. Wet places, esp on Great Plains; MT, WY. Asteraceae.

Composites Without Ray Flowers (Thistlelike)

The disk flowers that make up these flower heads range from drab and dry, in pussytoes, to showy and soft, in knapweed, but none of them are strap-shaped, which would make them rays. When showy, they are radially symmetrical—some actually look like little flowers, unlike most florets in composite flower heads.

WESTERN CONEFLOWER

Rudbeckia occidentalis (oxy-den-**tay**-lis: western). Disk purplish black, cylindrical to conical, up to 2¼" tall, surrounded by floppy green bracts; leaves alternate, long-pointed oval, often toothed; rank-smelling, coarse 2–6' tall plants often in clumps. Clearings; A–.

Though native, this black-headed plant is viewed as a pest on mountain meadows and clearcuts in Idaho, sometimes thriving to the exclusion of either trees or palatable forage. Cattle eat everything else in the meadow to avoid coneflower, enabling it to take over.

Most rayless composites are in the chicory tribe within the aster family. Outside that tribe, this species and a scattering of others (and local growth-forms) stopped producing rays—a strategic shift away from pollinators like bees that are drawn to bright colors. They may have rank smells that attract flies or ants instead.

DUSTY MAIDEN

Chaenactis douglasii (kee-**nac**-tiss: gaping rays; da-**glass**-ee-eye: for David Douglas, p. 65). Also false yarrow, hoary pincushion. Roundish heads of white to creamy-pink disk flowers so large you can see that each is a 5-lobed corolla with protruding 2-branched styles; involucre sticky-hairy; leaves lacy, fernlike; plants

years were sucked into a downward spiral of taxonomic splitting, leading to rejection by the scientific community, poverty, bitterness, and alleged mental instability, leading to even worse splitting, and so forth; you get the picture. After he died, most of his specimens were thrown out. No clue survived him as to how he came up with the name *Ratibida*.

CUTLEAF CONEFLOWER

Rudbeckia laciniata (rude-**beck**-ia: for Olof Rudbeck, Sr. and Jr.; la-sin-ee-**ay**-ta: cutleaf).

Woolly pussytoes   Rosy pussytoes   Umber pussytoes

gray-woolly, 2–20" tall; pine foothills, alpine gravels; A. Asteraceae. An adorable dwarfed alpine form is treated as either var. or species *alpina*.

PUSSYTOES

Antennaria spp. (an-ten-**air**-ia: antenna—). Disks dirty-white, soft-fuzzy, deep, surrounded by numerous scaly bracts (they aren't ray flowers); heads several, ¼" across; leaves woolly, mainly basal. Some species increase with over-grazing. Asteraceae.

Woolly Pussytoes

A. lanata (lan-**ay**-ta: woolly). Flowers dull white; basal leaves 1–4", pointed, linear, ± erect; stems 2–8", in clumps, often broad, but without runners. Alp/subalp; WA to WY.

Pearly Pussytoes

A. anaphaloides (an-af-a-**loy**-deez: looks like pearly everlasting). Disks white, soft-fuzzy, outer bracts white; heads in a cluster up to 2" wide; leaves narrowly elliptic, 1–8", woolly, with a few parallel veins; stems 8–20". Dry, rocky sites below tree line; A.

Littleleaf Pussytoes

A. parvifolia (par-vif-**oh**-lia: little leaf). Relatively showy because heads are larger and fewer (3–7); flowers often pure white, sometimes pink, greenish, etc.; leaves ⅜–1½", pointed; stems 1–5", in clumps, often broad, but without runners. Low to mid-elevs; A.

Rosy Pussytoes

A. rosea. Mat-forming, from leafy runners; sepal-like bracts bright pink to greenish-white; leaves less than 1"; stems 3–15". In sun, all elevs; A.

Dark Pussytoes

A. media. Also *A. alpina*. Mat-forming, from leafy runners; bracts blackish to dark green, sometimes white-tipped; leaves ¼–¾"; 2–5" tall. Alp/subalp; A.

Umber Pussytoes

A. umbrinella (um-brin-**el**-a: brown, small). Bracts light brown to white, sometimes pink-streaked; leaves ⅜–¾"; 2–8" tall. Up to tree line; A.

ALPINE FORMS OF many composites, including rosy and umber pussytoes, largely reproduce asexually, the ovules maturing into seeds without being fertilized by pollen. A patch is likely a female clone; it may be all interconnected by runner stems. Such plants are well adapted to a severe climate in which the blooming season all too often zips by in weather too nasty for small insects (the pollinators of minute flowers) to be out and about. Since each clone's flowers look alike over a wide area while differing from other clones, many were named as species. Weeding out unwarranted species names requires close taxonomic study.

Pearly everlasting

Tasselflower

Spotted knapweed

der lobes, coated with wool that rubs off; plants 1–5' tall, much branched. Rangeland, roadsides; A. Asteraceae.

Genus *Centaurea* ranges from garden bachelor's-buttons to several of the West's worst weeds. All are Eurasian. Spotted knapweed ranks as Montana's worst weed, infesting 7.2 million acres. It was originally brought here on purpose, as honeybee forage. Diffuse knapweed, *C. diffusa,* has spread as a tumbleweed onto 6 million acres. (The prototypical plains tumbleweed is Russian thistle, no relation to either knapweeds or thistles.) Yellow star-thistle, *C. solstitialis,* is prettier perhaps, but meaner, with long involucral spines. As knapweeds replace grasses, they increase erosion of soils and sedimentation of creeks. Their seeds stay viable in the soil for years. Goats and sheep can do a good job on them, at least locally. Many invertebrates that eat them in their native lands were identified and tested here in hopes of achieving biological control, with little success so far.

PEARLY EVERLASTING

Anaphalis margaritacea (an-**af**-a-lis: Greek name; marg-a-rit-**ay**-sha: pearly). Disks ± yellow, surrounded by tiny, papery white bracts; heads ⅜" across, in a convex cluster; leaves 2–5", linear, woolly (esp underneath), all on the 8–36" stem, lowest ones withering. Roadsides, burns, clearcuts; or alpine; A. Asteraceae.

Instead of ray flowers, everlasting has white, dry involucral bracts that persist everlastingly.

SPOTTED KNAPWEED

Centaurea stoebe (ken-**tor**-ia: plant of Chiron, the centaur; **stee**-be: Greek name). Also *C. maculosa, C. biebersteinii, Acosta maculosa.* Disk flowers pink-purple, showy, the outermost ones much larger, slender-lobed; involucres narrow-necked, ½" tall, with striped, black-tipped bracts; leaves dotted, cut in many slen-

TASSELFLOWER

Brickellia grandiflora (brick-**ell**-ia: for John Brickell; grand-if-**lor**-a: big flowers). Also large-flowered brickellbush, a misnomer. Flower heads 1–2" long, in pendent clusters, bell-shaped green involucres; disk flowers cream, tubular, with tassel-like style branches; leaves 1–5", opposite, triangular, edges toothed or scalloped; bushy herbs 1–3' tall. Blooming late (July–Oct) on rocky sites from low to subalpine; A. Asteraceae.

TRAIL-PLANT

Adenocaulon bicolor (a-den-o-**caw**-lon: gland stem). Also pathfinder. Flower heads tiny, white, several, in a very sparse panicle on a 10–32" stalk; leaves all on lowest part of stem, triangular, 4–6", dark green on top, white-fuzzy underneath. Moist forest; northerly. Asteraceae.

Trail-plant

Elk thistle

Eaton's thistle

Unlike most of the aster family, trail-plant is adapted to deep shade. The leaves are large, thin, and flat-lying—easy to recognize by their shape—while the flowers escape notice. (They attract small flies by their smell, not their looks.) The combination of weak leaf-stalks and high contrast between upper and lower leaf surfaces led to the common name: woodsmen tracking a large animal through the woods appreciated a conspicuous series of overturned trail-plant leaves. The sticky seeds also trail us by adhering to our legs. And then there are the tiny, circumlocuitous trails that leaf-miners (p. 472) etch across many trail-plant leaves by summer's end.

THISTLE

Cirsium spp. (**sir**-shi-um: dilated veins—an early medicinal belief). Thick-stemmed herbs; leaves have spine-tipped pinnate lobes; these native spp. have no spines on leaf faces, and often have cobwebby wool on upper stems. Asteraceae.

Elk Thistle

C. scariosum (scary-**oh**-sum: like thin paper, ref to inner bracts). Heads pink-purple to yellowish white, in a massive terminal cluster crowded by long leaves; bracts tipped with ⅛–½" spines; leaves long-hairy, often pale or partly white; stem extra thick, even near top, 1–6' tall, or no stem—just a rosette at ground level. Low-elev to subalpine meadows; A.

Eaton's Thistle

C. eatonii group (ee-**toe**-nee-eye: for Daniel C. Eaton). Heads off-white to pink-purple, nodding in some varieties; involucres have cobwebby hairs, bracts tipped with long spines; leaves variable in hairiness; 1–4' tall. High meadows and talus; sw MT to n NM. Dr. Jennifer Ackerfield discovered it is actually several species; she gave one of the new species the Southern Ute name *C. tukuhnikivatzicum*.

Wavyleaf Thistle

C. undulatum (un-dew-**lay**-tum: wavy). Heads pink, widely separated at branch tips and axils; each bract has a pale stripe and an outward-bent spine; leaves woolly beneath; stem spine-less, slender, 1–5'. Dry, low elevs; A.

Gray Thistle

C. griseum (**gris**-ium: gray). Also *C. clavatum* var. *osterhoutii*. Heads pale pink to dull off-white; leaves up to 14" long; a leggy thistle 1–5' tall; variable: leaves smooth or hairy, involucres hairless to thickly woolly. Mid to high elevs; southerly.

THISTLE TAPROOTS AND peeled stems are nutritious and tasty, highly rated in both

Wavyleaf thistle

Gray thistle

Western wormwood

Indigenous knowledge and survival-skills texts. Even horses may—very carefully—munch the sweet-nectared flowers. The plant responds with profuse branching. In Yellowstone, elk thistle is known as Everts' thistle after Truman Everts, who survived on thistle roots for a month in 1870 after being separated from his party, his horse, his gun, tent, and so forth. Today Yellowstone's grizzlies eat it. A biennial, elk thistle grows only a broad rosette of leaves the first year. In its second or third year, it may bear its flowers either atop the rosette or up on a stem.

Bull thistle, *C. vulgare,* and "Canada thistle," *C. arvense*—both actually from Europe—are noxious weeds. A weevil introduced to attack these weeds is now harming several native thistle species.

WORMWOOD

Artemisia spp. (ar-tem-**ee**-zia: Greek name, for the goddess Artemis). Also wormwoods, mugworts. Flower heads pale yellowish, within small cups of fuzzy bracts; leaves often spicy-aromatic. Asteraceae.

Western Wormwood

A. ludoviciana (lu-doe-vis-ee-**ay**-na: of the Louisiana Purchase area). Also white sagewort. Leaves silvery-fuzzy on both sides, linear or pinnate-lobed or with just a few irregular, fin-

gerlike lobes; 1–4' tall. Dry meadows, roadsides, all elevs; A.

Tarragon

A. dracunculus (dra-**cun**-kew-lus: little dragon, old name for it, from Arabic *tarchon*). Also dragonwort; *Oligosporus dracunculus.* Leaves not silvery, but sometimes long-hairy beneath, either odorless or with strong tarragon aroma, linear but can cluster to appear lobed; 1½–5' tall. Steppe, dry meadows, roadsides, low to mid-elevs; A.

Lemon Sagewort

A. michauxiana (me-show-zee-**ay**-na: for André Michaux). Flowers and bracts often red-tinged; leaves silvery beneath, dark green above and sticky, the biggest ones bi- or tripinnately lobed; 6–16" tall. Gravels, mid-elevs to alpine; A–.

Rocky Mountain Sagewort

A. scopulorum (scop-you-**lor**-um: of the Rockies). Involucres blackish-brown; leaves silky-fuzzy on both sides, predominately basal; 2–12" tall. Dry alp/subalp sites; NM, CO, UT, WY, MT.

Arctic Sagewort

A. norvegica (nor-**vedge**-ic-a: of Norway). Also *A. arctica.* Heads larger, to ⅜" across, nodding; bracts black-edged; leaves green, scarcely fuzzy, bipinnately lobed, basal ones the largest by far; 4–20" tall. Dry meadows, mid-elevs to alpine; CO, UT, WY, MT, WA.

NATIVE PEOPLES IN the West valued western wormwood above all other plants for ceremonial and purification purposes, right down

Rocky Mountain sagewort

to wiping their bowls with it after eating. Burned for their smoke, infused in grease or water, or just tied up to waft about, wormwoods drove away evil spirits, mosquitoes, weakness, snow blindness, colds, headaches, menstrual irregularity, tuberculosis, underarm odor, eczema, and dandruff.

Shrubs of this genus perfume half of the West. That would be sagebrush. Unlike sagebrush, these herbaceous wormwoods resprout after fire.

In Europe, use of *Artemisia* to repel midges led to the old name "mugwort," a *mucg* being a midge

Pale mountain-dandelion

Orange mountain-dandelion

in Middle English. Wormwoods of Europe include the intoxicating absinthe, the bitter *vermouth* ("wormwood" in German), and the culinary herb tarragon, which is native here as well (the Crow call tarragon "wolf perfume"). *Artemisia* has been used medicinally in China for thousands of years. More recently, its active ingredient, artemisinin, became the first-line therapy for malaria, achieving an impressive reduction in the disease. Time will tell how well that lasts, as the spirochete evolves resistance to artemisinin.

Composites Without Disk Flowers (Dandelionlike)

Dandelions—and all the flowers in this group—have colored, soft ray flowers all the way from the center of the flower head out.

MOUNTAIN-DANDELION

Agoseris spp. (a-**gah**-ser-iss: goat chicory). Head single; plants milky-juiced; leaves basal, ± linear (rarely with a few widely spaced teeth); stem not hollow. Asteraceae.

Pale Mountain-dandelion

A. glauca (**glaw**-ca: silvery pale). Head 1–1¾" across, yellow, drying pinkish; highly variable as to leaf shape, stature, and hairiness; 4–32" tall. Common, in sun, all elevs; A.

Orange Mountain-dandelion

A. aurantiaca (or-an-**tie**-a-ca: orange). Head 1" across, red-orange, drying purplish; 6–12" tall. In sun, mid-elevs to alpine; A.

AGOSERIS FLOWER HEADS often close up on hot days. They are more complex and interesting than dandelions, the stems are more solid, and the leaves usually lack teeth. Some *A. glauca* are toothed, with teeth much smaller than *Taraxacum* lobes.

DANDELION

Taraxacum officinale (ta-**rax**-a-cum: possibly "bitter herb" in Persian; o-fis-in-**ay**-lee: in herbal commerce). Head 1–1½" across, single, yellow; leaves basal, pinnate-lobed; stem hollow. Rampantly widespread Eurasian weed; A. Asteraceae.

Weedy European dandelions have invaded even remote alpine tundra here. We also have two native alpine dwarf *Taraxacum* species, but they're uncommon. Flimsy hollow stems distinguish *Taraxacum* from *Agoseris*.

HAWKWEED

Hieracium spp. (hi-er-**ay**-cium: hawk—). Flower heads ½–1" across; stems milky-juiced. In sun. Asteraceae.

Woolly hawkweed

Orange hawkweed

Scouler's hawkweed

Woolly Hawkweed

H. triste (**tris**-tee: sad or dull). Also *H. gracile, Chlorocrepis tristis*. Heads yellow, few or single; involucres finely dark-haired; leaves ± basal on long stalks, oval, smooth; stem 2–12", fuzzy. High meadows; A.

Orange Hawkweed

H. auranticacum (or-an-**tie**-a-cum: orange). Also king devil. Heads burnt-orange, clustered; plants 8–28", densely hairy. Disturbed soil; European weed; A–.

Scouler's Hawkweed

H. scouleri (**scoo**-ler-eye: for John Scouler). Also *H. cynoglossoides*. Heads yellow, several; leaves elliptic, lowest ones up to 8", trending smaller upward; involucres, stems, and leaves (at least beneath) with long white hairs; 12–40" tall. Foothills to subalpine; WA to UT and WY.

White Hawkweed

H. albiflorum (al-bif-**lor**-um: white flower). Also *Chlorocrepis albiflorum*. Heads white, sparse; plants 6–14", hairy. Steppe to subalpine; A–.

THE MILKY JUICE of hawkweeds and mountain-dandelions was dried and then chewed for pleasure or cleaning.

HAWKSBEARD

Crepis spp. (**crep**-iss: sandal, name used by Pliny for some plant). Flower heads yellow, ¾–1¾", several; involucre narrow, ½–¾" tall, of long, narrow inner bracts and short outer ones; plant milky-juiced. Asteraceae.

Western Hawksbeard

C. occidentalis (oxy-den-**tay**-lis: western). Rays 10–40 per head; involucre with blackish sticky bristles; leaves mainly basal, 4–12", gray-fuzzy, pinnate-lobed, lobes often slanted backward; stem leaves similar but small; 3–16" tall. Dry sites, low to mid-elevs; A–.

Dwarf Alpine Hawksbeard

C. nana (**nay**-na: dwarf). Also *Askellia nana*. Rays just 6–12 per head; involucres dark red or green; leaves mainly basal, spoon-shaped, sometimes a few teeth, ½–1½" long, hairless; 2–8" tall. Alpine; A–.

SALSIFY

Tragopogon spp. (trag-a-**po**-gon: billygoat beard). Also goatsbeard, oyster plant. Head single, 3–4" across, incl the spiky green bracts sticking out way past the ray flowers; in seed looks like a huge dandelion head, each seed with a parachutelike pappus ½" or wider; stem

Western hawksbeard

Dwarf alpine hawksbeard

Yellow salsify

Yellow salsify seed head

Purple salsify

swollen below the head, milky-juiced; leaves tough, grasslike, clasping the stem; 1–3' tall. Roadsides, rangeland; European weeds; A. Asteraceae.

Yellow Salsify
T. dubius. Rays yellow.
Purple Salsify
T. porrifolius (por-if-**oh**-lius: leaves like a certain onion). Rays purple.

PUREED WITH BUTTER, cultivated purple salsify is the most delicious root vegetable I have ever tasted. I failed to detect oysters, though. Sad to say, rumors of the far more abundant wild *T. dubius* being its equal are, well, dubious. Flora author H. D. Harrington reported searching far and wide in the Rockies and finding *dubius* always "small, fibrous, woody, and tough."

Flowers Without Petals or Sepals

Stinging nettle's four tiny sepals are so obscure that getting close enough to see them could inflict pain. Other than that, you should look closely for sepals before turning to this section. Find barely countable sepals or petals on mountain-sorrel (p. 191), pussypaws (p. 232), skunk-cabbage (p. 137), bunchberry (p. 235), biscuitroots (p. 225), and baneberry (p. 240).

LEAFY SPURGE
Euphorbia esula (you-**forb**-ia: good fodder, Roman name for some members of the genus, grotesquely untrue of this one; **ess**-you-la: Celtic

for "sharp," ref to acrid juice). Also *Tithymalus uralensis, T. esula.* Numerous pairs of yellow to green, round or heart-shaped bracts, ⅝" wide, flank 4-lobed, ⅛", conical greenish involucres, each holding one 3-tipped pistil and several stamens; no petals or sepals; linear 1–3" leaves densely clothe the many side branches; stems milky-juiced, 1–3'. Russian weed; rangeland; A. Euphorbiaceae.

Spurge, public enemy number one in eastern Montana, is allelopathic to other plants and toxic to cattle and people. It can grow back from bits of its roots as deep as 30 feet down; neither herbicides nor fire nor uprooting are much use against it after its first few years. It shoots its seeds explosively as far as 15 feet, and they float, so that streams help spread them. The best hope for controlling leafy spurge probably lies in bringing in the right insect to eat it. Nine species (including a fly called *Spurgia esulae*) have already been tested and released;

Leafy spurge

we await results. Sheep browse spurge contentedly, and may be of some help.

Genus *Euphorbia* includes spiny succulents often mistaken for cacti. This huge and diverse genus should be a juicy target for taxonomic splitters; most authorities keep it intact not because they like it humongous but because they think the paper accurately splitting it has yet to be written.

STINGING NETTLE

Urtica gracilis (**ur**-tic-a: burning; **grass**-il-iss: slender). Also *U. dioica.* Flowers with parts in 4s, tiny, pale green, many, in loose panicles dangling from the leaf axils, the panicles unisexual, with ♀ higher on the plant (our varieties); leaves opposite, saw-toothed, pointed/oval, 2–6"; stem and leaves lined with fine, stinging bristles; 2–6' tall from rhizomes. Moist forest, clearings; A. Urticaceae.

Herbal Invasions

I became a settler in Iowa twenty-two years ago and have seen great changes. . . . Mayweed and dog fennel, stinkweed and mullein have taken the place of purple flox and the mocassin flower.
—Edwin James, 1859

Non-native weeds have profoundly damaged huge portions of the West's open range-land. They displace both native flora and the native fauna that has evolved with it. They do enormous economic damage when they outcompete native species that are desired for forage or timber. In some instances, they sharply increase soil erosion or bring fire to formerly fire-resistant communities. In most species extinctions, competition from non-natives is a factor.

Climate change and pollution will favor weeds. Projections of the near-future climate predict increases in fire and forest pests—disturbances that weedy plants often exploit. Since weeds tend to be heavy nitrogen consumers, they also benefit from global nitrogen deposition, which originates as agricultural fertilizers and waste. Many invasive plants additionally can adapt to climate change by timing their growth and flowering in accord with temperature. Many temperate-zone natives time their season according to day length, which is less adaptive.

Most problem weeds are not particularly "weedy" or aggressive in their native lands. Without human help, species rarely travel around the globe very fast, and as long as plants migrate slowly, the fungi, bacteria, and grazers that consume them, as well as competitors, coevolve with them, maintaining a level of control so that a single species rarely takes over. Scientists seeking to control a weed look to that plant's native lands to see what attacks it there. They choose a few promising consumers (insects, usually) and set them up in quarantined tests for several years to see which ones could be introduced without undue risk to native plants. The process is slow, expensive, and far from foolproof.

Some see invasives as a grave threat to the biosphere. Promoting "free trade" based on political or economic theory alone, without considering biology, is so myopic it's deranged. Trade in unprocessed logs and wood chips introduces pests and diseases whose potential for economic harm dwarfs any potential profits. However, fighting them once they're here is often a lost cause. We may have to learn to live with them.

Stinging nettle

long sleeves and gloves (or socks) over your hands, lop them with your knife, and steam them limp in half an inch of water.

DWARF MISTLETOE

Arceuthobium spp. (ark-you-**tho**-bium: life on junipers). Tiny (1–4"), branching olive to dull blue-green or orange-green growths on conifer branches; flowers budlike, same color as plant; berries ± bicolored, with bluish-coated base and paler-tan outer ½; no recognizable leaves; A. Viscaceae.

Occasionally, you spot mistletoe on branches near eye level, but what you see more often are "witches' brooms"—big lumps in conifer crowns, consisting of massed, unhealthy, twiggy branches. These are induced by dwarf mistletoe, a parasite that extends threadlike roots inside tree limbs. Mistletoe sucks the tree's energy, stunting its growth and slowly killing it.

Dwarf mistletoe berries explode when ripe, shooting the single seed as far as 40 feet. Some seventy thousand seeds per year may shoot from one infested conifer. Gummy-coated, they stick if they hit another branch. When rain comes along, the gum dissolves, letting the seed slide down a needle to the twig. Grouse and various songbirds enthusiastically eat the berries and digest them beyond viability—unlike the bigger berries of Christmas mistletoes, which are disseminated mainly via bird droppings.

Nettles are back in style. "Really big in Oregon," writes chef John Gorham, "like double-green spinach with more chlorophyll and flavor . . . a blood and liver cleanser." In earlier times, they were also used for fiber: the mature stems were made into good twine, cloth, and paper, substituting for flax in wartime as recently as World War II. Nettle tea is widely used medicinally. Some Native people fashioned nettle-twine nets to capture ducks and fish; the sting itself helped stoical hunters stay awake through the night.

Like bee and ant stingers, nettle stingers are miniature hypodermic syringes. Scientists have struggled to figure out exactly which chemical they inject stings, partly because the stingers are so tiny it's almost impossible to isolate their contents. The most recent study I've seen blames oxalic and tartaric acids. Stinging must have evolved to defend against browsing, but it doesn't save nettles from fire-rim tortoiseshell and other caterpillars. Australia has large trees in the nettle family that bear much longer stinging hair syringes with either stronger toxins or higher doses; some victims require hospitalization.

However, thorough steaming or drying renders nettles harmless. If you camp in April or May near a bed of nettle shoots less than 2 feet high, cook them up. Wear

Dwarf mistletoe

Several dwarf mistletoe species look much alike, but specialize in one or two species of host tree. They pose a serious economic problem. After selective logging is used to mimic natural fire, mistletoe often spreads (actual fires don't spare so much mistletoe, since brooms are highly flammable). Sometimes the only cure is to log the infected tree species completely out of the stand. From an ecological standpoint, mistletoe fosters plant diversity and creates great nest sites for animals. Owls, accipiters, woodrats, and flying squirrels all benefit.

Dicots with Strongly Irregular Flowers

Irregular flowers are those in which the petals (or the sepals, if sepals are the showy parts) are not all alike. Some guides call them bilaterally—as opposed to radially—symmetrical, but some, like louseworts, have no symmetry.

Irregular flowers are highly specialized: each species matches a particular form and size of insect, for pollination.

Flowers are not placed here if the irregularity consists of modestly unequal, but otherwise similar, petals or lobes: for example, veronica (p. 231) and kittentails (p. 183), with four small petals; and cow-parsnip (p. 228), with five.

LOUSEWORT

Pedicularis spp. (ped-ic-you-**lair**-iss: louse—). Corolla fused, with 2 main lips, the upper ± long-beaked, often hoodlike, the lower usually 3-lobed; calyx irregularly 2-to-5-lobed; capsule also asymmetrical; leaves (exc on sickletop lousewort) fernlike, pinnately compound;

Sickletop lousewort

flowers many, on an unbranched stem. Orobanchaceae.

Sickletop Lousewort

P. racemosa (ras-em-**oh**-sa: bearing racemes). Also parrotbeak. Flowers (our variety) white, beak curled strongly sideways, so inflorescence looks pinwheel-like from above; leaves reddish, ± linear, fine-toothed, all on stem; 6–18". Openings near tree line, or alpine; A.

Elephant's Head

P. groenlandica (green-**land**-ic-a: of Greenland, a mistaken name). Flowers purplish-pink, in a dense spike, elephant-like (upcurved beak as the trunk, lateral lower lobes as the ears); leaves preponderantly basal; 8–16" tall. Rills and boggy meadows, alp/subalp; A.

Rocky Mountain Lousewort

P. sudetica ssp. *scopulorum* (su-**det**-ica: of the Sudeten Mtns, Czechia). Also *P. scopulorum*. Flowers magenta; beak curled sideways, so inflorescence looks pinwheel-like from above; leaves 1–3"; 5–12" tall; CO, NM (plus Canada and Alaska).

Elephant's head

Rocky Mountain lousewort

Fernleaf lousewort

Oeder's lousewort

Bracted lousewort

Coiled-beak lousewort

dark red, scarcely beaked, in a robust, cobwebby-haired spike; stem leaves as big as basal ones, aging purplish; 24–36" tall; lush meadows, mid-elevs to lower alpine; A. **Coiled-beak Lousewort** *P. contorta.* Flowers pale yellow (less often pink to purple), in a loose spike; beak semicircular, arching back into lower lip; leaves mostly basal; 8–14". Alp/subalp; northerly.

THE NAME "LOUSEWORT" dates from an ancient superstition that cattle got lousy by browsing louseworts. Each of the flower shapes suits the anatomy of one or more species of bumblebee—pollinators of the hundred or so species of louseworts as well as monkeyflowers, penstemons, and many other family members. The tip of the beak (or "trunk" in one case) positions the stigma to catch pollen from the bee's legs or abdomen. Louseworts are as curiously irregular and varied a genus of flowers as you could ask for.

Fernleaf Lousewort
P. cystopteridifolia (sis-top-ter-id-if-**oh**-lia: leaves like fragile fern). Flowers magenta; upper corolla lip a 1" hood; calyx 5-lobed; leaves give way upward to hairy bracts with a pair of side lobes but no fernlike incisions; 4–18" tall. Mid to subalpine elevs; s MT, n WY.

Oeder's Lousewort
P. oederi (**ö**-der-eye: for Georg C. Oeder). Also *P. flammea.* Flowers pale yellow, few, in a squarish cluster; ¾" corolla hood may be crimson or yellow; all leaves dissected; 4–7" tall. Alpine gravel; s MT, n WY.

Bracted Lousewort
P. bracteosa (brac-tee-**oh**-sa: with bracts). Flowers yellow or (increasingly n) purple or

PAINTBRUSH
Castilleja spp. (cas-til-**ay**-a: for Domingo Castillejo). Corolla a thin tube, usually dull green, barely emerging from very colorful floral bracts that may be red, pink, magenta, orange, yellow, or greenish white (often grading to green at their bases); calyx narrowly 4-lobed, same color as bracts; leaves (incl the colored floral bracts) often narrowly 3-, 5-, or 7-pronged, elliptical, their main veins appearing parallel; stems usually several, unbranched, from a woody base. In sun. Orobanchaceae.

Wyoming paintbrush

Scarlet paintbrush

Entire paintbrush

Rosy paintbrush

Harsh paintbrush

Western paintbrush

Wyoming Paintbrush

C. liniariifolia (lin-ee-air-ee-if-**oh**-lia: linear leaves). Bracts red-orange, hairy, narrow, with 2 linear side-lobes; leaves ¾–4" long, often yellowish, grayish, or purple-edged; 8–40+" tall. Low to mid-elevs; A. The Wyoming state flower.

Scarlet Paintbrush

C. miniata (mi-ni-**ay**-ta: cinnabar-red). Bracts scarlet (less often yellow-orange), deeply lobed; 8–30" tall. Widespread, foothills to tree line; A.

Entire Paintbrush

C. integra (in-**teg**-ra: bracts entire). Bracts scarlet; bracts and leaves usually without lobes or teeth; purplish stem with white hairs. Low woodland and brush; CO, NM, perhaps a few in UT.

Rosy Paintbrush

C. rhexiifolia (rex-ee-if-**oh**-lia: leaves like *Rhexia*). Bracts usually rose-purple, entire or shallowly lobed, oval; leaves entire. All elevs, but esp. subalpine; A.

Harsh Paintbrush

C. hispida (**hiss**-pid-a: bristly). Bracts red to orange near tips only, 3- or 5-lobed; most leaves deeply lobed, with stiff, white hairs. Dry forest or grassland up to mid-elevs; A.

Western Paintbrush

C. occidentalis (oxy-den-**tay**-lis: western). Bracts usually pale yellow, scarcely lobed, oval; leaves unlobed or a few with 2 small side lobes. Foothills to lower alpine; all exc OR, w ID.

Yellow Paintbrush

C. flava (**flay**-va: yellow). Bracts yellow, often orange-tinged, with 2 or 4 side lobes; flower cluster tall; leaves linear or with 2 or 4 small side lobes. Sagebrush; OR and nw MT to c CO.

Lovely Paintbrush

C. pulchella (pul-**kel**-a: beautiful). Bracts streaky pink-purple or yellow or white; leaves usually have 2 side lobes; foliage soft-hairy; 2–5" tall. Alpine; WY, MT, possibly ID.

Yellow paintbrush

Lovely paintbrush

Yellow owl's-clover

PAINTBRUSH IS EASILY recognized as a genus, but notoriously hard to identify to species. Hybrids abound. Neither color nor hairiness is reliably diagnostic, though within a given area the members of a species tend to match.

Paintbrushes are partially parasitic, typically on grasses or composites. (They also photosynthesize.) By late summer, they may seem to hog all the water, staying green and firm while their hosts wither. Ethnobotanist Rose Bear Don't Walk reports that the Salish people's name for this flower translates as "thunder's spark."

YELLOW OWL'S-CLOVER

Orthocarpus luteus (orth-o-**car**-pus: straight fruit; **loo**-tee-us: yellow). Corollas yellow, tubular, the closed upper and lower lips equal in length; flowers in a slender spike; floral bracts and upper leaves 3-lobed, sticky-hairy, lower leaves linear, all on the reddish 4–16" stem; annual plant. Full or part sun, low to mid-elevs; A. Orobanchaceae.

MONKEYFLOWER

Erythranthe spp. (air-ith-**ranth**-ee: red flower). Also *Mimulus* spp. Flowers snapdragonlike—corolla with a long (¾–2") throat, hairy inside, and 2 upper and 3 lower lobes; calyx 5-lobed; 4 stamens; leaves opposite; dense clumps grow from rhizomes. Seeps, wet talus, streambanks. Phrymaceae.

Purple Monkeyflower

E. lewisii (lew-**iss**-ee-eye: for Meriwether Lewis, p. 241). Flowers deep pink to violet; leaves stalk-less, pointed, 2–3"; plants 12–36" tall, sticky-hairy. A–.

Yellow Monkeyflower

E. guttata (ga-**tay**-ta: spotted). Flowers yellow, some less than ¾" long, throat often red-spotted, 5+ per stem; 4–30" tall. A.

Mountain Monkeyflower

E. tilingii (til-**ing**-ee-eye: for Heinrich Tiling). Flowers yellow, ¾–1⅜" (often much larger than

Purple monkeyflower

Yellow monkeyflower

Mountain monkeyflower

Primrose monkeyflower

Dwarf purple monkeyflower

ring many others to the oldest available name, *Erythranthe.* The new circumscriptions remain poorly understood, especially the yellow-flowered, seep-loving species including *E. tilingii.*

DWARF PURPLE MONKEYFLOWER

Diplacus nanus (**dip**-la-cus: 2 plates; **nay**-nus: dwarf). Also *Erythanthe nana, Mimulus nanus.* Corolla funnel-shaped, of 5 slightly unequal, rounded, flaring lobes, bright pink with 2 fine yellow lines (nectar guides) in throat; calyx 5-lobed; 4 stamens; leaves opposite; green parts sticky-fuzzy; annual plant up to 4" tall. Sagebrush and dry, sandy soils; WA to nw WY and sw MT. Phrymaceae.

After sufficient rains, this annual carpets the ground in pink.

DALMATIAN TOADFLAX

Linaria dalmatica (lin-**air**-ia: flax—). Also *L. genistifolia* ssp. *dalmatica.* Flowers yellow with an orange throat, in long spikes; corolla 5-lobed, 1–1⅝" long incl the long rearward spur; leaves 1–2", pointed oval, clasping the stems,

leaves), 1–4 per stem, throat red-spotted; calyx often purple-dotted; leaves ⅜–1"; 2–8" tall. A.
Primrose Monkeyflower
E. primuloides (prim-you-**loy**-deez: primrose-like). Flowers only ¼–⅝" long, yellow, often red-dotted; dwarf, mat-forming or erect, with long, straight, white hairs. OR, ID, sw MT, ne UT.

YELLOW MONKEYFLOWERS THRIVE both in and alongside small streams, including (at Yellowstone) hot-spring outlets that maintain warm, snow-free banks with green growth year-round. David Douglas coined the name *Erythranthe,* "red flower," for a bright red California flower, separating it from all its yellow, pink, and lavender *Mimulus* relatives; little did he know that the name would one day be transferred to exactly those relatives. Other botanists did not agree that the red flower needed its own genus, so the name was all but forgotten for more than a century. Then a 2012 paper broke up genus *Mimulus,* leaving only seven species with that name and transfer-

Dalmatian toadflax

Blue-eyed mary

Cyan penstemon

Alberta penstemon

smooth-edged; 1–3' tall. Disturbed soil in valleys; European; A. Plantaginaceae.

Though neither visually repulsive nor dangerous to stock, toadflax edges onto most top-ten lists of weeds threatening our region. The similar butter-and-eggs, *L. vulgaris* (with linear leaves that don't clasp the stem), is a garden flower now also invasive.

BLUE-EYED MARY

Collinsia parviflora (ca-**lin**-zia: for Zaccheus Collins; par-vi-**flor**-a: small flower). Also innocence. Corolla ¼–⅜"; blue to violet, the upper 2 lobes fading to white; lobes 5, the lower-central one inconspicuous, creased shut to enclose the stamens and style; calyx 5-lobed, green; flowers in axils of upper, often whorled leaves; lower leaves opposite, ± linear; plants annual, 2–14" tall. Meadows, up to lower alpine; A. Plantaginaceae.

PENSTEMON

Penstemon spp. (**pen**-stem-un: almost a stamen). Also beardtongue. Corolla swollen-tubular, with 5 (2 upper, 3 lower) short, rounded, flaring lobes; a broad, ± hairy, sterile stamen rests on the throat; fertile stamens 4, paired; sepals 5, hardly fused; leaves both opposite and basal; from ± woody bases. Rocky places and dry meadows, mostly below tree line. Plantaginaceae.

Cyan Penstemon

P. cyaneus (sigh-**an**-ius: sky blue). Flowers brilliant blue with variable amounts of purple, 1–1⅓", in widely separated, hairless whorls, sterile stamen yellow-haired; rosette of basal leaves much bigger than stem ones; 12–28" tall. WY, sw MT, ID. Similar *P. cyananthus* grows in the same area plus the Wasatch Range, and similar *P. cyanocaulis* grows in e UT and w CO.

Alberta Penstemon

P. albertinus. Flowers rich royal blue, ½–¾", in widely separated, sticky-hairy whorls at leaf axils; basal leaves bigger than stem ones; 4–16" tall. ID to c MT.

Waxleaf Penstemon

P. nitidus (**nit**-id-us: whitish). Flowers sky blue, purple near base, ½–¾" long; sterile stamen has long yellow hairs; leaves entire, grayish with a waxy coating that rubs off; stem leaves clasp the stem; 4–12" tall; blooms in May. ID, MT, WY.

Small-Flowered Penstemon

P. procerus (**pross**-er-us: tall and noble). Also *P. confertus* ssp. *procerus.* Flowers usually

Waxleaf penstemon

Small-flowered penstemon

Shrubby penstemon   *Rocky Mountain penstemon*   *Rydberg's penstemon*

Hall's penstemon

Fuzzytongue penstemon

Whipple's penstemon

blue-purple, ¼–¾", sticky-hairy, in 2–6 whorls; leaves narrowly elliptical, to 3" long; stem 4–24". Common, up to alpine; A.

Shrubby Penstemon

P. fruticosus (fru-tic-**oh**-sus: shrubby). Flowers blue to lavender, 1–2"; leaves crowded, minutely toothed or not, thick, evergreen, ¾–2½"; stems woody, 6–16" tall. Rocky places at all elevs; northerly.

Hall's Penstemon

P. hallii (**hall**-ee-eye: for Elihu Hall). Flowers blue and purple, ½–1", crowded; sterile stamen yellow-haired; leaves ± leathery, entire; 4–8" tall. Alp/subalp; CO only.

Rocky Mountain Penstemon

P. strictus (**stric**-tus: narrow). Flowers blue to purple, ¾–1¼", all ± on 1 side of stem; sterile stamen yellow-haired; leaves ± leathery, entire; 1–3' tall. From Wind River Range s.

Rydberg's Penstemon

P. rydbergii (rid-**berg**-ee-eye: for Per Axel Rydberg). Flowers purple, aging blue, ½–¾", in dense, round clusters; leaves entire, to 6" long; 8–20" tall. Lush, moist meadows—dominating some, to brilliant effect; A.

Whipple's Penstemon

P. whippleanus (whip-el-**an**-us: for Amiel Whipple, p. 217). Flowers a distinctive blackish purple (rarely cream, with fine purple lines), covered with white, sticky hairs, ¾–1⅛", in 1–4 whorls; leaves linear to elliptical; stems 8–24". Aspen groves to alpine; sw MT to NM.

Firecracker penstemon | Beardlip penstemon | Wyoming kittentails

Rocky Ledge Penstemon

P. ellipticus. Flowers lavender, 1–1½", in small clusters, all ± facing 1 way; leaves crowded, usually minutely toothed, some overwintering; stems woody, ± prostrate. Rock crevices; northerly.

Fuzzytongue Penstemon

P. eriantherus (airy-**anth**-er-us: woolly anthers). Flowers usually lavender, with dark guide lines, ⅝–1⅜", in 3–6 whorls; throat full of long hairs; leaves ± linear, with a few small teeth; plant finely hairy; stem 8–24". WA to n-most CO.

Yellow Penstemon

P. confertus (con-**fur**-tus: crowded). Flowers pale yellow, white, or pink-tinged, ¼–½", in 2–8 whorls; leaves elliptical, to 4" long; stem 8–24". c MT to OR.

Firecracker Penstemon

P. eatonii (ee-**toe**-nee-eye: for Daniel C. Eaton). Slender scarlet flowers are little more than tubes, as the lobes are minimal; stem leaves triangular, wavy-edged, at least 1" wide at base, which clasps stem; 1–3' tall. Sunny low elevs; UT, NM, sw CO, s WY, s ID.

Beardlip Penstemon

P. barbatus (bar-**bay**-tus: bearded). Also scarlet bugler. Slender tubular scarlet flowers with lip curled way back; stem leaves linear, widest near midpoint; 1–5' tall. Sunny low elevs; s CO, s UT, NM.

COLLECTIVELY, PENSTEMONS ARE the archetypal Rocky Mountain wildflower. With around two hundred species here, picking just fourteen was no easy task. The stems of many are woody at their bases, to varying degrees, but you're likely to see penstemons as herbs, so I've placed them all in this chapter. Bumblebees pollinate most penstemons; however, if the flowers are long and skinny, they are also red, a dead giveaway of hummingbird pollination.

Penstemon is usually translated as "five stamen," but "almost a stamen" makes more sense. The *pen* in "penultimate" means "almost," and similarly lacks a *t*; with so many five-stamened genera around, it would be strange to give the name "five-stamen" to the one with the appearance of four stamens.

BESSEYA KITTENTAILS

Veronica spp. (ver-**on**-ic-a: for Saint Veronica). Also *Besseya* spp., *Synthyris* spp. Erect flower spikes bristling with purple ⅜–½" stamens, 2 per flower; leaves thick, toothed; entire plant has fine gray wool. Compare featherleaf kittentails (p. 231). Plantaginaceae.

Wyoming Kittentails

V. wyomingensis. No petals; stamens provide all the purple; 2 or 3 small greenish calyx lobes; leaves elliptic, scalloped, both basal and (much

Alpine kittentails

smaller) alternate; 4–16" tall. Sagebrush, rock outcrops, all elevs; e ID, c MT to n-most CO.

Alpine Kittentails

V. besseya (**bess**-ia: for Charles E. Bessey). Also *B. alpina*. Corollas purple, tubular, 4-lobed; leaves basal; 3–6" tall. Alpine; mainly CO, rare in s WY and se UT.

SNOWLOVER

Chionophila jamesii (key-un-**ah**-fil-a: snow-loving; **james**-ee-eye: for Edwin James, p. 95). Flowers greenish white, variably tinged with purple, edges browning, inconspicuously 5-lobed, narrowly conical, barely 1" long, all on 1 side of stalk; leaves mostly basal, vertical, narrow, thick, entire; 2–7" tall. Alpine, where snow has melted recently; southerly. Plantaginaceae.

Alpine-flower aficionados love it when they manage to spot this odd, inconspicuous plant.

WOOLLY MULLEIN

Verbascum thapsus (ver-**bas**-cum: Roman name; **thap**-sus: ancient city near Tunis). Flowers yellow, in a massive spike atop a 2–7' conical, gray-woolly plant; corollas 5-lobed, slightly irregular; leaves fleshy, basal ones up to 16" × 5", smaller upward along stem; biennial, producing just a basal rosette in its first year. Roadsides, over-grazed range; noxious weed from Europe; A. Scrophulariaceae.

Mullein ("mull-en") has been in the herbal apothecary since Roman days, and spread in Native American medicinal use about as fast as it invaded. It contains rotenone, enabling Appalachian mountain folk to poison fish with it (for consumption). Prudent herbalists avoid it. Voyageurs called it *tabac du diable*—the devil's tobacco—and Texans call it "cowboy toilet paper."

NETTLE-LEAF GIANT HYSSOP

Agastache urticifolia (a-**gas**-ta-kee: many spikes; ur-tis-if-**oh**-lia: nettle leaf). Corolla purplish-pink to white, ½", tubular, shallowly 5-lobed, encased in the 5-pointed calyx; 4 stamens protrude; flower spike crowded; leaves opposite, toothed, heart-shaped to pointed-elliptic; stems square, 1–4' tall, often crowded. Moist soil, low to mid elevs; A–. Lamiaceae.

Snowlover

Woolly mullein

Nettle-leaf giant hyssop

MINT-LEAF BEE BALM

Monarda fistulosa (mo-**nar**-da: for Nicolás Monardes; fist-you-**lo**-sa: hollow). Also wild bergamot, horsemint; *M. menthifolia*. Showy rich-pink flowers whorled atop a mint-aromatic 1–2' plant; corolla 1–1½", tubular, hairy, 2-lipped; calyx 5-toothed; style and 2 stamens sticking out; leaves opposite, grayish, toothed, pointed-oval; stems square. Moist soil, low prairie to near tree line; A–. Lamiaceae.

Mint-leaf bee balm

Self-heal

While a bee dives deep into a bee balm tube for nectar, the stamens and stigma arch over its back to both deposit and pick up pollen.

The leaves' antiseptic and sedative powers were known to Native peoples, European

Showy locoweed

Lambert's locoweed

herbalists, and scientists alike. The aromatic oils perfumed Crow tresses, Kootenai sweat lodges, and Cheyenne horses. Chewed leaves anesthetized toothaches and skin irritations. A jug of tea induced sweating and, in combination with a sweat bath, was thought to cure fevers, colds, pneumonia, and most stomach and kidney complaints. This treatment got one of Lewis and Clark's men back on his feet overnight.

SELF-HEAL

Prunella vulgaris (pru-**nel**-a: quinsy, an ailment once treated with self-heal; vul-**gair**-iss: common). Also all-heal. Corolla blue-purple, ¼–¾", 4-lobed; upper lobe hoodlike, lower lobe liplike, fringed; calyx ½ as long; stamens 4 (rarely lacking); flowers bloom sequentially upward, in a crowded, broad-bracted spike of opposite pairs neatly offset 90°; leaves opposite, elliptical, 1–3"; stem squarish, 4–16" tall, or sprawling. Moist meadows up to subalpine; A. Lamiaceae.

The mint family is loaded with aromatic kitchen herbs, including catnip, pennyroyal, horehound, oregano, sage, savory, thyme, and of course mint. Other members are medicinal, or even toxic. Europeans once believed in self-heal as a panacea; in tribal and modern herbal lore, it's prescribed to heal sores, wounds, and chapped skin.

LOCOWEED

Oxytropis spp. (ox-**it**-ra-pis: sharp keel). Also crazyweeds. Flowers pealike, in ± egg-shaped clusters; leaves near-basal, densely silky-hairy, compound (pinnate exc in *O. splendens*); to 12" tall. Fabaceae.

Showy Locoweed

O. splendens. Flowers magenta (drying to blue), 12–40; plants heavily silver-hairy; leaflets crowded on all sides of leafstalk. Low to midelev, roadsides; MT, s WY, CO, NM.

Lambert's Locoweed

O. lambertii: (lam-**bert**-ee-eye: for Aylmer Lambert, a sponsor of David Douglas, p. 65). Flowers 5–20, magenta, drying blue, on leafless 2–8" stalks; leaves basal, hairy, 7–17 leaflets. Up to tree line; CO, NM, e WY, e MT.

White locoweed

Spiny milkvetch

Bessey's locoweed

Woollypod milkvetch

White Locoweed

O. sericea (ser-**iss**-ia: silky). Also white point-vetch. Flowers pale yellow to white, rarely pink-ish, 10–30; upper petal notched; calyx has fine black and white hairs; pod thick-walled, drying rigid; leaflets typically 7.

Bessey's Locoweed

O. besseyi (**bess**-ee-eye: for Charles E. Bessey). Also *O. nana* ssp. *argophylla*. Flowers purple; upper petal sharply notched; calyx with coarse, long, white hairs; leaflets 11–13. Dry sites; WY, MT, e ID, nw CO.

Field Locoweed

O. campestris (cam-**pes**-tris: of fields). Flowers pale yellow (or white), 6–12; upper petal has slight, rounded notch; calyx has fine gray hairs; pod very thin-walled; leaflets 11–25; stalks several, 3–15" tall. All elevs; A–.

WHITE LOCOWEED AND several other *Oxytropis* and *Astragalus* species are toxic, causing "locoism" in livestock if they eat quite a lot over a few weeks. Seen less often in deer and elk, locoism is neurological damage; symptoms include "aggression, hyperactiv-ity, a stiff and clumsy gait, low head carriage, salivation, seizures, apparent blindness, increasing miscoordination, weakness and death." It's the worst plant syndrome afflict-ing US livestock, since locoweed abounds in rangeland. The toxin is produced by fungi that live as endophytes within the plant leaves. Defensive toxins commonly emit aromas that warn herbivores to stay away, but unfortunately, this toxin's warning seems to be understood by rats, not hoofed mammals.

SPINY MILKVETCH

Astragalus kentrophyta var. *tegetarius* (a-**strag**-a-lus: Greek name for some legume; ken-tro-**fie**-ta: spur plant; teh-jet-**air**-ius: roof—). Also *A. k.* var. *implexus*. Flowers pealike, purple, ¼–⅜", 2-lipped, almost covering a mat of tiny (¼"), spine-tipped leaflets, 5–9 per pinnate leaf. Tundra, granite outcrops, clay badlands; e OR

Western sweetvetch

Utah sweetvetch

Yellow sweetvetch

and sw MT to NM. Species has low-elev varieties, mostly white-flowered.

WOOLLYPOD MILKVETCH

A. purshii (**pur**-she-eye: for Friedrich Pursh, p. 102). Flowers pealike, ⅜–1½", may be either magenta, pale lilac, or white with a purple spot; all green parts woolly; ± curved pods are covered in extra-thick white wool; prostrate, pinnate leaves up to 6" long; plants 2–5" tall. Sagebrush or juniper; WA to nw CO.

Close to a hundred *Astragalus* species grow in the region, mainly in grassland, steppe, and piñon-juniper communities.

SWEETVETCH

Hedysarum spp. (hed-**iss**-a-rum: Greek name for a sweet-smelling plant). Flowers pealike, ¾" long, in a ± 1-sided raceme; seedpods 2-to-6-seeded, flat, form long chains; leaves pinnate. Fabaceae.

Western Sweetvetch

H. occidentale (oxy-den-**tay**-lee: western). Flowers 20–80, pendent, magenta to faded pink, rarely white; upper 2 calyx lobes broader but shorter than the lower 3; 16–32" tall. Pine zone up to alpine; A–.

Utah Sweetvetch

H. boreale (bor-ee-**ay**-lee: subarctic). Flowers 5–25, magenta, not often pendent; calyx lobes all slender and similar; 6–24" tall. Low to mid-elevs; A.

Yellow Sweetvetch

H. sulphurescens (sulphur-**ess**-enz: yellowish). Flowers yellow to cream. All elevs; WA to WY.

YELLOW SWEETVETCH ROOTS provide the bulk of grizzly bear diets in some seasons in the Canadian Rockies. (Grizzlies rely similarly on glacier lilies at Glacier and on biscuitroot at Yellowstone.) The root-digging seasons can last from April to June, before green vegetation is plentiful, and again through the critical fall fattening-up season in years when the buffaloberry crop is poor. Indigenous peoples ate these "grizzly bear roots" too.

LUPINE

Lupinus spp. (lu-**pie**-nus: Roman name). Flowers pealike, usually blue to purple (in these spp.), small (½"), many, in racemes; calyx 2-lobed; pods hairy; leaves palmately compound; leaflets often center-folded. Fabaceae.

Silvery Lupine

L. argenteus (ar-**jent**-ius: silver). Back of upper petal hairless or with just a few silky hairs; leaves densely silvery-haired beneath, less so or smooth above; plant bushy, 5–40" tall; a

Silvery lupine

Bigleaf lupine

Dwarf lupine

dwarfed alpine form, var. *depressus*, just 5–10" tall, is treated as a full species by some. All elevs; abundant; A.

Silky Lupine

L. sericeus (ser-**iss**-ius: silky). Back of upper petal silky-haired; flowers in ± separated whorls, blue-purple less often ranging to pinkish white; leaves almost all on stem; plant slenderer, heavily covered with rusty to whitish silky fuzz, 14–32" tall. Lower elevs; A.

Bigleaf Lupine

L. polyphyllus (polly-**fill**-us: many leaves). Incl varieties Wyeth lupine, chokecherry lupine; *L. wyethii, L. prunophilus*. Petals hairless; flowers in a ± continuous 6–16" column; lower leaves long-stalked; plant often hairless exc for pods and sepals; robust, 1–5' tall. Moist forest and meadows; A.

Dwarf Lupine

L. lepidus var. *utahensis* (**lep**-id-us: charming). Also *L. caespitosus, L. lyallii, L. sellulus*. Back of upper petal hairless; leaves basal, leafstalks longer or taller than flower stalks; flower whorls crowded; plant 2–7" tall, silky-haired. Other varieties have taller stalks, more hair or less hair. All elevs; A–.

MASSED LUPINES CAN be intoxicatingly fragrant. A bumblebee in search of lupine nectar might be overwhelmed by too many choices, so the lupines help her out: the upper petal has white spots that act as nectar guides. As a blossom ages, the spots turn magenta. Bees learn to skip those blossoms, since their nectar is depleted. Efficiency is increased for both flower and bee.

Legume-family plants fertilize forests by putting nitrogen into the soil in a form all plants can use. Where forests became dense and shady due to fire suppression over the past century, silky lupine decreased in abundance, along with the nitrogen-fixers snowbrush and bitterbrush, and they've all been scarce for so long that a single fire (or thinning treatment) may not be enough to bring them back in strong numbers.

Lupine species hybridize promiscuously, resulting in controversial taxonomy and challenging ID. Garden lupines are hybrids bred primarily from bigleaf lupine; these hybrids are now invasive pests in the eastern US and in countries ranging from Sweden to New Zealand.

Dwarf lupine achieved media stardom in the 1980s at Mount St. Helens, pioneering all alone on a desert of new volcanic ash after the volcano's 1980 eruption. Its nitrogen-fixing symbiotic partners were one key to its success; another was its coat of silky hairs, which minimize water loss on hot, dry days by reflecting light and holding wind away from the leaf surface. (Either hair or waxy coatings can protect by reflecting ultraviolet radiation, which is strong at high elevations and can damage

Parry's alpine clover

Uinta clover

Yellow sweet-clover

Dwarf clover

tissue.) One lupine was able to photosynthesize at temperatures up to 104°F, while another lupine with much less hair gave up at around 86°F. Dwarf lupine also turns its leaves to face the morning and evening sun, maximizing photosynthesis at cooler times of day, and stays green through winter, making the most of early- and late-season sun.

Notice the way a little sphere of dew or rain is held on the center point of each leaf.

We don't know for sure why the Romans named these flowers after wolves. It may indicate an affinity they felt—remember their myth of a she-wolf as the mother, or at least the wetnurse, of Rome. Or they may have decried lupines as killers of sheep; lupines are toxic, to varying degrees. Sheep have died from grazing silky lupine. ("Lupine" means "of wolves" or "wolfish" in English too.)

ALPINE CLOVER

Trifolium spp. (try-**fo**-lium: 3 leaf). Flowers tubular, 2-lipped, ⅜–⅞"; calyx with 5 long, pointed lobes; compound leaves of 3 leaflets; flowers clustered (exc on *nanum*) on leafless stalks from a dense, leafy mat. Alpine. Fabaceae.

Parry's Alpine Clover
T. parryi (**pair**-ee-eye: for Charles Parry, p. 212). Flowers rose to purple; calyx hairless. A–.

Uinta Clover
T. dasyphyllum (daisy-**fill**-um: shaggy leaf). Lower petals rose and purple, upper petal pinkish-white or yellow; calyx hairy. MT, WY, UT, CO, NM.

Hayden's Alpine Clover
T. haydenii (hay-**den**-ee-eye: for Ferdinand Vandeveer Hayden, p. 217). Flowers magenta grading to almost white or pale yellow; calyx hairless; leaflets broad ovals. Absarokas, sw MT, ec ID.

Dwarf Clover
T. nanum (**nay**-num: dwarf). Flowers magenta, just 1 to 4 per cluster. MT, WY, UT, CO, NM.

SWEET-CLOVER

Melilotus spp. (mel-i-**lo**-tus: honey *Lotus*). Many-stemmed weeds with small (¼"), pealike flowers in curved, 1-sided racemes 2–5" long; compound leaves of 3 toothed, elliptical leaflets; plants 2–7' tall, sweetly fragrant, esp while drying. Disturbed ground; Eurasian. Fabaceae.

Yellow Sweet-clover
M. officinalis (o-fiss-in-**ay**-lis: medicinal). Flowers yellow. Lines thousands of miles of highway; A.

Golden banner

Lanzwert's sweetpea

Steer's head

Golden smoke

both "preferred forage" and "potentially toxic to stock," as well as "invasive."

GOLDEN BANNER

Thermopsis rhombifolia (ther-**mop**-sis: like lupine; rom-bif-**oh**-lia: rhomboid stipules). Also false lupine, golden or yellow pea or bean, buck; *T. montana*. Flowers bright yellow, pealike, ¾–1", in columnar racemes; leaves compound, alternate; 3 leaflets plus 2 smaller stipules clasping stem at leaf base; 6–40" tall. Meadows or dry grassland up to mid-elevs; A. Fabaceae.

This toxic color swath tends to increase with heavy grazing. The Blackfoot called it buffalo pea because it bloomed as the bison left theirwinter range.

LANZWERT'S SWEETPEA

Lathyrus lanszwertii (**lath**-er-us: Greek name perhaps meaning "very passionate," i.e., aphrodisiac; lanz-**wert**-ee-eye: for Louis Lanszwert). Also *L. leucanthus*. Flowers pealike, white with purple nectar guides and/or pink blush, or lavender overall, the upper petal largest, the lower 2 partly fused, creased, enclosing pistil and 10 stamens; stigma bristly on upper surface; fruit a pea pod; leaves pinnately compound, often tendril-tipped; leaflets oval or in some locales linear and 3" long; stems angled, 8–40" tall. Meadows, openings; all but MT; the white variety (sometimes treated as a species) is southerly. Fabaceae.

STEER'S HEAD

Dicentra uniflora (di-**sen**-tra: 2 spur). Bizarre, little pinkish flower resembles a horned skull, ⅝" long, with 2 petals fused (the muzzle) and 2 curved back (the horns); 2 sepals; 6 hidden stamens; stalks 1-flowered, 2–4"; leaves lacy, compound, on 2–4" stalks attached to flowers belowground; toxic. Blooms soon after snowmelt; moist gravels, steppe to tree line; northerly. Fumariaceae.

White Sweet-clover
M. albus (**al**-bus: white). Also *M. officinalis* ssp. *alba*. Flowers white. A.

SWEET-CLOVERS ARE BELOVED of bees, cows, highway departments, and herbalists. They contain blood anticoagulants related to warfarin, the well-known blood-pressure medication and rat poison. They get listed as

Mountain-sorrel

Monkshood

Canary violet

Goosefoot violet

GOLDEN SMOKE

Corydalis aurea (cor-**id**-a-lis: the crested lark; **aw**-ri-a: gold). Also scrambled eggs. Flowers yellow, in compact side clusters; petals 4, upper one with a round-tipped rearward spur; inner 2 ± hidden; sepals 2, falling off early; stamens 6; seeds black, in slender, curved 1" pods; leaves powdery-coated, pinnately 2× to 4× compound, fernlike; sprawling winter-annual or biennial plants, stems to 20" long. Burns, gravel bars, lower forests; A. Fumariaceae.

MOUNTAIN-SORREL

Oxyria digyna (ox-**ee**-ria: sharp—; did-**gy**-na: 2 ovaries). Also alpine sorrel. Flowers tiny, greenish, of 2 erect and 2 spreading lobes, 2 stigmas and 6 stamens, in rough spikelike panicles; fruit rust-red, tiny, 2-winged; leaves kidney-shaped, 1–2" broad, on long basal leafstalks, coloring brilliantly in fall; stalks several, 4–18" tall. Wet alp/subalp rocks and gravels; A. Polygonaceae.

Mountain-sorrel leaves offer a lemony note to salmon roe when boiled and pressed into cakes. Their tart acids include ascorbic (good for you) and oxalic (bad for you). Do snack, but not to excess.

MONKSHOOD

Aconitum columbianum (ac-o-**nigh**-tum: Greek name; co-lum-be-**ay**-num: of the Columbia River). Sepals 5, petal-like, blue-purple, the upper one hooding, helmetlike; 2 true petals small, hidden under hood; flowers ¾–1½" tall, in an open raceme atop the 1–7' stem; leaves 2–5" across, palmately deeply incised. Moist sites at all elevs; A. Ranunculaceae.

Bumblebees pollinate most blue irregular flowers. Monkshood's odd-shaped flower excludes from its nectary all insects except highly motivated, smart bumblebees, whose advantage to the plant is their fidelity: having been once well rewarded with monkshood nectar, they will visit only monkshoods, not wasting the pollen on a haphazard sequence of flower species. Monkshood is listed as toxic to humans and stock, but biologists found a lot of it in elk scats in the Blue Mountains.

VIOLET

Viola spp. (vie-**oh**-la: Roman name). Petals 5, lowest one largest, bulbous at its rear end, with dark-purple nectar guide lines; sepals 5; stamens 5, short; capsules split explosively to propel seeds. Violaceae.

Canary Violet

V. nuttallii group (nut-**all**-ee-eye: for Thomas Nuttall, p. 192). Includes *V. praemorsa, V. vallicola*. Flowers yellow; leaves blunt-tipped, elliptical; 1–5" stem. In sun, all elevs; A–.

Goosefoot Violet

V. purpurea (pur-**pew**-ria: purple). Also *V. utahensis*. Flowers yellow; leaves ± purple-veined, usually with distinctive blunt teeth; 2–6" stem. Dry sites at lower elevs; WA to nw CO.

Darkwoods violet   Canadian white violet   Hookspur violet

Darkwoods Violet

V. orbiculata (or-bic-you-**lay**-ta: round, ref to leaf). Flowers yellow; leaves scalloped, bluntly heart-shaped to nearly round, dark green, persisting through winter; 1–2" stem. Moist forest; WA to Wind River Range.

Canadian White Violet

V. canadensis. Incl *V. scopulorum.* Flowers ½" wide, several, ± white inside with yellow center, violet on back; leaves heart-shaped, pointed; stems leafy, 4–16" tall. Low to mid-elev woods; A.

Hookspur Violet

V. adunca (a-**dunk**-a: hooked). Also *V. labradorica.* Flowers violet or partly white, ½–¾" wide, with an upcurved rearward spur; leaves basal and on stem, ± heart-shaped; stems 1–4" when in flower. All elevs, favoring the high; A.

Northern Bog Violet

V. nephrophylla (nef-ra-**fil**-a: kidney-shaped leaf). Also *V. sororia.* Flowers violet, 1" wide; leaves strictly basal, heart-shaped, up to 2¾" wide. Wet places; A.

MANY VIOLETS BLOOM soon after snowmelt, often so early that pollinators are not yet on the wing. When they go unpollinated, violets respond with a second kind of flower: greenish, low, and inconspicuous, and able to pollinate themselves and produce seed without ever opening. Violets, including garden pansies, have tasty, nutritious leaves and stems, but poisonous seeds and roots.

Our Number-One Namer

Thomas Nuttall may have collected and named more new species from west of the Mississippi than anyone else. He came along at the ideal time: crossing the Rockies was easier and safer for him than for Meriwether Lewis or David Douglas, but plenty of conspicuous species remained to be described. Following early work on Great Plains flora, he wrote his magnum opus, *The Genera of North American Plants*, in 1818. In 1834, the ambitious entrepreneur **Nathaniel Wyeth** persuaded Nuttall to quit his prestigious chair at Harvard and join an expedition to Oregon. Nuttall collected all along the way, then sailed to Hawaii, California, and home via Cape Horn. Having exhausted his savings on the trip, he retired to an inherited estate in his native England.

Nuttall's enthusiasm inspired praise from botanists, and sometimes derision from, say, boatmen: "When the boat touches the shore, he leaps out, and no sooner is his attention arrested by a plant or flower, than everything else is forgotten. The inquiry is made, *ou est le fou?* where is the fool? . . . he is gathering roots."

Barbey's larkspur

Nuttall's larkspur

Low larkspur

LARKSPUR

Delphinium spp. (del-**fin**-ium: Greek name, from "dolphin"). Flowers deep blue to violet, ¾–1¼" across; sepals 5, petal-like, spreading, the upper one with a long nectar-bearing spur behind; petals 4, the upper 2 spurred (within the sepal spur), often much paler; leaves 2–5" across, narrowly palmately lobed and/or compound. Ranunculaceae.

Barbey's Larkspur

D. barbeyi (**bar**-bie-eye: for William Barbey). Also subalpine larkspur. Very tall racemes of 10–50 flowers; side sepals more forward than spreading; upper stems and stalks sticky-hairy; 3–5' tall, or even 7'. Abundant in subalpine meadows, aspens; southerly.

Western Larkspur

D. × *occidentale* (oxy-den-**tay**-lee: western). Many flowers, in a tall raceme, color incl a lot of white; upper petals white-edged; stem hollow, 3–7' tall. Dry meadows, all elevs; A–. A common hybrid between *D. barbeyi* and *D. glaucum*.

Low Larkspur

D. bicolor. 3–12 bright blue-purple flowers; upper petals (not sepals) ± white with blue lines; stem solid, 6–20" tall. Grassland, open woods, up to lower alpine; c CO n.

Western larkspur

Nuttall's Larkspur

D. nuttallianum (nut-tel-ee-**ay**-num: for Thomas Nuttall, p. 192). Like *bicolor,* exc stamens are exposed under the lower 2 petals, whose tips are split at least ⅛" deep; there's a surer difference in the roots, but please don't uproot a plant to ID it. Steppe, pine forest, sometimes alpine; A.

BLUE, IRREGULAR FLOWERS are generally bumblebee-pollinated, and Barbey's larkspur was by far the most popular flower among bumblebees in a study in central Colorado. Larkspur seeds, on the other hand, have for millennia been ground up to poison lice, and the plants are generally very toxic.

Regular Flowers

A "regular" flower is radially symmetrical: all its petals (or its sepals) are essentially the same size and shape.

Regular Flowers with 5 Fused Petals, 5 Stamens

Five-petaled flowers typically have five sepals and five, ten, fifteen, or more stamens in addition to their five petals or corolla lobes. Also included, on pp. 223–229, are flowers with just one set of conspicuous flower parts in fives—either petals or sepals but not both, and not the stamens. These variables tend to reveal family relations. Most Rosaceae have fifteen or twenty stamens; Polemoniaceae have five; Portulacaceae have two sepals; and Apiacieae lack sepals.

If you have a five-petaled flower that you can't locate here, try these leads: shrubs include many five-petaled genera, and some are very small. Irregular dicots (pp. 176–193) mostly have five dissimilar corolla lobes. Composites (pp. 155–173) have tight heads of tiny florets that are technically five-parted, and sometimes visibly so. Gentians (p. 236) have five petals on some blossoms, but usually six or seven on others nearby.

Petals are "fused" if a close look at their bases reveals that all the petals are all one piece: the seeming petals are actually lobes of the corolla.

JACOB'S LADDER

Polemonium spp. (pol-em-**oh**-nium: Greek name). Flowers light to purplish blue with yellow center; corolla lobes spreading, from a short, flaring tube; leaves of many small, hairy pinnate leaflets; may smell skunky (see sky pilot, below). Polemoniaceae.

Towering Jacob's Ladder

P. foliosissimum (folio-**see**-sim-um: most leafy). Flowers 1" wide, in small clusters; leaves on stem, 1–4" long, leaflets ¼–1"; 1–3' tall. Moist meadows, mid-elevs to subalpine; mainly south-erly. The spp. has varieties, mainly elsewhere, with white and pale-yellow flowers.

Showy Jacob's Ladder

P. pulcherrimum (pool-**ker**-im-um: most beautiful). Flowers ½" wide, not tightly clustered; leaves mostly basal, leaflets small, but at least ¼" long; 3–10" tall. Mid-elevs to alpine; A.

SKY PILOT

P. viscosum (vis-**co**-sum: sticky). Also skunk-flower; *P. confertum*. Flowers purplish blue, ⅝–1⅛", often densely clustered; corolla lobes flaring from a long tube; pinnate leaflets tiny, crowded, appearing whorled around the leaf-stalk; plants 4–14" tall, sticky-hairy, may smell skunky. Mainly alp/subalp; A.

Like most flowers, sky pilot produces a sweet fragrance in its nectaries to attract pollinators. In many individuals, this is overwhelmed by a skunky aroma emanating from sepals and bracts below the flowers. The preferred pollinators, bumblebees, like large and sweet-smelling flowers, which prevail at tundra elevations where ants are few and bees are many. Nectar-feeding ants slip past the pollen-bearing stamens on their way to robbing the nectary, and often destroy the pistil in the process. The skunky smell serves to repel ants; it wafts mainly from plants at elevations where ants abound. It offers a trade-off: no ant damage, but fewer bees, leaving flies to take up the slack. Candace Galen, who studied this species in Colorado for decades, found many individuals that flourished for thirty years.

SKYROCKET

Ipomopsis aggregata (ip-a-**mop**-sis: resembling morning-glory; ag-reg-**ay**-ta: clustered). Also scarlet gilia. Corolla trumpet-shaped, usually scarlet but some-

Towering Jacob's ladder

Showy Jacob's ladder

Sky pilot

Skyrocket

Nuttall's linanthus

times white or pink, the slightly flaring tube 2× as long (½–1¼") as the slightly recurved, pointed lobes; stamens borne near mouth of tube; leaves much-dissected, lobes linear; stems to 3' tall, many-flowered, or dwarfed (4" tall) at high elevs. Very widespread, dry meadows, disturbed ground; A. Polemoniaceae.

The long, tubular corolla and bright-red color are clues that this flower evolved with hummingbirds as pollinators. Most insects cannot see red, but hummingbirds crave it. From the the Wind River Valley south, many skyrocket flowers are white or light pink, perhaps with red speckles, instead of scarlet. In some populations, the same hillside—or even the same plant—switches from red blooms in July to pale ones in August. This happens where hummingbirds migrate through and are gone by August, leaving white-lined sphinx moths (p. 474) as the best available pollinators. At night, when the moths can barely see red flowers, they visit white ones.

The plant lives as a rosette of leaves for two to six years before producing one flowering stem and then dying.

ALPINE COLLOMIA

Collomia debilis (co-**lo**-mia: gluey; **deb**-il-iss: weak). Corolla trumpet-shaped, ½–1", blue to streaky white or light pink (or brilliant deep pink in WY); stamens sticking out, often blue-tipped; leaves alternate, crowded, ± sticky-hairy, usually oval, to 1"; dense cushions 2–3" deep. Unstable, high-elev gravels; WA to UT and w WY. Polemoniaceae.

Alpine collomia

NUTTALL'S LINANTHUS

Leptosiphon nuttallii (lep-to-**sigh**-fun: slender tube; nut-**all**-ee-eye: for Thomas Nuttall, p. 192). Also *Linanthus* or *Linanthastrum nuttallii*. Corolla white, with 5–6 flat-spreading lobes at the end of a hairy, straight tube; 3 stigmas and 5 stamens just emerge from tube opening; leaves seemingly linear, ¾" long, in big whorls, but actually these are leaflets of palmately compound leaves; numerous leafy stems, to 12" tall, from woody bases. Rocky slopes, mid- to high elevs; A. Polemoniaceae.

PHLOX

Phlox spp. (flocks: Greek name meaning "flame"). Corolla white, pink, or pale blue, with 4–6 flat-spreading lobes at the end of a straight tube; 3 stigmas usually just visible in the tube opening; stamens don't protrude; numerous leafy stems, from woody bases; leaves linear, pointed. Rocky, sunny spots. Polemoniaceae.
Cushion Phlox
P. pulvinata (pul-vin-**ay**-ta: cushion-forming). Also *P. caespitosa* ssp. *platyphylla*. Corolla white to light blue, ½–⅝" across; calyx sticky-hairy;

Cushion phlox

Spreading phlox

Spiny phlox

leaves ¼–½"; dense mats 1–3" thick. Alp/subalp; A–.

Spreading Phlox

P. diffusa (dif-**you**-za: spreading). Corolla pink, lavender, or white; leaves ¼–¾", pointed but not stiff, crowded in mats 2–4" thick. Mainly alp/subalp; WA to sw MT and nw WY.

Rocky Mountain Phlox

P. multiflora. Corolla usually white; leaves ½–1¼"; 2–4" tall. Up to mid-elevs; A–.

Spiny Phlox

P. hoodii (**hood**-ee-eye: for Robert Hood, p. 447). Corolla ⅜" across; calyx and leaf edges cobwebby-haired; leaves ¼", stiff, sharp; mats 1–3" thick. Lower elevs; MT, WY, CO, ID.

Longleaf Phlox

P. longifolia. Corolla usually light pink; leaves ¾–3"; plants 3–16" tall, not matted. Sagebrush and up to mid-elevs; A.

PHLOX COLONIZES HIGH ridges so windswept they fail to retain much snow, or much soil. Winter subjects plants that lack a snow blanket to ferocious, drying winds and to hundreds of freeze-thaw cycles. Though the rocks continuously break up, their particles tend to blow away as they reach soil size, preventing soil from accumulating. Alpine phlox species exemplify "cushion plants" adapted to this extreme environment.

Snow-Depth Specialties

Most alpine plants are adapted to particular snow depths and durations. These adaptations will partly control vegetation shifts as climate changes. Skip Walker's University of Colorado team measured snow over many winters at an array of Colorado sites, finding the following:

0–10 inches (optimal winter maximum depth): moss campion, four-nerve daisy spike woodrush, cushion phlox, alpine spring-parsley

10–20 inches: Uinta clover, Arctic sandwort, curly sedge, pixie goblet, Arctic harebell, Iceland-moss, map lichen, purple reedgrass, rockworm

20–40 inches: Rocky mountain sagewort, alp lily, Holm's Rocky Mountain sedge, alpine bistort, one-flower daisy, Arctic gentian, Rocky Mountain willow

40–80 inches: American bistort, alpine avens, marshmarigold

80–120 inches: sibbaldia, purple hairgrass, pygmy bitterroot, hair-tipped haircap moss, Western paintbrush, alpine buttercup, brown tile lichen

120–200 inches: Parry's alpine clover, Parry's primrose, Drummond's rush

Longleaf phlox

Sweet rockjasmine

Pygmyflower

These plants withstand the frost action whose powers of pulverization are potent on nearby rocks. The smooth cushion-plant form eases the wind on over with little resistance, especially if the plant contours itself in the lee of a rock or a crevice between rocks. The tiny, crammed leaves live in a pocket of calm partly of their own making, and there they trap windblown particles that slowly become a mound of soil. The stems grow to keep up with the soil depth—look for coarser gravel dammed against the upslope side of the mat and slipping past it on both sides while a downslope "shadow" area develops made of finer sediment that contrasts with the surroundings and becomes a seedbed for more-delicate plants.

Cushion plants often have really long taproots (say, 8 to 15 feet) to tackle soil drought. Ridgetops get their share of snow, but wind and the coarse, rocky substrate allow little of it to stay there.

ROCKJASMINE

Androsace spp. (an-**drah**-sa-kee: Greek name for some sea plant). Also fairy candelabra. Corolla white to cream yellow, aging pink, ¼" across; stamens hidden within tube; calyx cup-shaped, 5-lobed; flowers branch out from a single point; leaves a basal rosette, lanceolate. Primulaceae.

Sweet Rockjasmine

A. chamaejasme (cam-ee-**jaz**-me: dwarf jasmine). Also *A. lehmanniana*. Jasmine-fragrant flower has a yellow or orange center that turns pink after pollination; flowers in a compact cluster; leaves ¼–⅜", in dense mats; plant white-hairy, 1–4" tall. Alp/subalp, favoring limestone; MT, WY, CO, NM.

Pygmyflower

A. septentrionalis (sep-ten-tri-o-**nay**-lis: northern). Tiny white flowers in open clusters; annual or short-lived plant, 1–10", drying dark red. All elevs, esp on disturbed soil, common in alpine tundra, weedy in gardens; A.

SINCE 1991, COLORADO botanists have established an alpine "warming meadow" where they've electrically heated patches of ground to predict how plants will do in a warmer climate. Pygmyflower does very badly: few of the seedlings survive long enough to produce a good crop of seeds. Heating jacks up the seed-germination rate, which is actually a bad thing, because it uses up the seed bank, so that if eventually a year with better conditions comes along, there will be few seeds left to take advantage.

Dwarf primrose

Bonneville shooting star

Alpine primrose

Parry's primrose

Darkthroat shooting star

PRIMROSE

Primula spp. (**prim**-you-la: first to bloom). Flowers magenta, pink, or white, with yellow center; corolla lobes spread flat from a narrow tube; stamens within opening; leaves basal, succulent, bright green, hairless, entire or with a few teeth. Primulaceae.

Parry's Primrose

P. parryi (**pair**-ee-eye: for Charles Parry, p. 212). Flowers brilliant magenta, 3–12, ¾" across; leaves 1–12" long, fetid when touched; plants to 18" tall. Wet, wind-sheltered alp/subalp sites; from sw MT s.

Alpine Primrose

P. angustifolia (ang-gus-tif-**oh**-lia: narrow leaf). Flowers ⅝" across; leaves less than 1" long; ½–3" tall. Alp/subalp gravels; CO and NM.

SHOOTING STAR

Primula spp. Also *Dodecatheon* spp. Magenta 5-lobed corolla and calyx bent sharply back (i.e., upward), ½–1" long, fused "collar" portion usually yellow; stamens tightly clasping pistil or slightly spreading; flowers several, many facing down (i.e., the petals up) but all erect in fruit (a capsule); leaves basal, stalked; stem 2–18" tall. Primulaceae.

Darkthroat Shooting Star

P. pauciflora (paw-sif-**lor**-a: few flowers). Also *D. pulchellum*. Stamens are joined and columnar nearly ½ their length. Wet soils; A.

Bonneville Shooting Star

P. conjugens (**con**-jug-enz: joined). Stamens are joined and columnar less than ¼ of their length. Up to mid-elevs where moist, at least in spring; WA to all of WY.

DWARF PRIMROSE

Androsace montana (an-**drah**-sa-kee; mon-**tay**-na: montane). Also *Douglasia montana*. Corolla deep pink with yellow ring at center initially, ⅜" across, lobes shallowly notched at tip, spreading flat from narrow tube; stamens barely appear at opening (in contrast to moss campion, p. 211); flowers single on 1–4 tiny stalks per rosette; leaves linear, pointed, ¼", in rosettes aggregating in low, mosslike mats. Alpine, favoring limestone; northerly. Primulaceae.

Arctic harebell

Bluebells-of-Scotland

Ballhead waterleaf

THE KIND OF FLOWER SHAPE we call "shooting stars" and botanists call "reflexed" is also common in the nightshade family, which includes tomato and potato plants. These flowers are all buzz-pollinated: their stamens spew pollen in response to furious vibration. Bumblebees visiting the flowers land briefly on the robust column provided for that purpose, vibrate their wings, and are rewarded with a shower of pollen.

For the plant, there is efficiency in making bumblebees learn a flower-specific trick, because the bees will be faithful: on any given day, they visit one species of flower over and over again, and thus don't waste pollen by leaving it on the wrong kind of flower. The bee does exact a fee by eating a lot of the pollen, but apparently she's well worth the price. Lupine, lousewort, and monkshood win bumblebee fidelity another way—with robust, irregular flower shapes that only bumblebees of a particular size can unlock.

Fendler's waterleaf

BELLFLOWER

Campanula spp. (cam-**pan**-you-la: small bell). Also bluebells, but unrelated to *Mertensia*, below. Corolla pale blue, bell-shaped; pistil 3-forked. Campanulaceae.

Arctic Harebell

C. uniflora. Also *Melanocalyx uniflora.* Flower single, ¼–½"; leaves untoothed; 1–4" tall. Alpine; A–.

Bluebells-of-Scotland

C. rotundifolia (ro-tund-if-**oh**-lia: round leaf—a bit misleading). Flowers several, ⅝–1¼"; corolla lobes are less than ⅓ the length of corolla; stem leaves linear; basal leaves ± round on long stalks, often withering before flowers open; 6–30". Widespread, rocks, meadows; A. Similar Parry's harebell, *C. parryi*, has corolla lobes about ½ the length of corolla.

WATERLEAF

Hydrophyllum spp. (hyd-ro-**fill**-um: water leaf). Flowers in spherical heads bristling with stamens; corolla ⅜"; calyx bristly; leaves basal, pinnate; 5–11 leaflets. Boraginaceae.

Fendler's Waterleaf

H. fendleri (**fend**-ler-eye: for Augustus Fendler, p. 246). Flowers white, sometimes purplish, often rising above the 6–12" leaves; leaflets toothed, with soft white hair underneath; 8–32" tall. Often the most abundant herb in aspen groves; southerly, plus ID and OR.

Ballhead Waterleaf

H. capitatum (cap-i-**tay**-tum: flowers in heads). Flowers light blue; leaves 4–10" with a few lobes, hairy, taller than the 1–3" flower stalks. Sagebrush, aspen groves, moist, open forest; w CO to OR.

A Russian Ship Visits California

Estonian doctor **Johann Friedrich von Eschscholtz** studied insects while **Adelbert von Chamisso**, a poet from Berlin, hunted plants on Captain Otto von Kotzebue's Russian exploration of 1815–18. Count **Nikolai Rumiantzev** (or Romanzoff) financed the trip. The two collected much in Alaska and California, but sailed right past Washington and Oregon. Though he earned a fine reputation for his botany, von Chamisso is best known for writing a children's tale, *Peter Schlemihl's Miraculous Story*. His poetry was set to music by Robert Schumann, among others. Being nobility, the von Chamissos had fled the French Revolution when he was nine.

SITKA MIST MAIDEN
Romanzoffia sitchensis (roman-**zof**-ia sit-**ken**-sis: found at Sitka, Alaska, by botanists funded by Count Rumiantzev, see above). Flowers white, funnel-shaped with 5 round lobes, ¼–½", in loose racemes on leafless 2–10" stalks; leaves basal on long stalks, round, ½–1½" wide, with coarse, blunt teeth; resembles a more delicate brook saxifrage (p. 207), but has fused petals. Seepy rock ledges, mainly alp/subalp; far northerly. Boraginaceae.

PHACELIA
Phacelia spp. (fa-**see**-lia: bundle, referring to dense inflorescence). Flowers crowded, bristling with stamens about 2× as long as ¼–⅜" corollas; calyx hairy. Rocky places. Boraginaceae.

Silverleaf Phacelia
P. hastata (hass-**tay**-ta: halberd-shaped—not true of our varieties' leaves). Also scorpionweed. Flowers dirty-white to lavender; inflorescence often curled, uncurling as it blooms progressively upward; leaves gray-silky-coated, narrow elliptical, with a few prominent depressed veins; stems to 24", but often sprawling. All elevs; A.

Silky Phacelia
P. sericea (ser-**iss**-ia: silky). Flowers purple (to blue or white) in a ± tall spike; stamens pale-tipped; leaves silky-haired, mostly basal, pinnate-lobed almost to midvein, the lobes again divided; subalpine form 4–9" tall, densely hairy; mid-elev form 10–36" tall and less hairy. A.

Lyall's Phacelia
P. lyallii (lie-**ah**-lee-eye: for David Lyall, p. 62). Flowers blue-purple in a short spike; stamens

Sitka mist maiden

Silverleaf phacelia

Lyall's phacelia

Silky phacelia

Threadleaf phacelia

Baker's bluebells

pale-tipped; leaves with coarse, usually sticky hair, dark green (not silvery), coarsely pinnate-lobed, mostly basal, cushion-forming; stem 3–10". Alpine; ID, MT.

Threadleaf Phacelia

P. linearis (lin-ee-**air**-iss: leaves linear). Corolla lavender to blue, bowl-shaped, stamens don't stick out; sepals and bracts bristle with white hairs; leaves 1–4" long, entire or 3-lobed; annual, 4–20" tall. Low grasslands; WA to UT and WY.

MOST OF OUR *Phacelia* species have both compact alpine and leggy mid-elevation varieties. The silky is most striking, with ornate leaves and a compact, colorful inflorescence.

BLUEBELLS

Mertensia spp. (mer-**ten**-zia: for Franz K. Mertens, p. 120). Also lungwort. Corolla blue, pink-tinged at first, ½–¾", short-lobed, narrowly bell-shaped and pendent, stem leaves pointed-elliptical, often with bluish bloom. Boraginaceae.

Tall Fringed Bluebells

M. ciliata (silly-**ay**-ta: fine-haired along edge). Upper leaves stalkless; basal leaves, if present, not heart-shaped; robust plants, 1–5' tall, usually in clumps. Subalpine meadows; A.

Tall Bluebells

M. paniculata (panc-you-**lay**-ta: flowers in panicles). Upper leaves short-stalked; basal leaves heart-shaped, long-stalked; robust plants 1–5' tall. Lush subalpine meadows; northerly.

Tall fringed bluebells

Alpine mertensia

Baker's Bluebells

M. bakeri. Less than ¼ of calyx length is fused; leaves lanceolate; 5–12" tall. Alpine; CO.

ALPINE MERTENSIA

Mertensia tweedyi (**twee**-dee-eye: for Frank Tweedy). Also *M. alpina.* Flowers brilliant royal blue, face-up in small clusters; corolla wider than long (not bell-shaped), with flat-spreading lobes on a short tube; leaves narrow elliptical; 2–10" tall. Alpine, locally abundant; CO, NM, WY, s MT.

Forget-me-not

Alpine forget-me-not

leaves oblong, hairy-tufted, ⅜", crowded in a mat; stems ½–4" tall. Alpine; A. Boraginaceae.

> *The intervals of soil are ... covered with a carpet of low but brilliantly flowering alpine plants ... rarely rising more than an inch in height. In many of them, the flower is the most conspicuous and the largest part of the plant, and in all, the colouring is astonishingly brilliant. ... May the deep cœrulean tint of the sky, be supposed to have an influence in producing the corresponding colour, so prevalent in the flowers of these plants?*
> —Edwin James on Pike's Peak, 1820

FORGET-ME-NOT

Myosotis asiatica (my-o-**so**-tiss: mouse ear). Also *M. alpestris, M. sylvatica.* Flowers sky blue with yellow center, ¼" across, in clusters; corolla lobes spread flat from a short tube; leaves oblong, hairy, basal ones to 5" long, stem ones to 1½"; hairy stems 3–16" tall. Moist mid-elev to alpine soil; A–. Boraginaceae.

ALPINE FORGET-ME-NOT

Eritrichium nanum (er-it-**trick**-ium: woolly hair; **nay**-num: dwarf). Also *E. elongatum, E. aretioides.* Flowers sky blue with yellow center, ¼" across; corolla spreading flat from a short tube;

This dwarf, the official flower of Grand Teton National Park, holds the wildflower record for greatest intensity per millimeter of height.

PUCCOON

Lithospermum ruderale (litho-**sperm**-um: stone seed; roo-der-**ay**-lee: of waste places). Also western stoneseed, gromwell, or lemonweed. Flowers pale yellow, ⅜" across, in leaf axils; corolla lobes spreading flat from a narrow tube; leaves linear, 1–4", many, hairy, all on stem; stems several, rough-hairy, 8–28" tall. Dry meadows and open forest to mid-elevs; A. Boraginaceae.

Use of the roots for red dye or face paint (when mixed with grease) was so widespread that settlers called puccoon "Indian paint." The Okanagan dyed fishing lines, believing that puccoon masked human odors. Some women took puccoon to suppress fertility; lab work confirms that puccoon suppresses some reproductive hormones and organs.

FRINGED PUCCOON

Lithospermum incisum (in-**sigh**-sum: incised). Also plains stoneseed. Flowers yellow, ¾" across, showy;

Puccoon

Fringed puccoon

Miner's candle

Butte candle

Buckbean

corolla lobes strikingly ragged-edged, spreading flat from a narrow tube up to 3" long; leaves linear, ½–3", many, hairy, all on stem; stems numerous, 4–16". Sagebrush, piñon-juniper; all exc ID, OR. Boraginaceae.

OREOCARYA

Oreocarya spp. (oreo-**cair**-ia: mountain nut). Also *Cryptantha* spp. Corolla lobes spread flat from a narrow tube; calyx and leaves bristly with stiff white hairs; lower leaves spoon-shaped, upper ones linear. Rocky, dry grasslands, pine forest. Boraginaceae.

Miner's Candle

O. virgata (veer-**gay**-ta: wandlike). Flowers white, in an impressive columnar raceme bristling with narrow green bracts; 12–40" tall; biennial, with only a basal rosette in the 1st year. CO and s WY.

Butte Candle

O. glomerata (glom-er-**ay**-ta: clustered). Also cockscomb cryptantha; *O. celosioides*. Flowers white with yellow center, ½" across, in clusters initially round, sometimes candlelike; 3–20" tall. A–. We have many similar but smaller-flowered *Oreocarya* spp.

BUCKBEAN

Menyanthes trifoliata (men-**yanth**-eez: monthly flower; try-fo-lee-**ay**-ta: 3-leaved). Corolla white (often purple-tinged), ¼–½" long and wide, with hairy-faced, flat-spreading lobes on a straight tube; stamens purple-tipped; flowers in a ± erect, columnar raceme;

Bastard toadflax

3 leaflets elliptical, 2–5"; stems and leafstalks usually prostrate and submerged; from rhizomes. Bogs and ponds; A. Menyanthaceae.

BASTARD TOADFLAX

Comandra umbellata (co-**man**-dra: hairy men, ref to stamens; um-bel-**ay**-ta: flowers in umbels). Also *C. pallida*. Calyx lobes white to purplish, narrow, petal-like; lobes still conspicuous on the bluish to brownish, ¼", egg-shaped fruits; flowers in a round clusters atop unbranched but clustered 2–14" stems; leaves bluish-coated, somewhat fleshy, linear to narrow elliptical, untoothed, alternate, all up and down the stem; partially parasitic on roots. Sandy soil, foothills to timberline; A. Santalaceae.

Once you get past the name, there are things to like about this plant: Thoreau admired it, and the oily fruits are tasty when green, though said to be nauseating when eaten in bulk. There are also things to dislike, especially if you're a plant: it is a partial root parasite on nearby plants of many kinds, and it is an alternate host for *Cronartium coman-*

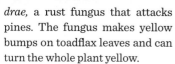
Spreading dogbane

Showy milkweed

Showy milkweed seed pod

drae, a rust fungus that attacks pines. The fungus makes yellow bumps on toadflax leaves and can turn the whole plant yellow.

SPREADING DOGBANE

Apocynum androsaemifolium (a-**pos**-in-um: Away, dog!; an-dro-see-mif-**oh**-lium: leaves like *Androsaemum*). Flowers pink, fragrant, ¼–⅜", bell-shaped with flaring lobes, clustered; leaves opposite, oval, ± glossy above, pale beneath, spreading flat or drooping; seeds cottony-tufted, in paired, slender, 3–6" pods; stems milky-juiced, smooth, often reddish, much-branched plants 1–2' tall, resembling shrubs. Dry, ± shady sites, up to tree line; A. Apocynaceae.

The Salish sewed hides together for teepees with a strong twine of Indian-hemp, *A. cannabinum;* they occasionally substituted dogbane. To humans and perhaps also to dogs, dogbane is purgative and diuretic and may cause heart irregularities. Butterflies and bees like it.

SHOWY MILKWEED

Asclepias speciosa (a-**skleep**-ius: Greek god of medicine; spee-see-**oh**-sa: showy). Strange, clawlike pink flowers in round clusters about 2" across; the 5 forward-curved pink hoods are structures on the stamens; behind those are 5 bent-back pink petals hiding 5 red/green-tinged sepals; hard-shelled, warty seedpods release seeds with copious white floss; leaves opposite, ± oblong, 4–8" long, with whitish veins; stems milky-juiced, gray-woolly, 1–4' tall. Lowland meadows, ditches, fields; A. Apocynaceae.

Milkweed's bizarre flower structure is an elaborate mechanism to avoid self-pollination. The flower snags an insect's leg on a V-shaped pair of pollen sacs, which the insect flies off with and leaves stuck on the next milkweed's sticky gland, on its pistil. Butterflies, bees, wasps, and ants are all known pollinators.

Milkweed leaves have toxins to keep butterfly larvae from eating them. Monarch butterflies evolved resistance to these toxins, enabling them not only to specialize in milkweeds but to accumulate enough alkaloids to make themselves toxic to predators. The toxins in the milky sap vary widely among the 107 species in the genus. Some are toxic enough to kill livestock, but showy milkweed is at the other extreme—non-toxic enough to have served as an important food for the Lakota and Crow tribes. Edible parts include the flower buds, young shoots, immature seedpods, and sap (dried, then chewed like gum). I don't recommend chewing it, considering the toxic relatives. During World War II, both sides experimented with milkweed sap as a source of rubber, and the United States launched commercial use of the seedpod floss as cushion stuffing. It is said to have higher insulative value than down. The sap is a folk remedy for warts and for poison ivy.

Felwort

Sibbaldia

Regular Flowers with 5 Separate Petals, 5 Stamens

A "regular" flower is radially symmetrical: all its petals (or sepals) are essentially the same size and shape.

FELWORT

Swertia perennis (**swert**-ia: for Emanuel Sweert, a 16th-century Dutch gardener; per-**en**-iss: perennial). Also star gentian. Petals ½", spreading flat, blue-purple variably streaked or speckled with white and/or green, with two bumps (glands) at the base of each; 5 large stamens; flowers in a spike, the lower ones in opposite pairs; leaves mainly basal, long-stalked lanceolate to oval, entire; 4–20" tall. Wet places, alpine down to mid-elevs; A. Gentianaceae.

Like its gentian relatives, felwort is a late bloomer, approaching star status in late August. It is remarkably disjunct globally: widespread in the Rockies and coastal Alaska, it grows in just a few spots elsewhere in North America, and then turns up in the high mountains of Europe and Asia.

SIBBALDIA

Sibbaldia procumbens (sib-**ahl**-dia: for Sir Robert Sibbald; pro-**cum**-benz: prostrate). Petals tiny, yellow, sitting on top of 5 slightly longer green bracts alternating with 5 much larger (up to ¼") green calyx lobes; leaves 3-compound; leaflets ½–1½", 3-to-5-toothed at the tip, white-hairy; leafstalks and flower stems rising 2–4" from rhizomes or prostrate stems. Alp/subalp; A. Rosaceae.

Blue flax

Wild sarsaparilla

BLUE FLAX

Linum lewisii (**lye**-num: Roman name; lew-**iss**-ee-eye: for Meriwether Lewis, p. 241). Also *Adenolinum lewisii, L. perenne*. Petals pale blue with yellowish base, ½–¾"; sepals ¼"; 5 styles; several flowers on each of the many stem branches; leaves linear, ⅓–1", all up and down the 6–30" stem; in alpine zone, petals, leaves, and plant are about ½ the above sizes. Dry meadows or open woods, all elevs; A. Linaceae.

WILD SARSAPARILLA

Aralia nudicaulis (a-**ray**-lia: Indigenous name; nu-di-**caw**-lis: naked stem). Flowers ¼" long, greenish-white, in a few open, round clusters ± hidden beneath the leaf on a 4–12" leafless stalk; berries ¼", insipid, dark purple, with 5 tiny styles; compound leaf rises on a 6–20" stalk from rhizomes; leaflets usually 9 or 15, oval, fine-toothed. Moist forest; WA, n ID, nw MT, plus e slope in CO, WY. Araliaceae.

Sarsaparilla was the great American "tonic" of the nineteenth century. The genuine item came from the Latin American

Brewer's mitrewort

Three-toothed mitrewort

Smallflower mitrewort

Smilax species, and then from sassafras. *Aralia* is unrelated to either of them, but its root puts out a similar flavor, and features in wildcrafter root beer recipes. It had its own reputation (learned from eastern Indigenous people in the 1600s) as invigorating, healing, sweat-inducing, and pimple-clearing. The roots survive fires to resprout.

GRASS-OF-PARNASSUS

Parnassia fimbriata (par-**nas**-ia: of Mount Parnassus, or mtns in general; fim-bree-**ay**-ta: fringed). Petals white, ¼–½", long-fringed near the base; flower single on a 6–16" stalk, in late summer; leaves heart-shaped, basal, long-stalked, plus 1 small leaf sessile halfway up stem. Wet meadows, streambanks, esp subalpine; A. Celastraceae.

The misnomer "grass of Parnassus" has withstood centuries of compleynts. Witness John Gerard writing in his *Herball* in 1597:

Grass-of-Parnassus

The Grass of Parnassus hath heretofore been described by blinde men; I do not meane such as are blinde in their eyes, but in their understandings, for if this plant be a kind of grasse, then may the Butter-burre or Colte's-foote be reckoned for grasses, as also all other plants whatsoever.

MITREWORT

Mitella spp. (my-**tel**-a: mitre—). Also bishop's cap. Petals branched, threadlike, sticking out between whitish calyx lobes; flowers 10–20 along a 6–14" stalk; leaves ± kidney-shaped, scalloped to toothed, basal, stalked. Moist sub-alpine forest or meadow. Saxifragaceae.

Brewer's Mitrewort

M. breweri (**brew**-er-eye: for W. H. Brewer). Also *Pectiantia breweri, Brewerimitella breweri.* Petals ± yellow-green, 5-to-9-branched; stamens aligned with calyx lobes; calyx saucer-shaped. Northerly.

Five-stamen Mitrewort

M. pentandra (pen-**tan**-dra: 5 stamens). Also *Pectiantia pentandra.* Like *breweri,* exc stamens aligned with petals. A.

Three-toothed Mitrewort

M. trifida (**trif**-id-a: 3-forked). Also *Ozomelis trifida.* Petals white or purplish, with 3 thick branches; calyx bell-shaped. Northerly.

Smallflower Mitrewort

M. stauropetala (stor-o-**pet**-a-la: cross petals). Also *Ozomelis stauropetala.* Petals white, with 3 slender branches; calyx bell-shaped. A–.

Naked mitrewort seeds

Redstem saxifrage

Rusty-hair saxifrage

Naked Mitrewort

M. nuda. Petals ± white, 9-branched; 10 stamens; calyx saucer-shaped. Northerly.

THE MITREWORT'S SEED capsule looks like a mitre—a bishop's tall, deeply cleft hat—until it splits open to resemble a chalice of caviar (ripe, they're much darker than in the featured photo). This splash cup invites a raindrop to knock the seeds out and away.

On a Japanese species, the slender, pinnate-lobed petals support landings by fungus gnats, its pollinators; this may turn out to be the case here as well.

Regular Flowers with 5 Separate Petals, 10 Stamens

SAXIFRAGE (TALL SPECIES)

Micranthes spp. (my-**cran**-theez: small stamens). Also *Saxifraga* spp. Pistil of 2 horns (rarely 3 to 5); several flowers in a loose raceme (exc *M. oregana*); often sticky-hairy near top; leaves all basal, somewhat fleshy. Wet rocks and gravels, low to subalpine elevs. Saxifragaceae.

Brook Saxifrage

M. odontoloma (o-don-ta-**lo**-ma: tooth fringe). Petals ⅛" long, nearly round exc for a short, slender neck, white with 2 yellowish-green spots near base; swollen white stamens; leaves 1–3", round to kidney-shaped, with large, even teeth; stem 6–12". A.

Redstem Saxifrage

M. lyallii (lye-**ah**-lee-eye: for David Lyall, p. 62). Petals white with 2 yellowish-green spots near base; sepals and pistil often crimson; stamens white, widened; leaves round, ½–1½", with large, even teeth; stem 4–8", leafless, often maroon. Wet gravels, esp alp/subalp; northerly.

Rusty-hair Saxifrage

M. ferruginea (fair-oo-**jin**-ia: rust-colored). Petals ¼" long, white, 3 wider petals each have a big yellow spot, 2 slenderer petals spotless; 10 slender white stamens with orange anthers; pistil with minute stigmas; some flowers often replaced by bulblets; leaves ½–4", oblanceolate to spoon-shaped, finely hairy, with 7–15 coarse, even teeth; flowers sparse on branched stalks 6–14". Northerly.

Western Saxifrage

M. occidentalis (oxy-den-**tay**-lis: western). Petals white; oval with a distinct, slender neck; yellow to dark-red anthers; leaves tapering to flattened leafstalks ½–2" long, leaf blades oval with tangled rusty hairs beneath, with coarse teeth; 2–8" stems may redden. WA to se WY.

Bog saxifrage

Bog Saxifrage

M. oregana (o-reg-**ay**-na: of Oregon). Also Oregon saxifrage. Petals white, small; flowers in a few round clusters on upper stem; leaves oblong, shallowly toothed, 4–7"; stems sticky-hairy, thick,

Diamondleaf saxifrage

Spotted saxifrage

Whiplash saxifrage

fleshy, ± seeds reddish, 10–30". Marshy meadows; CO, MT, ID, WA, OR. Similar but petalless *M. subapetala* grows in WY and s MT.

Diamondleaf Saxifrage

M. rhomboidea (rom-**boy**-di-a: diamond-shaped). Petals white, small; flowers in 1 (or 2) round clusters on upper stem; leaves diamond-shaped, 1–2"; stems sticky-hairy, thick, fleshy, ± reddish, 5–11". Dry meadows and tundra; A–.

Tufted alpine saxifrage

Purple saxifrage

SAXIFRAGE (MAT-FORMING SPECIES)

Saxifraga spp. (sac-**sif**-ra-ga: stone break). Pistil of 2 horns (rarely 3 to 5); leaves small, crowded, mat-forming. Wet rocks and gravels, alp/subalp. Saxifragaceae.

Spotted Saxifrage

S. austromontana (aus-tro-mon-**tay**-na: southern mountains). Also *S. bronchialis*, *Ciliaria austromontana*. Petals white, red-speckled on outer ½, yellow on inner; stems have a few small leaves; basal leaves sharp, linear, ⅛–⅝". Rock crevices; A.

Tufted Alpine Saxifrage

S. cespitosa (see-spit-**oh**-sa: growing in clumps). Petals spotless white; filaments and pistil often yellow; leaves of 3 narrow lobes. A.

Whiplash Saxifrage

S. flagellaris (flaj-el-**air**-iss: with whips). Also spider saxifrage; *Hirculus platysepalus*. Petals yellow; calyx red-tinged; all green parts bristling with sticky glands; leaves fleshy, ⅝", pointed-elliptic, clasping 3–6" stem and also in a basal rosette; red runners reach out across soil. s MT to NM.

Purple Saxifrage

S. oppositifolia (op-pos-it-if-**oh**-lia: opposite leaves). Petals pink to purple; flowers ± bell-shaped, ⅓", emerging singly from a mat of ⅛" leaves in scalelike opposite pairs; 1–2" high. Favors limestone rockfields; WA to Wind River Range.

Roundleaf alumroot

Tiarella

SAXIFRAGES IN THE Rockies seem to live up to their "rock-breaker" moniker, but the name actually derives from the medieval herbalists' doctrine of signatures: European species were prescribed for breaking up bladderstones, which their bulbs vaguely resemble.

TIARELLA

Tiarella trifoliata (tee-ar-**el**-a: crownlet; try-fo-lee-**ay**-ta: 3-leaved). Includes *T. unifoliata*. Also sugarscoop, laceflower, foamflower, false mitrewort. Flowers tiny, white, many, in a sparse raceme on an 8–22" stalk; petals threadlike, unbranched, less visible than the stamens; ovary and subsequent capsule of 2 very unequal sides; leaves mostly basal on short leafstalks, hairy, toothed, from 3-lobed to 3-compound and incised, 2–4". Dense forest; WA to nw MT. Saxifragaceae.

By midsummer, many tiarella leaves display pale curlicues, the tracks of leaf-miner larvae (p. 472). Look closely to see the larva at the front end of the track.

ROUNDLEAF ALUMROOT

Heuchera cylindrica (**hoy**-ker-a: for Johann von Heucher). Flowers in a crowded cylindric raceme, cup-shaped, creamy (sometimes pinkish), lacking petals; stamens hidden within the 5-lobed calyx; 2 styles emerge as seeds grow; leaves basal, oval with coarse lobes and teeth,

on long stalks; stem 6–32", leafless. Rock crevices, mainly mid-elevs; northerly. Saxifragaceae.

Alum is an astringent chemical long used to stop bleeding. European colonists learned from Indigenous people to use the roots of *H. americana* for that purpose, hence the name alumroot. Roundleaf alumroot roots were made into decoctions and poultices for bleeding, sores, rheumatism, sore throats, diarrhea, tuberculosis, and liver ailments.

TELESONIX

Telesonix heucheriformis (tel-e-**son**-ix: distant claw, ref to long-stalked petals; hoy-ker-if-**or**-mis: like alumroot). Also *Boykinia heucheriformis*. Petals violet, ¼", ± peeking out between the 5 calyx lobes; leaves kidney-shaped, doubly toothed, stalked, smaller upward; calyx and stem sticky-hairy; stems 1–several, scaly-based, 2–6". Alp/subalp limestone crevices; MT, WY, barely ID and UT. Saxifragaceae. Similar *T. jamesii*, it has bigger, showier petals and a very small range—just the CO Front Range.

SMALL-FLOWERED WOODLAND STAR

Lithophragma glabrum (lith-o-**frag**-ma: stone breaker; **glab**-rum: smooth, ref to leaves, not stem). Also starflower, rock

Telesonix

Small-flowered woodland star

Sticky geranium

Richardson's geranium

King's crown

Queen's crown

deeply 5-lobed, the lobes again split; basal leaves long-stalked, stem leaves small and stalkless; 1½–3' tall. Dry meadows, aspen groves, open forest. These two spp. cannot always be distinguished; their colors and other characters vary and overlap. Geraniaceae.

Sticky Geranium

G. viscosissimum (vis-co-**see**-sim-um: stickiest). Petals typically strong pink with purple veins; hairs white- or yellow-tipped (but not always sticky). A.

Richardson's Geranium

G. richardsonii (rich-ard-**so**-nee-eye: for Sir John Richardson, p. 136). Petals typically white, often with purple veins; sticky hairs purple-tipped. A.

KING'S CROWN

Rhodiola integrifolium (ro-dee-**oh**-la: rose—, ref to root fragrance; in-teg-rif-**oh**-lium: untoothed leaf). Also roseroot, midsummer-men; *Sedum integrifolia, Tolmachevia integrifolia, S. rosea*. Petals deep red, shorter than ¼"; stamens protruding; seed capsules 5 per flower, deep red; flowers several, crowded, ♂ and ♀ flowers on separate plants; leaves pale green, rubbery; stems 2–8", unbranched, in clumps, from rhizomes. Alp/subalp, often where wet; all but UT. Crassulaceae.

King's crown is closely related to *R. roseum*, the roseroot studied as a remedy for stress, fatigue, anxiety, depression, and cancer, and as an endurance performance booster. (King's crown itself has not been studied.) The two share some but not all of their essential oils.

star, prairie star, etc. Petals white, ¼", deeply 5-lobed—shaped ± like a maple leaf; 5-lobed calyx and stems often sticky-hairy; leaves deeply palmate-lobed, the basal ones stalked and much larger. Sagebrush, aspen groves, rocky slopes, all elevs; A. Saxifragaceae.

Look for minute maroon bulbs in the upper leaf and branch axils. They can germinate and grow after the stem topples—a way of cloning. Some specimens produce a lot of these and few or no flowers. Similar woodland stars without bulbils are either *L. parviflorum,* with three petal lobes, or *L. tenellum,* with five or seven. All three species are found throughout.

GERANIUM

Geranium spp. (jer-**ay**-ni-um: crane—). Also cranesbill. Flowers 1–1½" across, saucer-shaped, few to several; sepals and upper stems sticky-hairy; seedpods split open explosively; leaves

QUEEN'S CROWN

Rhodiola rhodanthum (ro-dee-**oh**-la; ro-**danth**-um: rose flower). Also rose crown; *Sedum rhodantha, Clementsia rhodantha*. Petals pink to white, ¼–⅜"; flowers have stamens and pistils; plants 2–14", much like king's crown (above) and may grow with it; *Colorado Flora* says you can tell queen's crown by a raised midrib on leaf underside. Mid-elevs to alpine, often where wet; c MT s. Crassulaceae.

Lanceleaf stonecrop

Wormleaf stonecrop

STONECROP

Sedum spp. (**see**-dum: seated). Petals yellow, pointed, ¼–⅜"; leaves alternate, fleshy, ± tubular, crammed together; plants spreading by rhizomes and/or runners. Dry, rocky places. Crassulaceae.

Lanceleaf Stonecrop

S. lanceolatum (lan-see-o-**lay**-tum: lance—). Also *Amerosedum lanceolatum.* Flowers in compact, broad-topped clusters; 5 seedpod cells form a narrow cone; leaves bluntly pointed, often reddening; 3–8" tall. A.

Wormleaf Stonecrop

S. stenopetalum (sten-o-**pet**-a-lum: narrow petals). Flowers in looser, flat-topped clusters; 5 seedpod cells spread flat like a star; leaves sharply pointed; axils of main upper leaves bear little clusters of smaller leaves or bulblets; leggy plants 8–12" tall. Northerly.

FAT SUCCULENT LEAVES, like cactus stems, maximize the ratio of volume to surface area, and hence the ratio of water capacity to water loss through transpiration. Even with leaves for backup water storage, stonecrops grow only while water is available and store water mainly to subsidize flowering and fruiting. You can squeeze water out of stonecrop leaves at temperatures that will then freeze it immediately, but the leaves resist internal freezing and do very well in the alpine.

MOSS-CAMPION

Silene acaulis (sigh-**lee**-nee: a Greek elf?; ay-**caw**-lis: stalkless). Also moss pink, cushion pink, carpet pink. Petals pink (occasionally

Moss-campion

white), separate, though they form an apparent tube, bent 90° to spread flat; styles and/or stamens often protruding; styles usually 3; calyx shallowly 5-lobed; leaves linear, pointed, thick, crowded, to ½"; mosslike mats 2" thick. Alpine; A. Caryophyllaceae.

The Pink family is so called not because the petals are pink but because they are pinked, or notched at the tip. Pinked petals do run in the family, but other family traits such as ten stamens and unfused petals are more reliable, and they distinguish moss-campion from phlox and dwarf primrose. (See pp. 195–197 on cushion plants.)

WHITE CATCHFLY

Silene parryi (**pair**-ee-eye: for Charles Parry, p. 212). Corolla white or lavender-tinged, ½" across, petals deeply 2- or 4-lobed, with 2 more small lobes on the throat—a seeming inner whorl of 10 petals; calyx long, sticky-hairy,

White catchfly

Ballhead sandwort

Arctic sandwort

Moths pollinate these blooms at night; they're at their best then. As "catchfly" implies, small bugs get caught on the sticky surfaces around the inflorescence. Research on tofieldia finds that sticky stems enable carnivory (p. 139).

SANDWORT

Eremogone and *Cherleria* spp. (air-em-**og**-un-ee: child of the desert?; cair-**lair**-ia: for Johann H. Cherler). Also *Arenaria* spp. Petals white, ¼–½"; sepals ⅛–¼"; 3 styles; leaves needle-thin but not prickly, ¾–1¼" long, often dense, basal plus 2–10 opposite leaves on stem; in tufts or mats. Rocky sites, esp steppe or alpine. Caryophyllaceae.

Fendler's Sandwort

E. fendleri (**fend**-ler-eye: for Augustus Fendler, p. 246). Also *Arenaria fendleri.* Petals ¼–⅜", barely longer than sepals; flowers appear scattered; 4–16" tall. Southerly.

Threadleaf Sandwort

E. capillaris (cap-il-**air**-iss: hair-leaved). Petals ¼–½"; flowers appear scattered; 2–10" tall. WA, OR, ID, w MT.

Ballhead Sandwort

E. congesta. Also *Arenaria congesta.* Flowers in a dense, round cluster; petals ¼–⅜"; 2–16" tall. All but NM.

Arctic Sandwort

C. obtusiloba (ob-too-si-**lo**-ba: blunt-lobed). Also *Minuartia* or *Lidia obtusiloba.* Petals ¼–½"; flower usually single; leaves ¼" long, opposite,

± red-striped; leaves narrow, 1–3", 2 or 3 pairs opposite, the rest basal; upper stem sticky-hairy; 6–24". Alp/subalp; WA to s WY. *S. menziesii,* with smaller flowers, extends s to CO and NM. Introduced white campion, *S. latifolia,* is common on roadsides. Caryophyllaceae.

Early Days in Estes Park

Charles C. Parry earned an MD, but preferred botany. After serving as botanist for the US and Mexican Boundary Survey in 1849–1853, he set about making Rocky Mountain botany his career, picking Colorado first, on the grounds that its wealthy mining towns offered amenities he wouldn't find elsewhere in the Rockies. While lodging with Joel Estes on the Big Thompson, he was enamored of the environs and predicted that this remote "Estes Park" homestead would be a successful resort someday. Less prescient was his assessment, after not quite conquering Longs Peak, that it was inaccessible. Parry presented a ground-breaking explanation of timberlines, noting that they get lower northward, and he gave us "alpine" as an ecological term that means the "terrain above tree line."

Field chickweed

Heartleaf pyrola

Greenish pyrola

linear, dense, on prostrate stems forming low cushions or mats. Abundant on alpine gravels, crevices; A.

FIELD CHICKWEED

Cerastium arvense (ser-**ast**-ium: horn—; ar-**ven**-see: of fields). Also prairie mouse-ear; *C. strictum.* Petals white, each with 2 round lobes; styles usually 5; calyx 5-lobed; flowers 1–6 per stem; leaves oblong, ¼–1", opposite, but often so crowded as to appear basal, in mats; 3" to (rarely) 10" tall. In sun at all elevs; A. Caryophyllaceae.

PYROLA

Pyrola spp. (**peer**-a-la: small pear, ref to leaf shape). Also wintergreen, shinleaf. Flowers 5–25 in a tall spike; style bends sharply down or sideways, then curves up at its tip; leaves evergreen, dark, basal on stalks, entire or finely toothed, 1–3" long; stem reddish, 4–16". Shady forest. Ericaceae.

Heartleaf Pyrola

P. asarifolia (a-sair-if-**oh**-lia: wild-ginger leaf). Petals spreading, pink to red, style strongly downturned; flowers 8–25 on a 6–16" stalk; leaves evergreen, dark, basal, long-stalked, round to heart-shaped, 1–3". A.

Greenish Pyrola

P. chlorantha (clor-**anth**-a: green flower). Also *P. virens.* Petals spreading, white to greenish or yellowish; style downturned; flowers 2–8; leaves basal, stalked, nearly round, ± fine-toothed, 1–2", often paler above than beneath; stem 3–8". A.

White-veined Pyrola

P. picta (**pic**-ta: painted). Petals spreading, pale, greenish to purplish, style downturned; flowers 5–20 on a reddish, 4–12" stem; leaves evergreen, basal, egg-shaped, 1–3", dark green, white-mottled along the ± pinnate veins (compare rattlesnake-plantain, p. 150). A.

One-sided Pyrola

Orthilia secunda (or-**thill**-ia: straight, small; se-**cun**-da: with flowers all to 1 side). Flowers greenish-white, bell-shaped, with long, straight style, 5–15 all facing ± the same way; leaves 1–2½", variably egg-shaped, running up the lower ½ of the 3–7" stems from rhizomes. Shady forest; A. Ericaceae.

THE NON-GREEN PLANTS on pp. 153–155 are heterotrophic, meaning they obtain their photosynthate nutrients (carbohydrates) from other plants that are autotrophic (creating photosynthates for themselves). A newer word, *mixotrophic,* was coined for plants that do a little of both. Pyrolas fall

White-veined pyrola

One-sided pyrola

One-flowered wintergreen   Western St.-John's-wort   Smoothstem blazingstar

into that category. One species, *P. aphylla,* has no leaves and is heterotrophic, and the other pyrolas are all so closely related to it that whether it was actually a species or just some degenerate specimen was debated for decades. Mixotrophy in pyrolas can be interpreted in either of two ways: they are hedging their bets, or they are at a transitional point in evolving, relying increasingly on their fungal partners underground and less on their green leaves. There's some truth in each.

ONE-FLOWERED WINTERGREEN

Moneses uniflora (mo-**nee**-sees: single delight; you-ni-**flor**-a: 1 flower). Also wood nymph, single delight. Flowers single, ½–1", waxy-whitish, fragrant; petals flat-spreading; pistil fat, straight, 5-tipped (like a chess rook); leaves oval, usually toothed, ½–1¼", from lower ¼ of 2–6" stem; often on rotting wood. Shady forest; A. Ericaceae.

This odd flower rewards looking closely. It was used medicinally by many Indigenous groups, and has shown activity against tuberculosis bacteria.

Regular Flowers with 5 Petals, 15 or More Stamens

ST.-JOHN'S-WORT

Hypericum spp. (hi-**per**-ic-um: over an icon). Flowers deep yellow, orange in bud; stamens showy, as long as the petals; capsule 3-celled; leaves opposite, oval to oblong, ± clasping-based, with tiny translucent spots ("perfora-tions") visible against light (fewer and less reliable in *H. scouleri*). Hypericaceae.

Western St.-John's-wort

H. scouleri (**scoo**-ler-eye: for John Scouler. Also *H. formosum*. Few flowers; sepals oval, less than 3× longer than wide—stems 4–24". All elevs; common subalpine; A.

St.-John's-wort

H. perforatum. Also Klamath weed. Many flowers, in a ± flat-topped cluster; petals often edged with tiny black dots; sepals narrow, 3× to 5× longer than wide—stems 1–3'. Roadsides, rangeland; European weed; A–.

THE OLD WORLD St.-John's-wort was gathered as a spell for St. John's Eve, June 23, and as a spell to place below images of the saints, hence the genus name "under the icon." It's now an abundant roadside weed in the West and a popular herbal treatment for several ills, notably depression—which it does alleviate, according to meta-analyses of controlled studies. As with other herbals, the dosage and side effects are poorly known. (Methodical testing of things like dosage is, after all, the chief difference between herbal medicines and prescription drugs, most of which originated as herbal compounds.)

St.-John's-wort contains photosensitizing toxins: they leave eyes and sensitive skin prone to severe reactions to bright sunlight. This made Klamath weed a killer weed, inflicting serious sheep and cattle losses in Oregon and California in the early 1920s and 1930s. After 1960, biological pest control got one of its first wins, using an introduced beetle to delete Klamath weed from lists of the "West's Ten Worst Weeds."

BLAZINGSTAR

Mentzelia spp. (ment-**zel**-ia: for Christian Mentzel). Also *Nuttallia* spp. Flowers bright yellow, numerous; numerous showy stamens at least ½ as long as petals, 5 distinctly widened sterile stamens alternate with petals; 5 tiny sepals; leaves pinnate-lobed, up to 6" long, alternate; leaves and stems usually rough-hairy; 1–3' tall, bushy. Roadsides. Loasaceae.

Smoothstem Blazingstar
M. laevicaulis (lee-vic-**aw**-lis: smooth stem). 5 petals 1–3" long. WA to just barely in CO.

Adonis Blazingstar
M. multiflora. Petals ½–1" long, seemingly 10, because 5 stamens are petal-like. Southerly.

Adonis blazingstar

THESE BRILLIANT NATIVE flowers get less love than they deserve, perhaps because they look weedy growing in weedy places like highway embankments. The Cheyenne and others used the roots medicinally.

Wood strawberry

STRAWBERRY

Fragaria spp. (fra-**gair**-ia: Roman name). Petals white to pinkish, nearly circular, ¼–½"; sepals apparently 10; stamens 20–25; berry up to ½" long; leaflets 3, toothed coarsely and ± evenly, on hairy leafstalks; 3–8" tall, with long reddish runners. Open forest and aspen groves to subalpine basins. Rosaceae.

Wood Strawberry
F. vesca (**ves**-ca: thin). Leaves minutely hairy on top, ± bulging between veins; tooth at leaf tip is as long as other teeth; seeds set on berry surface. A.

Blueleaf Strawberry
F. virginiana (vir-gin-ee-**ay**-na: of Virginia). Most leaves smooth and flat on top; tooth at leaf tip is shorter than other teeth; seeds embedded in pits on berry surface. A.

Blueleaf strawberry

COMMERCIAL STRAWBERRIES ARE a complex hybrid derived beginning 250 years ago from the two above species and *F. chiloensis* of the Northwest coast. None had big berries before the breeding began. Now you can buy strawberries the size of small apples. Ain't life grand?

The Okanagan used the leaves and the roots as poultices and as antidiarrheal teas.

PRAIRIE SMOKE

Geum triflorum (**jee**-um: Roman name; try-**flor**-um: 3-flowered). Also purple avens, old-man's-whiskers; *Erythrocoma triflora.* Flowers usually pendent, in 3s, dull reddish, 1" long,

Prairie smoke

Ivesia

purple-tinged) flowers of 5 true sepals alternating with 5 narrower bracts; leaves mainly basal, pinnately compound, the larger leaflets again deeply 3-to-5-lobed; stems rise 2–10" from mats of leaves. Alp/subalp; A. Rosaceae.

Alpine avens dominates great expanses of tundra, especially in southern Montana. Its unusually high storage capacity for carbohydrates in its roots helps it survive the occasional extra-poor (or heavily grazed) growing season. It is also one of the first plants to reestablish a claim on barren gopher blowouts. Broad adaptability surely contributes to its abundance: in the Colorado snow-depth study (p. 196), it was found growing in every depth category, with greatest frequency smack in the middle.

Alpine avens

Early cinquefoil

IVESIA

Ivesia gordonii (**ive**-zia: see p. 217; gor-**doe**-nee-eye: for Alexander Gordon, p. 142). Also Gordon's mousetail *Potentilla gordonii*. Yellow petals peek from between greenish-yellow sepals; flowers in round clusters; leaves basal, 1–4", pinnately compound, bottlebrush-shaped, leaflets ⅛–¼"; leaflets and sepals finely hairy; stem 2–7". Mainly alpine; A–.

CINQUEFOIL

Potentilla and *Drymocallis* spp. (po-ten-**til**-a: small but mighty, ref to medicinal uses; dry-mo-**cal**-iss: beautifying woodlands). Petals shallowly indented at tip; 5 true sepals alternate with 5 shorter bracts; leaves mainly basal, compound; leaflets 1–1½", coarsely toothed to narrow-lobed. Sunny sites at all elevs. Rosaceae.

Early Cinquefoil

P. concinna (con-**sin**-a: well made). Petals yellow; stem and leaves gray-woolly; basal leaflets 5 or 7, palmate; stems 1–6" but sprawling. A–.

vase-shaped, scarcely opening; petals yellow to pink, mostly hidden by the 5 reddish sepals and 5 bracts; seeds long-plumed; leaves mostly basal, finely cut (fernlike), hairy; 8–16" tall. Meadows at all elevs; A. Rosaceae.

ALPINE AVENS

Geum rossii (**ross**-ee-eye: for Captain James Ross, polar explorer, who first collected this flower in the High Arctic). Also *Acomastylis rossii*. Flowers 1–4, saucer-shaped, ½–1" across; yellow petals soon fall off, leaving green (or

The Army Went Exploring

Albert Bierstadt's too high and slim Rockies

Many scientific names honor US Army officers who commanded expeditions exploring the West in the 1850s and 1860s. While the primary task of these expeditions was to find routes for railroads to connect the East to California's gold, Congress also wanted to learn about the West, and sent scientists and artists along.

In 1849, Congress wanted to know about those rebellious Mormons who had set up camp by a big lake somewhere. They sent **Captain Howard Stansbury** and **Lieutenant John Gunnison** to report, which Gunnison did in a book, *The Mormons*. Gunnison died in a Native American attack in 1853. Stansbury also died rather young, due, his obituary claimed, to overexertion on the expedition.

Elsewhere in the Rockies, **Lieutenant Robert S. Williamson** survived an attack that killed his commanding officer in 1849. **Lieutenant Amiel Weeks Whipple** explored in 1853–54; **Lieutenant Joseph Christmas Ives** in 1857; **Colonel William F. Raynolds** in 1860; and **Captain Clarence King** in 1867.

En route to picking the first specimen of *Ivesia* while under Lt. Ives's command, John Strong Newberry descended the Colorado River through the Grand Canyon, making him the first geologist there, eleven years before John Wesley Powell. The Canyon failed to impress Ives: "It looks like the Gates of Hell. The region . . . is altogether valueless. Ours has been the first and will undoubtedly be the last, party of whites to visit the locality."

Clarence King was a geologist, not a soldier, yet at age twenty-five, on the strength of sheer charisma, he was commissioned as an army captain and invited to write his own orders for an expedition! His was the first army expedition to include a photographer (Timothy O'Sullivan), perhaps seeking to correct the melodramatic, romanticized image of the Rockies that easterners were getting from the painter Albert Bierstadt, who'd accompanied a previous army expedition. King fumed, "What has Bierstadt done, but twist and skew and distort and discolor and belittle and be-pretty this whole doggoned country? Why, his mountains are too high and slim; they'd blow over." King later became the first head of the US Geological Survey.

One day, a barefoot forty-two-year-old showed up in camp asking King for work. King took him on as a dishwasher and muleskinner. When the expedition's botanist fell ill, this man, **Sereno Watson**, took over plant-collecting duties. He turned out to be a Yale medical graduate who'd passed up careers in medicine and banking to wander the West. His collections on the King Expedition led him back to Harvard, where Professor **Asa Gray** groomed him to be his successor at the pinnacle of America's botany establishment. Even on the pinnacle, Watson remained a recluse. He named many of our plant species, but few are named for him.

The era of post-army exploration began with Professor **Ferdinand Vandiveer Hayden**, who first saw Yellowstone as Colonel Raynolds's geologist, but returned in 1871 and repeatedly thereafter, over a thirteen-year span. These trips were the Hayden Geological Survey, purely scientific explorations funded by the Department of the Interior.

Blueleaf cinquefoil

Snow cinquefoil

Tall cinquefoil

Blueleaf Cinquefoil

P. glaucophylla (glaw-co-**fil**-a: whitish leaves). Also *P. diversifolia*. Petals yellow; 5 or 7 palmate leaflets up to 1½" long, bluish, with a few long teeth near tip; 2–12" tall. Alp/subalp; A–.

Slender Cinquefoil

P. gracilis (**grass**-il-iss: slender). Petals yellow; 5–9 palmate leaflets at least 1¼" long, with long teeth all the way to base, grayish underneath; 6–18" tall. Mid-elev grasslands; A.

Littleleaf Cinquefoil

P. brevifolia (brev-if-**oh**-lia: short leaf). Petals yellow; basal leaves fine-cut in many tiny lobes, forming a dense cushion; 2–5" tall. Alpine; OR, c ID, w WY, sw MT.

Snow Cinquefoil

P. nivea (**niv**-ia: snow). Petals yellow; sepals, stems, and leaves gray-hairy; leaves mainly basal, 3 leaflets ⅜–1¼" long, with long, coarse teeth, greener above than below; 2–12" tall. A–.

Cliff Cinquefoil

D. pseudorupestris (soo-doe-roo-**pes**-tris: false rock cinquefoil). Also *P. glandulosa* var. *pseudorupestris*. Petals yellow to cream, usually longer than sepals; 5–9 pinnate basal leaves 1–6" long, deeply toothed; plant sticky-hairy, 3–16" tall. Rocky sites; A–.

Sticky Cinquefoil

D. glandulosa (glan-dew-**lo**-sa: with sticky hairs). Also *P. glandulosa*. Petals off-white to moderate yellow, often shorter than sepals; 5–9 pinnate basal leaves 4–12" long; plant sticky, usually hairy, 6–30" tall. WA, OR, ID, MT.

Tall Cinquefoil

D. arguta (arg-**you**-ta: sharp-toothed). Also *P. arguta*. Petals white to cream; flowers held tightly together on erect branchlets; leaves pinnately compound; 16–40" tall. MT and WY, mainly e of CD.

CINQUEFOILS ARE EASILY confused with buttercups until you look under the flower, where you'll see five sepals plus five bracts on a cinquefoil. Variation in leaf shape, hairiness, and habitat can make species ID challenging.

SILVERWEED

Argentina anserina (ar-jen-**tie**-na: silvery; an-ser-**eye**-na: of geese). Also *Potentilla anserina*. Petals yellow, sepals apparently 10; flowers 1" across, single on 1–5" leafless stalks from rosettes of pinnate leaves; tiny leaflets alternating with saw-toothed bigger ones; undersides (or rarely both sides) of leaves densely silver-haired; spreads by red trailing stems. Gravel bars, wet meadows, disturbed ground, foothills to subalpine; A (and worldwide). Rosaceae.

The Blackfoot used silverweed runners to tie up leggings and bundles. The long taproots were an important food.

Silverweed

BUTTERCUP

Ranunculus spp. (ra-**nun**-cue-lus: froglet). Petals glossy yellow; 5 sepals without 5 additional bracts; seeds in a conical head; stem 3–8". Ranunculaceae.

Sagebrush Buttercup

R. glaberrimus (gla-**bear**-im-us: smoothest). Leaves not toothed; basal ones elliptical, larger stem leaves 3-lobed. Sagebrush to (rarely) alpine; A.

Snow Buttercup

R. eschscholtzii (ess-**sholt**-zee-eye: for J. F. von Eschscholtz, p. 200). Leaves variably 3-lobed to 3-compound. Abundant in wet alp/subalp meadows; A.

Alpine Buttercup

R. adoneus. Like the preceding but with very finely cut leaf lobes less than ⅛" wide. Alp/subalp meadows; ID, WY, UT, CO.

Plantainleaf Buttercup

R. alismifolius (a-liz-mif-**oh**-lius: water plantain leaves). Leaves elliptical, smooth, or shallowly toothed, never deeply lobed or compound. Wet meadows; A.

SNOW BUTTERCUPS ARE solar trackers, bending to face the sun as it crosses the sky. Their sensors and bending mechanism are in the upper stem, which will bend with the sun even if the flower has been clipped off. Tracking to focus the sun's heat on the flower's sexual organs dramatically improves the pollen's fertility. Insects come to bask in the warmth, and leave dusted with pollen.

Buttercups are named for the peculiar waxy (cutinous) sheen of their yellow petals—your first clue for telling buttercups from cinquefoils (above). Buttercup-stem juices are blistering irritants. Indigenous people used them as arrowhead poisons and (cautiously, I imagine) as poultices for sore joints, toothaches, and so forth.

COLUMBINE

Aquilegia spp. (ak-wil-**ee**-ja: Roman name, from "eagle," ref to claw-shaped spurs). Flower shape unique: each petal is a front-facing cup with a long spur behind; petal-like sepals spread wide;

Sagebrush buttercup

Snow buttercup

Alpine buttercup

Colorado blue columbine | Yellow columbine (red-tinged hybrid) | Red columbine

leaves 9-compound with round-lobed leaflets, most basal, stalked. Alp/subalp moist crevices, meadows; mid-elev openings. Ranunculaceae.

Colorado Blue Columbine
A. coerulea (see-**rue**-lia: blue). Flowers huge, 2½–4" across, petals white, sepals blue, or sometimes both white or both blue; spurs 1–2" long. WY, sw MT, c ID. The Colorado state flower. In Utah's Wasatch, purple sepals could be Utah columbine, *C. scopulorum*.

Rocky Mountain Columbine
A. saximontana (sax-i-mon-**tan**-a: RM) Colored like the preceding, but flowers smaller (½–1½" across) and spurs hooked. c CO.

Limestone Columbine
A. jonesii (**jones**-ee-eye: for Captain William A. Jones). Alpine dwarf 2–5" tall; flowers light blue, spurs ⅜"; leaflets thick, grayish, fuzzy, tiny, very crowded in a mosslike cushion. Limestone terrain; MT, WY.

Limestone columbine

Yellow Columbine
A. flavescens (fla-**vess**-enz: yellowish). Flowers yellow. WA to nw CO.

Red Columbine
A. formosa (for-**mo**-sa: beautiful). Sepals and spurs red, petal cups yellow. WA to Yellowstone.

> *July 5, 1845. The flowers very large and beautyfully white, with varieties shaded a clear light blue—In my opinion It is not only the Queen of Columbines, but the most beautyful of all herbaceous plants—I never felt so much pleasure in finding a plant before.*
>
> *[Two months later] After a short search I found the Columbine I so much desired. . . . I collected seeds until it became dark, & commenced again the next morning as soon as it became sufficiently light. . . .*
>
> *It takes off a great deal of the pleasure of collecting knowing as I do that Mr. Geyer passed through this country in the seed season.*
>
> —Joseph Burke, (p. 142)

Oh, the fickle pleasures of a plant hunter!

The unique shape of columbine flowers is unmistakable, while the colors vary. White-lined sphinx moths flying at dusk see white best, and are the pollinator *A. coerulea* evolved for; their tongues are exactly long enough to reach the nectar at the tip of the spur. Where the same species is blue and white, it is hedging its bets, attracting both

Mountain hollyhock   *New Mexico checkermallow with bee fly*   *Scarlet globemallow*

sphinx moths and bumblebees. Bees nip the bulbous spur-tip to get the nectar, and also go to the front door for pollen. Hummingbirds and some bees are drawn to the red-yellow combination. Yellow columbine may be flushed with pink after messing around with red *A. formosa* where their ranges meet. Being interfertile, columbine species that occur together owe their separate identities to selecting different pollinators, via both color and the length of spurs—and tongues.

MOUNTAIN HOLLYHOCK

Iliamna rivularis (il-i-**am**-na: a mystery name; riv-you-**lair**-iss: of brooks). Also streambank globemallow. Flowers pale pink or lavender, 1½" across, in tall hollyhock-like clusters; leaves palmate-lobed, maplelike, 2–6"; robust herbs to 6' tall. Streambanks, ditches, disturbed clearings; A–. Malvaceae.

Growing from long-buried seeds, mountain hollyhock occasionally produces spectacular displays a couple of years after a severe fire.

NEW MEXICO CHECKERMALLOW

Sidalcea neomexicana (sid-**al**-sia: combo of 2 other mallow genera). Also saltspring checkerbloom. Flowers striated deep pink, 1–1½" across, in tall hollyhocklike clusters; numerous stamens basally fused into a tube around the pistil; leaves alternate, round in outline, lowest ones with many shallow lobes, upper with long narrow lobes with a few sub-lobes; white-hairy herbs 8–36" tall. Streambanks, sagebrush, ponderosa forest; OR to NM. Similar Oregon checkermallow, *S. oregana,* has pale to deep-pink flowers and grows from w WY and c UT w. Malvaceae.

GLOBEMALLOW

Sphaeralcea spp. (sfee-**ral**-sia: globe mallow). Flowers orange to brick-red, 1–1½" across, in short, spikelike clusters; petals heart-shaped; leaves alternate; plant hairy. Prairie, steppe, ponderosa forest. Malvaceae.
Scarlet Globemallow
S. coccinea (coc-**sin**-ia: scarlet). Calyx without bracts; leaves deeply narrow-lobed, or palmately compound; 4–14" tall. A–.
Munro's Globemallow
S. munroana (mun-ro-**an**-na: for Donald Munro, a friend of David Douglas). Calyx borne above 3 tiny linear bracts; leaves oval in outline, with teeth and shallow lobes; 8–30" tall. WA to nw CO.

Regular Flowers with 5 Petals, 2 Sepals

Five petals with two sepals is an unusual combination, represented here by just one family, Montiaceae.

Western springbeauty

Alpine springbeauty

SPRINGBEAUTY

Claytonia (clay-**toe**-nia: for John Clayton). Petals ¼–½", slightly notch-tipped; 5 stamens; 2 sepals; stem leaves 2, opposite, ½–3", lanceolate to elliptic. Montiaceae.

Western Springbeauty

C. lanceolata (lan-see-o-**lay**-ta: narrow-leaved). Petals white (rarely yellow—in e ID), usually with fine pink stripes; stems 3–6", ± succulent, hollow, weak, several-flowered, basal leaves few or none; from a bulbous root. Meadows and forest at all elevs, esp subalpine; A.

Alpine Springbeauty

C. megarhiza (meg-a-**rye**-za: big root). Petals pink to white; basal leaves numerous, fleshy, spoon-shaped; stem leaves small, occasionally lacking; stems numerous, ± recumbent, often reddish; from a long taproot. Alpine rocks; A.

Miner's Lettuce

C. perfoliata (per-fo-lee-**ay**-ta: perforated leaf). Petals white or pinkish; leaves basal, plus round stem leaves that the stem passes through. Moist forest; northerly plus UT.

WESTERN SPRINGBEAUTY BEGINS growing at its bulb tip in September, just when its neighbors are dying back. While snow insulates it at close to 32°F for the next eight months, the shoot inches up to the soil surface. Few plants—either Arctic or alpine—are active at such low temperatures. Without a snow blanket, growth would be impossible.

As soon as the snow melts away from the shoot in spring, springbeauty bursts to its full height of 3 or 4 inches, expending in a few days its large reserve of starches. It can even push through the last inch or two of snow by combusting some of the starch to melt itself a hole. It has two to four weeks to complete its life cycle—blooming, setting seed, and photosynthesizing like mad to store up starches for the next spring. Then it withers, existing only underground during late summer, the peak growing season for its neighbors.

Plants on this sort of schedule are called spring ephemerals. Many, such as blue-eyed grass, are common on semiarid land with just a few weeks of wet soil following snowmelt. Western springbeauty does well on such sites, but in moist subalpine meadows its timing has a different purpose—jumping the gun on bigger, leafier plants that monopolize the light later in the season.

The thin-fleshed, hollow stem acts as an internal greenhouse. When stored carbohydrates are burned off during the plant's growth spurt, some of the heat produced stays in the stem, making the internal air warm enough for photosynthesis even when the outside air is not. Waste CO_2 from respiration also stays inside, available for building new carbohydrates.

With their concentrated starches, western springbeauty bulbs are good edibles; Meriwether Lewis rated them top among Indian root vegetables. They taste radishy. But they aren't abundant, so leave them for gophers and bears, who also find a rich protein source in this plant's new spring leaves.

Miner's lettuce is so named because forty-niners learned to eat it (and similar

Toadlily

Tobacco-root

Sitka valerian

toadlily, below) to avert scurvy, the disease caused by vitamin C deficiency.

TOADLILY

Montia chamissoi spp. (for Giuseppe Monti; sha-**miss**-o-eye: for Adelbert von Chamisso, p. 200). Also water miner's-lettuce; *Crunocallis chamissoi*. Petals ⅛–½", white or pink with pink veins; anthers becoming pink; 2 unequal sepals; opposite leaves; stem 2–10", ± succulent, several-flowered; spreading by runners. Moist forest; A. Montiaceae.

Regular Flowers with a Total of 5 Petals or Sepals

If there's just one whorl, they are sepals, no matter how colorful or tender. Technically, then, no flowers have five petals and no sepals, but several appear to. Carrot family (Apiaceae) flowers have a vestigial whorl or fleshy ring—a calyx, with indetectable lobes—below their five petals. In valerian, sepals unfurl only as the flower goes to seed, so they aren't seen on the bloom.

VALERIAN

Valeriana (va-lee-ree-**ay**-na: strong—, ref to medicinal potency). Corolla white, 5-lobed; 3 stamens; calyx seemingly lacking; each seed grows a little parachute of plumes; foliage rankly aromatic, esp when drying out or frost-bitten in the fall; 1–4' tall. Caprifoliaceae.

Tobacco-root

V. edulis (**ed**-you-lis: edible). Flowers ⅛", flaring, in a tall, narrow, loose panicle; basal leaves 3–16" long, most of them simple and quite narrow, though a few may be divided into linear leaflets, as are the few pairs of small stem leaves. Open forest, meadows; A.

Western Valerian

V. occidentalis (oxy-den-**tay**-lis: western). Flowers ⅛", flaring, in 1 round-topped head ½–1¼" wide; leaves both basal and opposite, untoothed, either simple and pointed-oval or pinnately 3-to-11-compound. Diverse habitats up to low-subalpine meadows; A.

Sitka Valerian

V. sitchensis (sit-**ken**-sis: of Sitka, Alaska). Flowers ⅜", tubular, slightly asymmetrical, initially pink; terminal inflorescence rounded, 1–3" wide; lower ones smaller, in opposite pairs; leaves opposite, compound; leaflets usually 3 or 5, pointed-oval, vaguely toothed. Moist subalpine meadows; northerly.

VALERIAN ROOT HAS regained its ancient reputation as a tranquilizer and mood elevator. Sitka valerian fetches some of the highest prices paid to commercial foragers, leading to local overharvesting. "It is most effective," writes the herbalist Michael Moore, "when you have been nervous, stressed, or become an adrenaline basket case, with muscular twitches, shaky hands, palpitations, and indigestion." Thanks—I'll keep that in mind.

The Blackfoot and Okanagan spiked smoking mixtures with valerian leaves. Other peoples used tobacco-root roots as poultices on bruises and wounds, and also ate them. Poisonous raw, they became a favored vegetable after cooking for a day or two. Explorers and colonizers gave them mixed reviews.

Captain Frémont found them "rather agreeable," but his cartographer called them "the most horrid food he had ever put in his mouth." Tastes like chewing tobacco, he said, and smells like stinking feet.

As snowmelt releases subalpine meadows, deep-reddish shoots of Sitka valerian pop up abundantly. The redness disappears as the foliage matures, lingering in the budding flowers, only to disappear as they mature to white. Most redness in plants comes from anthocyanin, a complex-carbohydrate pigment that may be red, blue, or in between, depending on acidity (like litmus paper or hydrangeas).

American bistort

Alpine bistort

Anthocyanin may have several functions in high-elevation plants. First, it filters out ultraviolet radiation, which can be at least as hard on plant tissue as on human skin. Ultraviolet radiation is most intense at high altitudes, where there is less atmosphere to screen it, and most intense when sunlight peaks, at the summer solstice. In June, the high country is still snowbank-chilled, and plant tissues are young and tender, so that's where and when anthocyanin is brought out.

Second, while anthocyanin reflects ultraviolet radiation, it absorbs and concentrates infrared radiation, heating the plant. Third, anthocyanin is an interim form for carbohydrates on their way up from storage. To bloom and fruit early in their short growing season, high-country plants must store huge quantities of carbohydrates in their roots, and then move them up quickly after—or even before—snowmelt. (See springbeauty, p. 222.) In de-reddening, valerian stuffs itself with preserves from the root cellar.

BISTORT

Bistorta spp. (bis-**tor**-ta: twice twisted, referring to the rhizome). Also *Polygonum* spp. Flowers white, small, chaffy, fetid, in a dense head; 5 unequal calyx lobes; no petals; 8 unequal stamens; stem leaves few, ± linear, sheathing; basal leaves much larger (3–6", on 3–6" leafstalks), elliptical; stem unbranched. All elevs; but esp in alp/subalp meadows; A. Polygonaceae.

American Bistort
B. bistortoides (bis-tor-**toy**-deez: like bistort). Flower spike often ½" or wider, almost as wide as tall; 8–30" tall.

Alpine Bistort
B. vivipara (viv-**ip**-a-ra: giving birth to live young, ref to bulblets in place of flowers). Flower spike narrow, its width averaging about ⅓ of its height, or ⅓", with tiny pink bulbs in place of flowers in the lower portion.

THE FLOWERS STINK, luring flies as pollinators, but bears and elk eat the roots and shoots, respectively, and so did the Blackfoot and Cheyenne. The absurd scientific name—"bistort resembling a bistort"—came about by a mundane path. The scientist who named it saw it as a species of *Polygonum*, but resembling another genus, *Bistorta*, hence *Polygonum bistortoides*. A century later, further study found that in fact it belongs in *Bistorta*.

BISCUITROOT

Lomatium spp. (lo-**may**-shum: hem—, ref to seed margins). Also desert parsleys. Flowers tiny, in many ½" balls (like fireworks) on unequal rays from top of stalk; leaves lacy, much compounded. Apiaceae.

Cous

L. cous (cows: Nez Perce name). Also *L. montanum.* Flowers yellow; each ball-like flower cluster subtended by an involucre of often purplish, ± oval, ⅛–¼" bracts; leaves variable; plant 4–14" tall. Rocky slopes; OR to c WY.

Cous

Sandberg's Biscuitroot

L. sandbergii (sand-**berg**-ee-eye: for John H. Sandberg). Flowers yellow to orange; leaves ¾–3" long in total, with 3 branches each twice pinnate, with tiny ± linear leaflets; plant from almost ground-hugging to 12" tall. Alp/subalp gravels; northerly.

Sandberg's biscuitroot

Narrowleaf Desert-parsley

L. triternatum (try-ter-**nay**-tum: 3× divided in 3). Flowers yellow; leaves at least 3×3 compound, the leaflets threadlike, rather soft, ½–4" long; plant 8–16". Low to mid-elev dry meadows; A.

Chocolate-tips

L. dissectum (dis-**ec**-tum: finely cut). Flowers purple-brown or yellow; leaves at least twice pinnate, the leaflets very slender; stem hollow, often purple, 20–60". Piney or sage meadows, rarely up to subalpine; A.

Narrowleaf desert-parsley

IT'S EASY TO recognize the Carrot family: most members bear umbels, umbrella-shaped inflorescences in which many flower stalks branch from one point. The family includes carrots (called "Queen Anne's lace" when growing wild), parsnips, celery, fennel, and dill. Robust taproots in the family range in edibility from staple foods of many Indigenous groups to deadly poison hemlock. Identification of species within the family can be tricky, and a matter of life or death. The Okanagan reportedly ate young chocolate-tips roots, but also poisoned fish and lice with them. Scientific studies of their medicinal value are also ambiguous.

Chocolate-tips

Alpine spring-parsley

Snowline spring-parsley

SPRING-PARSLEY

Cymopterus spp. (sigh-**mop**-ter-us: wavy wings, ref to seed capsule). Flowers minute, in several ball-like clusters; leaves basal, twice compound, the leaflets minute, forming a mosslike cushion flanked by dried leafstalks of previous years. Rocky sites, foothills to alpine. Apiaceae.

Alpine Spring-parsley

C. alpinus. Also *Oreoxis alpina.* Flowers usually yellow; 1–5" tall. CO, NM, se UT, se WY.

Snowline Spring-parsley

C. nivalis (niv-**ay**-lis: of snow). Also *Aletes nivalis.* Flowers white to yellowish; 2–12" tall. OR, e ID, s MT, w WY.

TURPENTINE BISCUITROOT

Cymopterus terebinthinus (ter-ra-**binth**-in-us; turpentine tree). Also aromatic spring-parsley; *Pteryxia terebinthina.* Flowers minute, yellow, in widely spaced small clusters; leaves basal, 2–4× compound, the leaflets sparse, linear, aromatic;

Turpentine biscuitroot

5–16" tall. Sagebrush up to mid-elevs; WA to WY and nw CO.

The anise-scented leaves attract the anise swallowtail butterfly to lay her eggs.

AMERICAN THOROW-WAX

Bupleurum americanum (bew-**plu**-rum: Greek name, "ox rib"). Also *B. triradiatum.* Flowers yellow (or purplish), tiny, in several small balls on ray stalks from top of stem; each ball, and also the whole inflorescence, is subtended by a whorl of pointed-elliptical bracts; leaves simple, untoothed, narrow, up to 6" long × ⅜" wide, progressively smaller upward; 2–12" tall. Tundra, dry meadows, rock outcrops; OR to WY Bighorns, rare in CO and NM. Apiaceae.

LOVAGE

Ligusticum spp. (lig-**us**-tic-um: Roman name, ref to the region Liguria). Also licorice-root. Flowers white (rarely pinkish), tiny, in 1–3 flattish inflorescences 1–4" across; leaves mainly basal, compound, lacy; 2–4' tall, from a single thick taproot. Meadows, open forest; A–. Apiaceae.

American thorow-wax

Oshá

Yampah

Fernleaf Licorice-root
L. filicinum (fil-**iss**-in-um: ferny). Also *L. tenuifolium*. Leaflets very fine, linear, less than ⅛" wide. Foothills to subalpine; OR to CO.

Oshá
L. porteri (**port**-er-eye: for Thomas C. Porter). Also bear-root, love-root, Porter's lovage. Some leaflets more than ⅛" wide. All elevs, under aspen, rampant in many open subalpine meadows; CO, NM, UT, se WY.

OSHÁ ROOTS—BIG, RATHER hard, football-shaped tubers—contain Z-ligustilide, an effective anti-inflammatory according to test-tube and animal studies. Studies in humans are needed, but in the meantime brushpickers find a strong market for oshá roots. The most popular use today is said to be as a cold-and-flu remedy; ethnobotanical records find it used for almost anything: the Yuki used it to repel rattlesnakes. At least six western *Ligusticum* species, including fernleaf, are sometimes sold as oshá, and there is disagreement over whether *porteri* is the "true oshá." The word *oshá*, and the accent over the *a*, are apocryphal—said to be Indigenous, yet the *Oxford English Dictionary* says "no likely etymon has been identified."

YAMPAH

Perideridia gairdneri (per-id-er-**id**-ia: with a coat around it; for Meredith Gairdner). Also *P. montana*. Flowers white (rarely pinkish), tiny, in flattish inflorescences 1–3" across; seeds ribbed; leaves compound; stem leaves of 3–7 linear leaflets 1–6" long; basal leaves often withering by flowering time; aromatic plant 1–3' tall. Meadows, open forest; WA to nw CO. Apiaceae.

The mild, sweetish tubers of yampah were so popular that Blackfoot children would dig

Mountain sweetroot

them up and snack on them while playing; a shaman would carry them at all times and then fake digging them up, using a special bear claw, to proffer as a cure for virtually any ailment. The tubers were popular with Lewis and Clark's men, and with trappers and traders. They smell like parsnips but look quite different, tapering to both ends and often growing two or three per plant. The distinctive grasslike leaflets and caraway aroma make ID less risky than with other edible parsleys, but there is at least one toxic family member in our region with long, linear leaflets.

SWEETROOT

Osmorhiza spp. (os-mo-**rye**-za: aromatic root). Also sweet cicely. Tiny flowers in a loose, broad inflorescence; plant anise-fragrant, delicate, 6–48" tall. Apiaceae.

Mountain Sweetroot
O. berteroi (bare-**tare**-oh-eye: for Carlo Giuseppe Bertero). Also *O. chilensis*. Flowers white; seeds slender, with bristles that cling to

clothing; leaflets usually 9, oval, toothed, the larger ones often with a few lobes; stem solitary. Forest; A.

Dwarf Sweetroot
O. depauperata (de-pau-per-**ay**-ta: small). Like the preceding, 6–24" tall, and more abundant southerly. A.

Western Sweetroot
O. occidentalis (oxy-den-**tay**-lis: western). Flowers yellow; seeds slender, without bristles; leaflets 3, 9, or more, sawtoothed; stems often clustered. Open slopes, thickets; A–.

SWEETROOT SEEMS ABLE to grow in almost any sort of forest. The US Forest Service paper on Montana forest habitats at all elevations divides them into one hundred types. Mountain sweetroot was found in the sample plots of eighty-seven out of one hundred, but covered barely 1.5 percent of the ground.

Herbalists tout sweetroot as a fungicide, and the Salish used to smoke it. I would stay away from any internal use because of deadly relatives (especially hemlocks, below).

WATER HEMLOCK
Cicuta maculata (sick-**you**-ta: Roman name; mac-you-**lay**-ta: spotted). Flowers white (or greenish), tiny, in flattish inflorescences 2–5" across; leaves usually thrice compound; leaflets 5× longer than wide, 2–5" × ½–3", toothed, leaf

veins directed at the notches, not the tips; stem base forms a thick, hollow bulb with several horizontal partitions; stem 2–7'. Deadly poisonous. Streamsides, marshes, below 8,000'; A. Similar *C. douglasii* of the Pacific Coast reaches n ID and w MT. Apiaceae.

Known in England as cowbane. Per *Intermountain Flora*: "Early in the spring, when livestock that have been confined to a barn are first turned out to pasture, they may pull up and eat the whole base of the plant. One is enough." Serving as judge, jury, and executioner, this hemlock and the one (below) that killed Socrates have made careless plant-foraging a capital offense in the West. Even Indigenous children were occasionally poisoned when they used hollow hemlock stems, mistaken for cow-parsnip, as peashooters. While there are strong differences between the hemlocks and their sought-after relatives—cow-parsnip, biscuitroots, and yampah—the gravity of the risk demands absolute focus on plant identification (and stops me from recommending eating any wild plants in the Carrot family).

POISON HEMLOCK
Conium maculatum (**co**-ni-um: Greek name; mac-you-**lay**-tum: spotted). Flowers white, tiny, in flattish inflorescences 2–3½" across; larger leaves lacy, compound, commonly 6–12" long; leaflets typically pinnate-lobed; stem purple-splotched, at least near base, 2–8'. Deadly poisonous European weed. Roadsides, ditches; A. Apiaceae.

Poison hemlock inflicts more fatal poisonings than any other wild plant. To be safe, never eat any part of any wild "carrot-topped" plant with dissected or compounded leaves.

COW-PARSNIP
Heracleum maximum (hair-a-**clee**-um: Greek name, after Hercules). Also *H. sphondylium, H. lanatum.* Flowers white; 1 to several inflorescences nearly flat, 5–10" across, petals near edge of inflorescence much enlarged, 2-lobed; 5 stamens; 3 huge leaflets 6–16" wide, palmately lobed and

Cow-parsnip

Sharptooth angelica

Sharptooth Angelica
A. arguta (arg-**you**-ta: sharp-toothed). Leaflets less than 2× longer than wide, 2–5" × ½–3"; 2–7' tall. Northerly.

THE ARCHANGEL RAPHAEL allegedly revealed the true (European) angelica to humans as a remedy, hence its name, *Angelica archangelica*. It is cultivated today as a fragrance in cosmetics, as a seasoning, and to some extent for medicines. Our angelicas, whose flowers often swarm with beetles or flies, are not known to share the potency of their famous relatives, the true angelica and dong quai.

Regular Flowers with 4 Petals
Skunk-cabbage (p. 137) is a four-petal monocot. Leafy spurge (p. 173) has a tiny, four-lobed involucre but no petals or sepals. Stonecrops (p. 211) usually have flower parts in fives here, but occasionally in fours.

ALPINE WILLOW-HERB
Epilobium anagallidifolium (ep-il-**oh**-bium: upon pod; a-na-gal-id-if-**oh**-lium: with leaves like *Anagallis*). Also *E. alpinum*. Flowers deep to pale pink, nearly flat, ¼–½" across, usually 2–4 per stem, nodding when in bud; 4 petals, 2-lobed; 4 sepals and 4 stamens borne at the tip of a very slender maroon ovary; the ovary matures into a maroon seedpod that splits into 4 spirally curling thin strips, releasing tiny, downy seeds; leaves narrowly elliptical, ¼–1",

Alpine willow-herb

toothed; stems juicy, aromatic, hollow, 3–10'. Moist thickets; all elevs; A. Apiaceae.

Cow-parsnips, avidly eaten and widely eradicated by cows, are also browsed by wild herbivores including grizzlies. The fetid flowers draw flies. Some Native people ate the young stems, either raw or cooked but always peeled first to remove a weak toxin that causes skin irritation in reaction to sunlight; the stems taste milder and sweeter than the rank odor leads you to expect. Poultices and infusions of the leaves and the purgative roots held high medicinal repute.

ANGELICA
Angelica spp. (an-**jel**-ic-a: angelic). Flowers white (or pinkish), tiny, in flattish inflorescences 3–6" across; leaves usually twice compound; leaflets toothed; stem hollow. Moist to wet ground. Apiaceae.
Littleleaf Angelica
A. pinnata (pin-**ay**-ta: pinnately compound). Leaflets more than 2× longer than wide, ½–1½" long; 1–3' tall. Moist to wet ground; c ID to NM.

Fireweed

River beauty

Tufted evening-primrose

all on stem; stems red, 2–10", in sprawling patches from runners. Moist alp/subalp sites; A. Onagraceae.

FIREWEED
Chamaenerion spp. (ca-mi-**nee**-ree-on: dwarf oleander). Also *Chamerion* spp., *Epilobium* spp. Flower/seed structures as in alpine willow-herb, above. Onagraceae.
Fireweed
C. angustifolium (ang-gus-tif-**oh**-lium: narrow leaf). Also *C. danielsii.* Flowers pink to purple, nearly flat, 1–1½" across, blooming progressively upward in a tall conical raceme; leaves 3–6" × ¾"; plant 3–8' tall. Rampant in recent burns, avalanche tracks, clearcuts, roadsides, etc.; A.
River Beauty
C. latifolium (lat-if-**oh**-lium: wide leaf). Also *C. subdentatum.* Also dwarf fireweed, red willow-herb. Flowers deep pink to purple, nearly flat, 1¼–2" across; leaves 1–2" × ½–¾"; 3–24" tall. River bars to alpine talus; A–.

WITH ITS COPIOUS, plumed seeds that fly in the wind, fireweed is a major invader of burns, from Rocky Mountain clearcuts to bombed urban rubble. I still wouldn't call it a weed; it is native, beautiful, nutritious, and a poor long-term competitor, yielding its place to later-successional plants after a few years. Unlike many pioneer herbs, it is a perennial, and seems to require mycorrhizal partners. Lacking these, fireweed seedlings on deep Mount St. Helens ash in 1981 weakened and died without flowering.

In the far north, fireweed perseveres, beautifying vast areas of Alaska too severe for trees. Northerly peoples ate its inner stems as a staple food in spring. (Caution: may prove laxative.)

EVENING-PRIMROSE
Oenothera spp. (ee-**noth**-er-a: wine-catching). Flowers fragrant, blooming around sunset, then withering within 24 hours; petals heart-shaped; calyx lobes narrow, sharply bent back; seedpod woody, 1–2", splitting 4 ways. Onagraceae.
Tufted Evening-primrose
O. caespitosa (see-spit-**oh**-sa: in clumps). Also sand lily, moon-rose, rock-rose, etc. Petals 1–1½", white, turning rose-purple; flowers up to 5" high, reaching that height entirely on the basal tube of the calyx from an ovary below ground; leaves basal, oblanceolate, 4–8", raggedly toothed. Gravel roads and dry clay soils, mainly low to mid-elevs; A.

Tall evening-primrose

Ragged robin

Cusick's speedwell

Thymeleaf speedwell

Tall Evening-primrose

O. elata (ee-**lay**-ta: tall). Also Hooker's evening-primrose. Petals 1¼–2", lemon yellow, turning orange; leaves basal and on stem, oblanceolate, 4–21", dull green; plants 1–7' tall. Meadows and roadsides at lower elevs; CO, NM, UT, ID, OR, WA.

SPHINX MOTHS COME to pollinate these sensuous ladies of the night.

RAGGED ROBIN

Clarkia pulchella (for William Clark, p. 241; pool-**kel**-a: most beautiful). Also pink fairies, elkhorn. Dramatic, unique flowers: 4 fuchsia petals each have 3 long lobes from a slender base, their overall arrangement usually asymmetrical; calyx bent off to 1 side, 4-lobed; 4 fertile stamens plus 4 stunted ones; pistil resembles a small white 4-petaled flower near the center of the 4 petals; leaves alternate, linear, ¾–3"; annual plant 4–20" tall. Sagebrush and ponderosa; WA to w MT and disjunct in Teton County, WY. Onagraceae.

SPEEDWELL

Veronica spp. (ver-**on**-ic-a: for Saint Veronica). Corolla blue-violet with yellow center, unequally 4-lobed, ⅜" across; in a small raceme; leaves opposite; stem 2–12". Alp/subalp. Plantaginaceae.

American Alpine Speedwell

V. wormskjoldii (vormsk-**yol**-dee-eye: for Morten Wormskjold). Also *V. nutans*. Style and stamens shorter than petals; leaves lanceolate, ½–1½". A.

Cusick's Speedwell

V. cusickii (cue-**zick**-ee-eye: for William Cusick). Style and stamens longer than petals; leaves as above. w MT, ID, OR.

Thymeleaf Speedwell

V. serpyllifolia (sir-pil-if-**oh**-lia: thyme leaf). Also *Veronicastrum serpyllifolium*. Stamens about as long as petals; leaves elliptical to oval, ⅜–1"; stem may be prostrate. Wet places below tree line; A.

LINNAEUS APPROPRIATED THE vernacular name "veronica," long in use in English and several other languages: people thought markings on some veronicas resembled the imprint on Saint Veronica's handkerchief left when Jesus, carrying his cross, used it to wipe his brow.

SYNTHYRIS KITTENTAILS

Veronica spp. Corolla unequally 4-lobed, ⅜", in small racemes; 2 stamens protrude; leaves basal, 1–2". Plantaginaceae. Compare Besseya kittentails (p. 183).

Featherleaf kittentails

Mountain kittentails   *Pussypaws*   *Hairy clematis*

Featherleaf Kittentails

V. dissecta. Also *Synthyris pinnatifida.* Flowers deep blue-purple, stems to 8"; leaves deeply and finely dissected. Alpine; WY, sw MT, UT, c ID.

Mountain Kittentails

V. missurica (miz-**oo**-ric-a: of the Missouri River). Also *Synthyris missurica.* Flowers blue-lavender, stems to 12"; leaves round to kidney-shaped, with coarse, blunt teeth. WA, OR, c ID.

PUSSYPAWS

Calyptridium umbellatum (cal-ip-**trid**-ium: with a cap over the seed—not true of this species; um-bel-**ay**-tum: flowers in umbels). Also *Cistanthe umbellata, Spraguea umbellata.* Flowers rust-pink to white, in fluffy, chaffy heads on prostrate stalks reaching well past the basal rosette of tiny, narrow leaves; sepals 2, round, ⅛–⅜", sandwiching and nearly hiding the 4 much smaller petals and 3 stamens. Alp/subalp rock crevices, scree, disturbed sandy soil; WA to Wind River Range and Uintas. Montiaceae.

HAIRY CLEMATIS

Clematis hirsutissima (**clem**-a-tis: a vine cutting; her-sue-**tee**-sim-a: hairiest). Also sugarbowls, leatherflower, vaseflower; *Coriflora hirsutissima.* Unique, nodding, vase-shaped flowers, 1 per stem, consist of 4 heavy, recurved, petal-like sepals ranging from rich indigo to drab gray-brown; sepals, stems, and leaves all gray-haired; leaves much-dissected, fernlike; 8–20" tall, in clumps. Sunny lower elevs; A. (See p. 94 for woody-stemmed *Clematis* spp.) Ranunculaceae.

WESTERN MEADOW-RUE

Thalictrum occidentale (tha-**lic**-trum: Greek name; oxy-den-**tay**-lee: western). Sepals 4 (or 5), greenish, ¼"; petals lacking; flowers distinctive, many, in sparse racemes, ♂ and ♀ on separate plants, the ♂ being tassels of stamens, long yellow anthers dangling on purple filaments; ♀ less droopy, with several reddish pistils that mature into a starlike rosette of capsules; leaves twice or thrice compound, leaflets ¾–1½", round-lobed (similar to columbine); plants 20–40" tall. Open forests to subalpine meadows; A. Ranunculaceae.

Western meadow-rue female

Western meadow-rue male

Smelowskia   Heartleaf bittercress   Fendler's pennycress

The Blackfoot used meadow-rue seeds as both a spice and a perfume. Meadow-rues have found wide medicinal use, while some of their alkaloids have shown promise in labs.

I often see male meadowrues associating exclusively with other males, and females with females—a seeming inefficiency that I cannot explain.

SMELOWSKIA

Smelowskia americana (smel-**ow**-skia: for Timotheus Smelovskii). Also *S. calycina*. Flowers cream-white or purple-tinged, ⅜–¾" across, several, in roundish clusters; sepals fall off when flower opens; leaves crowded, basal, 1–4", pinnately compound or lobed, gray-fuzzy; low mat or cushion plants. Alp/subalp; A–. Brassicaceae.

HEARTLEAF BITTERCRESS

Cardamine cordifolia (car-**dam**-in-ee: Greek name; cord-if-**oh**-lia: heart-shaped leaf). Flowers white, ½–1" across, in broad clusters; leaves alternate, 2+" long, elongated heart shape with several blunt teeth; thin, 1–2" seedpods explode open to propel seeds. Wet soils of forest or meadow, up to subalpine; all exc MT. Brassicaceae.

FENDLER'S PENNYCRESS

Noccaea fendleri (no-**kee**-a: for Domenico Nocca; **fend**-ler-eye: for Augustus Fendler, p. 246). Also candytuft; *Thlaspi montanum, T. fendleri.* Petals white or rarely pink, ¼", spoon-shaped; 4 wee sepals; flower clusters at first broad-topped, becoming tall and open as first flowers go to seed; basal leaves spoon-shaped, ¾–2"; stem leaves much smaller, linear, clasping stem; hairless plants 1–12" tall, in small clumps. In sun, foothills to alpine; A. Brassicaceae.

LYALL'S ROCKCRESS

Boechera lyallii (**beck**-er-a: for Tyge Böcher; lye-**ah**-lee-eye: for David Lyall, p. 62). Also *Arabis lyallii.* Petals 4, purple, rounded, ¼–⅜"; 4 sepals, often purplish; 6 stamens; seeds in slender pods up to 2" long; leaves both alternate and basal, firm, the basal ones ⅝–1¼", pointed-oblance-olate; stem leaves smaller, with 2 earlike lobes at their bases; 4–10" tall. Alp/subalp dry meadows; WA to WY and UT. Brassicaceae.

Similar rockcress species occur throughout the West and are very hard to ID. At lower elevations, they can get quite tall and slender.

DRABA

Draba spp. (**dray**-ba: Greek name). Also Whitlow-grass. Flowers yellow, ¼", in small clusters; 6 stamens; seedpods elongated; leaves

Lyall's rockcress

Golden draba

Common twinpod

Sharpleaf twinpod

Western wallflower

and less matted. Telling them apart requires a lens strong enough to see the branching patterns of the hairs. Even scientific papers often punt on the ID.

TWINPOD

Physaria spp. (fiz-**air**-ia: bellows—). Also bladderpod. Flowers yellow, ± flaring, in small clusters, on short ± prostrate stems; seedpods hold 4 seeds; leaves whitish-felty, thick, basal ones in a spiraling rosette. Dry gravels. Brassicaceae.

Common Twinpod
P. didymocarpa (did-im-o-**car**-pa: twin fruit). Petals ½"; seedpods inflated into 2 lumpy lobes; spade-shaped leaves have 2–4 pronounced teeth. All elevs; WA to WY.

Sharpleaf Twinpod
P. acutifolia (a-cue-tif-**oh**-lia: sharp leaf, a misleading name). Petals ⅜"; seedpods inflated into 2 grape-shaped lobes; spoon-shaped leaves have 2–4 vague teeth, or none. All elevs, blooming very early at low elevs; A–.

Point-tip Twinpod
P. floribunda (flor-i-**bun**-da: many flowers). Many leaves have 2–6 longish lobes; seedpods modestly inflated. Sagebrush, up to mid elevs; w CO, UT, NM.

and stems hairy; few or no stem leaves; basal leaves crowded, paddle-shaped, mostly ¼–½"; above the mat of leaves rise a few 2–5" stems. Brassicaceae.

Few-seeded Draba
D. oligosperma (a-lig-a-**sperm**-a: few seeds). Upper stem and flower stalks hairless. Sagebrush to alpine rocks; A.

Golden Draba
D. aurea (**aw**-ria: golden). Upper stem and flower stalks hairy. Alpine; A–.

EASILY HALF A dozen species of draba fit the above description, and others yet are similar but white-flowered, or similar but taller

WALLFLOWER

Erysimum spp. (er-**iss**-im-um: Greek name). Petals round; flowers many, in a round-topped cluster; seedpods very narrow, splitting in 2; leaves 2–5", narrow, most in a basal rosette, a few on the stem, sometimes shallowly toothed;

Shy wallflower   Bunchberry   Northern bedstraw

plant grayish-hairy. Grasslands, meadows, all elevs; A. Brassicaceae.

Western Wallflower

E. capitatum (cap-it-**ay**-tum: flowers in heads). Also sand-dune wallflower. Petals ½–1", brilliant yellow, less often orange, lavender, or pink; 12–40" tall.

Shy Wallflower

E. inconspicuum. Petals ¼–⅜", yellow; 8–24" tall.

BUNCHBERRY

Cornus canadensis (**cor**-nus: horn). Also ground-dogwood; *Chamaepericlymenum canadense.* True flowers tiny, in a dense head ½–¾" across surrounded by 4 white bracts, ½–1" long, often mistaken for petals; berries red-orange, several, 1-seeded, ¼" across; leaves pointed, oval, 1–3", in a whorl of 6 beneath each inflorescence, and in whorls of 4 on flowerless stems (technically 4-leafleted basal leaves) nearby; stems 2–8", from rhizomes. Moist forest; A–. Cornaceae.

When you see the big, white floral bracts, you can't miss the oxymoronic resemblance between the dogwood tree and this 6-inch subshrub. The true flowers, on the other hand, are tiny but mighty: they explode pollen into the air on spring-loaded stamens. Researchers proclaim this snap action the fastest plant motion yet measured, and call the mechanism a trebuchet, referring to the compound-action medieval catapult with a flexible strap to fling the payload. Advantages include selecting for farther-traveling pollinators (those heavy enough to trigger the explosion) and blasting pollen onto body parts where they won't be able to eat it.

NORTHERN BEDSTRAW

Galium boreale (**gay**-li-um: milk—, ref to old use for clabbering cheese; bor-ee-**ay**-lee: northern). Also *G. septentrionale.* Corolla white, 4-lobed, flat, ¼" across; true flowers tiny, in a dense head ½–¾" across; 4 stamens out to sides, alternating with lobes; no sepals; flowers many, in a panicle; leaves in whorls of 4, lanceolate, 3-veined, 1–2" long; lower axils often bear branchlets with smaller leaves; stems square, 8–32" tall, from creeping rhizomes. Forest, thickets, meadows, to just above tree line; A. Rubiaceae.

FRAGRANT BEDSTRAW

Galium triflorum (try-**flor**-um: 3-flowered). Corolla white, 4-lobed, flat, up to ¼" across; sepals lacking; flowers usually in sparse 3s branching from leaf axils; leaves in whorls of 5 or 6, narrow-elliptical; stems minutely barbed, 4-angled, generally sprawling or clambering, dense. Widespread in forest and thickets; A. Rubiaceae.

The clinging tangles of weak bedstraw stems are irresistibly easy to uproot by the fistful; they are the plant's way of attaching its seeds to passing animals. Where other

Fragrant bedstraw

Rocky Mountain fringed gentian

Monument plant

plants have barbed fruits for this purpose, bedstraw barbs its entire stem, and then breaks off at the roots. It used to get put in straw mattresses to perfume them. In and near towns, it may be outnumbered by the Eurasian stickyweed, *G. aparine;* the latter differs in being an annual plant, but closely resembles our native bedstraw.

ROCKY MOUNTAIN FRINGED GENTIAN

Gentianopsis thermalis (jen-shun-**op**-sis: looks like Gentian; ther-**may**-lis: of hot springs). Also *G. detonsa, Gentiana elegans.* Corolla deep blue, its 4 lobes flaring, twisting, edges fringed, no pleats between lobes; calyx 4-lobed; leaves opposite, entire; annual plant 2–16" tall. Wet places; A–. Gentianaceae. The floral emblem of Yellowstone National Park.

MONUMENT PLANT

Frasera speciosa (**fray**-zer-a: for John Fraser; spe-see-**oh**-sa: showy). Also green gentian, elk-weed; *Swertia radiata.* Petals pale green, often purple-flecked, ½–1", oval, spreading almost flat; sepals linear, as long as petals; 4 stamens; flowers whorled in upper leaf axils, forming a massive column; stem leaves lanceolate, in whorls of 3–5; basal leaves oblong, 10–20"; flower stalks 2–5'; last year's thick, broken, dry stalks persist through summer. Sagebrush or dry meadows; A. Gentianaceae.

The monument plant starts as a rosette of few leaves and builds its resources year by year—sometimes over sixty years—until it is strong enough to bloom. Having bloomed once, it dies. Toppling over, it may cover its

own seeds, providing shelter that keeps the seedlings moist the next year; in years to come, its own progeny may reuse its nutrients. It shares this characteristic—post-mortem nurturing by decomposing parents—with salmon and praying mantids.

Regular Dicot Flowers with 3, 6, or Several Petals or Sepals

"Several" petals means a variable number between five and thirteen. Most apparently several-petaled flowers are in fact composites (p. 156) whose seeming petals are actually ray flowers. You'll find most three- or six-petaled flowers among the monocots (p. 138).

GENTIAN

Gentiana spp. (jen-tee-**ay**-na: Greek name, for King Gentius). Corolla with 4–7 (most often 5) shallow lobes, and fine teeth on the "pleats" between lobes; calyx lobed or not; leaves opposite. Gentianaceae.

Mountain bog gentian

Pleated gentian

Moss gentian

Arctic gentian

Mountain Bog Gentian

G. calycosa (cay-lic-**oh**-sa: cuplike). Also explorer gentian; *Pneumonanthe calycosa*. Flowers deep indigo-blue, single, 1½" tall; leaves blunt oval, ½–1½"; stems 3–12". Alp/subalp wet meadows; WA to UT and w WY.

Pleated Gentian

G. affinis (a-**fie**-nis: related). Also *Pneumonanthe affinis*. Flowers blue, 2–3 per stem, 1" tall; leaves lanceolate, pointed; stems 3–12". Diverse sites at all elevs; A.

Moss Gentian

G. prostrata. Also *Chondrophylla prostrata*. Flowers blue, wholly or partly pale, single, ½" tall, the lobes (most often 4) spreading wide in sun, but quickly closing when shaded; leaves ± oval, clasping, ¼"; stems 1–3", often partly prostrate; biennial plant. A–.

Arctic Gentian

G. algida (**al**-jid-a: cold). Also *Gentianodes algida*. Flowers single, oversized (1½–2"), ± white with purple streaks; leaves linear, may appear basal on low plants; 2–8" tall. MT to NM, but not ID.

OF ROUGHLY 360 species of gentian worldwide, many are collected for medicines and for flavorings in bitters and liqueurs, as well as for a soft drink, Moxie. I've seen few records of Indigenous medicinal use of these gentian species. (The Blackfoot, however, did use them "for attractiveness.") The flowers tend to close up when chilled or shaded by clouds, effectively keeping summer storms from splashing their pollen away.

BUCKWHEAT

Eriogonum spp. (airy-**og**-a-num: woolly joints). Calyx 6-lobed; petals lacking; 9 stamens; flowers in round clusters, small, on short stalks coming out of a conical involucre with or without lobes; leaves basal. Rocky, dry places, alpine to steppes. Each species may vary in hairiness or flower color, so ID is tentative. Polygonaceae.

Sulfur-flower

E. umbellatum (um-bel-**ay**-tum: flowers in umbels). Also umbrella plant. Flowers cream or reddish to bright yellow; outside of calyx hairless; stalklets of the several flower clusters meet at a whorl of leaves whose stem may meet others at a lower whorl; leaves usually almost smooth on top, fuzzy-white beneath; leaf blades 3× longer than wide; stems 4–12". In late summer, the leaf mats turn into lovely mosaics of red, green, white, and cream. Common at all elevs; A.

Sulfur-flower

Yellow buckwheat

Cushion buckwheat

Dense-flowered dock

Matted buckwheat

Matted Buckwheat

E. caespitosum (see-spit-**oh**-sum: in clumps). Flowers yellow, becoming rose-edged; leaves densely matted, grayish-green, fuzzy; just 1 flower cluster—cupped in 1 green involucre, with long, bent-back lobes—per 1–3" stem. All elevs; WA to WY and UT.

GENUS *ERIOGONUM* TURNS up frequently on lists of hosts favored by butterflies. It's one of North America's two or three largest genera—around 250 species, many of them in the Rockies. One of ours, sulfur-flower, has forty-one varieties.

DENSE-FLOWERED DOCK

Rumex densiflorus (**roo**-mex: Roman name, derived from sucking on it when thirsty). Also wild rhubarb. Flowers usually greenish, tiny, in 8–16" panicles, becoming all dark red in fruit; 6 sepals, but only 3 conspicuous ones; no petals; leaves heavy, narrowly arrowhead-shaped, the lower ones to 18" long; stout stems 10–40" tall. Moist sites, often forming dense patches, mid-elev to alpine; southerly. Polygonaceae.

The very young leaves are edible—and sour!—like French sorrel, *R. acetosa*. That non-native species may also turn up in the wild, even in the alpine zone. Our many *Rumex* species are hard to tell apart. Don't eat too much—the oxalic acid can upset your stomach.

Yellow Buckwheat

E. flavum (**flay**-vum: yellow). Flowers lemon yellow to cream, may shift to red-orange in heat; calyx hairy; sepals taper gradually to base; leaf blades 4–6× longer than wide; leaves usually hairy on both sides, forming wide mats; stems 2–10". All elevs; A–.

Cushion Buckwheat

E. ovalifolium (oh-val-if-**oh**-lium: oval leaf). Flowers cream, becoming rose-edged; leaves densely matted, silvery-white beneath or on both sides, ± succulent esp on alpine plants; 1 round flower cluster per 1½–7" stem, but it comes out of several tiny green involucres close together. All elevs but commonly alpine; A.

Yellow pond-lily

Marshmarigold

6–11, white, petal-like, ½–¾", often a few of them 2-lobed; petals lacking; stamens and pistils many; flowers usually 2, on a forked 3–10" stem; leaves basal, kidney-shaped, 2–4" wide, on 2–3" leafstalks, ± fleshy, edges ± scalloped, often curling. Alp/subalp streamsides, boggy meadows; A–. Ranunculaceae. Similar *C. leptosepala* grows in WA, c ID, nw MT; *C. chionophila* was included in *C. leptosepala* until recently.

ANEMONE

Anemone (a-**nem**-a-nee: Greek name, after Na'man, a name of Adonis). Also windflower. Sepals 5–9, petal-like; no petals; many stamens and pistils; usually 1 flower per stem, or more in *A. narcissiflora* and *A. multifida;* leaves palmately twice or thrice compound, mainly basal, long-stalked; 1 whorl of smaller leaves at midstem. Ranunculaceae.

Prairie Crocus

A. patens (**pay**-tenz: spreading). Also prairie pasqueflower; *Pulsatilla nuttalliana, P. patens, P. ludoviciana.* Sepals blue or purple outside, white inside, ¾–1½", flowering early, before leaves unfurl; stem later grows to 6–16"; styles grow to ¾–1½", feathery, in a round head, as seeds mature; plant covered with long white hairs. All elevs; A–.

Towhead Baby

A. occidentalis (oxy-den-**tay**-lis: western). Also white pasqueflower, western anemone; *P. occidentalis.* Sepals white, ½–1¼", flowering early,

YELLOW POND-LILY

Nuphar polysepala (**new**-fer: from the Arabic name; poly-**see**-pa-la: many sepals). Also wokas; *N. lutea.* 4 to 8 bright-yellow, heavy, roundish, petal-like sepals 1½–3" long; 4 smaller, green outer sepals; true petals and stamens numerous, much alike, crowded together around the large, parasol-shaped pistil; leaves heavy, waxy, elongated, heart-shaped, 6–18" long, usually floating. Widespread in ponds and slow streams up to about 6' deep; A. Nymphaeaceae.

Some Native peoples ate either the spongy rootstocks or the big, hard seeds—parched, winnowed, ground up, and boiled into mush—and some still do today. Oddly, Nancy Turner's Okanagan informants said the roots were considered poisonous, while the stems could be used to alleviate toothache. Many groups mashed the roots to make a pain-relieving poultice.

Recent studies suggest pond-lilies are among the closest living relatives of Earth's very first flowering plants.

MARSHMARIGOLD

Caltha chionophila (**cal**-tha: goblet; key-un-**ah**-fil-a: snow loving). Sepals

Prairie crocus

Towhead baby

Narcissus anemone

Fellfield anemone

Globeflower

before leaves unfurl; stem later grows to 1–2'; styles grow to 1–1¾" long, in a droopy head, as seeds mature; plant covered with long white hairs; basal leaves fernlike. Subalpine; northerly.

Narcissus Anemone

A. narcissiflora var. *zephyra*. Also *Anemonastrum zephyrum*. Sepals cream to white, sometimes blue on back, ¾" long, spreading flat; 40+ stamens form a yellow ring; leaflets in 3s, finely cut, stalkless. Moist, high meadows; CO and Bighorns.

Cutleaf Anemone

A. multifida (mul-**tiff**-id-a: many cuts). Includes var. *tetonensis*. Sepals 5–6, ⅜", color highly variable: from magenta inside and out to blue, cream, or pink, but most typically cream with some tint on outside; blooms after leaves unfurl; styles less than ⅛"; seed heads woolly, round, ⅜" across; stem ± silky-haired, 5–20", often 2-to-4-branched above leaf whorl. All elevs; A.

Fellfield Anemone

A. drummondii (drum-**on**-dee-eye: for Thomas Drummond, p. 136). Includes var. *lithophila*. Sepals 6–9, ⅜", ± white, often blue outside, flowering after leaves unfurl; styles ⅛"; seed

Cutleaf anemone

heads woolly, round, ⅜"; stem ± silky-haired, 4–10", not branched above base. Alp/subalp; WA to Wind River Range.

INDIGENOUS PEOPLE WERE aware of toxins in these bitter plants, using them to kill fleas. Some, however, inhaled their aroma or their smoke as a remedy for headaches and colds. Crushed leaves were used to stop nosebleeds or as a poultice for skin problems—which seems perverse, since they can blister skin.

The most strangely lovely of subalpine "flowers" is actually a seed head—a towhead baby. It looks like something Dr. Seuss dreamed up. The flower attracts less attention, blooming early when the plant is only a few inches tall.

GLOBEFLOWER

Trollius albiflorus (**tro**-lius: trollflower, German name; al-bif-**lor**-us: white flower). Also *T. laxus*. Sepals 5–12, petal-like, white or greenish, often aging yellow; petals lacking; many stamens and pistils; 1 flower per stem; dry seedpods form a tight, goblet-shaped head; leaves of 5 or 7 long, palmate lobes, irregularly toothed, mainly basal and long-stalked, plus 1 at midstem; stem hairless, 8–16". Wet soil near melting snow, mid-elevs and higher; A–. Ranunculaceae.

BANEBERRY

Actaea rubra (ac-**tee**-a: elder, for the similar leaves; **roob**-ra: red). Also *A. arguta*. Numerous ¼" white stamens are the showy part of the flower; the 5–10 white petals (occasionally lacking) are smaller; 3–5 petal-like sepals fall off as flower opens; flowers (and berries) in a ± conical raceme; berries ⅜" across, either white or (more often) glossy bright red; leaves 9-to-27-compound, leaflets pointed-oval, toothed, lobed, 1–3"; stem 16–40". Moist sites up to subalpine; A. Ranunculaceae.

These, our most poisonous native berries, are less than deadly. Sure, a handful could render you violently ill if swallowed—but that first taste would start you immediately spitting.

WILD GINGER

Asarum caudatum (a-**sair**-rum: Greek name; caw-**day**-tum: tailed). Calyx brownish purple, with 3 long-tailed lobes 1–3"; no petals; 12 stamens ± fused to the pistil; flower single on a prostrate, short stalk between paired leafstalks; leaves heart-shaped, 2–5", finely hairy,

Baneberry

spicy-aromatic, rather firm and ± evergreen, on hairy 2–8" leafstalks. Moist forest; WA, n ID, nw MT. Aristolochiaceae.

Meriwether and William, Naturalists

Lewis and Clark's voyage of 1804–1806 is so well known that it needs no lengthy description here. Less well known is that biological discovery, not exploratory heroics, was its great distinction. The explorer Alexander Mackenzie preceded them to the Pacific, following a Native grease-trading trail to Bella Coola in 1793, but he was interested in little but fur profits. In contrast, Meriwether Lewis and William Clark were briefed intensively on natural history and taxidermy before they set out, and they did a great job of recording natural history during their expedition. Lewis rivals Douglas and Nuttall in the number of first collections of Western plants credited to him, and surpassed both in writing down all he could learn from Indigenous people he met, including results from testing their herbal prescriptions on his men.

The medicine Lewis personally needed was an antidepressant. Near the Continental Divide, the apex of his superlative adventure, he turned thirty-two and pondered:

> I have in all human probability now existed about half the period which I am to remain in this Sublunary world. I have as yet done but little, very little, indeed, to further the hapiness of the human race or to advance the information of the succeeding generation. I view with regret the many hours I have spent in indolence.

Not true, Meriwether!

His mysterious death just four years later may have been murder, but his friend Thomas Jefferson believed it was suicide.

Wild ginger

Pygmy bitterroot

This odd plant we call "wild ginger" is unrelated to ginger, and even the tangy fragrance isn't really close, yet its stems have found favor as a seasoning with cooks of trapping, pioneering, and wild-food-stalking eras alike. The earthbound, camouflaged flowers are less fragrant. They attract creeping and crawling pollinators.

PYGMY BITTERROOT

Lewisia pygmaea (lew-**iss**-ia: for Meriwether Lewis, p. 241). Also *Oreobroma pygmaeum*. Petals 6–8, pink to white (sometimes greenish), ¼–¾"; 2 sepals, several stamens; leaves basal, linear, to 3" long, soon withering; stems mostly 1-flowered, 1–3", with a midstem pair of tiny bracts. Rock crevices, alp/subalp gravels; A. Montiaceae. Similar Nevada bitterroot, *L. nevadensis,* usually has white petals.

At least 7 percent of plant species, including cacti, *Lewisia* species, and many other succulents, share a nifty adaptation to drought. Plants have to open up their pores to take in CO_2 for photosynthesis, and open pores inevitably lose moisture. To conserve water during hot, dry days, plants close their pores (to varying degrees)—and give up on the opportunity to photosynthesize. Doing that too much can lead to death by starvation, since photosynthesis is how they make their food (how they make everybody's food, actually). What if plants could avoid the heat of the day and just open their pores at night? Since photosynthesis requires light and CO_2 at the same time, they can do so only if they take in CO_2 at night and store it until light returns. Bitterroots and their allies do exactly

that, storing CO_2 as malic acid and then photosynthesizing early the next morning.

BITTERROOT

Lewisia rediviva (ree-div-**eye**-va: revived). Flowers pink, apricot, or white, 2–2⅜" across, borne very low to the ground, 1 per stem in small clumps; 10–18 petals and 5–9 sepals all showy; leaves basal, linear, initially fleshy but usually withered by time of flowering. Arid gravels, mainly at lower elevs; A–.

Each spring when bitterroot was first dug, Salish and Kootenai bands held their great feast event, the First Roots ceremony. Prizing bitterroot above all plant foods, the Salish would happily trade a horse for a basketful of roots. The root is so small and the plant so sparse that it took a few days to fill a bag. They tempered the bitterness by digging bitterroot in spring when the leaves first appear, by digging only in locales known for relatively mild roots, by discarding the roots'

Bitterroot

cores, or by eating them after the roots had been a year or two in storage.

Meriwether Lewis tried some and wrote that they "had a very bitter taste, which was naucious to my pallate, and I transfered them to the Indians who had ate them heartily." He much preferred the root of springbeauty. Despite this bad first impression, he made a point, on his return to the river known for bitterroots, of collecting specimens for science. That sweet Montana river became the Bitterroot, along with the plant, which became the Montana state flower.

Several years later in Philadelphia, one of Lewis's dried roots got planted and watered. It sprouted green leaves and lived for some months. Privileged to name the plant for science, Friedrich Pursh (p. 102) determined that it was not just a new species but a new genus, which he named for Lewis, whose life by then had come to an early and bitter end. Pursh did not state whether by combining *rediviva* ("revived") with *Lewisia* he was hinting wishfully at bringing Lewis himself back to life. In the conventional view, *rediviva* refers simply to that undead dried specimen, reminding us that many plants received their names from botanists who never saw them bloom gloriously in their true place on Earth, but knew only a pressed, desiccated relic in a dim herbarium far away.

6

Ferns, Clubmosses, and Horsetails

Land plants that spread by means of spores, not seeds, are traditionally divided into those with and those without vessels, the tubes that conduct water and dissolved materials. Vessels are small, but their effects are large. Spore plants without vessels—mosses and liverworts—are only a few inches tall because without vessels, they can't raise vital fluids any higher. With vessels, on the other hand, spore plants grew to become tree ferns and tree clubmosses, which dominated the globe during the Mesozoic. Their modern descendants, typically between 6 and 60 inches tall, are still larger and palpably much more robust than mosses.

Spore plants are sometimes called cryptogams (meaning "hidden mating"). Yes, they have a sexual process—it's just less conspicuous than in seed plants, with their showy flowering and fruiting. More significantly, sex doesn't produce seeds capable of extended dormancy or travel. The traveling function is left up to an asexual stage in the cryptogam life cycle—a one-celled spore, more of a prototype of a pollen grain than of a seed.

Ferns

In ferns, the dustlike spores are borne in and released from sori—tiny clusters appear-ing as dark spots, lines, or crescents in patterns on the leaf underside. In some ferns, each sorus is shielded by a tiny membrane; in others, by a length of rolled-under leaf margin. Each fern frond and its stalk from the rhizome is one leaf; it is pinnately compounded or divided into pinnae. A pinna—a unit branching directly from the central leaf stalk—may be subdivided (compounded) an additional one to three times.

Evergreen Ferns

Few fern pinnae last for several years, as most conifer needles do, yet the following ferns at least retain green pinnae through their first winter. These pinnae may be noticeably more robust than those of deciduous ferns.

WESTERN POLYPODY

Polypodium hesperium (pol-y-**poe**-di-um: many feet, referring to the blunt rhizome branchlets; hes-**per**-ium: western). Leaves 3–12", green through winter, dark, smooth; pinnae broadly round-tipped, often not separated all the way to the stalk; sori large, ± round, exposed. Moist, shaded rock surfaces WA to MT; uncommon in s CO and NM. Polypodiaceae.

These dark, leathery leaves last all winter, but often fall short of lasting all summer. New leaves sprout with the fall rains. The roots taste of licorice, with a moment of

Underside of western polypody leaf

sharp sweetness, then a bitter aftertaste. The related licorice fern, *P. glycyrrhiza,* enters our range in Idaho. Licorice proper, *Glycyrrhiza glabra,* is not a fern at all, but shares some chemistry ("glycyrrhizin," naturally).

NORTHERN HOLLY FERN

Northern holly fern

Polystichum lonchitis (pa-**lis**-tic-um: many rows; lon-**kigh**-tiss: spear—). Leaves 6–24", dark, leathery, 1× compound, in clumps; pinnae asymmetrical at base, finely toothed, the teeth sharp and curved forward; stalk bases have long scales; sori round. Rock crevices, mid-elevs to subalpine; A–. Dryopteridaceae.

SWORD FERN

Polystichum munitum (mew-**nigh**-tum: armed). Leaves 20–60", dark, leathery, 1× compound, in huge clumps; each pinna asymmetrical at base, with an upward-pointing coarse tooth; stalks densely chaffy; sori round. Moist forest; WA to nw MT. Dryopteridaceae.

Underside of sword fern leaf

LACE FERN

Myriopteris gracillima (me-ree-**op**-ter-iss: myriad fern; gra-**sil**-im-a: slenderest). Also *Cheilanthes gracillima.* Leaves 3–10" tall, slender, evergreen, pale, at least twice compound, in clumps; leaflets tiny, appearing round from above due to strongly rolled margins, undersides covered in red-brown wool, hiding sori; stalks have narrow scales, not hairs. Igneous rock crevices; WA to nw MT. Pteridaceae.

Not all ferns are particularly moisture-demanding. The little lace ferns and lip ferns are drought-tolerant, living almost exclusively in cliff and rockpile crevices.

Lace fern

LIP FERN

Fee's Lip Fern

M. gracilis (**grass**-il-iss: slender). Like lace fern, exc stalks have sparse hairs as well as 2-toned

A Productive Winter in Santa Fe

Augustus Fendler arrived penniless in the United States from Germany, and lived much of his life wandering, working sometimes as a tanner or a lampmaker. After getting a tip that there were dollars to be scrabbled by collecting plants in the West, he taught himself botany and was granted passage with a troop of soldiers sent to occupy Santa Fe. On day hikes around Santa Fe over the winter of 1846–47, he collected one thousand plants. His work got high marks from the top botanists Engelmann (p. 70) and Gray, who exhorted him to return to the Rockies as soon as possible.

He tried. "My prospects for the first 8 days were most flattering and with every onward step the pleasure of collecting was heightened. . . . But it was the work of a few minutes to blast all my hopes and even to force me to abandon the undertaking." He turned back after a flash flood in Nebraska soaked his six months of provisions and all his collecting gear, killed two of his mules, and badly weakened the other two. He returned to St. Louis only to find all the possessions he'd left there destroyed by a fire. After another lampmaking stint, he collected plants in Venezuela and Panama on assignment from the Smithsonian, then tried settling in Germany again, St. Louis again, Delaware, and finally Trinidad—never again the American West. He did find time to write a book, *The Mechanism of the Universe*, using mathematics to prove the existence of God by way of gravitational force. Though not a great success at the time, it is in print today.

---

scales; pinnae undersides covered in thick brown wool. Calcareous rock crevices; A–.

Pinna on underside of Fendler's lip fern

Fendler's lip fern

Fendler's Lip Fern
M. fendleri (**fend**-ler-eye: for Augustus Fendler, see above). Like lace fern, exc stalks and pinnae undersides covered in coarse tan fibers. Crevices of various rocks; CO, NM.

CLIFF-BRAKE
Pellaea spp. (pel-**ee**-a: dark—). Leaves 2–8" tall, slender, evergreen; margins rolled under, covering the sori; stem shiny dark-brown, easily snapping off at any of several visible articulations near the bottom; leaves usually far outnumbered by snapped-off old stalk bases. Rock crevices. Pteridaceae.
Brewer's Cliff-brake
P. breweri (**brew**-er-eye: for William H. Brewer). Upper pinnae simple, oblong; lower pinnae deeply, unequally 2-lobed, mitten-shaped, or 3-lobed. From se WA to extreme nw CO.

Smooth cliff-brake

Smooth Cliff-brake
P. glabella (gla-**bel**-a: smooth). Nearly all pinnae simple, oblong. A–.

AMERICAN PARSLEY FERN
Cryptogramma acrostichoides (crypt-o-**gram**-a: hidden lines; acrostic-**oy**-deez: top row—) Also rock-brake; *C. crispa.* Vegetative leaves 3–8" × 2–4" broad, rather leathery, firm, yellow-green, at least twice compound, in clumps; fertile leaflets on 7–12" stalks, slender, tightly rolled. Dry, rocky sites in sun, all elevs; A. Pteridaceae.

American parsley fern

Parsley fern has spore-bearing leaves very different from its vegetative leaves—they're often twice as tall, but fewer, and not fine-toothed and parsleylike, as the sterile, vegetative leaves are.

Deciduous Ferns
Pinnae of the following ferns generally turn dry, crisp, and supine by midwinter, though a few may persist into milder winters.

BRITTLE FERN
Cystopteris fragilis (sis-**top**-ter-iss: bladder fern, ref to the covering of the sorus; fra-**jil**-iss). Also fragile fern, bladder fern. Leaves 5–14", twice or thrice compound; strung out along the rhizome rather than truly clumped; stalk slender, somewhat translucent; sori enfolded in a membranous sheath

a bit like a wall sconce (this is the positive way to tell it from woodsia, below). Common; often in partly shaded, moist rock crevices; A. Dryopteridaceae.

Fragile fern is all over the map. Not only does it grow all across the continent, but in the Rockies it can grow in sun or shade, on limestone or granite, and from foothill sagebrush to alpine talus. A cheat sheet on Rocky Mountain fern ID might say that any given fern is "Brittle fern, until proven otherwise."

WOODSIA
Woodsia (for Joseph Woods). Leaves 3–14" tall, twice compound; in clumps together with last year's broken bases; sori initially seen with starlike remnants of their membranous cover. Talus and other rocky sites; A. Dryopteridaceae.

Oregon Woodsia
W. oregana. Leaf undersides hairless; stalks tan to brown, scaly.

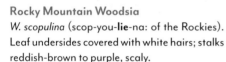
Rocky Mountain woodsia

Male fern

Rocky Mountain Woodsia

W. scopulina (scop-you-**lie**-na: of the Rockies). Leaf undersides covered with white hairs; stalks reddish-brown to purple, scaly.

SPINY WOOD FERN

Dryopteris expansa (dry-**op**-ter-iss: oak fern; ex-**pan**-sa: broad). Also shield fern; *D. austriaca.* Leaves 8–36" tall, broadly triangular, thrice compound, in small clumps; sori round-shielded; stalk bases scaly. Moist forest; northerly, rare in WY and CO. Dryopteridaceae.

MALE FERN

Dryopteris filix-mas (**fie**-lix-**mahss:** male fern). Leaves 15–40" tall, narrowing at both ends, twice compound; leaflet margins toothed; stalks and leaflet undersides ± covered with chaffy, hairlike scales; sori around a kidney-shaped shield. Moist sites at moderate elevs; A. Dryopteridaceae.

Medieval herbalists named the lady and male ferns, and prescribed their powdered roots and leaf infusions for jaundice, gallstones, sores, hiccups, and worms. Only since 1950 has male fern as a dewormer fallen from medical favor, replaced by synthetic drugs with a greater margin between the therapeutic and the toxic dose. Some Native peoples used them medicinally, and ate baked rhizomes of these and other ferns.

OAK FERN

Gymnocarpium dryopteris (gym-no-**car**-pi-um: naked fruit). Leaves 6–18" tall, broadly triangular, thrice compound, rising singly from runners; sori exposed; stalks pale, slightly scaly. Moist forest; dominates the herb layer in some western hemlock or redcedar stands; less often in full sun, where it tends to be yellowish and ± curled up. Conifer forest; all but UT. Dryopteridaceae. Oak fern appears to have 3 similar leaves on each stalk. Technically, this is a single leaf with 2 basal pinnae, left and right, each nearly as big and as dissected as all the remaining pinnae put together.

WESTERN MAIDENHAIR FERN

Adiantum aleuticum (ay-dee-**an**-tum: not wetted; a-**lew**-tic-um: of the Aleutians). Also *A. pedatum.* Leaf blades 4–16", fan-shaped, broader than long, twice compound; sori under rolled edges; stalks black, shiny, wiry. Saturated soil or rocks in shade; mainly northerly. Pteridaceae.

Western maidenhair fern

This is our easiest fern to identify, with its pinnae spreading fanlike and its striking, shiny black stalks splitting into two slightly unequal branches. Infusions of this fern were used to enhance the black sheen of Indigenous maidens' hair.

LADY FERN

Athyrium   filix-femina (ath-**ee**-ri-um: no shield; **fie**-lix **fem**-in-a: fern-woman). Also western lady fern; *A. cyclosorum.* Leaves 16–50" tall, narrowing at both ends,

twice or thrice compound; sori exposed (initially shielded on 1 edge); stalk base scaly. Forest streambanks, wet ravines; A. Athyriaceae.

ALPINE LADY FERN

A. distentifolium. Also *A. americanum, A.alpestre.* As above but smaller (8–32" tall) and more finely incised; sori not shielded by a membrane. Locally dominant in patches of wet subalpine talus; all exc NM.

BRACKEN

Pteridium aquilinum (teh-**rid**-ium: from a Greek name meaning "feather"; ak-wil-**eye**-num: eagle—). Also brake fern. Leaves 24–80" tall, ± triangular, twice to thrice compound, undersides fuzzy; sori under rolled edges. Widespread on ± sunny sites; A. Dennstaedtiaceae.

Bracken has the widest native distribution of any vascular plant in the world. No fern grows as tall or as fast: bracken has been measured at 16 feet in western Washington, and clocked at a growth rate of several inches a day. Though rarely much taller than 3 feet in the Rockies, it is no less aggressive here. It owes its success partly to allelopathy, or secretion of chemical compounds that are somewhat toxic to other kinds of plants, and partly to vegetative reproduction, sending numerous new shoots up from the rhizomes. In some burns and clearcuts, it seems able to hold off conifer reproduction for years.

Alpine lady fern

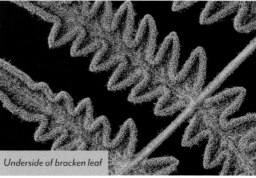

Underside of bracken leaf

Humans sometimes eat young bracken shoots, or "fiddleheads." They taste like asparagus with a dash of almond extract and an unnervingly mucuslike interior. But I don't recommend them. Try just one, if you must (being sure of your ID, as highly toxic monkshood shoots look similar). Bracken turns out to be an ancient traditional plant food that causes cancer; the carcinogen has been identified and proven—and is somewhat concentrated in the fiddleheads. A certain rare stomach cancer occurs more frequently in parts of Japan and Wales where bracken is popular. (Safer fiddleheads grow in the Northeast on the ostrich fern, *Matteucia struthiopteris.*)

Ranchers have long known bracken to be toxic, aside from being carcinogenic. Six hundred dry pounds of bracken consumed within a six-week period are enough to kill a horse via the enzyme thiaminase, which breaks down vitamin B_1. Vitamin B_1 works as an antidote. Thiaminase may have the same effect on humans, but humans are not known to stuff down that much bracken.

American moonwort Dense spikemoss Stiff clubmoss

AMERICAN MOONWORT

Botrychium neolunaria (bo-**trick**-ium: little bunch of grapes; ne-o-loon-**ar**-e-a: New World moon-shaped). Also new moon moonwort; *B. lunaria,* Single 4" leaf of 5–9 pairs of thick, fan-shaped pinnae, grass green, with fanned-out veins, wavy edge; 6–8" fruiting stalk bears a cluster (sometimes flagged to 1 side) of round spore capsules. Moist sites incl scree, mid-elevs to alpine; A. Ophioglossacee.

Moonworts are odd, very small ferns that often escape notice. Each has two stems that part ways an inch or so up: one for the pinnae and one for a taller cluster of spore-bearing spheres. Some species, like this one, have fan-shaped simple pinnae, while some others' pinnae are more fernlike (pinnate), but still small.

Clubmosses and Spikemosses

Like mosses, these fern relatives look shriveled and dead when they dry up, then quickly resurrect as soon as they're wetted. Having vessels (which mosses lack) enables their leaves to be more robust than moss leaves, so their closest semblance might be to heather or juniper. Most species have erect, fertile portions loosely termed "cones" terminating some of their branches. Cones are completely distinct from the leafy portions in clubmosses, though less so in spikemosses.

DENSE SPIKEMOSS

Selaginella densa (sel-adge-in-**el**-a: Roman name). Also *S. scopulorum, S. standleyi.* Spore-bearing portions of stems dull green like the sterile portions, but slenderer and neatly 4-ranked; spores ± orange; forms small mats or clumps. Rocky sites, all elevs; abundant in alpine sedge turfs, cushion plant mats; A. Selaginellaceae.

CLUBMOSS

Family Lycopodiaceae. Spore-bearing "cones" straw-colored, usually erect.

Stiff Clubmoss

Spinulum annotinum (**spin**-you-lum: little spines; an-oh-**tie**-num: marked yearly). Also *Lycopodium annotinum.* Cones ⅝–1⅜" tall, on crowded, mostly unbranched 4–8" erect stalks; leaves shiny, rather stiffly spreading, and long: ¼–½". Under conifers; all but UT.

Ground-cedar

Diphasiastrum complanatum (di-fay-zee-**ass**-trum: 2-sided; com-pla-**nay**-tum: flattened). Cones ⅝–1¼", on branched stalks; foliage flattened, cedarlike, leaves of 3 distinct shapes in the dorsal, ventral, and 2 lateral ranks. Sandy woodland soil; from Bitterroots n and w.

Running Clubmoss

Lycopodium clavatum (lye-co-**poe**-di-um: wolf foot; cla-**vay**-tum: club-shaped). Cones ¾–3" tall, on long, often branched stalks, from leafy runners. Forest; from Bitterroots n and w.

Spores are extremely fine, smooth, slippery, chemically nonreactive, water-repellent, and nonclumping. You may have noticed these qualities in pollen, which descended from spores through evolution. Pollen and spores need to be water-repellent and nonclumping to maximize air travel in the rain. Of all plant and fungal spores, clubmoss spores are the ones that have entered commerce, partly because they're easiest to collect. The cones were cut off, dried, pounded, rubbed, and sifted to collect spores, used for centuries to dust wounds, pills, and babies' bottoms, and also to make flash powder.

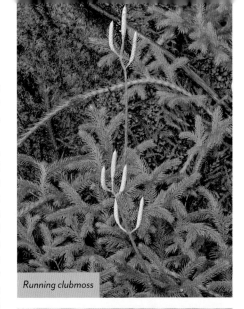

Running clubmoss

Horsetails

These plants bear spores in conelike structures at the top, as clubmosses do, but may be more closely related to ferns. Only one genus survives today from a group that was hugely important and diverse in the Paleozoic.

HORSETAIL AND SCOURING-RUSH

Equisetum spp. (ek-wis-**ee**-tum: horse tail). Thickets of hollow vertical stems with many sheathed joints; from blackish rhizomes. Equisetaceae.

Field Horsetail

E. arvense (ar-**ven**-see: of fields). Reddish-tan to almost white spore-bearing stems come up in early spring, soon wither; 1–3' green stems of summer have jointed, wiry, whorled branches you might mistake for leaves, and have 8–10 shallow, vertical ridges. Abundant in saturated ground at all elevs; weedy on roadsides; A.

Wood Horsetail

E. sylvaticum (sil-**vat**-ic-um: of forests). Our only horsetail with branches sub-branched; sterile stems 12–28", green; spore-bearing stems come up whitish and unbranched, later turn ± green

Field horsetail

and grow a few whorls of branches. Burns, bog edges, saturated meadows; northerly.

Smooth Scouring-rush

E. laevigatum (lee-vig-**ay**-tum: smooth). Also *Hippochaete laevigata*. Stems annual, unbranched or rarely with a few branches; most sheaths entirely green; stems smooth or weakly ridged; cone tips blunt. Wet ground, all elevs; A.

Rough Scouring-rush

E. hyemale (hi-em-**ay**-lee: of winter). Also rough horsetail; *Hippochaete hyemalis*. Stems evergreen, unbranched, ⅛–½" thick × 1–5' tall, with 18–40 fine, vertical ridges; most sheaths show a blackish ring; cones sharp-pointed. Wet ground, low to mid-elevs; A.

Long ignored for being primitive, common, and monochromatic, horsetails won their hour of media glory for sending the first green shoots up through Mount St. Helens's 1980 eruption debris. They can crack their way up through an inch of asphalt on highway shoulders. No wonder swimmers felt strong after scrubbing themselves with horsetails!

Leaves on *Equisetum* are reduced to sheaths made up of fused whorls of leaves, often straw-colored, growing from nodes at regular intervals along the stem. On horsetails, whorls of slender green branches grow just below the leaf sheaths, from the same nodes, producing a bottlebrush shape. The branches themselves have little nodes and sometimes little branchlets.

Indigenous peoples ate the new, fertile shoots and heads of common and giant horsetail eagerly. They were spring's first fresh vegetable—succulent beneath the fibrous skins, which were peeled or spat out. (Like bracken, horsetails cause thiaminase poisoning in cattle.)

Scouring-rushes (and some horsetails) have been picked worldwide for scouring and sanding—polishing arrow shafts, canoes, and fingernails, for example—thanks to silica-hardened gritty bumps on their skins.

7

Mosses and Liverworts

These spore-bearing plants lack vessels, leaving them with little ability to conduct water and dissolved nutrients up into their tissues. (In many of them, conduction takes place mainly along the outside of their stems, aided by surface tension.) They compensate with an ability to pass water through their leaf surfaces—either in or out—almost instantly. After a few dry days in the sun, a bed of moss may be grayish, shriveled, and brittle, appearing quite dead. But let a little dew or drizzle fall on it, or a little water from your bottle, and see how the leaves revive before your very eyes, softening, stretching out, and turning bright green. In freezing weather, they can give up their free moisture to crystallize on their surface rather than inside, where it would rupture cells.

Lichens (p. 286) share many of these characteristics with mosses and liverworts, and may lead similar lives. All three abound on trees and rocks, undergoing countless alternations between their dried-out and their moist, photosynthetically active states. When growing as epiphytes ("upon-plants"), they use another plant to hold them up off the ground, but draw little or no material out of it.

Mosses that grow on the forest floor stake out a seasonal niche as much as a spatial one, living their active season in spring and—where there is no snowpack—winter. Quick recovery from nightly frost is essential. By summer, these mosses are shaded out by perennial herbs and deciduous shrubs, and go largely dormant until fall.

The sexual life cycles of spore-bearing plants evolved before those of seed plants; they are more primitive, but not simpler. In fact, these cycles are so complex and varied that I won't even try to describe them here. Many mosses and liverwort species propagate vegetatively from fragments more commonly than from spores. Some produce multicelled, asexual propagules called gemmae just for this purpose. In mosses and seed plants both, adaptation to Arctic and alpine climates often entails skipping sexual reproduction in favor of cloning, either vegetatively or via unfertilized spores or seeds.

Mosses, Upright, Fruiting at the Tip

The "fruits" of mosses are spore capsules, usually borne on slender, vertical fruiting stalks. To release spores, most open at the tip after shedding first an outer cap (the calyptra) and later an inner lid (the operculum). Most guides to mosses first separate the peat mosses and one other odd family, and then divide moss families into two groups based on where on the stem the fruiting stalks attach:

Bristly haircap moss

Common Haircap Moss

P. commune (com-**mew**-nee: common). Leaves finely toothed (under 10× lens); capsule held horizontal; shoots often branching, 2–6" tall (rarely up to 18") + 1–3" stalk; ♂ cups brownish. Less common here, low to mid elevs; A but NM.

the first group fruits from the tip of the leafy shoot, and typically grows upright in crowded masses or small tufts.

The season of fruiting varies between species, but can last several months. Positive identification of most mosses requires not only fruiting specimens but also a microscope and a technical key. The following pages offer tentative identifications of a few common species, with or without the help of a 10× or 12× hand lens or monocular. A basic hand lens is cheap, and the plant forms it reveals are gorgeous.

HAIRCAP MOSSES

Polytrichum spp. (pa-**lit**-ric-um: many hairs). Stems wiry, rarely branched, vertical, in dense colonies; leaves ⅜" avg, narrow, inrolling when dry; sheathing the stem; stem and stalk (or sometimes entire plant) rich reddish; capsule single, initially cloaked in a densely long-hairy cap, 4-angled (after dropping the hairy cap); ♂ stalks tipped with splash cups like little flowers. Polytrichaceae.

Bristly Haircap Moss

P. piliferum (pil-**if**-er-um: hair-bearing). Leaves untoothed, with whitish, translucent hair-tips; shoots smaller: ¼–1¼" + ¾–1¼" stalk; capsule very short, tilted; ♂ cups wine-red. Dry, rocky soil; A.

Juniper Haircap Moss

P. juniperinum. Leaves straight, often bluish, untoothed, with short reddish hair-tips; leafy shoots 1–4" plus the 1–2½" red fruiting stalk; ♂ cups greenish to reddish. Sunnier sites, disturbed soil, up to alpine; A.

Juniper haircap moss

HAIRCAP MOSSES ARE palpably more substantial than other true mosses. Their stems contain a bit of woody tissue, as well as primitive vessels; they can even store carbohydrates in underground rhizomes, like higher plants. The leaves, too, are more complex and thicker than most mosses' translucent, one-cell-thick leaves. Several species of haircaps have translucent leaf margins that, when they dry out, curl inward to protect the chlorophyllous cells. This adapts the haircaps to sunny sites, as well as to human use—the tough stems were plaited for baskets or twine, and the whole plants used for bedding. Linnaeus reported sleeping well on a haircap-moss mattress on a trip to Arctic Scandinavia.

ALPINE HAIRCAP MOSS

Polytrichastrum alpinum (pa-lit-ric-**ast**-rum: somewhat like *Polytrichum*). Also *Polytrichum alpinum, Pogonatum alpinum.* Stems wiry, reddish, ¾–4" tall + ¾–2" stalk, vertical, in dense colonies; leaves heavy but fairly soft, narrow, tapering to a short brown hair-tip, fine-toothed their entire length; capsule cylindrical or ovoid (not angular) underneath a densely hairy cap, pale green, drying tan. Moist soil or rocks in forest or alpine tundra; A. Polytrichaceae.

FALSE HAIRCAP MOSS

Timmia austriaca (**tim**-ia: for Joachim Christian Timm; aus-**try**-a-ca: of Austria). Stems 1–5", red to orange, unbranched, erect; leaves ¼"

long, narrow, fine-toothed most of their length (under 10× lens), often orange-tinged at base and/or at tip; spore capsules nodding to inclined, sparse, without a hairy cap. Usually in shade, common on logs, cliff ledges; all exc UT. Timmiaceae.

Since these leaves are just one cell thick (like most mosses, but unlike haircap mosses), they are rather translucent, and they crumple up against the stem when dry.

FRAYED-CAP MOSS

Racomitrium canescens group (ray-co-**mit**-ri-um: ragged hat; cay-**nes**-enz: gray-white). Also hoary fringe-moss; *Niphotrichum* spp. Stems 1–3", sprawling in large mats; spore capsules vertical, on ½" stalks that twist counterclockwise when dry; leaves taper to whitish, translucent hair-tips that give the whole mat an ash-gray color when dry. In sun, on tundra, rocks, sand, or gravel outwash; A–. Nearly identical *R. elongatum* (illustrated on this page) is northerly. Grimmiaceae.

Beds of frayed-cap moss spend a lot of time all dried out, utterly crisp, but when moisture falls, they absorb it and spring back to life almost instantly.

BLACK MOSS

Grimmia montana (for Johann F. K. Grimm). Stems ½"; leaves less than ⅛", yellowish or bluish-green ranging to blackish, with white hair-tips that sometimes turn the entire cushion hoary; spore capsules uncommon; fine moss forms dense cushions in rock crevices. Mid to high elevs; A. Grimmiaceae.

Whitish, fine tips on moss leaves, like the silvery hairs all over some higher plants, conserve water by reflecting solar heat. On one *Grimmia* species, clumps with their hair-tips clipped off lost 50 percent more moisture in a day than the unclipped controls. Alpine rock surfaces can be ferociously hot—exceeding 150°F—on some summer afternoons.

Frayed-cap moss

Black moss

ERECT-FRUITED IRIS MOSS

Distichium capillaceum (dis-**tick**-ium: 2-ranked leaves; cap-ill-**ay**-see-um: hair—). Stems ½–3" tall, rarely branched, densely packed in clumps; leaves needle-slender, tiny, clasping stem in 2 ranks (the shoot is flattish); lower stem has fine reddish-brown root hairs; capsules erect, smooth, on ⅜–¾" stalk. Wet rocks, rotten logs, or streambanks at all elevs; A. Ditrichaceae.

FORKED BROOM MOSS

Dicranum scoparium (die-**cray**-num: pitchfork; sco-**par**-ium: broom). Stems 1–4", in tufts; lower stem has fine white-to-brown root hairs; glossy ½" leaves typically all arc, sicklelike, in 1 direction; capsules strongly bent, on 1" stalks, in spring. Forests, often on logs or tree bases; northerly or A– (it's debated). Dicranaceae

CRISPING CUSHION MOSS

Hymenoloma crispulum (hi-men-o-**lo**-ma: membranous fringe; **crisp**-you-lum: leaves crisping). Also mountain pincushion; *Dicranoweisia crispula.* Stems ⅜–1" tall, dense, forming round cushions or flat carpets; leaves ⅛", yellow-green to dark green, occasionally blackening, needle-thin and tapered, twisted when dry; capsule is nearly straight and erect, yellow-brown; stalk ½–1", yellow-brown. On high-elev rocks (esp granite) or trees; A. Dicranaceae.

BIGLEAF ROSETTE MOSS

Rhizomnium magnifolium (rye-zoam-**nigh**-um: moss with rhizoids; big leaf). Stems 1–3" tall, forming loose mats; leaves egg-shaped, up to ⅜" long × ¼" wide near top, smaller downward, from bright to milky-green, with a conspicuous midrib; capsules in summer on 1" stalks. Wet rocks or soils in shade; all exc UT and NM. Mniaceae.

These are among the largest leaves you'll see on moss. At 12× magnification, held up against light, their cells can just be discerned, while a distinct border of different cells around the edge should be clear. They shrivel when dry.

STAR MOSS

Syntrichia ruralis (sin-**trick**-ia: fused hairs). Also *Tortula ruralis.* Stems ½–4" tall, with few or no branches; leaves often orange-tinged, oblong with reddish hairlike tip, form stars when wet but shrivel and press against stem when dry; spore capsule reddish, skinny, cone-tipped, on ⅜–¾" stalk. Common, dry sites at all elevs, sometimes cow pies; A. Smaller red-haired screw moss, *S. norvegica,* has red leaf-tip hairs and likes wetter habitats. Pottiaceae.

PEAT MOSS

Sphagnum spp. (**sfag**-num: Greek name). Robust mosses typically in deep, spongy mats, the stems crowded, supporting each other; leaves tiny, mostly crowded on ¼–¾" branches, the branches (these 3 species) mostly tufted at the top; fruiting stalks short, several per shoot tip, each bearing 1 ± spherical blackish capsule, which releases spores all at once, explosively; typical color in shade is pale, glaucous green, paler and bluer than other mosses; where

Star moss

Bigleaf rosette moss

A mix of peat mosses

growing in sun, plants can be brilliant scarlet (*S. capillifolium*) or green mixed with wine red (*S. warnstorfii*); *S. fuscum* is brown-tinged above, straw-colored beneath. Typically floating in slow-moving water, or terrestrial on seeps; all elevs. A. Sphagnaceae.

The most important mosses, both ecologically and economically, are peat mosses. Estimates have them covering 1 percent of the earth's land—among the most extensive of all dominant plant types. They can take over by changing their environment to suit themselves and to discourage others. Growing in cold, slow-moving water, they draw oxygen and nutrients from the water and replace them with hydrogen ions, creating uronic acids. The water then becomes too acidic and too poor in oxygen and nutrients to support most kinds of plants. The new, peat-dominated plant community is a mire or peat bog, as opposed to a non-acidic fen or marsh. (Though bog peats thrive in the acids they create, they die in the sulphuric acids resulting from acid rain.)

In a bog, the bacteria that normally perform decomposition duties underwater are suppressed—by cold, lack of oxygen, and antibiotics produced by the mosses. Very little decomposition takes place. The floating mass of peat moss lives and grows at the top, in the air, and dies bit by bit just below. The dead part, failing to decompose, gets thicker and thicker beneath the waterline. In some places (western Ireland, northern Minnesota), this can go on indefinitely, the dead peat compressing and becoming a concentrated deposit of biomass suitable for fuel—the leading economic use of peat, at least historically. (I'd argue that it's high time to stop exploiting peat both as a common heating fuel and in gardens. Peat is an excellent carbon sink and arguably an endangered ecosystem.) Peat's inimitable contribution to whiskies occurs when peat fires dry the malted barley, flavoring it with both peat and smoke.

Peat bogs are not always an ecological dead end. The peat moss surface may grow or float high enough above water to become a seedbed for dry-land plants. That's one way for glacial-cirque tarns to end up as forests. Most cirques, though, have plenty of streamflow to avoid ever creating a bog: tarns silt up and turn into glorious meadows after a marshy (not boggy) transitional phase. Toward the south we have few bogs; we have low-acidity fens instead. Fens have peat mosses too, though fewer of them.

Sphagnum species specialize. *S. fuscum* forms hummocks in bogs. You may see brown *fuscum* on top of a hummock, scarlet *capillifolium* (northerly) on the sides, and green species in the soggy hollows between hummocks. *S. warnstorfii* grows in fens and not bogs, often along with sedges, which don't tolerate bogs. Robust pale-green *S. squarrosum* can grow on moist, shady forest floor in the Northwest, but in Colorado it's in fens.

To some Native peoples, peat moss was an irreplaceable material. Its phenomenal water-absorbing capacity made it perfect for diapers, cradle lining, and sanitary napkins. Expectant mothers gathered quantities of peat moss, sometimes lining the entire lodge for a birthing event. Other mosses were used for sponging, padding, and wiping in areas without peat mosses. Most native languages

Hoar moss | Yellow starry fen moss | Pipecleaner moss

don't distinguish types of moss other than peat moss, but among peat mosses they knew pink ones made the best diapers, while red ones were to be avoided.

Mosses Fruiting from Midstem

Mosses in a second group of families fruit from midpoints along the year's new, leafy shoot, which typically arches, trails, or hangs.

HOAR MOSS

Hedwigia ciliata (for Johann Hedwig; sil-ly-**ay**-ta: with fine hairs on edges). Stems sprawling, forming shaggy mats; leaves often whitish-tipped, bright yellow-green wet, dull gray-green or even blackish dry; capsules almost spherical when moist, light brown with red mouth, growing directly—without any stalk—from a stem tip, but seeming to grow from midstem because a branch extends from the same point, ± hiding capsules. On granite and similar rocks, low to subalpine; all exc UT. Hedwigiaceae.

YELLOW STARRY FEN MOSS

Campylium stellatum (cam-**pill**-ium: curved, referring to capsules; stel-**ay**-tum: star-shaped). Stems 1–4", ± sprawling, irregularly branched, often forming deep mats or tufts; leaves broad-based and tapering, coarse and chaotically spreading, resembling star points, green to russet-gold; spore capsules rare, curved, on 1" stalks. A dominant among mosses on seeps and marshes; all exc UT. Amblystegiaceae.

BEAKED MOSS

Eurhynchiastrum pulchellum (yew-rink-i-**ast**-rum: true beak—; pool-**kell**-um: adorable). Also *Eurhynchium* spp. Stems creeping, like fine yarn, much branched, with rhizoids all along attaching them to substrate; leaves minute, relatively broad, pale green; spore capsules horizontal to inclined, on ½" vertical stalks that mature red-brown. Forests, esp on tree bases and logs; A. Brachytheciaceae.

PIPECLEANER MOSS

Rhytidiopsis robusta (rye-tiddy-**op**-sis: looks like genus *Rhytidium*). Also rope moss. Yellow-green to brownish moss in loose mats, the strands looking thick and ropy due to close-packed leaves and sparse branching; stems yellow-green; leaves ¼", irregularly and deeply wrinkled (under lens), tending to curve all to 1 side of stem; fruiting stalks 1", red-brown; capsules often sharply crooked downward. Subalpine forest floor; mainly northerly. Hylocomiaceae.

BIG SHAGGY MOSS

Rhytidiadelphus triquetrus (rye-tid-dy-a-**del**-fus: wrinkled brother, ref to genus *Rhytidium*; try-**kweet**-rus: 3-ranked). Also rough gooseneck

Big shaggy moss

Stairstep moss

Knight's-plume

moss; *Hylocomiadelphus triquetrus.* Light green moss in coarse mats; stems red-brown, with conspicuous right-angled branching; leaves triangular, ¼ × ⅛", faintly but neatly pleated (under 12× lens), sticking out on all sides of stem; fruiting stalks few, 1"; capsule ± bent, maturing in autumn. On humus, logs, rocks, or trees; northerly; infrequent s to CO. Hylocomiaceae.

STAIRSTEP MOSS

Hylocomium splendens (hi-lo-**coe**-mium: forest hair; **splen**-denz: lustrous). Also fern moss; *H. alaskanum.* Glossy gold to brownish-green moss in a stepwise form—each year's growth, shaped like a much-compounded tiny fern, rises from a midpoint on the previous year's stem, growing vertically at first and then arching to horizontal; a stem may show 10+ such steps, but only the upper 1–3 look very alive; fruiting stalks ½–1" tall, red, not twisted; capsules ⅛", horizontal; leaves minute. Luxuriant mats on rocks, logs, and earth; boreal forest to alpine slopes; northerly; infrequent s to CO. Hylocomiaceae.

BIG REDSTEM MOSS

Pleurozium schreberi (ploo-**row**-zi-um: side branch; **shreb**-er-eye: for J. C. Schreber). Stems 4–6", red-brown; shoots ± featherlike but much less even and perfect than fern moss or knight's-plume; leaves minute, blunt tips visible with 10× lens; spore capsules curved, ± horizontal, on 1–2¼" red-brown stalks;

forms golden-green carpets. Conifer to aspen forest; all exc UT. Hylocomiaceae.

KNIGHT'S-PLUME

Ptilium crista-castrensis (**till**-ium: feather; **cris**-ta cas-**tren**-sis: a plumed military crest for wear when in camp). Also ostrich-plume. Stems 1–4", finely, closely, evenly, flatly branched (like a feather) and evenly tapering to tip; leaves minute, all curving with point toward stem base; spore capsules few, inclined to horizontal, red-brown, on 1–2" red-brown stalks; forms golden-green carpets in moist forest, or orange-green in sun. WA to CO. Hypnaceae.

PALM-TREE MOSS

Climacium dendroides (cli-**may**-shum: ladder—; den-**droy**-deez: treelike). Stems 1½–3" tall, erect, like tiny Christmas trees with many mostly upswept, leafy branches from the upper ½; in loose clumps or ± single, connected by rhizomes; capsules scarce, ± erect. Swamps, seeps, streambanks; A. Hypnaceae.

Palm-tree moss

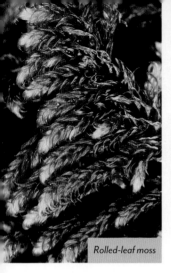
Rolled-leaf moss

Common liverwort

ROLLED-LEAF MOSS

Hypnum revolutum (**hip**-num: "moss" in Greek; rev-o-**loo**-tum: with margins rolled under). Also *Roaldia revoluta*. Fine golden-green moss, often glossy; stems either creeping, in a thin mat, or semierect, 1–2" tall, in a loose mat; leaves minute, sickle-shaped, narrowing to slender points in long arcs to 1 side and down; from above (under 10× lens), the shoot looks like a braid; the distinctive, rolled-under leaf margins may demand a microscope; fruiting stalks few, ⅜–1", capsules curved. Abounds on boulders, logs; A. Hypnaceae.

Liverworts

The first two liverworts below are in order Marchantiales, the thallose liverworts, which look quite different from their leafy brethren in order Jungermanniales.

COMMON LIVERWORT

Marchantia polymorpha (mar-**shahn**-ti-a: for Nicolas Marchant; po-ly-**mor**-fa: many forms). Green, ± leathery, flat lobes textured with close, regular rows of pale bumps; lobes regularly branching in 2-way equal splits; surface often bears conspicuous cups holding a few gemmae (vegetative propagules); tall (¾–4"), fruiting stalks like umbrellas, the ♀ ones 9-ribbed, present only briefly, in spring. Streambanks, burned or disturbed earth, very widespread, weedy in greenhouses; A. Similar, paler green *A. alpestris* is alp/subalp. Order Marchantiales.

This type of liverwort bears no obvious resemblance to either leafy liverworts or mosses. It's more likely to be confused with foliose lichens, but the liverlike textural pattern distinguishes it from the lunglike branching ridges of lungwort or the patternless black speckles on freckle pelt lichens (p. 292). Liverwort, lungwort, and dog lichen are all names bestowed by medieval herbalists, who found resemblances to human body parts in the plant world, and took them to be God's drug prescriptions. The bumps that give liverworts their liver texture are air chambers, each opening to the outside by a tiny pore. They offer a favorable environment for photosynthesis.

Rarely do ecologists pinpoint a plant's habitat as boldly as Alsie Campbell describing *Marchantia*'s streamside habitat: "Occasionally a semiporous barrier will be deposited upon a slightly sloping, non-porous surface across which water seeps all year, such as a small log impeding drainage from an almost flat rock. If there is no disturbance, organic matter and extremely fine inorganic particles build up and form an aqueous muck." Campbell concluded that *Marchantia* has "practical" importance in treacherously hiding slick footing.

SNAKEWORT

Conocephalum salebrosum (co-no-**sef**-a-lum: cone head; sal-eb-**ro**-sum: rough). Also *C. conicum*. Similar to common liverwort; no gemma cups; surface bumps very large and close-packed; fruiting stalks mushroomlike, the coni-

Snakewort

Paw-wort

cal head fringed with spherical spore capsules; aromatic when crushed. Moist earth, streambanks; A. Order Marchantiales.

Conocephalum is named for a female "cone head" that rests on the liverwort's surface for several weeks in spring, waiting for liverwort sperm to swim to it through a film of rain or dew or in the splash of a raindrop. When the spores are ripe and weather permits, a stem of special cells that don't grow or multiply (they simply balloon lengthwise and upward) raises the head a few inches into the air in the space of a few hours. This unorthodox manner of growing a stalk offers speed but not durability. You'll be lucky if you ever catch one of these cute but short-lived fruitings.

PAW-WORT

Barbilophozia lycopodioides (bar-bil-o-**foe**-zi-a: bearded tuft-branch; lie-co-po-di-**oy**-deez: like clubmoss). Also *Lophozia lycopodioides.* Leaves frilly, more than 1 mm wide, four-lobed, each lobe coming to a tiny point (barely visible under

10× lens); in 2 ranks, overlapping like shingles; stems 1–2½" long, sparingly branched; capsule black, egg-shaped. Moist subalpine forest, alp/subalp meadows.

Leafy liverworts—mosslike growths of branched, flattened, ribbonlike strips (about ⅛ inch wide) of leaves neatly overlapping in two, four, or five ranks—are easily mistaken for mosses. The most dramatic difference is in the spore-bearing stalks: on liverworts, they are watery, insubstantial, and rarely seen, as they last only a few spring or summer days. From close up, liverwort foliage is more distinctive; contrasted with moss foliage, it reminds me of small plants flattened by winter snows.

Fungi

A mushroom is not a fungus—at least not by itself. Neither is a puffball, or a shelf fungus on a tree. Each is instead a fruiting body (what we might loosely call a "fruit") of a fungus organism, which is many times larger and longer-lived. The fungus as a whole is a network of tiny, tubular hyphae too fine to see or to handle except when they aggregate in bundles or in fruiting bodies. The fungus produces mushrooms or other fruiting bodies to disseminate its spores, as an apple tree produces apples to carry its seeds.

Picture a "fairy ring"—a circle or partial circle of mushrooms that comes up year after year. These mushrooms are fruits of one continuous fungus body whose perimeter they mark; it expands year by year as the fungus grows within the soil. In some years, the mushrooms may fail to appear, but the fungus is still alive and growing; that year, it merely didn't get the moisture or other conditions for fruiting. The largest fairy rings on record are over 600 feet in diameter. Dividing that by the observed rate of growth yields a likely age of five to seven centuries. Larger outlines, made by honey mushrooms (p. 269), are visible from airplanes over forests: they are circular dents in the tree canopy, consisting of sick and dying trees attacked by the fungus. Some map lichens (p. 288) have been calculated—extrapolating from diameter—to be four thousand years old, rivaling the oldest trees. Each of these fungal circles (including any and all fruiting bodies produced) is a clone; the genes throughout are the same,

so some call it an individual, though it's not quite analogous to a mammal individual. In any case, the mushroom is not an individual.

A fungus is also not a plant. It's more closely related to an animal. Fungal cell walls, like insect skeletons, get their strength from chitin; they contain no cellulose, the characteristic fiber of the plant kingdom. Also in common with animals, fungi obtain their complex hydrocarbon foods from photosynthetic (green) organisms. Animals do it by eating, fungi via any of four nutritional modes:

- **Mycorrhizal:** connecting to plant roots for exchange of nutrients and water
- **Lichenized:** forming a symbiotic composite organism with algae, cyanobacteria, and other organisms
- **Saprobic:** decomposing (i.e., rotting) dead organic matter
- **Parasitic/endophytic:** drawing nutrition from living plant or animal matter

The first two are often mutualistic symbioses, meaning that they benefit both the fungal and the green partners in the relationship. The third and fourth modes benefit biotic communities, even when they harm some individuals. Rotting is necessary for clearing away dead material, recycling nutrients, and—one might add—for keeping the world from bursting into flame due to excess plant-produced oxygen in the air. Even parasitic fungi, usually considered diseases, are often beneficial when viewed from a whole-systems perspective. Fungi inside plants are called parasites when they are pathogenic to the plant, but endophytes when they are harmless or beneficial. A given fungus can shift between different modes with different plants, or at different times in its life.

Some saprobic fungi are actually omnivorous, in that they supplement their diet of dead things by catching microscopic living animals (p. 272).

The first thing to understand about fungus-plant relationships is that they're what makes the world go around—at least the terrestrial part of the world. When life first moved from the sea onto the land, it did so by means of partnerships between primitive fungi and algae. The algae evolved into the first plants, and plants coevolved with fungi up to the present day. Mycorrhizae and primitive lichens each count among the very oldest multi-celled terrestrial fossils ever found. About 90 percent of plant species are known to form mycorrhizae, and the remaining 10 percent are rife with other trans-kingdom mutualisms. Sedges have relatively few mycorrhizae, but are infested with endophytes, which protect them from pathogenic microorganisms. Mosses—the broadest plant group not yet known to be heavily mycorrhizal—have endophytes and also cyanobacteria that fix nitrogen for them, as they do for lichenized fungi.

One study found that only a single gene mutation is required to turn a pathogenic micro-fungus into a beneficial one; such an adaptation could benefit the fungus, which presumably can live longer in a healthy host than a dying one. No wonder fungal endophytes are proving to be extremely common in leaves. Conifers, the backbone of Western ecosystems, are 100 percent mycorrhizal, as well as commonly infested with needle endophytes that help protect them from grazers and pathogens, and often festooned with lichens, some of which fix nitrogen.

Of an estimated 1,500,000 species of fungi in the world, 5 percent have been named. Classification of fungi into families and orders is in a similarly primitive stage of development, so I won't be naming those higher taxa.

THE SYMBIOSIS OF FUNGI AND ROOTS

The "web of life" is not a metaphor: it's out there, and it's made of fungi. Simply put, the job of root hairs in plants is actually performed mostly by fungi in symbiosis with these plants.

Soon after a plant germinates from seed, its rootlet is likely to meet up with its fungal counterpart, a hypha. Indeed, some plants, including most orchids, won't germinate until they're penetrated by a hypha. The plant and fungus each contain hormones that stimulate and alter each other's growth, forming a joint fungus-root organ that immediately goes to work as a nutrient loading dock. This fungus-plant organ is a mycorrhiza, a word coined in 1885 that simply means "fungus root."

In one type of mycorrhiza, hyphae grow as a net around the root tip. In another, they digest root cell walls, and penetrate and inhabit the cells, which eventually digest them. Either way, the hyphae hormonally suppress the formation of plant root hairs, but provide a preestablished network of root hair surrogates finer and more efficient than the plant's own. (A root hair is about 1 millimeter in diameter; a hypha is about 0.003 millimeter. A hypha can thus provide about three hundred times as much absorptive surface area for a given investment in biomass.) In addition, mycorrhizal fungi deliver nitrogen and phosphorus that would otherwise be unavailable to plants. They can even bore into solid rock to extract nutrients.

Mycorrhizal fungi depend on the host plant for their food, or carbohydrates. Few plants are equally dependent on fungi, since most plants have the genetic information for making root hairs to obtain water and minerals. Planted in a rich, moist, nutritious substrate, a seedling may do just that, and repel mycorrhizal infection. But given real-world challenges, nonmycorrhizal seedlings often fail to thrive. For example, on a hot, sunny, logged-over slope, Douglas-fir seedlings proved more likely to survive if planted right next to Heath-family shrubs that partner with suitable fungi. In another study, seedlings survived only if they were inoculated with a teaspoonful of forest soil. The active ingredients in that elixir are poorly known, but mites, nematodes, bacteria, and other microorganisms are present in great numbers and variety, and surely share credit with the 16 miles or so of hyphal tubes in a tablespoon of soil.

Most plants have many mycorrhizal species as potential partners. Our conifers each have hundreds or thousands. A tree forms mycorrhizae with several species at a time, and grows best if it has a range of them available for changing conditions, even changing them with the seasons.

While the typical transfers are carbohydrates from the plant, and water and minerals from the fungus, there are additional benefits. Root rots (parasitic fungi) have a harder time attacking roots mantled in healthy mycorrhizal hyphae, which form both a physical and a chemical barrier. Many fungi compete with other fungi by secreting selective toxins, and they secrete antibiotics that fight bacteria, including some plant pathogens. Mycorrhizal fungi exude carbohydrates into the soil, making the soil more cohesive, more porous, and better aerated (and sequestering carbon while they're at it).

Then there's the all-important nitrogen cycle. Some fungi have the ability to break proteins into amino acids and deliver these to their plant partners. With this help, some plants can meet their nitrogen needs in substrates where it had been thought that all the nitrogen was organically bound up, unavailable to plants. Other fungi attack and digest live animals—a nitrogen source available to few plants. John Klironomos and Miranda

Hart measured nitrogen transfer to white pine seedlings from springtails (tiny soil insects) killed and consumed by mycorrhizal partners. These seedlings were getting as much as 25 percent of their nitrogen this way. The two researchers had started out intending to observe springtails eating fungi. The reversal was "as shocking as putting a pizza in front of a person and then having the pizza eat the person."

Finally, the fungal web sometimes carries carbohydrates from plants to other plants. Beneficiaries might include small plants growing in shade where they cannot photosynthesize enough to survive. The now-illustrious Suzanne Simard measured belowground transfer of carbon to Douglas-fir seedlings from birches above them; the flow increased when the seedlings were starved by shading them. Herbs, as well as tree species, may participate. In eastern Canada, trout lilies were found both to receive sugars from sugar maples in summer when the lilies store sugars in their bulbs, and to lose sugars to the maples in spring when the lilies are leafed out but the maple leaves are just emerging.

That said, media enthusiasm for trees taking care of each other goes beyond current scientific evidence. Yes, nutrients pass from plant to plant, but it's very hard to rule out other paths these nutrients may take—directly through soil moisture, or via naturally occurring root grafts. And so far, the detected quantities of resources transferred from plant to plant are small. As for the reach and the durability of hyphae, the few studies trying to measure them established that a hyphal connection can reach at least a few yards—not hundreds—and that hyphae frequently break, while new ones grow. To imagine that everything in the forest is literally connected by one fungal web is to get ahead of your skis.

Fungi did not evolve to be servants to plants. They convey nutrients for their own benefit. Plant-fungus symbioses that usually

To Eat Them, or Not?

A small minority of mushrooms is seriously poisonous. A greater number may make you sick in the stomach, or uncomfortable somewhere else. Still more are considered edible by most who have tried them, but their reputations are tainted by a few reports—allergic reactions, possibly. Even the supermarket mushroom, *Agaricus bisporus,* upsets some tummies. Many other mushrooms go unrecommended on the grounds of flavor or texture. Again, one person's "edible and choice" is another's "Bleccch!"

Though the odds favor mushroom eaters, the risks are extreme. There are old mushroom hunters, the saying goes, and there are bold mushroom hunters, but there are no old, bold mushroom hunters. Actually, there was one: Captain Charles McIlvaine lived sixty-nine years while routinely tasting mushroom species on first encounter. He claimed to have tried well over a thousand and liked most of them.

Mushroom identification is more subtle and technical than most plant identification, often requiring chemical reagents and a microscope, or even DNA sequencing. Since this book describes only a few fungi and doesn't use those tools, I restrict the "edible" recommendation to some that either lack paper-thin gills or lack a distinct stem: you can separate these recommended species from any poisonous ones based on a careful look at field charactersistics. All the same, you must assume responsibility for your own results, both gastronomic and gastrointestinal:

- Try a small nibble and spit it out. The test nibble can be raw, as long as you don't swallow any. Cook all fungi before eating.
- When eating a species for the first time, don't mix more than one species, and give your stomach twelve hours to fully test.
- Very small children are more susceptible to mushroom poisoning and allergies, and should not eat wild mushrooms.
- Excessive bugginess or worminess occasionally causes stomach upset. If the stems but not the caps show larval boreholes, leave the stems behind so they won't infect the caps.
- Carry mushrooms in paper bags, not plastic. Keep them as cool as possible.
- Deep chomp marks from deer or rodents are not evidence of safety.
- To see spore color, make a spore print by laying a cap flat on a piece of white paper until spores appear. White spores won't show up on white paper, so try wiping for them with your finger. Don't mistake moist stains for spores.
- Gentle, thorough sautéing on medium heat, without a lid, is rarely a bad way to cook a fungus. Scrambled eggs, toast, or crackers rarely fail to complement.

Is mushroom harvesting sustainable? As of thirteen years into one unpublished study, chanterelles were fruiting slightly *more* abundantly where they're harvested year after year. In the longer term, who knows? The side of the debate in favor of harvesting likes to say mushrooming is like picking the apples from an apple tree: mushrooms are the fruits, not the organism. The side opposed points out that fungi wouldn't have evolved to disseminate staggering numbers of spores if that trait didn't have significant value to them. Some proportion of regeneration from spores may be essential to the long-term health of the population.

Truffle harvests in Italy and France have declined by around 95 percent. Though the decline follows centuries of ardent truffle-hunting, expert opinion attributes it more to environmental change.

Fly amanita

Death cap

benefit plants have been seen harming them; the symbiosis can turn from mutualistic to parasitic.

That said, you can easily see proof that the web does exist and does provide plant-to-plant transfers. Just look at the non-green herbs (p. 153), epiparasites that make a livelihood of sponging off this very web.

Mushrooms with Stem, Cap, and Gills

A cap with paper-thin gills on its underside, atop a stem, is simply the "normal" mushroom form, like the ones in the supermarket.

FLY AMANITA AND WESTERN PANTHER AMANITA

Amanita spp. (am-a-**night**-a: Greek name for some fungus on Mt. Amanus). Cap usually sprinkled with whitish warts; stem white, with skirtlike ring near the top and a bulbous (not quite cuplike) base; gills and spores white.

Fly Amanita

A. muscaria group (mus-**car**-ia: of flies). Cap typically yellow to orange, bright red, tan, or white; bulb at stem base tapers upward, with about 4 concentric rings. On soil, Aug until hard freeze.

Western Panther Amanita

A. pantherinoides (pan-ther-in-**oy**-deez: panther—). Cap typically tan (ranging from brown to nearly white); bulbous base has a single, heavy, rolled upper edge that may or may not pull away from stalk. Common after summer rains under aspen or Douglas-fir; Jul until hard freeze.

THE GORGEOUS FLY amanita is the archetypal mushroom—or perhaps the archetypal malevolent toadstool (a British word for any inedible mushroom). William Rubel and David Arora use it as a case study of the biases of field guides, which tend to err on the side of caution (as this one does) when it comes to mushroom toxicity. Rubel and Arora consider *A. muscaria* a fine edible, and regularly share it with dinner guests. They also concede that it is toxic, but they detoxify it by boiling it, draining and discarding the water, and then cooking it. I don't recommend this. Symptoms reported from inadequately cooked mushrooms include hallucinations, ataxia, hysteria, myoclonic jerking, hyperkinetic behavior, stupor, seizures, and coma. Nevertheless, some individuals sign up for this potluck in hopes of hallucinations. Certain Siberian tribes ate them ritualistically.

Dogs and cats are more sensitive; some have died. To kill flies in times past, fly amanitas were left around the house in saucers of sugared milk—toddlers beware! No harm to flies has been verified experimentally; the mode of fly death may have been drowning while inebriated.

The fly amanita has several equally toxic close relatives with warts on the cap, fruiting in various seasons and with caps ranging from white through blushing yellow.

Whatever the species, trailside specimens are a scenic resource to be left untouched.

DEATH CAP

Amanita phalloides (fa-**loy**-deez: phalluslike). Cap skin bronze to pale silvery-green, sometimes white; flesh, gills, and stem white; stem cylindrical or slightly smaller upward, with a tattered, skirtlike ring (sometimes missing) and a ± bulbous base in a thin white cup (requires careful excavation to see; the cup often barely emerges from ground, and may be dirty); spores white; becomes fetid with age; do not taste.

Funeral bell

Lovely but monstrously poisonous, the death cap and several relatives called destroying angels are our most dangerous fungi. Though reports of any of them in the Rockies are exceedingly rare, it's worth knowing their characteristics. Thought to have arrived from Europe on the roots of trees for planting, the death cap seemed at first to grow only under non-native trees, but it's now spreading out from towns. It is established in Boise.

All amanitas have white spores and more or less white gills. Most have a definite ring around the stem, the remnant of a "partial veil" that extended from the edges of the cap, sealing off the immature spores to keep them moist. Most emerge from a "universal veil," an additional moisture barrier that wraps the entire young mushroom from under its base to all over its cap. As a young button, each amanita mushroom in its universal veil is egg- to pear-shaped, resembling a puffball but with the outline of cap and gills visible in cross-section. Remnants of this veil usually persist as a cup or lip around the base of the stem, and/or on top of the cap in the form of warts, crumbs, or broad patches. But absence of these remnants doesn't disprove any amanita, since the stem easily breaks off above the cup, and the cap remnants can wash off.

All white-spored, white-gilled mushrooms should be carefully examined for ring and cup. Genus *Amanita* contains good edibles, some popular in Europe, but the chance of misidentifying a deadly one leads American guides to caution against all amanitas. (This book doesn't recommend eating *any* stemmed, gilled mushrooms without consulting much more complete mushroom resources elsewhere.)

Most mushroom deaths in America follow eating these amanitas. That said, the media have exaggerated the fatality rate, which may actually be lower than 11 percent. Hospitalization using best practices (including IV fluids) could probably reduce it to less than 5 percent. Some people survive even without treatment, but few dogs do—and dogs do eat amanitas.

FUNERAL BELL

Galerina marginata (gal-er-**eye**-na: helmeted; mar-gin-**ay**-ta: edged). Also deadly skullcap; *G. autumnalis.* Caps mostly ¾–2" across, slightly tacky and deep yellow-brown when wet, dull tan when dry, radially striped near edge when moist; gills pale, becoming brown with spores; stem 1–3", thin, darkening toward base, with a thin whitish ring; rusty-brown spores may dust the ring. Clustered on rotting (sometimes buried) wood, often among mosses; A.

This unprepossessing "LBM" (little brown mushroom) contains the same amatoxins as the deadly amanitas, and is just as lethal. It is less of a danger because it's so small, and people don't normally bother with LBMs—unless they're looking for *Psilocybe* species.

Rosy waxy cap

Shoestring root rot or honey mushroom

WAXY CAP

Hygrophorus spp. (hi-**groff**-er-us: moisture-bearing). Also waxgills, woodwax. Mushrooms 2–4" across; gills white, clean, slightly waxy-feeling, rather thick.

Alpine Waxy Cap

H. subalpinus. All-white to cream-tinged; cap 2–5", sticky when wet, becoming flat or slightly concave by maturity; stem stout (often less than 2× taller than wide) with a slight ring (at least initially), the remnant of a veil from edge of cap. Spring, often near melting snow; A–.

Rosy Waxy Cap

H. pudorinus (pew-dor-**eye**-nus: ashamed, or blushing). Pretty, warm tan, often with a pinkish blush; cap sticky when wet, remaining shiny when dry, maturing ± flat but with downturned edge; gills don't change color where bruised; stalk fairly slender, often curved, with minute fibers (lower) and scales (higher); no veil. Fall, under conifers, esp spruce; A.

AMONG THE SEVERAL mushrooms that often come up near melting snow, the subalpine waxy-cap is relatively common, and may also be ecologically significant in offering food to bears at a nutritionally critical time of year, soon after they emerge from hibernation.

The waxy quality is subtle, but useful in ID once you get to know it. Brush a fingertip across the gills, and notice a slight stickiness and moisture, together with relative strength and thickness.

SHOESTRING ROOT ROT OR HONEY MUSHROOM

Armillaria solidipes (ar-mil-**air**-ia: arm-banded; sa-**lid**-ip-eez: solid foot). Also *A. ostoyae*, *Armillariella mellea*. Cap yellow-brown to brown, with coarse fibrous scales radiating from the center; gills white, stained rusty in age; stalk has stringy white pith and a substantial, up-flaring, brown-edged ring; clustered on wood or ground, caps often coated with spores where overlapped; spores white; coarse, black, thread-like rhizomorphs often visible around base or netting across nearby wood, often under bark, in fall; A.

Some *Armillaria* species live as saprobes; some become deadly parasites where the trees are stressed—in highly competitive stands, in logged and roaded regions, and during drought, which means that climate change could favor them. Close study has found that this mushroom with many faces is actually several species. They vary in pathogenicity to trees and in edibility, and are suspected of some degree of toxicity.

Broadly, *Armillaria* is a leading agent of white rot in wood. In ecological terms, that makes it a valuable recycler of nutrients and facilitator of animal habitats. White rot breaks down both cellulose and lignin, in contrast to brown cubical rot, which breaks down only cellulose, turning dead wood into, yes, those rotten brown cubes we often see.

The evolution of the first white rot fungi was a big deal for life on Earth. For 60 million years, plants ("tree ferns" in particular) had been evolving to grow much bigger and more quickly, thanks to lignin in their stems. There were no organisms yet that could decompose lignin, so masses of lignin were accumulating, eventually getting buried and turning into coal. (That geologic period is named Carboniferous, or "coal-bearing.") The removal and burial of so much carbon was making the air carbon-poor and oxygen-rich, which led to ice ages and presumably would, if unchecked, have eventually caused the land biota to catch fire. I was much taken with a hypothesis that Earth was saved from that fate, and ushered out of the Carboniferous Period, by the arrival of white rot fungi and their lignin-decomposing skills. This appealing hypothesis fell into disfavor for several reasons, though no alternative savior has been identified.

The "shoestrings" are the black rhizomorphs distinctive of this genus—bundles of hyphae with tough, protective black sheaths that enable them to reach across nutrient-poor stretches in search of the next nutrient-rich one, usually the roots of another stump or stressed tree. Rhizomorphs enhance *Armillaria*'s ability to spread, which has inspired a modern legend. The media anointed the honey mushroom the world's largest living thing in 1992, when Michigan mycologists found a honey mushroom that had spread over (or rather under) 91 acres. Other researchers then mapped one under 1,500 acres in Washington, and another under 2,395 acres in eastern Oregon's Strawberry Mountains. (We aren't talking about Godzilla mushrooms, but about networks of tiny hyphal tubes.) The Oregon researchers matched the *Armillaria* genes at widespread spots in the forest, and found that a single mushroom clone had spread, all starting several thousand years ago, perhaps from one spore. Since testing hyphal DNA across broad areas is rarely done, it's highly unlikely that the world's largest clone has been measured—or that all the hyphae in a vast clone remain interconnected.

WESTERN MATSUTAKE

Tricholoma murrillianum (trick-**ahl**-a-ma: hair fringe; mur-rill-ee-**ay**-num: for William Murrill). Also *T. magnivelare.* Cap and stem firm, initially very white, with brown fibers or bruises; cap 2–5", margin inrolled when young; gills fine and close, initially covered by a velvety veil; stem ½–1" thick, white above the thick ring; spores white; uniquely aromatic. In duff—often barely emerging—under conifers, mainly lodgepole; all exc UT.

The mycologist Lorelei Norvell assures us we can identify matsutakes by their "delicious" fragrance, but *Mushrooms of Northeast North America* grumbles that the emperor has no clothes: "odor like dirty gym socks but usually described as spicy-sweet, aromatic, fruity, or fragrant." To my nose, yes, a little like dirty socks but somehow bewitchingly sexy. Once you've smelled it, you have no trouble identifying this mushroom anywhere.

The original matsutake, *T. matsutake,* has been prized in Japan almost forever, but has become rare there, so the Japanese pay prices that have exceeded $200 a pound to import its close relatives. Those prices peaked years ago, before China discovered the matsutake market. Western matsutakes for export are still a mainstay of commercial pickers, but

Western matsutake

Shaggy mane

Inky cap

breaking through asphalt) to subalpine meadows; spring to late fall; A.

Hard to mistake for anything else, the shaggy mane is edible if picked young and cooked soon, before it starts to inkify (see next species).

INKY CAP

Coprinopsis atramentarius (co-prin-**op**-sis: looks like shaggy mane; at-ra-men-**tair**-i-us: black ink—). Also *Coprinus atramentarius*. 1–3" tall; cap taller than wide, pale gray, then soon lead-gray; furrowed, smooth, or with modest scales; gills and then the whole mushroom turn black and slimy, eventually becoming a black puddle. Clustered, on or near rotting wood (or chip piles); spring or fall; A.

The shocking deliquescence, or transformation of the gills into an inky mess, is an adaptation for rapid dissemination of spores. Many mushrooms that do this were formerly assumed to be related, and combined in the genus *Coprinus,* but DNA study finds that the trait evolved convergently in widely separated families. The ink was used as black ink in times past.

Inky caps are edible but contain a sort of natural Antabuse. If you eat them and consume any alcohol within thirty-six hours, the combination will make you very sick, using a different chemistry than Antabuse to provoke more or less the same symptoms: flushing, tingling, racing heart, and vomiting. Basically, it's alcohol poisoning: the toxins stop the liver from converting the alcohol into more-benign compounds.

mainly near the coast. In the Rockies, they are widely scattered but nowhere abundant.

SHAGGY MANE

Coprinus comatus (co-**pry**-nus: growing on dung; co-**may**-tus: woolly). Also lawyer's wig. 2–8" tall; cap taller than wide, long, egg-shaped becoming bell-shaped, whitish, covered with coarse white scales that darken later; gills fine, white, maturing pinkish; spore print black; stalk has a loose ring; a cut stalk is chalk white, not browning; with age, first the cap margins, then the gills, then the whole mushroom turn black and slimy, eventually becoming a black puddle. Clustered, from parking lots (commonly

SHORT-STEMMED RUSSULA AND LOBSTER MUSHROOM

Russula brevipes (**russ**-you-la: red—; **brev**-ip-eez: short stalk). Also brittlegill, duff-humper. White throughout, may bruise brown or yellowish; cap 3–10" across; dry, quite concave, margins usually rolled under; gills fine, crowded; spores white to cream; flesh odorless, bland to peppery; stem short, cylindrical, rigid, brittle: it snaps like chalk, leaving a rough, fiberless

Short-stemmed Russula   *Lobster mushroom*

surface. Mycorrhizal with conifers; scattered, usually ± buried in forest duff; fall; A.

These two species put a twist on the Frog Prince story: as a humble mushroom morphs into the object of our desire, it turns from virginal white to enflamed and pimply. Our story begins with a rather attractive mushroom that, due to its abundance and blah flavor, was voted "most boring mushroom" at one mycophagist convention. It likes to sturdily push thick duff up from underneath, and usually has a great many crumbs on its face. It belongs to a huge and abundant genus nicknamed JARs ("just another *Russula*"). In several species, the brittle, pure-white flesh and gills contrast with a cap skin—often quite peelable—that may be carmine red, green, or black.

Some short-stemmed Russulas host a parasitic fungus, *Hypomyces lactifluorum*. Think of it as a bright-orange mold. The mold prevents paper-thin gills from developing; alters the graceful form into a crude knob and the delicate, brittle flesh into spudlike firmness; and wraps the whole thing in a scurfy deep-orange skin. Voilá, it's a marketable delicacy: a lobster mushroom. It also keeps longer than normal mushrooms, apparently killing or repelling bugs and rot.

Is it safe to eat? Some supermarkets deem it safe enough to sell, but Michael Beug, a top expert in mushroom toxins, rules out lobsters for his personal consumption due to a few reports of severe symptoms.

Mushrooms with Cap and Gills But No Stem

Stemlessness helps make the oyster the rare gilled mushroom that can be rated safe for beginners to eat. (Oyster mushrooms occasionally have a short, semihorizontal stem from one edge of the cap—never from the center.) Beware, though, of undesirable species that have both stemless and off-center-stemmed fruits, and show at least one of these traits: tough, leathery, thin flesh; sawtoothed gills; yellow gills or brown spores or a dark, drab cap; or an intensely bitter or peppery taste. (Take small nibbles and spit them out.)

OYSTER MUSHROOM

Pleurotus spp. (ploor-**oh**-tus: side ear). Also *P. ostreatus*. Fan-shaped, usually stemless mushrooms growing in clusters from the side of fallen, dead, or wounded trees; mildly fragrant, pleasant-tasting.

Aspen Oyster Mushroom

P. populinus (pop-you-**lie**-nus: of poplars). Cap cream white to light gray; spore print pale buff. On wood of aspen, rarely cottonwood, spring to early summer; A.

Summer Oyster Mushroom

P. pulmonarius (pull-mun-**air**-i-us: like lungs). Cap cream to medium gray-brown; spore print white to lilac-gray. On wood of cottonwood, rarely conifers, spring to late fall; A.

Summer oyster mushroom

CARNIVOROUS PLANTS LIKE Venus flytrap may be notorious, but carnivorous fungi are *news.* Tiny nematode worms have long been observed tunneling into mushrooms for dinner. Many saprobic fungi, it now turns out, turn the tables and "eat" their nematodes. *Pleurotus* has specialized cells that splash poison onto any nematodes that touch them; hyphae then seek, invade, and digest the immobilized worm. Without nitrogen-rich foods such as these animals, a saprobic diet can be poor in nitrogen.

Oyster mushrooms are especially prodigious spore producers. "Mushroom worker's lung" is an affliction common among those who inhale spores in commercial oyster-mushroom grow houses. The classic oyster mushroom, *P. ostreatus,* is grown commercially not only for food but for medical research, for bioremediation of contaminated soil, and now for its mycelium as a biodegradable packing material grown within forms, yielding complex shapes like those of styrofoam packing.

The role of thin gills was rather mysterious until 2016, when high-speed cameras showed that water evaporating from gill surfaces chills the air, creating sharp downdrafts and then convective eddies that push the spores not only outward but even upward several inches within seconds. This explains how mushrooms benefit from losing water

rapidly, and why they wait for rain before appearing.

Mushrooms with a Wrinkled Undersurface

On both true chanterelles and the unrelated woolly chanterelle, the spore-bearing underside of the cap has rounded ridges (not paper-thin gills), sometimes with cross-wrinkles; there's also no sharp division between the cap and stem.

CHANTERELLE

Cantharellus spp. (canth-a-**rel**-us: small vase). Vase-shaped mushroom, cap margin often wavy or irregular; peppery aftertaste when raw. On soil, often nearly buried in duff, summer and fall.

Rainbow Chanterelle

C. roseocanus (rosy-oh-**cay**-nus: pinkish-grayish). Ranges from pale to strong yellow-orange; the wrinkled underside a stronger, deeper color than the cap surface (in contrast to the reverse in the coastal *C. formosus*); edges of cap often have a pinkish bloom; spores ochre; sweet (apricot?) aroma. Under lodgepole pine or spruce-fir; A.

White Chanterelle

C. subalbidus (sub-**al**-bid-us: almost white). Cap white, bruising yellow to (eventually) rusty orange; often quite stout and short; spores white. WA, OR, nw MT, n ID.

A gorgeous, rich color, easy and safe identification, abundance, resistance to bugs,

Rainbow chanterelle

Woolly chanterelle

Hedgehog mushroom

Shingled hedgehog

and an established place in French cuisine all boost the popularity of chanterelles.

We used to call all our yellow chanterelles *C. cibarius,* but it turns out the West has several species. *C. roseocanus* predominates in the Rockies, with *C. cibarius* also present. All are fine edibles. The only poisonous mushroom sometimes mistaken for a chanterelle (aside from the unrelated "woolly," below) has paper-thin gills.

WOOLLY CHANTERELLE

Turbinellus floccosus (tur-bin-**el**-us: small top; flock-**oh**-sus: woolly-tufted). Also scaly chanterelle; *Gomphus floccosus.* Orange, trumpet-shaped fungus, often deeply hollow down the center; inside surface roughened with big, soft scales; outside surface paler and irregularly, shallowly wrinkled (no paper-thin gills); spores ochre. On soil, late summer and fall; A.

Over the years, mushroom guides have batted this pretty species back and forth between the "edible" and "poisonous" lists. We'll call it "highly questionable."

Mushrooms with Fine Teeth Under the Cap

Tooth mushroom undersides are covered with fine teeth all pointing down; like gills, they are organs for releasing spores.

HEDGEHOG MUSHROOM

Hydnum repandum group (**hid**-num: truffle; rep-**and**-um: wavy-edged). Also sweet tooth mushroom, *lengua de vaca* (cow's tongue) in Spanish; *Dentinum repandum;* many new cryptic-species

names pending. Cap yellow-orange to buff or nearly white, only vaguely differentiated from stem; underside teeth of mixed lengths (avg ¼") in lieu of gills; spores white. Mycorrhizal with Douglas-fir; on ground, summer through late fall; A.

Eyes scouring for chanterelles' soft, gold caps may jump at this one, but fingers will be in for a surprise when they reach underneath and find a soft, spiny texture. Don't be disappointed—this mushroom is edible, and even harder to confuse with anything poisonous. Some toothed relatives, however, are bitter. The tiniest taste test will tell you.

SHINGLED HEDGEHOG

Sarcodon imbricatus (**sar**-co-don: fleshy tooth; im-brick-**ay**-tus: shingled). Also hawkwings, scaly urchin; *Hydnum imbricatum.* Cap large,

2–8", gray-brown, dry, with coarse, dark scales in a concentric pattern; underside covered with fine teeth in lieu of gills; spores medium brown; stalk stout, light brown inside and out. Mycorrhizal with conifers, summer and fall; A.

Some people are disconcerted by the idea of eating this coarse, blackish species, and a few have reported stomach upsets or bitterness—though I've enjoyed it. David Arora recommends prolonged cooking to mellow its strong flavor. The only unsavory similars will warn you with a bitter taste test.

King Boletus

Mushrooms with a Layer of Close-Packed Tubes Under the Cap

"Bolete" is a casual term for a group of mushrooms that, in place of gills under their caps, have a spongy mass of vertical tubes called pores. Like gills, they are organs for releasing spores. If you slice through the cap, you'll see that the pore layer is separate—and peelable—from the smooth cap flesh above it. Though not every bolete is safe, boletes feature in any listing of edible mushrooms safe for beginners because of that one easily seen character—pores. The two largest genera are *Boletus* and *Suillus*, but there are many others. Shelf fungi also have pores underneath, but their pores are so fine you can barely detect them without a lens.

Ruby Boletus

White king Boletus

BOLETUS

Boletus spp. (bo-**lee**-tus: clod). Cap 3–12"; flesh ± white, firm; pores white, becoming olive to tawny yellow with age or bruising, less often blue; stem often very bulbous, white with ± brown skin, upper part finely net-surfaced; spores olive-brown. Mycorrhizal with conifers; late summer to fall.

King Boletus

Boletus edulis group (**ed**-you-lis: edible). Also porcini, cepe, steinpilz. Cap skin golden or tan to red-brown, often redder just under surface, ± bumpy but not fibrous, often sticky; stem white with ± brown skin. Often under spruce; A, more common n.

Ruby Boletus

B. rubriceps (**roo**-bri-seps: reddish mushroom). Cap skin deep red, greasy or sticky; pores whitish, aging toward olive. Often under spruce; southerly.

White King Boletus

B. barrowsii (**bear**-owes-ee-eye: for Chuck Barrows). Cap skin off-white, dry, not sticky;

stalk yellow-tinged white; spores olive brown. Typically under pines; June–Aug; A.

TO MY PALATE, these are the most delicious fungi of all. Just don't leave them to sit around at room temperature, as they're notoriously vulnerable to maggots and molds.

Most boletes range between choice and not quite worth eating, but one or two red-orange-pored species are potentially deadly. The small number of modestly toxic species can be ruled out rather simply, leaving you to safely eat the rest without necessarily identifying them to species. Here are the features to avoid:

- Red-orange or pinkish-tinged tube openings.
- Bitter, burning, or peppery tastes (check with a tiny nibble-and-spit of the cap).
- Cap flesh that turns blackish, gray, pink, or blue when cut or cooked, on an orange-skinned bolete under aspens (these boletes can cause prolonged flulike symptoms).
- Tubes that turn deep blue within a minute where handled. A few species showing this sharp reaction cause stomach upset. Good edibles may also bruise bluish—but slowly—so expert mycophagists are undeterred by bluing unless it is intense and quick.
- Mushrooms totally riddled with tiny larval holes. Boletes can get gross pretty quickly, and gross ones are hard on the stomach.

SUILLUS

Suillus spp. (sue-**ill**-us: piglet).

Western Painted Suillus

S. lakei (**lake**-eye: for E. R. Lake). Cap rough with flat red-brown scales, yellow and ± sticky under the scales (or can be slimy instead after rain); flesh and tubes ± yellow, maturing orange, may bruise reddish; tubes angular and radially stretched when mature; stem has whitish ring when young; base bruises blue-green; spores

Blue-staining slippery jack

Western painted suillus

Hollow suillus

brown to cinnamon. Under Douglas-fir, late spring to fall; A. Edible.

Hollow Suillus

S. ampliporus (amp-li-**por**-us: big pores). Also *S. cavipes, Boletinus cavipes*. Similar to *lakei*, but with a hollow stem, and growing under larch; northerly. Good to mediocre edible, or somewhat bitter.

Blue-staining Slippery Jack

S. tomentosus (toe-men-**toe**-sus: woolly). Cap dry, with fine, fibrous scales (but may become smooth in age, and then slimy when wet); flesh pale yellow to orange-buff, bruising blue ± slowly; stem covered with brownish specks; no veil or ring; spores dark brown. Under 2- or 3-needle pines; similar species under 5-needle pines is *S. discolor;* A. Edible, often tasty, but be careful with ID if the blue stain is rapid and strong.

Dotted-stalk Suillus

S. subalpinus. Also *S. granulatus.* Cap tacky- to slimy-skinned, the skin peelable, tan to cinnamon, faintly mottled; flesh white to cream; tubes tan to yellow, very small, "dewy" in youth, staining or speckling brown with age; no veil or ring; stem white, developing brown dots or smears with age; spores cinnamon to ochre. Under 5-needle pines; A. Edible.

Short-stalk Suillus

S. brevipes group (**brev**-i-peez: short stalk). Like *S. subalpinus,* but cap starts deep red-brown and pales to ochre with age, and the short stem remains pure white exc, rarely, for faint dots in great age; and pores don't stain. Under pines; A. Choice edible.

SUILLUS SPECIES ALMOST all live mycorrhizally with conifers. No American suillus is reported as poisonous, aside from warnings about allergic reactions (often blamed on the cap's skin, which you can peel before cooking). No single field trait perfectly distinguishes suillus from boletus; if you're going to try them, read the bullet points under boletus (above). Many suillus have at least two of the following traits: yellow tubes; tube mouths stretched out and aligned radially from the center; a partial veil that persists, at least for a while, as a ring; a glutinous skin on the cap that is slimy when wet, tacky when dry; or glandular dots on at least the upper part of the stem at maturity.

These mushrooms often seem mushy or slimy for a "good" edible, but that can be remedied by drying and then reconstituting them. Dried suillus powder—a cheaper substitute for porcini powder—adds umami flavor to many commercially prepared foods.

Mushrooms with a Distinct Cap Without an Underside

In these, the cap and stem differ sharply in color and in texture, but the cap has no underside covered with gills, wrinkles, or pores.

MOREL

Morchella spp. (mor-**kel**-a: from the Old High German name for morel). Cap conical or rod-shaped to egg-shaped or round, gray-brown or somewhat yellowish; exterior honeycombed with narrow ridges with deep pits in between; flesh brittle; interior hollow; no gills; cap is attached at its bottom edge to the stalk (exc in cottonwood morel); stalk smooth or wrinkled, whitish; odor slight or pleasant. Abundance peaks in late spring.

Black Morel

M. elata clade (ee-**lay**-ta: tall). Cap 1–3"; color variable, cap usually darker than stem, ridges may blacken with maturity; main ridges are all vertical. Montane conifer forest, esp in the 1st or 2nd spring after a fire.

Yellow morel

Black morel

Yellow Morel

M. americana. Also *M. esculentoides.* Cap 1–5" or more, typically tan; ridges more erratically oriented (not so strongly vertical), not darkening much with maturity. Under cottonwoods.

Fuzzy-foot Morel

M. tomentosa (toe-men-**toe**-sa: fuzzy). Also gray morel. Cap 1–4"; young specimens dark gray, coated with fine velvety hairs; fades to any of several shades with age, but with cap and stalk a roughly similar shade at any time; ridges vertical. Conifer forest from late spring through summer, a year or two after fire.

Cottonwood Morel

M. populiphila (pop-you-**lif**-il-a: poplar-loving). Also half-free morel; *M. semilibera.* Cap 1–2", fairly dark, conical, skirtlike, i.e., separate from the stem for part of its length. Under cottonwoods. In the *elata* clade.

FUZZY-FOOT AND SOME black morels are phoenicoid, rising phoenix-like from forest-fire ashes. Exactly how they do this is a fun puzzle for scientists. (One paper is titled "Where Are They Hiding? Testing the Body Snatchers Hypothesis in Pyrophilous Fungi.") Do they blow in as spores and grow rapidly because the competition has been decimated? Do they grow from deeply buried morel lumps called pseudosclerotia? Or did they survive as endophytes in plant roots?

Burn morel tissue has been detected inside plants as diverse as juniper and cheatgrass; they help cheatgrass to menace our ecology, but such endophytes can also benefit corn crops if inoculated into corn. Endophytic growth in plant roots is distinct from mycorrhizae. Black morels apparently don't form mycorrhizae, but yellow morels do, notably with cottonwoods in the Rockies. Some insect-killed forests see rich fruitings of yellow morels. Black morels popped up on barren ash at Mount St. Helens, and where they rotted back into the ash, algae and mosses soon flourished.

Both black and yellow morels are tricky to spot, camouflaged against their respectively burnt and unburnt forest floors. All morels are somewhat toxic raw. Most people can safely eat them cooked. They are the most lucrative class of wild mushroom shipped from the Rockies. To make sure they are morels, sniff their pleasing odor and slice one lengthwise to see how the honeycombed cap is attached at its bottom—no underside, only the full-length hollow interior shared with the stem. Often sandy, morels benefit from washing.

Morels are giving modern taxonomy a hard time. Like other fungi, they have many "sexes" (mating types). Unlike typical mushroom fungi, their cells have many genetically distinct nuclei, with the result that adjacent morels in the same cluster are genetically distinct, even when they arise from the same mycelium. In just the last fifteen years, molecular study has led to the publication of literally dozens of new and resurrected species names in genus *Morchella*. There are now at least twenty-two species in North America, up from half a dozen in 1980; even if their names become more set-

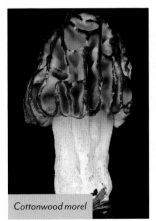

Cottonwood morel

Fuzzy-foot morel

tled, many pairs will be cryptic, meaning no one can identify them without sending a tissue sample to a lab. Some experts hold out hope that they'll learn to ID some species, provided they can watch a lot of caps as they develop over several days. Luckily, we have a workaround: we can identify the clades (branches of the family tree; see p. 14.) Black and yellow morels are each now well established as clades. In the yellow clade, only *M. americana* is common in the Rockies, so that's an easy ID.

SNOWBANK FALSE-MOREL

Gyromitra montana (jye-ro-**my**-tra: round hat). Also *G. gigas, G. korffii*. Cap dull yellow-brown, convoluted, lacking gills, flesh in cross-section thin, white, brittle; stem white, largely hidden by cap, nearly as big around as the cap and with similarly convoluted, thin, brittle flesh enclosing many irregular hollow spaces. Spring or early summer under high-elev conifers, near melting snow.

Gyromitra toxicity was long a mystery. Countless people over the centuries happily ate the so-called edible false-morel, *G. esculenta*. A few of them died. The toxin (after decomposing in your body) is monomethylhydrazine, a rocket fuel that sent astronauts into space—now a well-studied chemical known to be both poisonous and carcinogenic. However, it is volatile enough to be removed by either drying or boiling, and it typically produces no symptoms until a victim exceeds a certain threshold, presumably by eating a lot of inadequately boiled *Gyromitra* over a few days. Now that the explanation is known, mushroom guides say "POISONOUS" right under the name *esculenta,* which means "good to eat." Some people still eat *G. esculenta* after scrupulously boiling it.

As for the snowbank false-morel, chemical testing has generally found either no toxin or just a trace. Consequently, some guides call it toxic and compare eating it to Russian roulette, while others call it a good edible that requires parboiling as a precaution. I have eaten it happily. Steve Trudell and Joe Ammirati hedge wisely: "We recommend that it be cooked well and not eaten frequently or in large amounts."

Fungi Without Cap-and-Stem Form

Many familiar fungal fruiting bodies do not look at all like a mushroom. Think of coral fungi, shelf fungi on trees, and little odd-shaped jellies and ears on rotting logs. Then there are truffles—abundant but unfamiliar, because they are buried in soil.

NORTHERN RED BELT CONK

Fomitopsis mounceae (foam-i-**top**-sis: tinder lookalike; **moun**-see-ee: for Irene Mounce).

Snowbank false-morel

Northern red belt conk

Medicinal Shrooms

Ötzi, the Bronze Age man who froze to death on a glacier in the Alps and was found 5,200 years later, had three kinds of shelf fungi in his bag. He was probably using one of them, a species of *Fomitopsis*, to treat his intestinal worms. We deduce that because it is one of dozens of mushrooms with specific medicinal uses that are well known to ethnobiologists, and because Ötzi had whipworms. Studies find cholesterol-lowering, anti-cancer, antioxidant, anti-fungal, and anti-microbial properties in many mushrooms, including turkey tail, bear's head, witch's butter, hedgehog, and king bolete. Mushroom-based medicines are marketed in China, and one is officially approved in Japan, yet they remain on the alternative-medicine shelf in North America. (Many of the studies do not meet FDA standards.) This seems odd when you recall it was a fungus, *Penicillium*, that practically invented Western pharmacology. Go figure.

Also *F. pinicola, Fomes pinicola.* A dense, long-lasting shelf fungus on conifers, 3–12" across and about ⅓ as deep; initially a ± white knob, later hoof-shaped; becoming an inverted shelf (flat bottom, sloping top), blackish-brown at the base, with concentric arcs of bright, often shiny red-browns, ocher, or white toward the edge. Likely A.

The handsome red belt conk is a serious parasite on conifers. It causes "brown cubical rot," leaving long-lasting cubes of lignin in the soil, since it decomposes cellulose but not lignin.

Fomitopsis and aspen bracket both used to be included in a huge genus, *Fomes,* but that genus was split several ways, with only the true tinder conk remaining in it. *Fomes* was for many centuries cut up, soaked in saltpeter, dried, and then used as matches; what made that work was its unlayered, very long internal spore tubes. *Fomitopsis* species, in contrast, have short tubes because they start a new layer of them every year. The fruiting bodies grow for decades; up to seventy annual layers have been counted.

TURKEY TAIL

Trametes versicolor (**tram**-a-teez: woven; **verse**-a-color: multi-colored). Also *Coriolus v.* Dense

Turkey tail

shelf fungus lasting several months on dead wood or stumps, 1–4" across, often in huge clusters, thin and leathery, with beautiful concentric arcs of red-brown and deep brown shades, pale at outer edge, where newest; underside is a layer of minute white pores. A.

ASPEN BRACKET

Phellinus tremulae (fel-**eye**-nus: cork—; **trem**-you-lee: of trembling aspen). A dense, long-lasting shelf fungus very common on aspens, 1–4" across; initially shelf-shaped, later hoof-shaped, meaning too steep on top to set objects on; blackish-brown above, cracking with age; underside appears velvety, yellowish to dark brown. On aspen trunks; A.

Aspen bracket

Dyer's polypore

DYER'S POLYPORE

Phaeolus schweinitzii group (fee-**oh**-lus: dusky—; **shwigh**-nit-zee-eye: for Lewis David de Schweinitz). Also velvet-top fungus. Fleshy rosettes 3–16" across, flaring out and upward from a thick central point or short stalk; bands of color initially include rich yellow or orange near margin, but darken to browns; surface woolly or velvety; underside of fine greenish-yellow pores; flesh soft and moist in youth, later brittle and light. Grows in fall, persists to spring; resembles a shelf fungus and sometimes is one, on stumps or trees, but more often from soil near its conifer host; A.

Dyer's polypore is a common cause of heart rot in conifers, yet it isn't a terrible scourge like some diseases. It's basically normal for trees to get heart rot if they live long enough (some species rot sooner than others), and then many trees live with it for a long time. Eventually the trunk snaps.

This is one of many mushrooms and lichens that make beautiful dyes, and have been used in that way since biblical times.

CONIFER CHICKEN OF THE WOODS

Laetiporus conifericola (lee-tip-**or**-us: bright pores; conifer-**ick**-a-la: living on conifers). Also *Polyporus sulphureus* (sul-**few**-ri-us: sulphur-yellow). Thick, fleshy, shelflike growths 2–12" wide, typically in large, overlapping

Conifer chicken of the woods

Most shelf fungi live as heart rot in either dead or (as with aspen bracket) living trees. In either case, the cells they attack are dead cells, since only the thin outer layers of a tree trunk are live cells. That makes them saprobes technically, but they kill trees indirectly by weakening them to wind breakage, and they spoil the lumber, making themselves economic pests in timber country. From an ecological point of view, killing a tree here and there in the forest is beneficial: it provides habitat to many animals, lets in light, and altogether enhances plant and animal diversity. In the case of aspens, you can be sure that whenever an old stem dies, there are living roots all around, ready to send up new stems. You can imagine that in aspen's evolution, tolerating this common recycler of old stems would not harm the population; it would continually rejuvenate it. In our modern world, that may not always work perfectly because of a wild card: hoofed browsers.

Alpine jelly cone

Tree ear

clusters; orange above, yellow below. On conifers, summer through fall (long-lasting); northerly.

Witch's butter

Shelf fungi grow simple fruiting bodies that release spores through tiny pores on their smooth undersides. The pores are never separable from the cap flesh. Dense, rather woody flesh makes these fungi much longer-lasting than mushrooms, but it doesn't do a thing for their edibility. The one relatively tender exception is this "chicken," considered edible when young, soft, and bright yellow. A mycophagist convention once voted it the top edible mushroom, but one mushroom book calls it too sour (it's lemony) and one mushroom-toxin expert questions its safety.

ALPINE JELLY CONE

Guepiniopsis alpina (gwe-pin-ny-**op**-sis: looks like *Guepinia*, a similar fungus). Also poor man's gumdrop; *Heterotextus alpinus*. Top-shaped yellow-to-orange gelatinous bodies ⅛–1" across. On dead conifer wood or twigs soon after snowmelt, at higher elevs; A.

Orange-peel fungus

TREE EAR

Auricularia americana (or-ick-you-**lar**-ia: little ear). Also *A. auricula*. Brown (drying blackish), ear-shaped or cupped shelves; texture earlike (rubbery tough) when moist, hard when dry; 1 side is spore-bearing and usually concave, often facing down; sterile side is silky-hairy. On dead conifer wood; common near melting snow; A.

Tree ears are widely eaten in Chinese soups and other dishes; for Anglos, their rubbery texture may be an acquired taste. Witch's butter is sometimes substituted, but its flavor is missing in action. Tree ears are saprobes—rotters of already dead wood.

WITCH'S BUTTER

Tremella mesenterica (trem-**el**-a: trembling like jelly; mez-en-**tair**-ic-a: like intestines). Mass of leaflike lobes, golden yellow to orange and gelatinous but tough when moist; hard, small, dull, and inconspicuous when dry; can become a shapeless blob when old and wet. Spring to fall, on rotting broadleaf trees, where it lives by parasitizing a rot fungus; A.

ORANGE-PEEL FUNGUS

Aleuria aurantia (a-**loo**-ri-a: flour; aw-**ran**-ti-a: orange). Bright orange, thin-fleshed cups ½–4" across; underside paler; no stalk; brittle, not tough. Gravelly soil (e.g., road shoulders)

Scorched-earth cup

where a forest burned and no scorched-earth cups turned up afterward, forest failed to reestablish. Their mycelium helps prevent erosion of the scorched earth—a benefit that we lose out on when we salvage-log after fires. As saprobes, they also contribute nutrients by recycling them.

YELLOW CORAL FUNGUS
Ramaria rasilospora (ra-**mair**-ia: branched; ra-sil-a-**spor**-a: shaved spores). Somewhat cauliflowerlike structure, with massive white stems branching many times before reaching the buff, cream, or pale yellow tips; odor and flavor mild; spores pale yellow. Under conifers, mainly spring; A.

or grass or moss, often grouped; summer and fall; likely A.

This common fungus, edible but insipid, can release visible puffs of spores.

SCORCHED-EARTH CUP
Geopyxis carbonaria (geo-**pix**-iss: earth goblet). Also fire-following fairy cup, stalked bonfire cup. ½", smooth, light-brown cups with pale, perhaps ruffled, rims; usually on very short stalks (often not visible until picked); no gills. Abundant in the 1st to 3rd year after fire, spring to fall; A.

Hot tip: these tiny cups come up in many of the same spots as black morels—typically in fire-thinned duff very near fire-killed trees—and often before the morels. In Sweden, scientists found that in the rare instances

Coral-like branching fungi grow on the ground or on rotten wood. They include a few stomach-upsetting species with translucent, gelatinous flesh at their cores. *R. largentii,* coming up in summer and fall in Colorado, ranges to deep yellow-orange, while *R. formosa* usually has a peachy-pink cast and strong yellow tips. Still more beautiful are fairy fingers, *Alloclavaria purpurea,* deep purple and found on boggy trailsides.

CROWN CORAL FUNGUS
Artomyces pyxidatus (ar-to-**my**-seez: narrow fungus; pix-i-**day**-tus: with goblet shapes, a poor description of the crowns). Also *Clavicorona pyxidata.* Crowded clusters of upright 2–5" stalks,

Fairy fingers

Yellow coral fungus

Crown coral fungus

and the "scenic resource" categories; trailside specimens should be left alone for others to see. The fragrance and flavor are lovely. The larger branches are tough, demanding longer, gentle cooking. Cut into the mass discreetly, removing only modest quantities of such parts that are pure white and have few or no boreholes from insect larvae. Soak them in a pot of water to float any beetles out from the crevices. Then mince and sauté gently.

each topped with a 3-to-6-pointed crown unique to this species; pale cream, aging dull ocher, tan, or pinkish; from dead aspen or other hardwoods.

Back when fungus taxonomy was based on what the fruiting bodies look like, coral fungi were an obvious group. Molecular study has scattered the corals to the four winds; this one's closest relatives are *Russula* spp. (p. 270).

PUFFBALL

Lycoperdon and *Apioperdon* spp. (lie-co-**per**-don: wolf fart; ay-pee-o-**per**-don: tip fart). Stalkless balls, lacking any kind of gills; upper skin covered with fine, crumblike bumps, turning light brown; flesh initially pure white, marshmallow-textured, later turning first brown then green and soggy.

Gemmed Puffball

L. perlatum (per-**lay**-tum: pearl-studded). Upper skin covered with fine, crumblike bumps, turning light brown. In large clusters on soil; A.

Pear Puffball

A. pyriforme (pie-rif-**or**-me: pear shape). Also *Lycoperdon pyriforme*. Few or no crumblike bumps. Rotten wood or in wood-rich soil; A.

BEAR'S HEAD

Hericium abietis (her-**iss**-ium: hedgehog; ay-bih-**ee**-tiss: of fir trees). Large, 6–12", or rarely up to 30" high × 16", white to cream fungus consisting of many branches and sub-branches ending in fine teeth all pointing down, lengthening with age; rounded overall, with crowded, short branches from a massive, solid base; spores white. On dead conifer wood or tree wounds, late summer, early fall; northerly. A similar fungus on dead cottonwood would be coral tooth, *H. coralloides*.

Bear's head's distinctive beauty puts it in both the "safe for beginners"

Bear's head

Gemmed puffball

Warted giant puffball

A MATURING PUFFBALL either splits open or opens a small hole at the top, and spores come puffing out by the millions, mostly when raindrops strike. Spores, like pollen, resist wetting, so their flight is undamped by rain or fog. Using raindrops as the trigger is a brilliant tactic in the Rockies, where summer raindrops so often coincide with thunderstorms and strong updrafts. Fungal spores are a hundred times smaller than pollen grains. One good-sized gemmed puffball produces around four billion spores. Squeeze a ripe puffball on a stormy day, and you may send a few spores around the world.

Some Indigenous peoples used puffball spores to dress wounds. If you try that, take care not to inhale clouds of spores, which can infect the lungs.

Young deadly amanitas can be mistaken for puffballs, so before eating any puffball, slice it through its center vertically, and eat only those that are undifferentiated white and marshmallow-soft inside, from tip to toe. Amanita buttons in cross-section reveal at least a faint outline of developing stem and gills. Other telltales for the discard pile would be a green-, brown-, or yellow-stained center (a puffball starting to mature into a bag of spores) or a punky center distinct from a smooth, ⅛-inch outer rind (a *Gastropila* or a *Scleroderma*). Puffballs take a couple of weeks to mature, so you stand a good chance of catching them young. They benefit from treatments that add texture and flavor: grilling, batter-frying, and ample butter and garlic. They have some of the highest protein content of any vegan food item.

WARTED GIANT PUFFBALL

Calbovista subsculpta (cal-bo-**vis**-ta: bald fox fart; sub-**sculpt**-a: somewhat sculptured). Stalkless, slightly flattened ball 3–6" across, white, patterned with brownish raised polygons. Scattered among subalpine grasses or under conifers, midsummer.

Most puffballs are smaller than golf balls, but *Calbovista* is baseball-sized, and its *Calvatia* cousins can grow to basketball size. (That would be the smooth-skinned edible *C. booniana,* which reaches our region in low-elevation grasslands.) The name *Calbovista* is a compound of two other genera: *Calvatia* is Latin for "bald head," while *Bovista* ("fox fart") is old German vernacular for a puffball.

If it's to be your first taste of puffball, just pick a small sample. You may hate it. The flavor is mild and pleasantly fungal, while the texture is at once slippery and marshmallowy. Frying hot enough to brown the surface helps. My finest trailside lunch was a warted giant puffball sandwich: fry half-inch slices in butter with garlic, then place on crackers thinly layered with anchovy paste and cheese.

FUZZY TRUFFLE

Geopora cooperi (ge-o-**por**-a: hole in the earth; **coop**-er-eye: after J. G. Cooper). Roundish lumps ¾–3" across; exterior shallowly furrowed, covered with light- to dark-brown fuzz; interior a convoluted mass of white to cream or tan folds. Buried or in the duff layer, or—if partially buried—often partly eaten; under other conifers, or sometimes willow or aspen in the north, Apr–Oct; possibly A.

Truffles live mycorrhizally just like many mushrooms, but instead of giving their spores to the wind to carry, they use distinctive aromas to entice mammals to eat them.

(See pp. 329–330.) Some get by with tiny fruits that get moved around accidentally by burrowing animals. Hundreds of trufflelike (i.e., aromatic, underground-fruiting) species abound in our forests. Most are rarely seen, but the fuzzy truffle tends to reach the surface as it ages, perhaps when a squirrel digs it up. If you interrupt a squirrel messing around in the duff, check what it was messing with. If you interrupt a snail dining on one, expect slime.

This is not an exorbitantly prized truffle. Expert opinions vary: Larry Evans describes the fuzzy's aroma as "fruity and morel-like at once," but Michael Kuo counters that it's "sour and reminiscent of bad apple cider."

Fuzzy truffle in cross-section

Dogs and sow pigs have been used to sniff out truffles, finding them at their aromatic peak. Both need rigorous training, though sows come partially preprogrammed: some truffles share an aromatic compound with boar pheromones. However, sows may eat the truffle before they can be restrained, and they damage the mycelium. Truffling with pigs is a thing of the past—banned in Italy since 1985. Without four-legged help, a rake is the usual tool, but careless raking also damages mycelium and wastes truffles, finding them before they ripen into anything worth eating.

The king of Prussia commissioned mycologist Albert B. Frank to cultivate truffles in the 1880s. Professor Frank failed to grow truffles, but he looked so closely at how they work that he made arguably the greatest single advance in mycology: the discovery and naming of mycorrhizae.

Lichens

A lichen is a sophisticated organism in which one or more kinds of fungi enclose one-celled photosynthetic organisms—algae, cyanobacteria, or both. The lichen produces organs, tissues, and chemicals that are never seen in nonlichenized fungi, algae, or cyanobacteria. Unlike mushrooms, lichens grow in the open and are tough, durable, and perfectly able to revive after drying out. The lichen symbiosis seems to have evolved separately in seven branches of the fungus family tree. Its original occurrence may have been the breakthrough that life needed in order to move from the sea onto the land.

Lichen fungi get their food from their photobionts—the algae or cyanobacteria—which benefit by being able to grow on sites too severe for their non-lichenized relatives. This process has been compared to farming: a farm's irrigation enables plants to grow in climates too dry for them, while barns and hay let cattle live in climates that would otherwise be too snowy. The lichen offers physical shelter and also chemical protection: lichen chemicals repel herbivores, microbes, and competing plants. You could say these are counterparts to a farmer's greenhouse, pesticides, herbicides, and deer repellent. Most lichenologists agree to call this symbiosis a beautiful mutualism, but some have floated other views: that the fungus is slowly killing the alga, or that the alga hijacks the fungus to build a perfect house for it, like a gall wasp inducing a plant to build a gall for its larvae.

Lichens exemplify a common theme in the diversity of life on Earth: simpler organisms living as symbionts inside bigger organisms are the rule, not the exception. For example, photosynthesis is performed by chloroplasts, which originated as cyanobacteria that survived inside bigger, one-celled organisms that tried to consume them. Most green plants benefit from fungal or bacterial endophytes in their leaves, or mycorrhizae invading their roots. Then there's the human body: nine out of ten cells in it are bacteria, most of which benefit us. But lichens go beyond all the above in that symbionts join up to form a common body.

The photobiont, with only about 7 percent of the tissue mass of a typical lichen, is the junior partner. It was once assumed that a lichen spore can't begin to farm unless it happens to land near a free alga of the right type. Alternative speculation holds that a spore can grow into a slight smudge called a prethallus, and then survive meagerly for years waiting for the right alga to turn up—and then for the two to form a lichen.

To varying degrees, different lichens do release spores that reproduce. Spores are produced in the fruiting disks you'll find in the species descriptions. While sexual propagation via spores is important for genetic diversity, it is far less common than propagation from fragments that include both partners together. These may be just any old fragment that happens to break off, or they may be specially evolved propagative bun-

Air Pollution Indicators

Lichens have discriminating tastes for the airborne solids and solutions that, after all, they live on. They accumulate air pollution in their tissues along with natural aerosols, and it can kill them. Since different lichens tolerate different amounts and types of pollutants, scientists can use them to monitor an airshed's pollution by analyzing tissue samples in the lab, or simply by checking which species are still present. Compared to testing air samples directly, this has the advantage of averaging pollution levels over several years.

On the other hand, a few lichens thrive on some kinds of pollution. Concerned about ancient rock art fading rapidly in Hells Canyon, archaeologists asked the lichenologist Linda Geiser to find out if air pollution could be the culprit even in that remote wilderness. The telltale clue she spotted was not a missing lichen species, but abundant ones—a belt of bright-orange sunburst lichens (p. 293) low on the canyon walls near rapids. Those lichens need extra nitrogen or calcium, and typically get them, at least in wilderness, from urine or guano, thus marking marmot or pika perches. Here, nitrogen probably came down the river from Idaho's feedlots, fertilized fields, and fish farms, and went airborne as nitrate, feeding the lichens and eating away the petroglyphs.

dles (propagules) in the form of powders, grains, or tiny protuberances.

Of all familiar, visible organisms, lichens have perhaps the most modest requirements—a little moisture, any amount of sunlight from minimal to extreme, and solid materials that even the cleanest air carries. From minimal input comes modest output; growth rates measured in lichen species range from 0.0012 to about 4 inches per year.

Success in lichen niches is a matter of making hay while there's moisture—rain, snow, dew, fog, or even just humidity—and then drying up, suspending all activity. When dry, lichens can survive temperatures from 150°F (typical of soil surfaces in the summer sun) down to near absolute zero (in laboratories). Drying out and suspending respiration is the only way they can get through hot days in the sun. At 32°F or a bit colder, they remain active and unfrozen, thanks to complex chemistry that includes alcohols. Our alpine lichens get some of their photosynthesizing done while buried in snow, which keeps them moist and lets sunlight in. While alpine rocks with their extremes of sunlight

and temperature are no problem, far more of our lichen biomass and much faster growth rates occur in an environment with moderate temperatures and reduced light levels: tree bark in shady forests. There, too, the lichens dry out many times a year when the rain and fog lapse. Most lichens need to dry out periodically, or they will die.

Lichens that colonize bare rock may, after centuries of life, death, and decomposition, alter the rock enough for plants to grow, initiating the long, slow succession of biotic communities. This scenario has captivated countless naturalists, from Linnaeus to our high school biology teachers, though there are certainly other common paths of primary succession on barren, newly exposed ground. (Associations with nitrogen-fixers are typical: mosses loosely coated with cyanobacteria pioneer on rock, and nitrogen-fixing plants like lupine and alder sprout from barren ash or even just crevices that hold a bit of fine gravel or dust.)

As for crumbling rocks, frost riving is the main process, but lichens help. Their hyphae penetrate rock and secrete acids that dissolve minerals, creating pore spaces. Their gelatinous gripping surfaces expand and contract as they moisten and dry out, slowly crumbling some kinds of rock. (In a lab, plain gelatin can replicate this effect on rocks.)

Lichens have been suggested as candidates for restarting succession, or even evolution, after a global ecocatastrophe. This idea may have been inspired by Arctic lichens that accumulated radioactive fallout and survived, even as they passed harmful amounts of it into the bones of lichen-eating caribou and caribou-eating Inuit and Saami. There may be something to this myth. Lichen lineages are so old that they must have survived three of the big five mass extinctions.

Lichens are informally sorted by growth form into crust lichens, leaf lichens, and shrub lichens.

Crust Lichens

Crust lichens are thin coatings or stains found on or in rock or bark surfaces—so thin that you could not pry them up.

MAP LICHEN

Rhizocarpon geographicum (rye-zo-**car**-pon: root fruit). Chartreuse-yellow patches broken by fine black lines and surrounded by a wider black margin. On rocks; abundant in exposed situations; A.

Crust lichens grow slowly. Because they live for thousands of years and are ubiquitous on Arctic and alpine rocks, map lichens dominate the science of lichenometry, which uses lichen growth to measure time—for example, the time since glaciers retreated from rocks that now bear lichens. While lichenometry always requires local calibration, a typical

Biological Crusts

In alpine and cold steppe environments, soil surfaces are often a structured mix of very small lichens, cyanobacteria, mosses, and fungi. The resulting crust, ranging from an eighth of an inch to over an inch thick, often goes unnoticed, but is vital to plants. Estimated to cover 12 percent of the global land surface, biocrust stays relatively moist in the sun, resists erosion, and often fixes nitrogen.

Biological or cryptobiotic ("hidden life") crust is easy to damage and slow to repair itself. Rocky Mountain air is measurably dustier than it used to be, primarily because cattle broke up much of the biocrust of the Great Basin. A footprint in biocrust can last for decades. If you hike off-trail above tree line, watch for nubbly, crusted, sandy earth and avoid walking on it. Detour onto snow or bare rock, ideally; raw gravels or dense sedge turf are second best. In ungrazed steppe or semidesert, alternatives to crust may be hard to find; staying on the trail is best.

Biologists are trying to learn how to foster biocrust growth by seeding the soil surface with appropriate cyanobacteria. In a recent study, inoculated crusts grew much faster where they were shaded under solar panels. Solar farms could become nurseries for inoculum cultures. Crusts develop in ways analogous to forest succession, with various community members facilitating—or sometimes discouraging—the growth of others. For example, certain cyanobacteria evolved chemicals that act as a sunscreen—protecting them from ultraviolet damage—while also darkening the crust, so that it absorbs more heat. If those darker bacteria take over a crust, it can get too hot for certain other crust cyanobacteria. Climate warming itself threatens biocrusts.

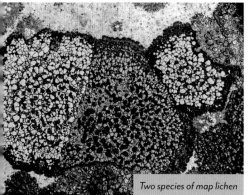

Two species of map lichen

Brown tile lichen

radial growth rate for map lichens might be an inch per century at first, then slowing.

GOLD COBBLESTONE LICHEN

Pleopsidium flavum (plee-op-**sid**-ium: full faces, ref to a microscopic part; **flay**-vum: yellow). Brilliant yellow, shiny, pebbly crust; fruiting disks brownish, common but inconspicuous. Abundant on basalt, all elevs; common on other non-calcareous rocks at high elevs; A.

Recognizable from a distance, this lichen paints landscape-scale murals across cliffs of Columbia River basalt; in the Rockies, it's less obvious but still widespread.

BROWN TILE LICHEN

Lecidea atrobrunnea (les-**id**-ia: small dish; at-ro-**brun**-ia: black-brown). Pebbly crust of black (fruiting disk) and dark brown tiles with thin grayish edges, ± mixed near the center, all brown toward the outsides. Common on mainly igneous rocks; A.

BLACK-AND-GOLD TILE LICHEN

Calvitimela armeniaca (cal-vit-im-**mee**-la: naked honey; ar-men-**eye**-a-ca: of Armenia). Also *Tephromela armeniaca*—and expect the genus name to change again. Pebbly crust mostly of dull-black and honey-yellow tiles, and some brownish. On high-elev, non-calcareous rocks; A.

GOLDSPECK LICHEN

Candelariella vitellina (can-del-airy-**ell**-a: small brilliant; vit-el-**eye**-na: egg yolk—). Bright

Gold cobblestone lichen

Black-and-gold tile lichen

Goldspeck lichen

Zoned dust lichen

Spraypaint lichen

Sulfur dust lichen

SPRAYPAINT LICHEN

Icmadophila ericetorum (ick-ma-**doff**-ill-a: mud-lover; er-iss-e-**tor**-um: among heaths). Also fairy barf, candy lichen. Pale blue-green (moist) or nearly white (dry) granular crust dotted with white-rimmed, slightly raised, pink fruiting disks; the disks are on short stalks if you look closely enough. On rotting wood, or growing over mosses; A–.

Dust Lichens

This degenerate lichen growth form (also known as imperfect lichen or powdery paint lichen) is just a powdery layer of fungi and algae bundled together in little clumps called soredia. Many lichens bear soredia as reproductive propagules. They're the "granular" or "powdery" stuff in several species descriptions. Apparently dust lichens evolved repeatedly (in several unrelated lichens) when certain soredia learned the trick of propagating more soredia without bothering to propagate the rest of the lichen. Lacking the continuous cortex layer that most lichens have, dust lichens are more vulnerable to drying out, but better able to absorb moisture

egg-yolk-yellow crust (not as orange as sunburst lichens, p. 293) of small bumps; fruiting disks same yolk color, with raised edges. Common on granitic rocks, sometimes wood; A.

from the air: they live by absorbing humidity from the air, but they repel raindrops and drips, being unable to absorb liquid water. They specialize in rain-deprived substrates like the undersides of limbs, and rock overhangs. They like foggy climates, and go crazy in shaded waterfall spray zones.

Early naturalists filed dust lichens under *Lichenes Imperfecti* within Deuteromycetes, two taxonomic dustbins revealing their inability to categorize fungi that had never been seen with sexual organs. Modern molecular techniques allow for them to be transferred from these dustbins to their rightful places throughout the fungus family tree.

Dog lichen

Membranous dog lichen

ZONED DUST LICHEN

Lepraria neglecta (lep-**rair**-ia: leprous, i.e., scurfy). Speckly bluish/pale-gray coating; may form concentric rings. All elevs; abundant on alpine soil or (often blackened) moss; A.

Zoned dust lichen is ecologically important for stabilizing alpine soils. It frosts bare soil or moss with white, typically producing a nubbly texture of 2-inch mounds.

SULFUR DUST LICHEN

Chrysothrix chlorina (**cris**-o-thrix: gold thread; clor-**eye**-na: greenish-yellow). Bright greenish-yellow, scurfy, granular coating. Spectacular on moist, shaded cliffs; northerly.

Leaf Lichens

Leaf lichens range from salad-like heaps attached to their substrate in only a few spots to almost paint-like sheets with distinct, thick margins that can be pried from the substrate.

DOG LICHEN

Peltigera spp. (pel-**tidge**-er-a: shield bearer). Also veined lichen, pelt, frog's blanket. Big sheets with lobed and curled-up margins; conspicuous tan to dark red, ± tooth-shaped fruiting bodies on lobe edges; underside of some spp. netted with branching veins, raised from surface, bearing coarse, hairlike rhizines.

Dog Lichen

P. canina (ca-**nye**-na: of dogs). Gray or brownish, covered with minute hair visible with hand lens; lobes often more than ¾", margins downcurled; rhizines tufted. Forest soil; A.

Field Dog Lichen

P. rufescens (roo-**fes**-enz: reddening, a misleading name). Gray or brownish, covered with minute hair; lobes less than ¾" wide, margins ± upturned, often whitish; rhizines become densely matted toward the center. Abundant in sun, from dry forests to alpine crusts; A–.

Membranous Dog Lichen

P. membranacea. Gray or brownish, covered (at least near edges and on underside veins) with minute hair; lobe margins downturned, with

Ruffled freckle pelt

Lungwort

Ragbag lichen

thick, dark veins grading to white near edges, with tufts of unmatted rhizines. On soil or rotten wood; A–.

Freckle Pelt

P. aphthosa (af-**tho**-sa: blistered). Bright green above when wet, pale green to tan when dry; upper side freckled with raised, dark bumps; lobe edges ± flat; rhizines matted. On soil and rocks, both sun and shade; A–.

"FROG'S BLANKETS," MY favorite name for these lichens, is a translation of what British Columbia's Gitxsan people called them. The European name dates from the medieval Doctrine of Signatures, which looked in nature for semblances of body parts and read them as drug prescriptions writ in God's own handwriting. In the dog lichen's erect fruiting bodies they saw dog teeth, so they prescribed dog lichen for dog-bite disease: the decoction for rabies, right up until the last century, was ground dog lichen and black pepper in milk. The doctrine was a wrong turn in the history of herbal medicine. Lichens do have medicinal value, and they were used more appropriately in pre-Christian Europe.

Dark-gray lichens like the dog lichen have nitrogen-fixing cyanobacteria, often of genus *Nostoc*, as their chief photosynthesizing partners. Green species like freckle pelt employ green algae throughout, adding cyanobacteria only within their "freckles."

LUNGWORT

Lobaria pulmonaria (lo-**bair**-ia: lobed; pull-mun-**air**-ia: lung—). Also lettuce lichen. Big "leaves" olive to pale brown above when dry, green when wet, with a netlike pattern of ridges sprinkled with pale "crumbs" (on underside, these ridges are valleys between domes); underside paler, mottled with brown in the furrows; margins deeply lobed, curling; fruiting disks rare. On trees or mossy rocks—either in very old forest or near cottonwood or aspen; WA to nw MT.

Cyanobacteria in lichens pull nitrogen out of the air; the nitrogen moves on in the nutrient cycle when it leaches out in rainwater, or

conspicuous, separate rhizines often ⅜" long. Moist forest floor; A–.

Ruffled Freckle Pelt

P. leucophlebia (loo-co-**flee**-bi-a: white veins). Bright green above when wet, with many warty, dark bumps; pale green to tan when dry; lobe edges crisped or ruffled; white underside has

when insects or decomposing bacteria consume the lichen. To intercept nutrients before they wash away, cottonwoods and alders may extend rootlets among the lichens and mosses on their own bark.

"Lungwort" derives from the medieval Doctrine of Signatures, which prescribed this lichen for lung ailments because of its textural resemblance to lung tissue. *Mertensia* and at least six other green plants were also called lungwort and prescribed for lung ailments.

Elegant sunburst lichen

RAGBAG LICHEN

Platismatia glauca (plat-iss-**may**-sha: broad—; **glaw**-ca: whitish). Fluffy wads of small sheets (up to 1" wide) with strongly ruffled edges; pale greenish gray above; underside patchy with white, greenish brown, and black; surface often granular-coated. On trees in moist forest; northerly, plus a few CO canyons.

Ragbag lichen tolerates pollution and is common on small urban trees.

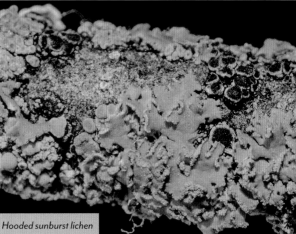

Hooded sunburst lichen

It's a cosmopolitan lichen, native to six continents. Like the tube lichens (p. 300), it grows throughout young conifer forests, but shifts into the mid-canopy as the forest ages; in ancient forests, you see it mainly where it has fallen to the ground.

ELEGANT SUNBURST LICHEN

Rusavskia elegans (**el**-eg-enz: elegant). Also elegant orange lichen; *Xanthoria elegans*. Bright orange patches adhering tightly to rocks, easily mistaken for a crust lichen, but with slightly raised flakes and distinct, lobed edges; fruiting discs deeper orange, small, concentrated near center of patch. On sunny rocks fertilized with nitrogen or calcium; A.

Elegant sunburst lichen marks habitual perches of rockpile-dwelling pikas and marmots, where it is fertilized by nitrogen in urine. More conspicuous than the pika

itself, this orange splash may help us spot the source of that little "eeenk" sound.

The orange pigment is a sunblock protecting the lichen algae from ultraviolet light. It's so effective that this lichen survived being experimentally subjected to unearthly levels of ultraviolet plus cosmic radiation in space, at the International Space Station. Elegant sunburst lichen grows from seashores up to 23,000 feet in the Karakoram Range—possibly the greatest elevation range of any macroscopic species.

HOODED SUNBURST LICHEN

Xanthomendoza fallax (zanth-o-men-**doe**-sa: yellow, for Fernando Fernández-Mendoza; **fal**-ax: deceptive). Also *Xanthoria fallax*. Deep-orange clumps of lobes, often quite small, with relatively large, raised, orange fruiting disks. On

Orange chocolate-chip lichen

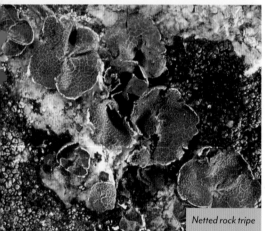

Netted rock tripe

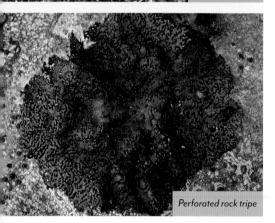

Perforated rock tripe

orange, showing at many upturned edges, finely woolly; with brown veins, sparse rhizines. High-elev soil; A.

This showy lichen excels on late-lying snowbed sites, often alongside mosses that get started there with the help of nitrogen-fixing cyanobacteria on their leaves. The lichen can then incorporate those same bacteria, but may remain quite small for years before the right green algae blow in and enable it to realize its full potential.

ROCK TRIPE

Umbilicaria spp. (um-bil-ic-**air**-ia: navel—). Dark-gray to light-brown lichens attached to rock by a single, central "umbilical" holdfast; tough and slippery when wet, hard and brittle when dry; ± lobed and curling at the edges; fruiting disks uncommon in first 3 described. Abundant on non-calcareous rock.

Netted Rock Tripe

U. decussata (de-cuss-**ay**-ta: crossed). Dark gray, crossed with delicate white ridges in an almost roselike pattern; underside smooth, black. A.

Blistered Rock Tripe

U. hyperborea (hy-per-**bor**-ia: of the far north). Dark gray, pebbly, and irregular all over; underside smooth, tan to black. Abundant; A.

Grizzled Rock Tripe

U. vellea (**vel**-ia: velvety). Pale or brownish-gray; underside with a patchy coat of minute, pale rhizines interspersed with short, black rhizines. High elevs.

Perforated Rock Tripe

U. torrefacta (tor-a-**fac**-ta: roasted). Medium brown with numerous black speckles (fruiting disks) and fine perforations near edges; underside light brown with tiny plates. A.

COASTAL PEOPLES SOMETIMES added boiled rock tripe to herring eggs, a prized delicacy, to make the eggs go further. In Asia, people fry it up to eat like potato chips, or relish it tender in salad or soup, always drying and boiling it first to leach out dark and bitter flavors. After boiling it for three hours,

twigs or bark of deciduous trees and shrubs, incl sagebrush; A.

ORANGE CHOCOLATE-CHIP LICHEN

Solorina crocea (so-lor-**eye**-na: sunny; **cro**-sia: crocus, i.e., saffron). Upper surface bright green (wet) to dull pale green (dry), with dark-brown fruiting disks; underside bright

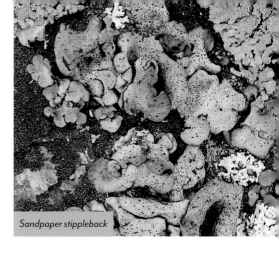
Sandpaper stippleback

Ernest Thompson Seton damned it with faint praise: "the most satisfactory of all the starvation foods." He credited it with saving Franklin and Richardson (p. 447) on their desperate Arctic trek in 1821: "Their diet was varied with burnt bones when they could find them and toasted leather and hide; but the staple and mainstay was rocktripe."

Scots scraped rock tripe from rocks for use as a dye that ended up in Harris Tweeds and litmus paper. Many lichens yield yellow, olive, or brown dyes when boiled in water; the shift to purple and red occurs when certain lichens are cured for weeks in warm ammonia. For at least two thousand years, the ammonia used was stale human urine. The dyed fabric came out with a fine, long-lasting fragrance. Dye-lichen gathering rarely flourishes for long in one locale, since lichens grow so slowly that they are soon depleted.

STIPPLEBACK

Dermatocarpon spp. (der-mat-o-**car**-pon: leather fruit). Also leather lichen. Gray to tan, leafy lichen attached to a rock by a single, central holdfast-like rock tripe (above) but thicker and finely speckled; ± lobed; tough and slippery when wet, hard and brittle when dry, or may take a balled-up, vagrant or tumbleweed form (see sidebar, below).

Vagrant Lichens

Since lichens live on air, dust, sunlight, and atmospheric moisture, they don't necessarily need to be rooted at all. In some wide-open habitats, you may find perfectly healthy lichens utterly on the loose. Their closest relatives can be either crust, leaf, or shrub lichens, and they range from fully vagrant species, to erratic growth forms, to mere accidents—that is, formerly attached individuals that got knocked off their holdfasts.

When a specialized growth form does occur, the branches are balled up, a little like a tumbleweed—with the difference that tumbleweeds don't tumble until they're dead. It's tempting to imagine dispersing vast distances on the wind as the main draw of vagrant life, but most likely the "normal" propagules—spores and lichen bits—blow farther. Many vagrant lichens specialize in small habitats like pans where rainwater persists. Noting a geographic correlation with pronghorn range, Roger Rosentreter speculates that a vagrant lichen's best hope of traveling far in the American West lies in catching on a pronghorn's fur.

Pronghorns enthusiastically eat tumbleweed rock-shield, and might disagree with the Wyoming agencies that blame the same lichen for mass poisonings of sheep and elk. A Persian village was once saved from wartime starvation when a windstorm covered it with vagrant lichens, *Lecanora esculenta,* which they recognized as edible and baked into a sort of bread. The same story is the most plausible explanation for the manna from heaven that saved Moses and company.

Green rock posy

Tumbleweed rock-shield

Green starburst lichen

Colorado rock-shield

Blushing Stippleback

D. miniatum (mi-ni-**ay**-tum: scarlet, not normally but in response to a chemical test). Undersurface ± brown, smooth to net-patterned. On limestone or basalt; A.

Sandpaper Stippleback

D. reticulatum (ret-ic-you-**lay**-tum: netted). Undersurface ± black, sandpapery rough. On non-calcareous rocks; A.

GREEN ROCK POSY

Rhizoplaca melanophthalma (rye-zo-**play**-ca: rooted plate; mel-un-off-**thal**-ma: dark eye). Also rock-brights, rock-brooches, manna; *Lecanora melanophthalma*. Small (1") but sturdy, round gray-green lichen largely covered with ± darker-centered, white-rimmed fruiting disks; a few small lobes showing at edges; may take a balled-up, vagrant form (see sidebar, p. 295). On rocks or old, debarked wood; A.

The white-rimmed fruiting disks in this genus almost dwarf the vegetative lobes, suggesting a bouquet of flowers.

GREEN STARBURST LICHEN

Parmeliopsis ambigua (par-me-lee-**op**-sis: shield lookalike). Pale yellow-green rosette of branched, radiating, narrow (less than ⅛") lobes; fruiting disks few or none; underside ± black with sparse rhizines; tiny mounds of pale, granular soredia are scattered on top, mainly near center. On bark or dead wood; A.

COLORADO ROCK-SHIELD

Xanthoparmelia coloradoensis (zan-tho-par-**meal**-ia: yellow shield—; col-o-rad-o-**en**-sis). Pale greenish-gray rosette of branched lobes,

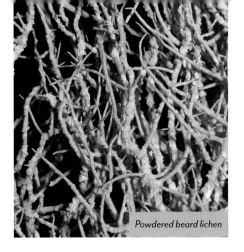
Powdered beard lichen

their tips raised from the substrate; underside tan with tan rhizines; center of rosette usually decorated with brown fruiting disks. On rock; A. Similar *X. wyomingica* has narrower lobes and is found typically on alpine gravel, occasionally vagrant.

TUMBLEWEED ROCK-SHIELD

Xanthoparmelia chlorochroa (clor-o-**cro**-a: green ocher). Pale, greenish-gray, narrow branches rolled almost into tubes, forming a cushion or a messy tangle; no fruiting disks; underside smoky gray-brown, blackish in part, with numerous black rhizines not attached to substrate. Vagrant (p. 295), free to blow about on high plains or semidesert; A–.

Navajo rug and basket weavers collect this lichen for a reddish-brown dye.

Shrub Lichens

Shrub lichens are three-dimensional and often much-branched, the branches ranging from slender hairs to robust clubs, either erect or pendent.

OLD-MAN'S-BEARD

Usnea spp. (**us**-nee-a: Arabic term for some lichen). Pale greenish-gray tufts on trees; when stretched gently, the thicker branches (unless very dry) reveal an elastic, pure-white inner cord inside the brittle, pulpier, skin—like wire in old, cracked insulation.

Powdered Beard Lichen

U. lapponica (la-**pon**-ic-a: of Lapland). Somewhat pendent tufts 1½–3" long; many fine, perpendicular side branches; under 10× lens, see scattered, concave, oval ruptures full of mealy propagules. A.

An Explorer Who Aspired to High Places

Captain **John C. Frémont** explored the West throughout the 1840s. A civil engineer by training, he became enthusiastic about plant collecting, inspired by watching the diligent botanist Karl A. Geyer (p. 142) in Iowa in 1841. The admiration was unrequited: Geyer wrote that Frémont didn't "understand anything about Botany" and grumped that Frémont was able to collect many new species because, traveling with plenty of soldiers to protect him from attack, he could go places where lone botanists—well, at least Geyer himself—dared not tread.

One of the most exciting places Frémont visited was the heart of the Wind River Range, where with Kit Carson for a guide he made a point of becoming the first white man astride the highest pinnacle (now named for him—but actually only second highest). Atop Fremont Peak, he boasted of another first when he captured "*bromus*, the humble bee. . . . It is certainly the highest known flight of that insect." Sorry, Captain: humble bees did become bumblebees over time, and they still visit the tops of the Rocky Mountains, but they were always *Bombus*, and *Bromus* was always a grass.

A grandiose man, Frémont conquered Spanish California on his own initiative, and ran for president as an ardent abolitionist and westward expansionist.

Pitted beard lichen

Edible horsehair

Exploding beard lichen

Nevertheless, there are cottage industries producing lichen extracts for medicinal and cosmetic use. Usnic acid is named for this genus; the acid extracted from it and several others is used as an antibiotic. You can find it in first-aid salves in Sweden, in teas and throat lozenges in Switzerland, and in the deodorant in my kit. I vouch for it!

HORSEHAIR LICHEN
Bryoria spp. (bry-**or**-ia: moss—). Also tree hair, black tree lichen. Blackish, 2–18" festoons of fine, weak fibers, often with pale, powdery specks. Abound on trees, krummholz, rarely on rocks; clean-air indicator.

Edible Horsehair
B. fremontii (fre-**mont**-ee-eye: after John C. Frémont, p. 297). Typically dark reddish brown, relatively long (to 18"), occasionally yellow-specked; under 10× lens, thicker strands are grooved and twisted; most branchings are ± perpendicular. All but UT and NM.

Dusky Horsehair
B. fuscescens (fus-**kess**-enz: darkening). Also *B. lanestris*. Dull brown, paler toward base, copiously white-speckled; most branchings are at acute angles. A.

Pitted Beard Lichen
U. cavernosa. Pendent tufts up to 12" long; many fine, threadlike branches, but main branches are much thicker and conspicuously pitted and ridged. In the 3 more easterly ecoregions.

Bristly Beard Lichen
U. hirta (**her**-ta: hairy). Bushy tufts 1¼–2" long; mealy-coated; no fruiting bodies. A.

Exploding Beard Lichen
U. intermedia. Bushy tufts 1–1¾" long, with long, perpendicular branches, adorned with large, thin, long-fringed fruiting discs of the same gray-green color. sw CO, NM.

BEARD LICHENS OF this genus abound in boreal regions, inspiring schemes for industrial conversion of lichen starches into glucose for food, or alcohol for fuel or drink. However, even abundant lichens grow too slowly to sustain a starch industry.

THESE FIBROUS FESTOONS achieve surreal loveliness in moist western-larch forests near the Idaho-Montana border, draping darkly from larch limbs with their bright-green or (in the fall) bright-yellow foliage. Higher up, they add drama to weather-bonsai'd timberline trees. Don't, however, confuse them with ugly, old snow mold, *Herpotrichia nigra* ("black creeping hair"), a pathogenic fungus that turns mountain hemlock and subalpine fir branches into mats of smutty needles, mainly during late-snowmelt years. Horsehair, like other lichens, is neither pathogenic nor parasitic.

Edible horsehair is considered edible when it lacks the lichen acids that defend its relatives from herbivores. For flying squirrels and woodland caribou, it is a winter staple; for deer and elk, it is a minor food that helps them get more nutrition out of conifer needles. Indigenous people ate horsehair lichen where carbohydrate foods were scarce in winter, and some bundled it to make shoes. Reports vary as to whether it was delicious or starvation fare, bitter or soapy or sweet or bland.

Taste tests distinguish a bitterer, yellow form that does have defensive acids. DNA sequencing determined both forms to be *B. fremontii;* but Trevor Goward, a coauthor of that study, independently expressed qualms about such rigidly DNA-based determinations: "I personally think [the Indigenous peoples who see two different lichens] are on to something, and would like to see professional lichenologists continue to follow suit."

His wish spurred his friend Toby Spribille on to a breakthrough in lichen science, finding a crucial difference between the edible and bitter forms. Both have an additional fungus in them, a single-celled fungus (yeast) unrelated to the main fungus. There's ten times more of this yeast in the bitter form than in the edible form. Spribille went on to find the yeast's relatives similarly hiding in most of the lichens where he looked for them. Some lichenologists question whether the yeasts are truly part of the lichen symbiosis or just trivial hangers-on; however, the correlation with producing different lichen acids gives Spribille a strong case.

Witch's hair

Some lichens have two kinds of yeasts and also several kinds of bacteria; of course, humans contain symbiotic bacteria as well, but unlike in humans, bacteria in lichens (these hidden yeasts, not just the conspicuous cyanobacteria) may yet prove to help determine the appearance and structure of the lichen. This could shake up lichen taxonomy, which since 1950 has been governed by a rule that lichen genus and species names are those of the one, primary fungus. Spribille says that a lichen is not a species but a "microbial conversation."

WITCH'S HAIR

Alectoria sarmentosa (al-ec-**tor**-ia: rooster?; sar-men-**toe**-sa: twiggy). Pale gray-green festoons on trees; wispily pendulous, 3–30" long; when pulled, a moist strand snaps straight across (no rubbery central strand). On conifers; northerly.

When you cross Lolo Pass or many other ridgelines near the tipping point of marine climate influence, you see an abrupt transition from witch's hair on the wet west side to horsehair lichen on the dry east side.

Forking bone lichen

Hooded tube lichen

V fingers

TUBE LICHEN

Hypogymnia spp. (hyp-o-**jim**-nia: naked underneath). Also bone lichen, puffed lichen. Tufts of hollow branches, sharply 2-toned, pale greenish-gray above with black specks, blackish-brown beneath, lighter brown at the tips.

Forking Bone Lichen

H. imshaugii (imz-**how**-ghee-eye: for Henry A. Imshaug). Tufts relatively airy, erect, with tubular branches forking dichotomously; interior ceiling white; numerous yellowish-brown fruiting cups on upper sides avg ⅛". Common on small limbs; northerly.

Hooded Tube Lichen

H. physodes (fie-**zo**-deez: bladder). Tubes flattened, ⅛–¼" wide, yet raised from the bark; interior ceiling white, floor dark; tips look exploded, full of granular soredia; fruiting disks rare. On bark. Northerly, rare in CO.

Lattice Tube Lichen

H. occidentalis (oxy-den-**tay**-lis: western). Tubes flattened, ⅛" wide, crowded, pressed against bark; tube tips often punctured; interior surfaces dark. Northerly.

V FINGERS

Allocetraria madreporiformis (al-o-set-**rair**-ia: different *Cetraria*; ma-dreh-por-i-**for**-mis: resembling *Madrepora,* a coral). Also *Dactylina madreporiformis*. Clumps of pale yellow-green, hollow fingers, many of them branched in Y shapes, up to 1⅜" tall and ⅛" thick. Alpine sod; NM to sw MT.

ICELAND-MOSS

Cetraria islandica (set-**rair**-ia: shield—; iss-**land**-ic-a: of Iceland). Clumps 1–3" tall × 2–8" wide, brown to olive when dry, consisting of narrow, flattened lobes with sparsely fringed edges sometimes rolled into tubes, with white specks (under 10× lens) scattered

Iceland-moss · Yellow ruffle lichen and snow lichen

across back surfaces. On soil, mostly alpine; NM, CO, WY, MT, WA. Similar *C. ericetorum* (er-iss-e-**tor**-um: among heaths) has white specks aligned under edges, which are rolled tubewise. A–.

The time-honored misnomer "Iceland moss" comes from Europe, where this is the best-known lichen that people eat. The Inuit also eat it. Calling it *brødmose* ("bread moss"), Norwegian sailors used to bake it into their bread to extend its shelf life at sea. It had to be parboiled with soda before milling, or it would have been unspeakably bitter.

SNOW LICHEN

Cetraria nivalis (niv-**ay**-lis: of snow). Also *Flavocetraria nivalis*. Like the preceding but pale yellow to off-white; white specks beneath margins; lobes strongly ruffled all over. Alpine; NM, CO, WY, MT.

In the High Andes, herbal healers among the Kallawaya people brew snow lichen into a tea to alleviate altitude sickness.

YELLOW RUFFLE LICHEN

Cetraria juniperina. Also *Vulpicida juniperinus, V. tilesii*. Like snow lichen but brighter yellow-orange with black specks, and less strongly ruffled; rhizines and fruiting bodies rare. On calcareous soil, incl tundra; A–.

WOLF LICHEN

Letharia spp. (leth-**air**-ia: death—). Brilliant sulphur-yellow, stiff tufts, the profuse branches cylindrical or ± flattened, often pitted, black-dotted. On trees; abundant on pine and juniper; clean-air indicator.

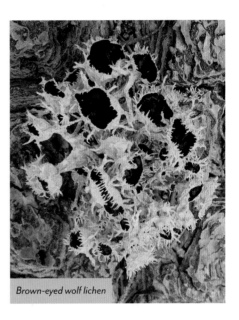

Brown-eyed wolf lichen

Brown-eyed Wolf Lichen
L. columbiana (co-lum-be-**ay**-na: of the Columbia River). Black-speckled; has conspicuous fruiting cups with brownish disks avg ¼" across (up to ¾"). Northerly.
Wolf Lichen
L. lupina (loo-**pie**-na: of wolves). Also *L. vulpina*. Black-speckled; fruiting cups usually absent. A.

SOMEWHERE THERE WAS a tradition of collecting this intense-colored lichen to make a poison for wolves and foxes; my American sources say this used to be done in Europe, while a British source writes it off as an American barbarism.

Wolf lichen imparts its chartreuse color to fabrics, as a dye. Before they had cloth, Indigenous people used the dye on

Woolly foam lichen

1½–4" tall. In dense patches on soil but barely attached to it.

Reindeer Lichen
C. arbuscula (ar-**bus**-cue-la: little tree). Also *Cladina mitis*. Pale yellowish green, sometimes with bluish bloom; tips split into 3 or 4 points. Low to mid-elev soil; in Yellowstone, it's found only in geothermal areas; A–.

Gray Reindeer Lichen
C. rangiferina (ran-jif-er-**eye**-na: of reindeer). Ash-gray, or slightly greenish when moist; tips split into 2 or 3 points. Thin soils, often with moss; WA to nw MT.

THOUGH FOUND AT all elevations here, reindeer lichens are best known for covering vast areas of Arctic tundra, where they are the winter staple food of caribou (known in Europe as reindeer). They aren't really very digestible or nutritious; even caribou prefer green leaves when they can get them, but occupying tundra habitat, for caribou, has entailed adapting to wintering on lichens. This requires ceaseless migration, since the slow-growing lichens are eliminated where grazed for long. The Saami people of northern Scandinavia harvest lichens for their reindeer herds. Some Inuit relish a "saladlike" delicacy of half-digested lichens from the stomachs of caribou killed in winter.

Reindeer lichen, left, and gray reindeer lichen, right

moccasins, fur, feathers, wood, porcupine quills for basketry, and their own faces.

WOOLLY FOAM LICHEN
Stereocaulon tomentosum (stereo-**call**-un: solid stem; toe-men-**toe**-sum: woolly). Low (1½–3" tall), nearly white clumps of delicate, ± coral-like branches covered with minute wool; tiny brown fruiting bodies on ends of side branches. On soil, often with mosses; A–.

REINDEER LICHEN
Cladonia spp. (cla-**doe**-nia: branched). Also *Cladina* spp. Profusely fine-branched lichens

Closer to home we encounter them, dyed green and softened in glycerine, as fake trees and shrubs in architectural models. They also supply extracts for commercial uses, including perfume bases and antibiotics. Along with *Usnea* species, they are a chief source of usnic acid, an antibiotic potent against ailments ranging from severe burns and plastic surgery scars to bovine mastitis—and potentially MRSA, a dangerous staph infection.

Pixie goblet

Mealy pixie goblet

CLADONIA

Cladonia spp. (cla-**doe**-nia: branched). Clustered greenish-gray fruiting stalks ½–1¼"tall, from a mat of small, leafy lobes. On soil or well-rotted wood.

Pixie Goblet

C. pyxidata (pix-ih-**day**-ta: explained on next page). Stalks golf-tee-shaped, coated with coarse scales and/or globular granules; "baby tees" may sprout from the rims of main ones. A.

Mealy Pixie Goblet

C. chlorophaea (clo-ro-**fee**-a: dusky green). Stalks golf-tee-shaped, coated with fine white granules. A.

Ladder Lichen

C. verticillata (vert-iss-il-**ay**-ta: tiered). Also *C. cervicornis.* Stalks like multi-storied narrow trumpets, the 2nd- and 3rd-story trumpets rising from the centers (not rims) of lower ones. A.

Ladder lichen

Powderhorn Lichen

C. coniocraea (cony-o-**cree**-a: pale cone). Stalks very slender, tapering to a point, coated with granules, often curvy, up to 1" tall, from a patch of green, leafy lobes ¼" wide—larger than on most cladonias. A.

Sulphur Cup

C. sulphurina (a misleading name, as it is only slightly, if at all, yellower than other cladonias). Stalks yellow-tinged gray-green, irregular, ± cupped at top, mealy-coated. A.

Brown Helmets

C. cariosa (carry-**oh**-sa: rotten). Stalks very ragged and irregular, often Y-branched, tipped with smooth, rounded, saddle-brown fruiting bodies. A.

Brown helmets

Red-crowned sulphur cup

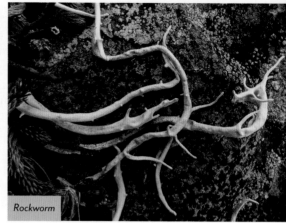
Rockworm

Red-crowned Sulphur Cup

C. deformis. Stalks ± yellowish, mealy-coated, mostly cupped at top, often with small scarlet "fruit" along some of the cup rims. A–.

THOUGH GOLF TEES are the obvious resemblance, goblets for pixies are hinted at by the name *pyxidata,* which actually refers to a pyxis, a goblet-shaped, lidded container in ancient Greece. The shape likely evolved because it makes raindrops splash farther, disseminating tiny propagules from the goblet.

The huge genus *Cladonia* tends to have a two-part growth form. The primary growth is a mat of flakes (squamules) resembling some leaf lichens, only finer. After a while, fruiting twigs may rise from this mat. The reindeer lichens (p. 302) lack squamule mats and were long treated as a separate genus.

ROCKWORM

Thamnolia subuliformis (tham-**no**-lia: bushy, a misleading name; sub-you-li-**for**-mis: needle shape). Also *T. vermicularis.* White, tapering tubes 1–2½" tall, without fruiting cups, typically in lackadaisical clumps of standing tubes with some reclining tubes and a few branched tubes. Alpine; A.

The rockworm has neither spore-bearing organs nor any sort of vegetative propagule, so it apparently reproduces only from broken bits of stalk—hardly an effective or speedy mode of intercontinental travel, one would think, yet it is found on every continent except Africa. It likely evolved hundreds of millions of years ago on Pangaea and spread throughout suitable habitat; plate tectonics later split it up among six continents.

Mammals

Do we even need an introduction to mammals? We *are* mammals. We're well schooled in the salient characteristics. Most mammals give birth to live young; yet egg-laying platypuses are mammals and live-bearing snakes are not. We are all warm-blooded: we maintain a constant relatively warm body temperature. But so do the birds and, arguably, the bees.

In mammals only, a coat of keratinous hairs serves to insulate the body, so hair may be the definitive mammalian body part, since only half of mammals (the females) produce milk. Whales, armadillos, and other mammals even less hairy than we are all have at least a few hairs at some developmental stage.

Mammal coats come mainly in shades of brown and gray. Even on the few mammals, like skunks, with "showy" rather than cryptic coloring, no fur is bright blue or green, as some feathers are. Most mammals instead need to be inconspicuous: they wear drab colors, are quiet and elusive, and are nocturnal. That makes mammal-watching a less popular pastime than birdwatching, and makes mammal tracks, scats, and other signs the key to being woods-wise to mammals. Mammals learn to avoid popular hiking trails. I've often begun to see more of them as soon as I got even a hundred yards off trail. Motor noise, of course, is worse. Surprisingly distant noise is enough to increase animal mortality: among other things, it masks the sounds of predators—demanding a heightened level of vigilance, which in turn means less time for foraging and can lead to starvation.

The mammals we see most often each have their reasons to be unafraid: porcupines are spiky, bears and bison are big, squirrels go up, marmots and pikas go down, pronghorns go fast, and National Park–dwelling mountain goats enjoy protected status—all the more reason for caution around them!

Small mammals are especially limited to the murky corner of the spectrum. The shade of a species' coat may vary not only by region but also for different seasons, for juveniles versus adults, for infraspecific varieties and color phases, for the paler underside and darker back, and between the hair tips, underfur, and guard hairs. Among closely related mammals, populations in dry regions are often paler than their moist-habitat relatives, each tending to match the ground's hue to be less visible to predators (especially owls) that hunt from above in dim light. (Dry country has pale, dry dirt, whereas moist forests have dark humus and vegetation.)

By some calculations, the global biomass of terrestrial wild mammals is down 85 percent from the early days of humans, and outweighed nine to one by both human biomass and cattle biomass.

Positive identification of small mammals utilizes the number and shape of molar teeth, and caliper measurements of skulls and penis bones (a feature of most male mammals). The creature must first be reduced to a skeleton. Don't worry, this chapter will not go into molar design or the meaning of "dusky," but will use size, ratio of tail to body length, form, habitat, and sometimes color to facilitate educated guesses as to small-mammal ID. Our typical glimpse of a shrew, mouse, or vole, unless we trap it or find it dead, is so fleeting that we can only guess based on habits and habitat.

Order Eulipotyphla (Shrews and Moles)

The order containing shrews and moles has gone through some recent renamings and reshufflings; it now includes Old World hedgehogs. Eulipotyphla (you-lip-a-**tiff**-la:) translates—inaccurately—to "fat and blind."

SHREW

Sorex spp. (**sor**-ex: Roman name). Mouselike creatures with very long, pointed, wiggly, long-whiskered snouts, red-tipped teeth, and ± naked tails. Soricidae.

Wandering Shrew

S. vagrans (**vay**-grenz: wandering). 2½" + 1¾" tail; fur dark-gray-frosted brown on back, pinkish on sides, and pale on belly. Meadows; northerly.

Dusky Shrew

S. monticola (mon-**tic**-o-la: mountain dweller). Also montane shrew; *S. obscurus*. 2¾" + 1¾" tail; dark brown. Moist habitats; A.

Western Water Shrew

S. navigator. Also *S. palustris.* 3½" + 3¼" tail; blackish above, paler beneath; tail sharply bicolored. In or near high marshes and lakes; A.

Masked Shrew

S. cinereus (sin-**ee**-rius: ashen). 2¼" + 1¾" tail (ratio 3:2); brown with pale-gray underside. Various habitats, common; A.

Pygmy Shrew

S. hoyi (**hoy**-eye: for Philo Romayne Hoy). 2¼" + 1" tail (ratio 2:1); varies from dark brown or gray to coppery. WA, ID, MT, CO.

SHREWS, OUR SMALLEST mammals, lead hyperactive but simple lives. As the shrew expert Leslie Carraway writes, "*S. vagrans* exhibits some behavior that tends to indicate it does not perceive much that transpires in its microcosm." Day in and night out, shrews rush around groping with their little whiskers, sniffing, and eating most everything they can find. This goes on from weaning, between April and July of one year, until death, generally by August of the next.

Shrews eat insects and other arthropods (often as larvae), earthworms, and a few conifer seeds and underground-fruiting fungi; they have been known to kill and eat other shrews and mice. They have cycles of greater and lesser activity, but as a rule they can't go longer than three hours without eating, and the smaller species eat their own weight

Western water shrew

Masked shrew

equivalent daily. As with bats and humming-birds, such a high caloric demand is dictated by the high rate of heat loss from small bodies: at two grams, masked shrews approach the lower size limit for warm-blooded bodies. Baby shrews nurse their way up to this threshold while huddling together so that the combined mass of the litter of four to ten easily exceeds two grams. Before the young learn to navigate well, some form a train with the mother as engine, each shrew biting and holding the tail of one in front of it. Whereas bats and hummingbirds take half of every day off for deep torpor (p. 318), shrews do not. Nor do they hibernate. It's hard to imagine how our shrews meet their caloric needs during the long snowy season when insect populations are dormant, and heat loss all the more rapid. But they do—or at least enough of them do to maintain the population.

A sort of fatal shrew shock syndrome seems to be triggered, for example, by capture or a sudden loud noise. Some scientists relate it to the shrew's extreme heart rate (1,320 beats per minute have been recorded) and others to low blood sugar caused by even the briefest shortage of food. At any rate, it is common to see little shrew corpses on the ground. With poor eyesight and hearing, shrews are ill-adapted to evade predators. Their defense is simply to be unappetizing. Owls, Steller's jays, and trout are on the short list of predators known to have a taste for shrews.

Water shrews, the ones most likely to tempt trout, spend much of their time going after aquatic insects and larvae, snails, leeches, and so on. Their fine, dense fur traps an insulating air layer next to the skin, giving them enough buoyancy to scoot across the water's surface for several seconds. When they dive and swim, they must paddle frenetically to stay under, aided by stiff, hairy fringes on the sides of their hind feet. As soon as they stop, they bob to the surface. Underwater, they have a silvery coat of little bubbles coming out of their fur.

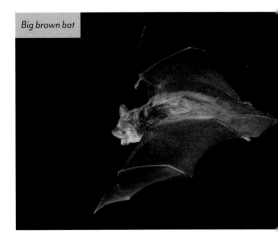
Big brown bat

Our more terrestrial shrews may run on the ground or even climb trees, but more of the time they are subsurface. Dusky and vagrant shrews specialize in the duff layer; others use existing vole runways.

Shrews are ferociously solitary. In order to mate, they calm their usual mutual hostility with elaborate courtship displays and pheromonal exchanges—a real-life "taming of the shrews."

Order Chiroptera (Bats)

Among mammal orders, bats include the greatest number of individuals.

BAT

Family Vespertilionidae. Mouselike bodies with large, membranous wings in place of forelegs; nocturnal, most often seen at dusk, swooping back and forth in the air.

Big Brown Bat

Eptesicus fuscus (ep-**tee**-sic-us: a flier?; **fus**-cus: brown). Fur medium to dark brown; 4–5" long, with 9–14" ws. Often roosts in buildings; widespread; A.

Western Small-footed Myotis

Myotis ciliolabrum (my-**oat**-iss: mouse ear; silly-o-**lab**-rum: hairy lip). Fur ocher, with blackish muzzle, ears, and wings; 3–4" long, with 9–10" ws. Mainly hunts a few feet from the ground; roosts in rock crevices, caves, sometimes buildings; A.

AS EVENING GETS too dark for swifts and nighthawks to continue their feeding flights, bats and owls come out for theirs. Though our largest bats and smallest owls overlap in size—5½ inches long, 16-inch wingspread—bats are easy to tell from owls by their fluttering, indirect flight. Our species never venture out in the daytime, so you are unlikely to see one well, and I won't identify more than the two species.

Bats probably take a bigger slice out of the insect population than any other type of predator. Bats catch flying prey either in the mouth or in a tuck of the small membrane stretched between the hind legs. Each wing is a much larger, transparently thin membrane stretched from the hind leg up to the forelimb and all around the four long "fingers." Since the wing has no thickness to speak of, it is less effective than a bird wing or airplane wing at turning forward motion into lift. To compensate, bats generally have much greater wing area per weight than birds, and use a complex stroke resembling a human breaststroke to pull themselves up through the air. Bats are more agile in flight than birds, able to pursue flying insects rather than just intercepting them. Some are also fast: the Brazilian free-tailed bat, common in southern Colorado and New Mexico, has been tracked at over 100 miles per hour in short bursts.

Bats roost upside down, hanging from one or both feet, often in large groups. Daytime roosts during the active season are in caves for some species, tree cavities and well-shaded branches for others. Colonies larger than a hundred thousand bats may return to the same cave day after day—famously Carlsbad Caverns, New Mexico, but also the Orient Mine in Colorado. They sleep all day and hibernate all winter upside down. (A few species fly south instead.) Our bats typically hibernate in caves, which provide stable above-freezing temperatures and high humidity. Some males mate very aggressively, others so sneakily that the female appears unaware that anything is going on.

Our bats typically mate in autumn, then delay fertilization and bear a single pup in spring. Still hanging upside down, for a few weeks the mothers nurse almost constantly except while out hunting. Coming in from the hunt to a colonial roost, they are fairly accurate in picking out their own pup.

Bats may carry rabies, but even rabid ones rarely bite people.

Since 2006, a fungal disease called white-nose syndrome has afflicted North American bats, nearly wiping out entire cave populations. Victims fail to make it through hibernation after they wake up repeatedly, using up their fat stores at a time of year when they cannot be replenished. From a beachhead in the east, the fungus spread westward year by year, reaching Devils Tower, Wyoming, and then hopping over to the Washington Cascades. It appears to be a Eurasian pathogen that Eurasian bats are resistant to, most likely brought over by humans. It is an existential threat to many of North America's bat species, possibly already killing more individuals than any other disease in mammals. The big brown bat appears to be resistant, and vaccines have been developed. Losing most bats would be a terrible blow to the agriculture industry, which owes bats billions of dollars a year for pesticide services.

Wind turbines strike and kill large numbers of bats, especially a few species that are positively attracted to turbines. The sounds and the lights of turbines were an early guess as to the attraction, but a review of studies concludes towers are sites for scent-marking. While this research may be difficult, it is crucial to find a way to diminish the turbine's allure for bats. Going forward, we need a healthy bat population along with plenty of turbines to meet our energy needs.

Order Lagomorpha (Rabbits)

Females in this order are larger than males. All have four upper incisors, as opposed to two in rodents. As in rodents, these teeth grow continuously throughout life, replacing the tooth material lost through incessant chewing. The order has just two families: pikas run, rabbits and hares hop.

PIKA

Ochotona princeps (ock-o-**toe**-na: Mongolian name; **prin**-seps: chief). Also cony. 8" long if stretched out, but a thickset 5–6" long in typical postures; tailless; brown; ears round, ½". On or near talus (slopes comprised of large rocks); A. Ochotonidae.

Pikas peek in most western national parks, almost always from rocks in a talus field. They aren't exactly a snap to see, but their distinctive "eeenk" draws our attention, while alerting other pikas to our arrival. They quickly drop out of sight, but if we keep watching, one or more may re-emerge not far away. You might mistake them for rodents, but the posture is rabbitlike—pikas are indeed akin to rabbits, not rodents.

Pikas don't just peek out from behind a rock; they "peek" vocally. Their name, originally the Siberian Tungus people's word for them, is an imitation of that legendary "eenk," so it ought to be pronounced "peeka." That pronunciation prevails in Canada, but south of the 49th, I mostly hear "pike-a," perhaps due to confusion with "pica" or "piker."

Pikas run around clipping vegetation near talus and then ferrying it back to pile in "pika haypiles" to cure for winter storage. You may spot haypiles under small rock overhangs; others—up to bathtub sized—are out of view down in the rockpile. Since they don't hibernate, most (but not all) pikas depend on this hay to get them through the winter.

To keep their hay from rotting, pikas harvest highly phenolic (mildly toxic) plants that inhibit rot. The phenols degrade with age, allowing the cured hay to be eaten. One favorite food plant,

the alpine avens, now produces more toxins than in 1992, thanks to CO_2 fertilization. This makes the haypiles keep better, so it may be just fine with the pikas.

Pikas are intensely antisocial, attacking other pikas that infringe on their territories (except, of course, when they're ready to mate). "Eeenk" can express either alarm or territoriality—the two salient features of a pika's so-called social life. This zealous territorial defense seems warranted, since kleptoparasitism—theft from each other's haypiles—is rife.

As nonhibernating warm bodies in cold climates, pikas have to wear warm fur coats. The downside is that being out and about for more than a few minutes on a hot day could kill them. So they don't do that. They stay cool by retreating deep into the talus, where it is never hot, even when a forest fire sweeps through. In other words, they regulate their body temperature not through vulgar bodily functions like sweating or panting, but by retiring to an air-conditioned chamber. They choose their habitat for that characteristic—thermally moderate rockpiles with ample food nearby. So while pikas appear to be dependent on talus or, less commonly, lava beds, the pressing question is whether they are dependent on cold climates and deserve

Pika

Coprophagy

The high fiber content in a diet of leaves, twigs, and mosses is a challenge to digestive systems. Cud-chewing grazers meet the challenge with multiple stomachs; they can chew away at their leisure on food that their first stomach has already partially broken down. Pikas, rabbits, and several other small mammals have a way of rechewing food that's already been in their stomachs, without having to squeeze extra stomachs into their small frames. They defecate soft, lumpy pellets of lightly digested food and eat them—sometimes immediately or sometimes in winter after leaving them stored. Twice-digested waste, in contrast, comes out as neatly formed, hard pellets. (Pikas' hard pellets are spherical, a bit bigger than peppercorns, and often aggregated in huge latrines.) Pikas also eat marmot scats, and sometimes store them in their haypiles.

The benefits go beyond shorter chew times. Soft pellets commonly have higher nutrient value than the foods that go into them, suggesting that symbiotic bacteria have been working their magic. There are some nutrients that pikas can't absorb until they've been worked on first in the stomach, then in the caecum (a pocket between the stomach and the intestine), and then in the stomach again. Some coprophagists also inoculate food with a bacterium that completes its digestive work when exposed to sunlight.

their poster-child status on the climate-change scene.

Simply being able to drop out of sight and chill doesn't quite mean they're home-free in a warming world. If they spend too many of the summer daytime hours underground, they may not have time to harvest enough hay to survive the winter. It's true that a few Nevada mountain ranges have lost their pikas within recent decades, but many pika researchers see an adaptable creature with a complex prognosis. "Are pikas going extinct? It depends," I was told by one, Thomas Rodhouse. "They're at risk of extirpation in certain regions for very clear, understandable, climatological reasons. But not everywhere."

Idaho's Craters of the Moon lava beds are one of the hot, moderate-elevation pika habitats. Pikas prefer pahoehoe lava, the smooth, ropy kind riddled with tunnels where cold air sinks into dead ends. Rodhouse's group planted thermometers a meter down in pahoehoe holes, finding that some were colder than others at all times, both summer and winter. It seems a pika home can have both a cool room and a warm room. They do need warm rooms: winter cold limits their range just as surely as summer heat does.

Pikas even live close to sea level in the Columbia River Gorge in Oregon. Many summer afternoons there see pika-lethal heat, but pikas only expose themselves to it for a few minutes at a time. They've adapted to these local conditions with a locavore diet heavy on moss, which grows luxuriantly within inches of their cool refuges.

Adapting to the future within the Rockies will be a challenge. Pikas of anomalously low habitats did not necessarily migrate there from the alpine. More likely, all of today's pika habitats are remnants of a much larger habitat at the height of the last ice age—continuous across much of the West—and their range has been shrinking ever since.

Young pikas dispersing from their natal talus must find another good site quickly. They disperse and make their own hay their first summer, even if they're only half-grown. A dispersing youngster's prospects are favorable only when there's room for them somewhere on that rockpile they call home—typically near the pile's center, since dominant individuals have already claimed the choice outer sites near the meadow. Youngsters dispersing beyond the natal rockpile risk death. Picture youngsters driven out of their parents' homes, exploring unfamiliar forest alone, some of them escaping the notice of predators; if the afternoon gets hot, they absolutely must find refuge underground, but where is there talus?

The biologist Embere Hall described adaptive behavior within typical high-mountain habitat. On one Wyoming mountain, during the hottest days, some pikas stayed underground for longer midday periods than others, and yet they gathered equally nutritious food. If this smart trait can be passed along to offspring, either genetically or by teaching, it can help pikas keep up with climate change. (But it won't work where it gets to be too hot all day, all summer. For pikas, night foraging is so rare that there clearly must be a good reason they avoid it; becoming nocturnal won't be the solution.) Now Hall is looking closer at those pikas that do not adjust their routines, to see if they have a physical trait that helps them tolerate heat, or if they don't and will likely die out over time. Either one (a superior trait, or dying off in favor of more flexible individuals) would be a way for a population to adapt.

To sum up their prospects, broad surveys across the West find decent numbers of pikas still hanging on in the overwhelming majority of their known sites. Several long-held assumptions about pikas' limitations—dependence on talus, high elevations, haypiles, and mild afternoon heat—all turn out to be overdrawn. Pikas show impressive adaptability in some areas. That said, as cool-loving animals, they are still potentially vulnerable to warming. It all depends on how much warming.

MOUNTAIN COTTONTAIL

Sylvilagus nuttallii (sil-vi-**lay**-gus: forest rabbit; nut-**all**-ee-eye: after Thomas Nuttall, p. 192). 13" + 1½" tail; pale gray-brown; white under tail; ears 2½" (shorter than head); nocturnal. Sagebrush with rock outcrops, open woodland; A. Leporidae.

The mountain cottontail ventures out from its refuge under sagebrush, bitterbrush, or rabbitbrush to graze on grasses. As steppe animals go, it is only marginally adapted to drought, and has learned to climb juniper trees—a rare ability among lagomorphs—to nibble foliage at night, largely for the dew.

WHITE-TAILED JACKRABBIT

Lepus townsendii (**lep**-us: Roman name; town-**send**-ee-eye: after John Kirk Townsend, p. 312). 20" + 3½" tail; white tail, black ear tips; otherwise gray-brown in summer; in winter, most

Mountain cottontail

White-tailed jackrabbit

high-elev populations are white with some buff around face; nocturnal. Grassland from plains to alpine; A. A jackrabbit in NM and s CO lowlands would more likely be the somewhat darker, black-tailed *L. californicus*. Leporidae.

Jackrabbits (genus *Lepus*) are not rabbits but hares, born fully furred and ready to run and eat green leaves within hours of birth. Rabbits, in contrast, are born naked and blind, and must be nursed for ten to twelve days before leaving the nest. Physical differences between the two are quite arcane.

This is one of the biggest, fastest jackrabbits. It can sprint up to 45 miles per hour, and leap as high as 20 feet. It eats grasses and some herbs; in winter, it nibbles on shrubs too. A grassland beast, it is more common on plains than in the mountains, but it has been found at all elevations. Its numbers are declining due to habitat loss, extermination as a pest, a virus, and climate change. One problem with climate change is color mismatch—that is, being white in winter where there's no longer much snow. Whiter genetic lineages will likely die out in areas where snow cover gets less consistent, and browner lineages should move from the lowlands to replace them. A deadly virus, rabbit hemorrhagic disease, has been spreading since 2020 in our region's black-tailed jackrabbits and cottontails. More than half of infections in domestic rabbits with this virus are fatal. It is too early to assess the effects on wild populations.

Townsend's Agonies and Ecstasies

John Kirk Townsend came west in 1834 with Nathaniel Wyeth and Thomas Nuttall (p. 192). Young (twenty-four) and enthusiastic, Townsend wrote the most vivid naturalist's account of the highs and lows of exploring the West:

> In the morning, Mr. N. and myself were up before the dawn, strolling through the umbrageous forest. . . . None but a naturalist can appreciate a naturalist's feelings—his delight amounting to ecstacy—when a specimen such as he has never before seen, meets his eye, and the sorrow and grief which he feels when he is compelled to tear himself from a spot abounding with all that he has anxiously and unremittingly sought for.

Of portaging his canoe in heavy rain, Townsend wrote:

> It was by far the most fatiguing, cheerless, and uncomfortable business in which I was ever engaged, and truly glad was I to lie down at night on the cold, wet ground, wrapped in my blankets, out of which I had just wrung the water. . . . I could not but recollect . . . the last injunction of my dear old grandmother, not to sleep in damp beds!!!

Occupational hazards brought several explorer-naturalists to untimely ends. Townsend devised his own formula to keep pests from eating his stuffed specimens: it contained arsenic; he died at age forty-two of chronic arsenic poisoning. Townsend's letters reveal that he was biased against Indigenous people, and dug up and dismembered skeletons for the purpose of pseudoscientific measurement. In consequence, his name is being expunged from the common names of birds.

SNOWSHOE HARE

Lepus americanus. Also varying hare. 16" when stretched out, plus 1½" tail; gray-brown to deep chestnut-brown (incl tail) in summer; in winter, high, most races turn white, with dark ear-tips; ears slightly shorter than head; nocturnal and secretive. Widespread in forest; A. Leporidae.

Snowshoe hare

Thanks to their "snowshoes"—large hind feet with dense, stiff hair between the toes—these hares can be just as active in the winter, on the snow, as in summer. They neither hoard nor hibernate, but molt from brown to white fur, and go from a diet of greens to one of conifer buds and shrub bark made all the more accessible by the rising platform of snow. They make the animal tracks we see most often while skiing. They also become the crucial staple in the winter diet of several predators: foxes, great horned owls, golden eagles, bobcats, and especially lynxes. Though the hares' defenses (camouflage, speed, and alertness) are good, the predator pressure on them becomes ferocious when the other small prey have retired beneath the ground or the snow. Hares can support their huge winter losses only by being greater summer prodigies of reproduction, the proverbial "breeding like rabbits." Several times a year, a mother hare can produce two to four young. She mates immediately after each litter, and gestates thirty-six to forty days.

Drastic population swings in eight-to-eleven-year cycles are well known in Canada, but not in the Lower 48. The other kind of "varying" the hare is named for is from summer brown to winter white pelage. The semiannual molt is triggered by changing day length. In years when autumn snowfall or spring snowmelt comes abnormally early or late, the hares find themselves horribly conspicuous and have to lay low for a few weeks. Permanently brown races have evolved in lowland areas that fail, year after year, to develop a prolonged snowpack. These brown hares stand out during the occasional snows. The rest of the year, they make the most of their camouflage, foraging when they can best see without being seen—by dawn and dusk, and sometimes on cloudy days in deep forest. Snowshoe hares don't use burrows, but retire to shallow depressions, called "forms," under shrubs.

Though infrequently vocal, snowshoe hares have a fairly loud aggressive/defensive growl, a powerful scream perhaps expressing pain or shock, and ways of drumming their feet as their chief mating call. A legendary courtship dance, in which they may literally somersault over each other for a while, appears to crescendo out of an ecstatic rapture of foot-drumming.

Order Rodentia (Rodents)

Out of the 4,000-plus species of mammals living today, almost 1,700 are rodents, and almost 1,300 of those, or 32 percent of all mammal species, are myomorphs, or mouselike rodents. As with insects, songbirds, grasses, and composite flowers, such disproportionate diversification indicates great success in recent geologic times.

CHIPMUNK

Neotamias spp. (ne-o-**tay**-me-us: New World storer). Also *Tamias* spp., *Eutamias* spp. Rich

Yellow pine chipmunk

Colorado chipmunk

Uinta chipmunk

Yellow Pine Chipmunk
N. amoenus (a-**me**-nus: delightful). 4½" + 3½" tail; usually has 2 gray and 2 white stripes. WA to n UT.
Colorado Chipmunk
N. quadrivittatus (quad-re-vit-**ay**-tus: 4 stripes). 5" + 4" tail; 4 white stripes. CO, NM, e-most UT.
Uinta Chipmunk
N. umbrinus (um-**bry**-nuss: brownish). 5" + 4" tail; 4 white stripes; tail horizontal while running. CO to e-most ID.
Least Chipmunk
N. minimus. 4¼" + 3⅜" tail; 4 white stripes; holds tail vertically while running, often flicks it up-down, but not left-right; may "chip" incessantly. Sagebrush to tundra; A.

brown with four pale stripes and five dark stripes down the back; striped from nose through eyes to ears; species are very hard to tell apart. Sciuridae.

CHIPMUNKS ARE CONSPICUOUS—DIURNAL, noisy, and abundant. Their diverse chips, chirps, poofs, and tisks can be mistaken for bird calls.

Our chipmunks forage terrestrially for seeds, berries, bulbs, flowers, buds, caterpillars, and—increasingly toward winter—underground fungi. To facilitate food-handling, they have an upright stance (like other squirrels and gophers) that frees the handlike forefeet. They store huge quantities of food, carrying it to their burrows in cheek pouches.

Their winter strategy varies with food supply conditions, but commonly they rely on stored food alone, without fattening up. They pass the winter with a series of torpid bouts at only moderately depressed body temperatures; every few days, they get up to eat stored food and excrete, and on milder days at lower elevations they go out and forage. You may hear their chatter even in midwinter. Their torpor is shallow, taking about an hour (far less than most squirrels) to recover from fully. Born naked, blind, and helpless, they mature fast enough to disperse and make their own nests for their first winter. Though winter burrows are belowground, nests in tree cavities may be used as summer homes.

The previous four chipmunks occupy similar ecological niches, but tend to exclude each other. (If you have two different stripy critters in your campground, one may be a ground squirrel.) In Glacier National Park, yellow pine chipmunks take the lower forests, least chipmunks the alpine tundra. Almost the reverse holds in central Idaho and eastern Oregon, where yellow pine is king of the mountains, leaving low steppe for least. However, where they get a mountain range to themselves, either least or yellow pine can occupy all elevational zones.

Golden-mantled ground squirrel

GOLDEN-MANTLED GROUND SQUIRREL

Callospermophilus lateralis (cal-o-sper-**mah**-fil-us: beautiful seed-lover; lat-er-**ay**-lis: sides, referring to stripes). Also copperhead. 7" + 4" bushy tail; medium gray-brown with 2 dark and 1 light stripe along each flank; head and chest tawny to yellow-brown, and not striped; "milk-bottle" pose typical while looking around. In sun, all elevs, even tundra; A. Sciuridae.

These stripy campground scavengers are sometimes mistaken for chipmunks. They are also close to chipmunks in ecological niche, but may avoid direct competition by being less arboreal and by hibernating for months. They eat a smorgasbord of seeds, nuts, leaves, shoots, and fungi. Ground squirrel cheek pouches—the mucus-lined mouth interior that extends nearly to the shoulders—can hold several hundred seeds. Their alarm call is a single, sharp whistle, higher-pitched than a marmot's (and used less often).

Our many kinds of ground squirrels have a common pattern to their annual life cycle:

Gopher Teeth

To see some top-notch work from nature's design shop, check out a cleaned rodent skull. Carefully slide out one of the four incisor teeth. It's a long skinny arc and slides out of an arcuate socket the entire length of the skull or the jaw. Throughout life, new material is added at the tooth's back end and hardens as it moves forward, continually making up for wear. The outer side of the arc gets thicker enamel, resisting wear differentially to create a self-sharpening chisel edge. Iron oxides further harden the enamel, while staining it orange.

Porcupines wear away roughly 100 percent of the tooth length every year. That's a lot of gnawing. (No wonder that the root word in rodent means "gnaw.") These are rootless teeth, neither fixed nor ever done growing. Some rodents have rootless molars as well as incisors, but porcupines do not: chewing on trees for their entire lives inflicts heavy wear on their molars, limiting their life spans. Occasionally, a rodent dies of malocclusion: an upper incisor and a lower incisor don't quite meet, so they keep growing and either pierce the skull or lock the jaw.

Thirteen-lined ground squirrel

they are serious hibernators. Most individuals are active for only three to four months. The colony stays in view longer—from late March, say, to early September—because they are on staggered calendars. Adult males emerge first, and go into estivation (summer torpor) in July when the herbs and grasses begin to dry up, making it harder to put on body fat. Females are a few weeks later to do both. As soon as they emerge (or sometimes earlier, still underground), breeding begins, replete with male combat. Gestation takes twenty to twenty-four days, and the young remain in their natal burrow as many days again, not emerging until mid to late May; they're the ones you still see through August. Ground squirrels can perform estivation and hibernation (winter torpor) consecutively without noting any difference, and in fact there is none.

During mating season, males suffer wounds and tend to severely deplete their energy. As yearlings, they have to disperse, and this, too, reduces their foraging success. In consequence, males have shorter lives, and colonies are preponderantly female.

THIRTEEN-LINED GROUND SQUIRREL

Ictidomys tridecemlineatus (ic-**tid**-a-miss: weasel-mouse; try-des-em-lin-ee-**ay**-tus: 13-lined).

Also *Spermophilus tridecemlineatus*. 7" + 3" tail; ears scarcely visible; 13 dark/light stripes run length of back; each dark stripe contains light spots in a row. Lower-elev grasslands, mainly e of CD. Sciuridae.

These squirrels—conspicuous in High Plains towns, where they are called "gophers"—eat far more insects than most squirrels do.

WYOMING GROUND SQUIRREL

Urocitellus elegans (oo-ro-sigh-**tel**-us: tail squirrel; **el**-eg-anz: elegant). Also *Spermophilus elegans*. 8½" + 2¼–4" tail; rather evenly gray with white flecks above, tawny beneath. Extreme sw Montana, the e ½ of the ID RM, and disjunctly from the s end of the Wind River Range s through CO. Sciuridae.

UINTA GROUND SQUIRREL

Urocitellus armatus (ar-**may**-tus: armed). Also *Spermophilus armatus*. 9" + 2–3" tail; head, shoulders, and underside of tail medium gray; rest of body may be gray or brown to cinnamon. w WY, e ID, sw MT.

Uinta females are asocial and aggressively territorial. They are virtually as big as the males, and tend to dominate them after the mating season. Territoriality in this animal is believed to protect a "garden" of previously cropped grass, whose new replacement shoots are more nutritious than uncropped grass.

COLUMBIAN GROUND SQUIRREL

Urocitellus columbianus (co-lum-bee-**ay**-nus: of the Columbia River). Also *Spermophilus columbianus*. 10" + 4" bushy tail; back gray, ± white-flecked; face, throat, forelegs, and tail reddish tawny. Meadows; ID, w MT, OR, WA.

This, the largest ground squirrel, is the most social and the most associated with subalpine meadows. In these ways, it approaches the marmot, for which it is sometimes mistaken, much as the small, stripy ground squirrels seem to approach the chipmunk. (For ID, remember that marmots

Columbian ground squirrel

squirrels dive into their burrows. A long, steadier whistle signals a surface threat; responding squirrels stand and scan the horizon.

The males are combative at mating time; several in the colony may die of their wounds, and few will exceed half the life span of an average female. The female is receptive for only a few hours, but manages to squeeze several suitors into her busy schedule. Thenceforth she is hostile to all adult males, but will associate and bed with her female kin. For the rest of their active year, the males focus on putting on weight for next spring's bouts.

are considerably larger, Richardson's are somewhat smaller, and neither has a reddish forehead.)

Social bonds are reinforced by "kissing"—sniffing around each other's muzzles, where scent glands are located. By scent, ground squirrels can tell close relatives from distant relatives from nonrelatives. Their protective behavior, including alarm calls, seems to favor the close relatives but not the distant ones. Females protect access to food, whereas the big males, rubbing their scent glands on the ground to mark territory, guard access to females whose territories they overlap. As with marmots, there's a lot of growly squabbling and chasing. Among youngsters, it's play; between adults, it's territory; and adults of both sexes chase yearling males out of the colony, forcing them to go find new home ranges.

RICHARDSON'S GROUND SQUIRREL

Urocitellus richardsonii (rich-ard-**so**-nee-eye: for Sir John Richardson, p. 447). Also *Spermophilus richardsonii*. 9" + 3" tail; evenly gray to somewhat tawny above; tawny beneath; tail fluttering. Meadows, prairie; MT.

Richardson's ground squirrel has a piercing alarm whistle. If the whistle is short and descends in pitch, it signals a raptor, and the

MARMOT

Marmota spp. (mar-**moe**-ta: French name). Also rockchuck, whistle-pig. Heavy-bodied, thick-furred, large rodents of mountain meadows and talus, known for their piercing "whistles." Sciuridae.

Yellow-bellied Marmot

M. flaviventris (flay-vi-**ven**-tris: yellow belly). 16" long + 6½" tail; yellowish-brown, often gray-grizzled, with ± yellow throat and belly; feet darker brown. All elevs, prefers high; A.

Yellow-bellied marmot

Hoary Marmot

M. caligata (cal-i-**gay**-ta: booted). 20" long + 9" tail; grizzled gray-brown; black feet; ± white belly and bridge of nose. Alp/subalp; WA to nw MT.

THE MARMOT ANNOUNCES your approach with a sudden shrill shriek (not truly a whistle, since it's made with the vocal chords). This warning may send several other marmots lumping along to their various burrows. They'll pause en route, perhaps standing up like big milk bottles, to look around and see how threatening you actually appear. A scarier predator than you, such as a red-tailed hawk, would have elicited a shorter, descending whistle conveying greater urgency. In your case, they easily become nearly oblivious to you, or even quite forward and interested in your goods. Or you may get to watch them scuffle, box, and tumble, or hear more of their vocabulary of grunts, growls, and chirps.

Marmots need their early-warning system because they're slower than many other prey, and count all the large predators as enemies—including humans (i.e., the Mountain Shoshone in the Wind River Range). To protect themselves from the phenomenal digging prowess of badgers, yellow-bellied marmots locate their burrows in rockpiles. Occasionally, a bear will dig out and eat a whole hibernating family. These days, humans rarely hunt marmots.

Torpor and Hibernation

Most warm-blooded species have normal body temperatures about as warm as our own 98.6°F, but many spend much of their time at sharply reduced metabolic levels: they become torpid. Torpor is deep sleep, with very slow breathing (one breath per minute) and heartbeat (four to eight beats per minute) at body temperatures close to the ambient temperature, down to a limit of a few degrees above freezing. Its purpose is to conserve calories at times when they are hard to come by. It follows several patterns:

- **Seasonal** torpor is usually called either hibernation (from the Latin for "winter") or estivation (for "summer"); the animal may or may not wake up occasionally to excrete, stretch, eat stored food, or go out and forage.
- **Daily** torpor is seen in bats by day and in hummingbirds by night; ordinary sleep would waste too many calories through heat loss from tiny bodies like these in temperate climates.
- **Occasional** torpor is used by many kinds of small mammals in response to calorie shortage, water shortage, or even to shock, as in "playing possum."

Torpor helps animals adapt to severe environments. Some species of jerboas include subarctic races that hibernate, desert races that estivate, and in-between races active year-round. Some chipmunks vary from year to year, as well as with elevation and latitude, as to whether they will hibernate or forage through the winter, and they may dehibernate if the weather turns better in midwinter. Ground squirrels and marmots of steppe country go into estivation in the summer and don't come out until they dehibernate in early spring.

Hibernating is costly. The animal must put on a lot of weight—33 to 67 percent on top of its midsummer weight. Since obesity slows animals down, increasing their vulnerability to predators, they squeeze their hyperphagia (overeating) into a short season, after the

Hoary marmot

As befits the largest of all squirrels, marmots take their hibernating seriously. They put on enough fat to constitute as much as half their body weight, and then bed down for more than half the year, the colony snuggling together to conserve heat. Resist the temptation to think of their seven-month hibernation as a desperate response to an extreme environment. It is one of several strategies that work here; small subalpine grazers like the pika and the water vole stay active beneath the snow at a comfortable, constant 32°F, while long-legged browsers forage above the snow, a few staying subalpine, most migrating downslope. Different wintering strategies go with different kinds of food, so that species rarely compete directly for one food resource. Marmots prefer perennial broad-leaved herbs, and actually benefit from sharing their meadow with hoofed grazers, whose selection of grasses allows more broad-leaved herbs to grow, up to a point: seriously overgrazed meadows begin to have less forage for either class of eater.

carbohydrate-rich seeds and berries ripen, and then go to ground as soon as they can. Much of the tissue gained is brown fat, which can oxidize to produce heat directly, without muscular activity. But burning off fat draws water out of the bloodstream, whereas burning off muscle adds water, and dehydration is a big problem during hibernation; muscle mass gets added, only to be burned off again. To conserve water, pituitary hormones suppress urine formation, but then with minimal urination, toxic urea accumulates in the blood, requiring another special process to convert the urea into harmless compounds.

For roughly one day per month, some long-term hibernators raise their temperature while remaining asleep and apparently dreaming. Their brain waves indicate REM sleep (associated with dreaming), whereas in deep torpor they show little brainwave activity. They probably do this to maintain brain function.

Triggers for hibernation include scarcity of food, abundance of fat, cold outside temperatures, short days, and absolute internal calendars. Once it's hibernating, a rodent is hard to rouse. To wake up, most species spend several hours raising their body temperature by shivering violently. Marmots hibernate in heaps, and the shivering of one will trigger the others to join in a group shiver.

Bears, in contrast, lower their body temperature only a little, can rouse into full activity quickly, and commonly do rouse in winter. Some people conclude that bears are "not true hibernators." A more accurate statement would be that bears are not squirrels. Bears reduce their metabolic rate only about 75 percent (vs. 95 to 98 percent in small mammals), but they can do some things squirrels can't: they can go without urinating or defecating for months, and can maintain 100 percent of their muscle mass by recycling metabolic waste into protein. Perhaps bears are simply so much bigger and better-insulated that lowering their body temperature is difficult, and would serve no purpose.

Subalpine meadows lacking marmots may simply be too grassy.

Marmots squeeze their whole year of eating into a brief green season. A hoary marmot mother is hard put to fatten enough for her nursing litter's hibernation as well as her own; she is infertile in alternate years, restoring her metabolic reserves while loosely tending her yearlings. A dominant hoary marmot male typically keeps two mates, impregnating the fertile one and leaving her to run a nursery burrow while he shares a burrow with the infertile one and her yearlings. The colony is like an extended family: subordinate adults are "aunts and uncles" surrounding the dominant "alpha ménage à trois" and this year's and last year's litters. As summer wears on, parents increasingly work at chasing their male yearlings away from the colony. So while many marmot tussles you see are simply play between youngsters, those that end in a one-sided chase may be adults making yearlings unwelcome. Young and yearling marmots suffer heavy casualties to both predation and winter starvation.

Yellow-bellied males don't stop at two mates, but will tend harems as large as they can get by chasing other males out of a territory that overlaps the territory of several fertile females. The females maintain matrilineal territories, behaving amicably with their daughters and sisters while chasing non-kin away.

NORTHERN FLYING SQUIRREL

Glaucomys sabrinus (**glawk**-a-miss: silvery-gray mouse; sa-**bry**-nus: of the Severn River). 7" + 5½" tail; tail broad and flat; large flap of skin stretching from foreleg to hind leg on each side; eyes large; red-brown above, pale gray beneath. Conifer forest; UT and w WY to WA. Sciuridae.

Active in the hours just after dark and before dawn, these pretty squirrels are rarely seen. On a quiet night in the forest, you might hear a soft, birdlike chirp and an occasional thump as they land low on a tree

Northern flying squirrel

trunk. They can't really fly, but they glide far and very accurately, and land gently, by means of lateral skin flaps, which triple their undersurface. They can maneuver to dodge branches, and almost always land on a trunk and immediately run to the opposite side—a predator-evading dodge that includes a feint of the tail. Large owls preying on them often pick off and drop the tail, so we sometimes find jettisoned tails on the ground.

Flying squirrels usually nest in old woodpecker holes, and have their young gliding at two months of age, around midsummer. They don't hibernate, nor do they store great quantities of food for winter. Ours eat mainly truffles (underground-fruiting fungi), augmented in winter with horsehair lichens, which also insulate the nest.

Sabrinus was a river nymph in Roman myth, but the squirrels didn't get their species name for being river nymphets. The type, or first-described specimen, lived by Ontario's Severn River, named for England's Severn River, named Sabrinus by the Romans.

One night in 2017, while mucking around the Wisconsin woods with an ultraviolet LED flashlight, Professor Jon Martin became the first human to see flying squirrels as they see themselves: fluorescent Barbie pink. The pale parts of their fur fluoresce pink in ultraviolet light, which they perceive. On further investigation, his team found this trait in all *Glaucomys* species—even 130-year-old dried specimens—but in no other American rodents. They did find pink fluorescence in several owl species that prey on squirrels, and also in lichens on trees, suggesting that it may serve the squirrels either to camouflage them against trees or to repel a predator by passing as one.

ABERT'S SQUIRREL

Sciurus aberti (sigh-**oo**-rus: shadow tail; **ay**-bert-eye: for John James Abert, Frémont's boss). Also tassel-eared squirrel. 10–12" + 8–10" bushy tail; large squirrel, very striking with its tall ear tufts (taller in winter) and its varied coloring, either salt-and-pepper, black, pale brown, or dark brown; white belly; some have rust-red patch on nape. Around ponderosa pines; CO, NM, se UT. Sciuridae.

The various coat colors in Abert's squirrels can occur in the same litter, but different regions also have different typical coloring. Their beauty makes them a favorite critter to many people, including hunters. Hunting has depressed their numbers at times. Their dietary mainstays grow on ponderosa pines—seeds, buds, pollen cones, and cambium—but they also feed on other conifers, and on fungi, larvae, and shed antlers. They are rather sol-itary overall; if you see two in an extended chase, that's courtship.

RED SQUIRREL

Tamiasciurus spp. (tay-me-a-sigh-**oo**-rus: chipmunk squirrel). Also pine squirrel, chickaree. 8" + bushy 5" tail; variably gray-brown to dark red above; white beneath, ± graying in winter; a black line usually separates the 2 color areas; white eye ring. Conifer forest. Sciuridae.

American Red Squirrel
T. hudsonicus (hud-**so**-nic-us: of Hudson's Bay). UT and nw WY to WA.

Southwestern Red Squirrel
T. fremonti (**free**-mont-eye: for John C. Frémont, p. 297). Usually gray-brown, not red. CO, NM, s WY.

NOISY SPUTTERINGS AND scoldings from the tree canopy call our attention to this creature who, like other tree squirrels, can afford to be less shy and nocturnal than most mammals thanks to the easy escape offered by trees. A "chirr-rr-rr" that begins with a slight, high-pitched "peep" lets neighborhood squirrels know there's a dangerous animal nearby. Once red squirrels know about you, they don't consider you much of a threat. There

American red squirrel

Abert's squirrel

Southwestern red squirrel

Gopher cores revealed by spring snowmelt

are predators that take squirrels in trees quite easily—martens, goshawks, and large owls—but apparently they were never common enough to put a dent in the squirrel population, and are scarcer than ever today, in retreat from civilization (unlike squirrels). Extended chatter often asserts territory near a food cache. A "buzz" call is used in courtship chases.

Sometimes in late summer and fall, the sound we know red squirrels by is green cones hitting the ground. Grand- or white-fir cones are designed to open and drop their seeds while still on the tree; if you see those on the ground, they are a squirrel's harvest. The squirrel runs around in the branches nipping off cones—as rapidly as one every three seconds—then runs around on the ground carrying them off to cold storage. True-fir cones are too heavy to drag or carry, so it gnaws away just enough on the outside to reduce a cone to draggable weight, leaving the seeds still well sealed in. Someday, it will carry the cones back up to a habitual feeding-limb and tear them apart, eating the seeds and dropping the cone scales and cores, which form a heap called a "midden." The squirrel may use either the center of a midden or a hole dug in a streambank to store cones for one to three years. There, the cool, dark, moist conditions keep the cone from opening and losing its seeds. A midden can be maintained for many generations and grow to several cubic meters in size. Other species, from martens to grizzly bears, have learned to poach from middens, especially those that hold whitebark pine nuts.

Mushrooms, which must be dried to keep well, are festooned in twig crotches all over a conifer, and later moved to a dry cache such as a tree hollow. With such an ambitious food-storage industry, this squirrel has no need to hibernate. For the winter, it moves from a twig-and-cedar-bark nest on a limb to a better-insulated spot, usually an old woodpecker hole. This year's young winter in their parents' nest (unlike smaller rodents such as mice and chipmunks) since they need most of a year to mature.

The proportion of all conifer seeds that are consumed by rodents, birds, and insects is huge, exceeding 99 percent in some poor cone-crop years. Foresters have long regarded seed-eaters as enemies. But the proportion of conifer seeds that germinate and grow is infinitesimal anyway, and of those, a significant percentage succeeded because animals harvested, moved, buried, and then neglected them. Trees coevolved with seed-eaters in this relationship, and may depend on it. Additionally, all conifers are dependent on mycorrhizal fungi, many of which depend in turn on these same rodents to disseminate their spores. Conifers limit squirrel populations by synchronizing their heavy cone-crop years. A couple of poor cone-crop years bring the number of squirrels way down, and when the trees next produce a bumper crop, there are way too many seeds for the reduced population to harvest.

NORTHERN POCKET GOPHER

Thomomys talpoides (**thoe**-mo-mis: heap mouse; tal-**poy**-deez: molelike). 5½" + small, ± naked, 2½" tail; variable gray-brown tending to match local soil; front claws very long, eyes and ears very small, large incisors protrude. Sporadic in open areas with loose soil; A. Geomyidae.

Pocket gophers spend their lives underground, and have much in common with moles—powerful front claws, heavy shoulders; small, weak eyes; small hips for turning around in tight spaces; and short hair with a reversible "grain" for backing up. But moles are predators of worms and grubs, and gophers are herbivores. They can suck a plant underground before your very eyes, making hardly a dent on the surface. They get enough moisture in their food that they don't even go to water to drink. In fact, only two occasions always draw them into the open air. One is mating, in spring, which takes only a few

Sex Versus Survival, Underground

A pattern affects all three big genera of burrowing sciurids—marmots, ground squirrels, and prairie dogs: as you go farther north or higher into the mountains, you find species or populations with larger bodies, stronger social organization, more delayed dispersal of the young males, more delayed or intermittent fertility in females, and smaller litters. For example, young yellow-bellied marmots disperse (leave their maternal care and burrow) at the end of their first or, more often, second summer. Hoary marmots, with a shorter, colder active season, mature very slowly for rodents, dispersing only in their third summer when their mother's subsequent litter arrives, and breeding in their fourth summer. With the shorter, colder growing season, more energy must go into weight gain, and less is left over for reproduction.

Nutritional energy has to be allocated between survival (i.e., fattening up) and reproduction. Natural selection favors those that reproduce the most; but those that overallocate for reproduction won't survive winter and can't reproduce the next spring. Or they may just survive, but with no fat left on their bodies to sustain mating efforts, since many populations here emerge from hibernation and/or do their mating when snow still covers the ground, and little green forage is available.

For the male in many species, such as the yellow-bellied marmot, increased reproductive effort consists of patrolling a larger territory, to include more mates. For the female, it consists of bearing and nursing larger litters, and doing so without skipping years. (Hoary marmots normally skip alternate years, but may skip two years in a row if they are of low rank in the harem and/or when food conditions are poor.) As long as the females are lactating, they lose weight; they only begin fattening after weaning, which is why females return to hibernation later than males. By that time of year, vegetation is drier and less nutritious. In years with exceptionally late snowmelt, most mothers of new litters fail to fatten up enough, and will die during the winter. Nevertheless, life expectancies for males are much lower than for females, mainly due to exposure to predators, both while dispersing in youth and while patrolling and defending in adulthood.

minutes and draws out only the males; afterward they return to mutually hostile solitude, plugging up burrow openings behind them. The other is evicting young gophers from their mothers' burrows.

Though badgers and gopher snakes are well equipped to take them, the fact that gophers sustain their numbers with just one small litter per year reflects the overall safety of life underground. It also reflects enormous energy costs: burrowing 10 feet takes about a thousand times more energy than walking 10 feet. Tracking that consumption on down the food chain, we find gophers eating the most calorie-rich part of the meadow—the roots—and eating way more than their share.

In the winter, gophers can, without exactly going out, eat aboveground foods like bark and twigs by tunneling around through the snow. In spring and summer, you find "gopher eskers"—3-inch-wide, sinuous ridgelets of dirt and gravel that came to rest on the ground during snowmelt. (Voles leave similar but smaller cores.) Shortly before snowmelt, the gopher resumed earth burrowing, and used the snow tunnels it was abandoning as dumps for newly excavated dirt. Summer "tailings" heaps are fan-shaped.

Gophers churn the soil in a big way, with big ecological effects. Churning fosters diversity in mature prairies. It sped plant recovery near Mount St. Helens by bringing soil and seeds to the ash-buried surface. On alpine tundra, on the other hand, where high winds are nearly constant and soil development is glacially slow, churning wastes a lot of soil. As soon as gophers churn a patch, the wind starts carrying soil particles off to the lowlands. Soon you have a "gopher blowout," a sharp-edged basin whose floor is rocky, with little soil or vegetation left. It may take many years for a stable meadow to return. Churning can also wipe out pine seedlings in clearcuts—but good forestry can avoid this problem by not growing the weeds that gophers like.

A pocket gopher's pocket is a cheek pouch used, like a squirrel's, to carry food. Unlike a squirrel's, it opens to the outside. Fur-lined and dry, it turns inside out for emptying and cleaning. "Gopher teeth," big protruding incisors, are used (at least by gophers) for digging. The lips close behind them to keep out the dirt.

WHITE-TAILED PRAIRIE DOG

Cynomys leucuris (**sigh**-na-mis: dog mouse; loo-**cure**-iss: white tail). 12" + 2" tail; yellowish tan with whitish underside and tail tip; dark patch above and below the eye. Shrub steppe up to 11,000'; s MT s to Ouray, CO; Mar–Oct. Slightly smaller Gunnison's prairie dog, *C. gunnisoni*, replaces it from c CO s. Sciuridae.

The best-known prairie dogs are the black-tailed, who live on the Great Plains, forming massive towns. Our montane species live in relatively sparse and mobile clans. All of the prairie dog species have been decimated by the loss of the bison ecosystem, the hostility of ranchers, and finally by sylvatic plague, a flea-borne disease that reached North America in 1903. White-tailed populations, though small, have stabilized and do not appear threatened today, despite the apparently permanent plague

White-tailed prairie dog

and virtually unrestricted shooting and even some ongoing poisoning in all four states. Their range is projected to get somewhat drier overall, which will worsen conditions for them after 2040.

Ranchers and farmers consider prairie dogs enemies because they eat and dig up crops, compete with livestock for forage, and riddle the earth with hazards for hoofed mammals. For many decades, the US policy on prairie dogs was basically to exterminate them. Wyoming still classifies them as an agricultural pest.

Prairie dogs have been called keystone species because, when they decline, several other species decline with them. Burrowing owls are one of many species that use their burrows. Black-footed ferrets prey on prairie dogs exclusively, and are now America's rarest mammal. (They were feared extinct in the wild before the 1981 discovery of one Wyoming population associated with white-tailed prairie dogs.) Ferret reintroduction programs favor white-tailed prairie dog colonies, which are typically more remote than black-tailed colonies, despite an affinity for town edges and disturbed land.

White-tailed prairie dogs eat grass, flowers, sometimes sagebrush, and even weird shrubs like saltbush. They spend the majority of their time belowground, in extensive burrow systems. Researchers excavated one burrow that held 100 feet of passages, and rooms specialized for sleeping, hibernating, and giving birth. So many males die young that, as adults, females outnumber them two to one.

AMERICAN BEAVER

Castor canadensis (**cas**-tor: Greek name). 25–32" + 10–16" tail; tail flat, naked, scaly; hind feet webbed; fur dark reddish brown. In and near slow-moving streams; beaver signs frequently seen include beaver dams and ponds, and beaver-chewed trees and saplings; A. Castoridae.

American beaver

The beaver is by far the largest North American rodent alive today, and was all the more so ten thousand years ago, when there was a giant beaver species the size of a black bear. Of all historical animals, the beaver has had the most spectacular effects on North American landforms, vegetation patterns, and Anglo-American exploration.

The sound of running water triggers a dam-building reflex in beavers. They cut down poles with their teeth and drag them into place to form, with mud, a messy but very solid structure up to a few yards wide and several hundred yards long. A beaver colony maintains its dam, pond, canals, and berms for generations, adding new poles and using their feet to puddle fresh mud into the interstices.

Most beaver foods are on land, so the aquatic life is primarily for safety: the pond is a large foraging base on which few predators are nimble. Living quarters, in streambank burrows or mid-pond lodge constructions, are above the waterline, but their entrances are all below it, as are the winter food stores—hundreds of poles cut and hauled into the pond during summer, now submerged by waterlogging. Beavers favor cottonwood, willows, and aspen for poles. They can chew the bark off underwater without drowning, thanks to watertight closures right behind their incisors and at their epiglottis. They maintain breathing space just under the ice by letting a little water out through the dam. They are well insulated by a thick fat layer just under the skin, plus an air layer just above it, deadened by fine underfur and sealed in by a well-greased outer layer of guard hairs.

The typical beaver lodge or burrow houses a pair, their young, and their yearlings. Two-year-olds must disperse in search of new watersheds, where they pair up, commonly for life, and build new lodges. While searching, they may be found far from suitable streams.

When beavers dammed most of the small streams in a watershed, as they once did throughout much of the West, they stabilized river flow more thoroughly, subtly, and effectively than concrete dams do today. Beaver ponds also have dramatic local effects. First off, they drown a lot of trees. Eventually, if maintained by successive generations, they may fill up with silt, becoming first a marsh and later a level meadow with a stream meandering through it. Remaining saturated year-round, beaver pond-and-canal systems are immune to wildfire, showing up as bright-green refugia within the most intense burns.

Beavers were originally almost ubiquitous in the United States and Canada, outside of desert and tundra. Many Native groups considered beavers kin to humans, showing them respect in lore and ritual, while also hunting them for fur and meat. Europeans, in contrast, trapped Eurasian beavers (*Castor fiber*) nearly to extinction just to obtain a musky glandular secretion for use as a perfume base. There was a busy trade in this castoreum for centuries before the beaver hat craze hit Europe in the late 1700s and the demand for dead beavers skyrocketed, not letting up until the supply of American beavers became scarce around 1840. Beaver-pelt profits provided the main impetus for westward exploration and territorial expansion, including the Louisiana Purchase.

Changing fashions dropped beaver far below carnivore furs in value. (Castoreum,

though, is still used in perfumes.) Populations recovered considerably, but will never foreseeably regain their natural level because so much beaver habitat has been lost, and an agricultural economy does not easily tolerate the beaver's considerable ability to rebuild it.

Beavers are outstanding allies in both mitigating and adapting to climate change; we should accommodate them as much as we can.

If we judge the charisma of megafauna by the number of places named after them, then in the US beavers come in third, after bears and deer. (A zoology journal published a paper on geographic name counts.)

BUSHY-TAILED WOODRAT

Neotoma cinerea (nee-**ah**-ta-ma: new cutter; sin-**ee**-ria: ash-gray). Also packrat. 8" + 6½" tail covered with 1"-long fur, hence more squirrel- than ratlike; brown to (juveniles) gray above, whitish below; whiskers very long; ears large, thin; can either run or, if chased, hop like a rabbit. Widely but patchily distributed in non-alpine sites with rock outcrops or talus; A. Cricetidae.

For some strange reason, packrats love to incorporate human-made objects, shiny ones especially, into their nests. Possibly some predators are spooked by old gum wrappers or gold watches, or perhaps the packrat's craving is purely aesthetic or spiritual. She is likely to be on her way home

Bushy-tailed woodrat

with a mud pie or a fir cone when she comes across your Swiss Army knife, and she obviously can't carry both at once, so there may be the appearance of a trade, though not a fair one from your point of view. Packrats have other habits even worse than trading, often driving cabin dwellers to take up arms. They spend all night in the attic, the woodshed, or the walls noisily dragging materials around—shreds of fiberglass insulation, for example. Or they may mark unoccupied cabins copiously with foul-smelling musk. The males mark rock surfaces with two kinds of smears: one dark and tarry, the other calcareous white and crusty.

Our common species, the bushy-tailed, is an active trader. It nests in crevices of trees, talus, cliffs, mines, or cabins, and sometimes assembles piles of sticks to cache food in. In caves in arid climates, packrat nests and piles can last for millennia, held together and perhaps preserved by rock-hard calcareous deposits of dried urine, enabling paleobotanists to study vegetation shifts as far back as the last ice age. Bushy-tailed woodrat middens are studied in Yellowstone.

DEER MOUSE

Peromyscus maniculatus (per-o-**miss**-cus: boot mouse; ma-nic-you-**lay**-tus: tiny-handed). 3½" + 3" tail (length ratio near 1:1); brown to (juveniles) blue-gray above, pure white below incl white feet and bicolored, often white-tipped tail; ears large, thin, fully exposed; eyes large. Ubiquitous. Cricetidae.

The deer mouse is far and away the most widespread and numerous mammal on the continent. Not shy of people, it makes itself at home in forest cabins, farmhouses, and many city houses in the Northwest. It remains less urbanized than the house mouse, *Mus musculus,* which spread from Europe to cities worldwide together with its relatives the black rat and Norway rat. (Those three invasive urbanites are easily distinguished from our native mice, rats, and voles by their scaly, hairless tails.)

Deer mouse

Piñon mouse

The deer mouse builds a cute but putrid nest out of whatever material is handiest—Kleenex, insulation, underwear, moss, lichens—in a protected place such as a drawer or hollow log. In winter, it is active on top of the snow. Omnivorous, it favors larvae in the spring and seeds and berries in the fall. Such a diverse diet adapts it well to burns and clearcuts; as the forest matures, more-specialized fungus-eating voles gradually displace deer mice. After the 1980 eruption of Mount St. Helens, deer mice were common in the blast zone so soon that they are presumed to have survived the blast in their homes, and to have found food under the ash.

PIÑON MOUSE

P. truei (**true**-eye: for F. W. True). 4¼" + 4" tail (length ratio near 1:1); buff, with paler feet and bicolored tail; ears 1" (extra large), thin. Piñon and juniper; w and s CO, UT, NM. Cricetidae.

Piñon mice are major consumers and disseminators of juniper berries. Piñon and deer mice can both carry hantavirus, the flulike symptoms of which can turn deadly if left untreated for a few days. While piñon mice don't come indoors much, the chief danger to humans is in dry, dusty cabins with deer mice, where people can get infected by breathing mouse-contaminated dust. Avoid handling mice, and wash your hands promptly and thoroughly if you do handle one. If you must clean up a mousy corner, spray first with bleach solution, and wear rubber gloves, goggles, and a HEPA-filtered respirator.

WESTERN JUMPING MOUSE

Zapus princeps (**zay**-pus: big foot; **prin**-seps: a chief). 4" + 5" tail (tail longer than body); back has broad, dark-brown-to-black stripe; sides ± washed with dull lemon-yellow; belly buff-white; hind feet several times longer than fore-feet; ears small. Thickets and meadows near streams, May–Sep; A. Zapodidae.

The jumping mouse normally runs on all fours, hops along in tiny hops, or swims, but if you flush one it's likely to zigzag off in great bounding leaps of 3 to 5 feet. This unique gait gives you a good chance of recognizing them, even though they're mainly nocturnal and not all that common. The oversized feet are for power, and the long tail for stability: jumping mice that have lost or broken their tails tumble head-over-heels when they land

Western jumping mouse

from long jumps. While none of our other mice or voles hibernate, this one hibernates deeply for more than half the year. It eats relatively rich food—grains, berries, and tiny (¼-inch or less) underground fungi that grow on maple roots.

MEADOW VOLE

Microtus spp. (my-**cro**-tus: small ear). Mouselike creature, yellowish brown in summer, brownish gray in winter; eyes small; ears barely visible; feet pale; tail bicolored. Cricetidae.

Western Meadow Vole

M. drummondii (for Thomas Drummond, p. 136). Also *M. pennsylvanicus*. 5" + 1½" tail (length ratio 3:1); relatively dark brown. Moist to wet meadows; A–.

Montane Vole

M. montanus. 4¾" + 1¼–2" tail (length ratio less than 2:1). In vole runways in grass; WA to sw MT and s to NM.

Long-tailed Vole

M. longicaudus (lon-ji-**caw**-dus: long tail). 4½" + 2½" tail (length ratio less than 2:1); fur may be reddish. Diverse grassy habitats at all elevs; A.

MEADOW VOLES ARE among the world's most prolific mammals. The female gestates for three weeks, and breeds again immediately after bearing her litter of three to ten young ones, who nurse for the first two weeks of the next pregnancy. At that rate, the numbers really add up. With, say, an average litter of five and a cycle of twenty-two days, a vole pair could generate billions of descendants in a year. Needless to say, mortality keeps pace—meadow voles are the dietary staple of many predator species, both winged and four-legged—and prevents the numbers from actually going there. But they do go quite high, and then plummet, cyclically. Drastic population cycles three to five years long characterize many of the vole species in this large New World genus. Each species's cycle stays synchronized over much or all of its range. Hormonal and behavioral mechanisms, rather than either starvation

Montane vole

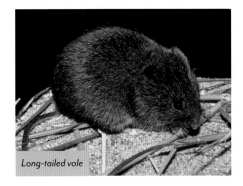
Long-tailed vole

You might think of this critter as a small muskrat rather than a big vole. However, it spends time in water for refuge, rarely for forage. It dines on our favorite lush wild-flowers—lupine, valerian, glacier lilies, and such—eschewing grasses and sedges. In winter, it digs up bulbs and root crowns of the same flowers, or eats buds and bark of willows and heathers, while tunneling around under the snow. Look for mud runways running straight to the water's edge from its burrow entrances, which are up to 5 inches wide.

RED-BACKED VOLE

Clethrionomys gapperi (cleth-ree-**on**-a-mis: keyhole mouse; **gap**-per-eye: for Anthony Gapper). Also *Myodes gapperi*. 4" + 1¾" tail; gray-brown, with distinct rust-red band down length of back; tail ± bicolored similarly; active day or night. Common in mature conifer forests; A. Cricetidae.

Red-backed voles should be respected as gourmets. They dine largely on underground-fruiting fungi—which few of us are aware of except when, as "truffles," they are imported from France or Italy at $300 a pound. It's no accident that truffles are uniquely fragrant, delicious, and nutritious. These fungi have no way of disseminating their spores other than by attracting animals to dig them up, eat them, excrete

or predation, curb the population explosions somewhere short of mass starvation—though not always soon enough to prevent damage to grain crops.

Increasing drought has extirpated western meadow voles from central New Mexico. Further losses at the dry margins are likely.

Montane voles and heather voles spend winter tunneling through the crumbly bottom layer of the snowpack, called "depth hoar." They maintain vertical tunnels to the surface here and there, perhaps for ventilation or for help in escaping weasels. Occasionally they venture to the surface, where they make short, shallow tunnels to rest and warm up in.

WATER VOLE

M. richardsoni (richard-**so**-nye: for Sir John Richardson, p. 447). Also water rat; *Arvicola richardsoni*. 6" + 2¾" tail (length ratio 2:1); fur long and coarse, ± reddish dark-brown above, paler beneath. In and near high bogs, streams, and lakes; northerly plus UT.

Red-backed vole

the undigested spores elsewhere, and preferably thrive on them over countless generations.

Many forest rodents are avid participants in this scheme, but none are more dependent than red-backed voles. Since the fungi stop fruiting if their conifer associates die, red-backed voles disappear after a clearcut or destructive burn. They can't switch from fungi to fibrous grasses because their molars have lost the ability to keep growing throughout life to replace unlimited wear and tear. Rodent incisors, or front teeth, keep growing lifelong (see p. 315), but *Clethrionomys* molars are irreplaceable—like ours. Several *Microtus* voles, in contrast, can grind up grasses to their hearts' content. Red-backed voles are the dietary mainstay of martens in parts of our range.

HEATHER VOLE

Phenacomys intermedius (fen-**ack**-o-mis: impostor mouse). 4¼" + 1" tail (length ratio 3:1); gray-brown above, paler beneath; tail distinctly bicolored; feet ± white, even on top. Sporadic in forests and alpine meadows; A. Cricetidae.

Above timberline, you may spot a heather vole's nest shortly after snowmelt: it's a 6-to-8-inch ball of shredded lichens, moss, and grass, typically not far from a big heap of rust-colored dung pellets. Both the winter nest and the dung heap were in a snow burrow atop the soil. The summer nest is underground.

SAGEBRUSH VOLE

Lemmiscus curtatus (lem-**iss**-cus: small lemming; cur-**tay**-tus: short, referring to the tail). Also *Lagurus curtatus*. 4" + ⅞" tail (ratio more

Rodents and Fungi

Underground-fruiting fungi—truffles, broadly speaking—are a mainstay of rodent diets in Western conifer forests.

Rating the nutrition in these morsels has proven tricky. Laboratory analysis finds them high in protein, very low in fats, moderate in carbohydrates, and very high in some vitamins and minerals. But their digestibility turns out to be abysmal—after all, the principal contents are spores, and the passage of spores through the rodent's digestive tract intact and viable is the whole point of the relationship from the fungus's point of view, so the spores have to be pretty indigestible. After adjusting for digestibility, truffles seem barely worth eating. Two factors tip the scales in their favor: they fruit in late fall and winter, when green foods are least nutritious; and finding them consumes fewer calories than finding plant foods because it's done with the nose, a small mammal's most effective sense organ.

Moisture content is another big plus: eating fungi is often less taxing than making the trip to a stream or puddle to drink, and reduces exposure to predators. Moist fungi thus enlarge the fungivores' habitat by freeing them from having to live near creeks. (On the other hand, some squirrels hang fungi up to dry, preserving them for winter.)

Fungal spores aren't the only potent stowaways in vole and squirrel droppings. Nitrogen-fixing bacteria, which live in the truffles, also pass unharmed through the rodents' bowels, as do yeasts, which contribute nutrients these bacteria need to fix nitrogen. Since nitrogen fertility is often a limiting factor on conifer growth, the conifers may be as dependent on the bacteria and yeasts as on their mycorrhizal partners, the truffles. It adds up to a five-way symbiosis, including the rodents that disseminate all four other partners.

Sagebrush vole

Muskrat

than 4:1); pale gray with whitish buff on ears, nose, feet, belly, and underside of tail. Arid habitats at all elevs; OR, c ID, s MT, UT, WY. Cricetidae.

Like other voles, the sagebrush vole lives on greenery, but it's a rare vole in braving near-desert habitats, and in wearing sandy-colored fur to match. (Other small rodents of the semidesert rely more on insects and seeds.) The sagebrush vole is active year-round, at any time of day or night, with some preference for dusk and dawn. It doesn't expose itself much, spending most time in burrows or surface runways through dense grass or under snow. Sagebrush voles cram a lot of reproduction into spring and fall, when moist greenery is more abundant. Within twenty-four hours of giving birth, the mother leaves her two to thirteen newborns in their nest while she steps out to look for her next mate.

MUSKRAT

Ondatra zibethicus (ahn-**dat**-ra: Wendat term for muskrat; zi-**beth**-ic-us: civet- or musk-bearing). 93" + 72" tail; tail scaly, pointed, flattened vertically; fur dark glossy brown, paler on belly, nearly white on throat; eyes and ears small; toes long, clawed, slightly webbed; voice an infrequent squeak; largely nocturnal. Scattered, in or near slow-moving water up to mid elevs; A. Cricetidae.

It's tempting to think of the muskrat as an undersized beaver, but its anatomy unmasks it as an oversized vole. Leading similar aquatic lives, beavers and muskrats grow similar fur, which was historically trapped, traded, marketed, and worn in similar ways. (In both cases, the guard hairs are removed, leaving the dense, glossy underfur.) Several million muskrats are still trapped annually—more individuals and more dollar value than any other US furbearer. In the South, the meat also finds a market as "marsh rabbit."

Muskrats are smaller than beavers—half the length and rarely even a tenth the weight—and their teeth and jaws aren't robust enough to cut wood, so they build no dams or ponds. Soft vegetation like cattails, rushes, and water-lilies makes up the bulk of their diets, their lodges, and the rafts they build to picnic on, and yet they stray from the vegetarianism typical of the vole family, eating tadpoles, mussels, snails, or crayfish. Their interesting mouths remain shut to water while the incisors, out front, munch away at succulent stems underwater. They can take a big enough breath in a few seconds to last them fifteen minutes underwater; under winter ice, they are thought to extend their range by releasing bubbles to create an air pocket they can use on the next outing. To a mammal this small, ice water's chill is a great challenge; to recoup body heat, they huddle with other muskrats, and burn off calories through "non-shivering thermogenesis."

Neatly clipped sedge and cattail stems floating at marsh edges are a sign of muskrats. Both sexes, especially when breeding, secrete musk on scent posts made of small grass cuttings.

PORCUPINE

Erethizon dorsatum (er-a-**thigh**-zon: angering; dor-**say**-tum: back). 28–35" long, incl 9" tapered tail; large girth; blackish with long, coarse yellow-tinged guard hairs and very long whitish quills; incisors orange. Anywhere, but more common around conifers; A. Erethizontidae.

Porcupines' bristling defenses permit them to be slow, unwary, and much too blind to obtain a driver's license. This is both good and bad news for you. Though they're mainly nocturnal, you stand a good chance of seeing one some morning or evening, and perhaps of hearing its low, murmuring song. (And you stand a good chance of having your equipment eaten by one during the night; more on that below.)

Porcupine

Quills and spines are modified hairs that have evolved separately in many kinds of rodents, including two separate families each called porcupines. On American porcupines, they reach their most effective form: hollow, only loosely attached at the base, and minutely, multiply barbed at the tip. The barbs engage instantly with enough grab to detach the quill from the porcupine, quickly swell in the heat and moisture of flesh, and work their way farther in (as fast as 1 inch per hour) with the unavoidable twitchings of the victim's muscles. Though strictly a defensive weapon, quills can occasionally kill, either by perforating vital organs or by starving a poor beast that gets a noseful. More often, they work their way through muscle and out the other side without causing infection, since they have a greasy antibiotic coating.

So, just maintain a modestly safe distance while the porky most likely retreats up a tree. "Throwing their quills" is a myth; in fact, porcupines just thrash their tails, whose range sometimes takes people by surprise.

Since quills are their only defense, porcupines have to be born fully quilled and active in order to survive infancy in good numbers. This requires very long gestation (seven months) and small litters (one or rarely two). But still, how to get the little spikers out of the womb? Answer: the newborn's soggy quills are soft, but harden in about half an hour as they air-dry. And then again, how to get mama and papa close enough to mate? Or baby close enough to nurse? Solving these problems has given porcupines their Achilles' heel, or rather, soft underbelly. With a quill-less underside including the tail, a porcupine can safely mate with her tail scrupulously drawn up over her back. (They mate in late autumn but are otherwise solitary.) The soft underbelly actually has very tough skin, to hold up through months of dragging up and down tree bark. Both sexes protect their genitalia from abrasion by withdrawing them entirely behind a membrane, making it hard to tell the sexes apart.

If it weren't for that unarmed belly, predators would have no place to begin eating. And predators there are, mainly fishers and cougars. The soft underbelly is safe as long as porky is alive enough to stay right side up. Fishers attack the head. Follow their example, with a heavy stick, if you ever find yourself lost, truly starving, and near a porcupine. Some states protect porcupines on the grounds that they are easy edible prey for unarmed humans lost in the woods. They are not choice fare. Indigenous people sometimes ate them, and turned the quills into an elegant art medium on clothing and baskets.

Porcupines eat leaves, new twigs, and aspen catkins in summer, and tree cambium, preferably pine, in winter. They select cambium near the treetop, where it is sweetest. Bright patches of stripped bark high up in pines are a sign of porcupine. When they kill a tree's top by girdling it, the tree turns a branch upward to form a new main trunk, leaving the tree kinked and diminished in lumber value. So foresters are alarmed at the increasing porcupine populations that have resulted mainly from human persecution of predators.

Like red tree voles and blue grouse, porcupines pay a nutritional price for dining at this all-you-can-eat cafeteria. They inexorably lose weight in winter, even while keeping their big guts stuffed with bark. In summer, trees offer tender new foliage, and defend it with toxically high potassium levels. For the porcupine kidney to keep up with the task of eliminating potassium, the porcupine is driven to find low-acid, low-potassium sources of sodium. Your sweaty boots and packstraps fit the bill, as do the coolant hoses, electrical insulation, and even tires that abound in trailhead parking lots. For similar reasons, marmots also may consume car parts, deer and elk eat the mud around soda springs, and mountain goats go after human urine. It's a good idea to always sleep by your boots, and to make your packstraps hard to get at too. Fair warning.

Wildlife managers seem to agree that porcupine populations are declining across the West, but there are no firm numbers and no conclusive explanation. Drought may be implicated, as well as increasing numbers of the known killers—cougars, fishers in a few states, and cars.

Order Carnivora (Carnivores)

Members of six families in the order Carnivora follow. All six are household words: dog, cat, raccoon, bear, skunk, and weasel. Carnivora also include seals and walruses.

RED FOX

Vulpes vulpes (**vul**-peez: Roman name). Also *V. fulva.* 25" + 17" tail; shoulder height 16" (terrier size); commonly red-orange with black legs and ears and white belly (red phase); other color phases are "black," "silver" (white-tipped black fur), "frosty" (creamy pale gray, at high elevs in Yellowstone area), and "cross" (like red, but with dark brown down back and across shoulders); all typically have white-tipped tail. Widespread. Canidae.

Foxes are little seen here: they're nocturnal, shy, elusive, and alert. Though they can bark and "squall," they rarely make a concert of it. Their tracks and scats are hard to tell from those of small coyotes, and their dens are most often other animals' work taken over without distinctive remodeling. Rocky areas are preferred for denning, and mixed brush/grassland areas for habitat.

Native Rocky Mountain foxes run the gamut of color phases. Red is most common at most elevations, but the unique "frosty" form is said to predominate at the very highest elevations in the Yellowstone ecosystem. Even littermates may differ, just like blond, dark, and redheaded human siblings. Foxes of the plains and open valley country probably descend partly from fur-farm escapees, inheriting a substantial genetic component from European and

Red fox

to tawny dog, grayer and thicker-furred in winter; runs with tail down or horizontal. Ubiquitous. Canidae.

In pioneer days, wolves ruled the forests, leaving coyotes to range over steppes, brushy mountains, and prairies. Then, during the nineteenth century, guns and traps tipped the scales in favor of coyotes by targeting the bigger predators—cougars and grizzlies, as well as wolves. Where the big predators were eliminated, coyotes, bobcats, and eagles assumed the mantle of apex predators, also inheriting mankind's hatred even though all are too small to significantly limit deer numbers. Coyotes are America's most bountied, poisoned, and targeted predator: USDA Wildlife Services kills more than sixty thousand annually. Yet coyotes uncannily persist, increase, and spread. Returning wolves may halt that trend, but reports are mixed as to whether that is happening yet in Yellowstone.

More often than you'll see coyotes, you'll hear them howling at night. You can guess that those hair-filled scats in the trail are likely theirs, especially when placed smack dab on a stump, hummock, or footlog over a creek; in an intersection; on a ridgetop; or any combination thereof. Coyote feces and urine are not mere "waste," like yours, but scent posts, like graffiti signatures, full of olfactory data that later canine passers-by, including other species, can read. In Chinook myth, Coyote consulted his dung as an oracle! The male canine habit of fiercely scratching the ground after defecating deposits scents from glands between the toes. The long noses in the dog family really are "the better to smell you with, my dear," as the large olfactory chamber is loaded with scent receptors. Coyotes can detect the passage of other animals a mile or two away, or days earlier. They read scent posts to learn each other's condition and activities (territoriality not being a

Eastern subspecies. But all races are considered the same species.

Red foxes mate in midwinter, bear their litters (of four to seven) in early spring, and commonly stay mated. In contrast to the cat and bear families, canids make good fathers. Foxes eat insects, earthworms, fruit and seeds, and some birds' eggs in addition to the preferred mice, voles, hares, frogs, and squirrels. They hunt with devious opportunism and stealth, often culminating in a spectacular aerial pounce. Waving the huge plumey tail in mid-pounce can perfect the aim of pounces of up to 15 feet. The tail also warms the face when a fox curls up. The large paws are good on snow.

As wolf range shrank over the last century, coyote and red fox range expanded; red foxes displaced wolves as the world's most widely distributed rural wild-mammal species. Where wolves were removed and coyotes thrived especially, they tended to chase out the foxes; today, where wolves return, they chase out coyotes, and foxes recover.

The slender, tan, catlike swift fox, _V. velox_ (**vee**-lox: swift), of the Great Plains, may sometimes reach the eastern edge of our range. It is rare and endangered in Canada and Montana. The similar kit fox, _V. macrotis,_ of the Great Basin and Southwest deserts, barely reaches our range in Idaho.

COYOTE

Canis latrans (**cay**-niss: dog; **lay**-trenz: barking). 33" + bushy 14" tail; shoulder height 16–20"; medium-sized, pointy-faced, erect-eared, gray

primary intent). Mates will even scent-mark together on the same spot.

When you're identifying scats, predator specimens will stand out from those of herbivores or kibble-fed canids by being full of hair, while size separates coyotes from, say, martens. However, scat ID is trickier than you might expect: In one study, biologists collected 404 predator scats in a national park. (They were finding out who was eating the marmots. Answer: mainly coyotes.) They field-identified the scats, then confirmed the ID using DNA. Their field accuracy rate was only 63 percent, when with only three candidates—cougar, bobcat, and coyote—random guessing would have yielded 33 percent.

As for their spooky nocturnal cacophony, most people hearing it feel that it, too, conveys something above and beyond mere location—though helping a family group relocate each other is its best-understood function. Often it's hard to tell how many coyotes we are listening to; the Modoc used to say it's always just one, sounding like many. Coyote choruses mix up long howls with numerous yips; wolves howl without yipping.

Though preferring small mammals and birds, coyotes also eat grasshoppers, berries, mushrooms, and scavenged carrion. Stalking mice, they patiently point like a bird dog, then pounce like a fox. Against hares they use the fastest running speed of any American predator. Though they commonly try to pick off fawns or elk calves, they hunt adult deer only rarely, locally, and usually in later winter when they gain an advantage from crusted snow as well as the winter-weakened condition of the prey. Coyotes usually hunt alone or pair up, one mate perhaps decoying or flushing prey to where the other lurks. An unwitting badger, eagle, or raven may be briefly employed to do the decoying or flushing.

Pairs bond for years, and display apparent affection and loyalty. To say they mate for life would be about as euphemistic as saying that Americans do. In years when coyotes are abundant and rodents are unusually scarce, as many as 85 percent of the mature females may fail to go into heat, and those who do so will bear litters smaller than the typical five or six pups. On the other hand, they reproduce like crazy when their own populations are depleted. Some coyotes remain with their parents as yearlings, helping to raise the new litter. While the most typical pack is a family, there's a range from larger groups down to solitary or paired individuals.

Wolves, coyotes, jackals, and domestic dogs are all interfertile and beget fertile offspring. Yet these species are dramatically different, physically and behaviorally, despite having long occupied overlapping ranges. Wild "coy-dog" hybrids occur, yet fail to establish reproducing populations. The fierce antagonism of wolf packs toward outsiders (including coyotes) maintains wolves and coyotes as distinct species. At Yellowstone, wolves kill (but don't eat) coyotes they catch trying to scavenge wolf kills, but this doesn't stop other coyotes from trying. Coyote numbers have fallen sharply there since wolves returned.

Coyote

Coyote the Trickster is a ubiquitous, complicated figure in Indigenous lore, a devious intelligence undermined by downright humanoid carelessness, greed, conniving lust, and vulgarity. In some tales, Coyote exemplifies the bad, greedy ways of hunting that destroyed a long-gone Edenlike abundance. In others, he brings the poor, starving people rituals and techniques they need, like catching salmon. Some Indigenous, on learning about Jesus, saw him as the white man's Coyote. The opposing stories reflect the likely history: the first people in North America found an Edenlike abundance, but when a local population reached a saturation point for simple hunting and foraging, the people had to learn to store foods.

In other stories, Coyote is the Creator. No wonder the world doesn't always work perfectly.

GRAY WOLF

Canis lupus (**loop**-us: Roman name). 52" + bushy 20" tail; shoulder height 26–34"; big, erect-eared gray to tan dog with massive (not pointy) muzzle; long legs; majority of pups are black, turning pale by adulthood; fur very thick in winter; runs with tail down or horizontal. Native in A; southerly populations pending restoration.

> *At night they are so fearless as to come quite within the purlieus of the camp, and there sit, a dozen together, and howl hideously for hours.*
>
> —John Kirk Townsend in Wyoming, 1834

Today, the eerie howling of a wolf pack is music to the ears of urban and suburban North Americans—an amazing sea change in a very few years, matching the strong recovery of wolf populations in the Northern Rockies. Wolves captured in Canada were reintroduced into central Idaho and Yellowstone in 1995–96. Others had already been slipping across the Canadian border into Montana and Washington. In 2020, Wyoming packs entered two Colorado counties, and in 2023 additional wolves were flown into the state in accord with an initiative passed by Colorado voters in 2020.

Few western ranchers share city-dwellers' love of wolves. Wolves do prey on calves and lambs when those prey are the most convenient, and some packs acquire a preference for them. In the nineteenth century, Euro-American culture unanimously despised and feared wolves. In much of today's media, wolves have gone from cruel to cuddly. Both views are exaggerations. Wolves are wild carnivores. Statistically, they rarely attack people but, like black bears, they can do so, and are more likely to if they are habituated to people or if they have rabies. Wolf-dog hybrids and wolves that have been raised by people are especially dangerous.

Sometime during the last ice age—probably many times, in many places—some wolves began hanging around humans to scavenge food scraps. At some villages, this strategy fed them well; their descendants thrived, or the mellower-tempered ones did while the aggressive ones were driven away or killed. Over the generations, humans came to like these self-selected dogs while often finding ways to exploit them—for meat or wool or to haul loads. At some point, dogs began to participate in a hormonal response known for strengthening trust and bonding between human mothers and infants: if a dog and its owner gaze into each other's eyes, oxytocin levels rise in both of them. This doesn't happen in wolves, even ones raised by humans. The wolves that stayed wild back then continued to evolve, so that modern wolves are not as closely related to dogs as ancient wolves were. Dogs are thus not wolves, even though they are officially *C. lupus*. (Linnaeus named dogs *C. familiaris*, but current taxonomists don't regard domestication as speciation.)

Wolves can be born either black or a gray patterned with brown. With age, they often lighten: gray wolves to white, black wolves to gray. Black coats may have entered the gene pool via interbreeding with Inuit sled dogs somewhere in northern Canada four thousand to seven thousand years ago. Black wolves remain uncommon in the East, and very rare in Eurasia. Biologists at Yellowstone found that opposites attract (gray wolves mate with black) to an unexpected degree, and infer that heterozygous black wolves (carrying genes for both traits, with black dominating) are coincidentally advantaged with some kind of disease resistance, while also being less aggressive on average.

Gray wolf

Perhaps the most exciting thing about wolves at Yellowstone is that they allow biologists (as well as tourists) to study them. Previously, most observations of wolf behavior were of captive wolves thrust into pens together, leading to a wildly false stereotype of wolves engaged in ceaseless aggression over alpha status. The wolf biologist David Mech made a breakthrough by spending thirteen summers on Ellesmere Island up close with wolves who had never encountered humans and were unwary. His conclusion: a pack is a family, and aggression is infrequent. The kids wouldn't think of challenging Dad's or Mom's alpha status.

Mix of colors typical of Yellowstone wolf packs

Study of forty-three packs (and counting) at Yellowstone reveals complexity and variation. Still, bottom line, packs are extended families, and display impressive teamwork. Monogamy prevails (sometimes it's lifelong), but versions of polygamy are also common. Large packs sometimes "adopt" an outside male, who may breed with a subordinate female. Only the breeding adults scent-mark. While dominance is roughly equal between the alpha female and male, some Yellowstone researchers see a little more leadership from the females.

Around age two or three, wolves may try dispersing from the pack, but some never disperse. They have several potential paths to breeding: without dispersing, wait to replace a breeder who dies; disperse and join another pack at subordinate rank; disperse and then kill a breeder of another pack; or, most often, join other dispersers to found a new pack. As overall wolf density increases, so, too, does aggressive competition between packs—the main setting of wolf-on-wolf fatalities. While bigger packs hold a clear edge in these fights, a less obvious advantage accrues to packs with elders (ages six and over), who apparently can share the battle wisdom they've acquired.

A major virtue of pack sociality is having a lot of help in feeding and training the pups. This kind of help characterizes less than 2 percent of mammal species. Hunting as a pack facilitates taking prey as large and dangerous as elk, moose, or bison. Actually, though, lone wolves sometimes take moose. The pack advantage may center less on the hunt than on holding ravens at bay afterward until the meat is consumed.

When the saber-toothed tigers and other Pleistocene megapredators went extinct, leaving only wolves, grizzlies, and humans to prey on the large herbivores, it destabilized North American ecology. Regional extirpations of wolves threw things further out of balance, with no regular predator at all for the large herbivores that grizzlies rarely select. This affects not only prey species and the prey's food plants, but also competitors and the competitors' competitors, like foxes, which come back in when wolves chase coyotes out. Yellowstone has seen at least modest recovery of beleaguered aspen, willow, and beaver populations; exactly how much credit the wolves deserve for that, and through what mechanism, is debated.

Increasing numbers of Yellowstone wolves are infected with the protozoan *Toxoplasma gondii,* with interesting effects. This parasite causes an increase in testosterone and hence in risky behavior. The parasite benefits in that rodents it infects are more likely to get eaten by a cat, and the parasite has to infect a cat of some kind in order to reproduce. Similarly, infected hyenas are more likely to get eaten by a lion. Our wolves, in contrast, have no predators, so infection may actually benefit individual males: pathological boldness encourages exploiting new territory, and infected male wolves were found forty-six times more likely than uninfected males to become pack leaders. The downsides are increased contact with humans and highways, leading to mortality, and fetal and infant mortality from infected mothers.

RINGTAIL

Bassariscus astutus (bas-a-**ris**-cus: little fox). Also miner's cat, civet cat, ringtailed cat, bassarisk. 24–34" long in total, with slightly more than ½ that in the huge, bushy tail with 7 or 8 light/dark rings; low and slender; buff overall, hairs black-tipped; erect ears; light/dark goggles around the big, dark eyes; 5 clawed toes; call a sharp bark; nocturnal. Canyons, piñon-juniper; southerly. Procyonidae.

Ringtail

Ringtails are true omnivores: pine nuts, juniper berries, insects, berries, spiders, reptiles, toads, songbirds, and small mammals all play substantial roles in their diets. They typically find all the water they need in their food, conserving it by concentrating their urine more than any other carnivore. Excellent climbers, they can stem up cliff crevices, ricochet up or down if the crevice gets wider, and rotate their hind feet 180 degrees to descend trees.

They are related to raccoons—visibly so. Raccoon sightings are likely more frequent here; raccoons (*Procyon lotor*) venture higher up, farther north, and of course much closer to human habitation. But ringtail sightings are more fun.

BLACK BEAR

Ursus americanus (**ur**-sus: Roman name). 4' long (4" tail inconspicuous); 3½' at shoulder; typically jet black, with a tan nose (less common phases are "cinnamon" red-brown, tannish brown, "blue" gray, and even white); facial profile ± straight; no shoulder hump; claws dark, 1–1½". A. Ursidae.

Perhaps even more than the sneaky coyote, smart raven, and industrious beaver, the bear has always been seen by humans as somehow our kin. Though the bear's mirror to human nature would seem darker and grander than the coyote's or beaver's, it's hard to pin down the essence. I concede that bears are among the most humanoid of animals in terms of their diet and their feet, which are five-toed, plantigrade (putting weight on the heel as well as on the ball and toes), and about as big as ours, so that the prints—especially the hind print—look disturbingly familiar.

Their diet seemingly includes almost anything, and varies enormously by season, region, and individual. Plant foods predominate, starting in spring with tree cambium, horsetails, grass, bulbs, and all kinds of new shoots, and working up to berry gluttony in fall, the fattening-up season. Prey includes small mammals, fawns, fish snatched from streams, and insects or larvae where they can be lapped up in quantity—anthills, grubby old logs, wasp nests, and beehives, preferably dripping with honey. An adult bear can chase predators from their kills, but is less adept at hunting for itself, and will only eat large animals as carrion. Some bears learn to rob woodpecker nests, grain crops, fallen orchard fruit, garbage dumps, or hiker camps. During heavy berry-eating, bear scats become semiliquid like cow pies, and show a lot of fruit seeds, leaves (blueberry), and skins (apple). Earlier in the season, they are thick, untapered cylindrical chunks, perhaps showing animal hair, but often resembling horse manure. Fresh, they are often as jet-black as the beast itself.

Trees show several different kinds of bear markings. To eat conifer cambium, bears peel huge strips of bark, typically from around 5 feet down to the ground, with irregular incisor gashes. (Though not all bear-stripped trees die, bears are serious tree-killers; during times when cambium seems like the best food available, a bear may strip sixty trees a day.) To assert territory, bears bite trees or claw them with long, parallel slashes 5 to 9 feet from the ground (compare with cat scratches). They rub against rough bark—to scratch an itch, to scent-mark, or to apply tree volatiles as insect repellent—leaving diffuse abrasion and a lot of bear hair 2 to 4 feet up. They also leave claw marks when they climb trees to eat catkins, nuts, or fruit. Cubs are often sent into trees for safety while Mom forages, or whenever a threat (such as a hiker) is perceived.

Black bears, especially the females, are aggressively territorial, but they tolerate

Bear claw marks

Black bear, cinnamon phase

each other's company occasionally at banquets, such as salmon runs or great berry crops. Once the ripe fruits are all gone in the fall, there is little bears can do to fatten up. They begin to slow down even before hibernating, and appear listless while preparing their dens—for example, building a nest of fluffy stuff like cedar bark, which stays resilient a long time. The bear sleeps curled up in a ball, with the crown of its head down. With insulative nest padding as well as thick fur, the bear loses little heat to the air of its den, and maintains a body temperature of about 88°F, even while its heart rate may reach an astonishing low of just eight very weak beats per minute (p. 318).

Bears are rather nearsighted; when they stand on their hind legs, it may be to see farther. They have excellent hearing and a phenomenal sense of smell, able to detect carrion many miles away.

Two or three cubs are born around January. The mother wakes up to give birth, then nurses them mostly in her sleep for the next few months. A den of cubs nursing emits a loud buzz or hum, like a beehive, only deeper. About 25 percent of litters include half-siblings, indicating that females as well as males are promiscuous. Cubs are smaller at birth, relative to their adult weight, than almost any mammals short of marsupials. To nurse them to viable size, the mother may lose 40 percent of her weight during hibernation, whereas males lose only 15 to 30 percent. The next year, she will hibernate with her yearlings rather than breeding again. The ferocity of black bear mothers accompanying cubs likely protects cubs from males of their own species. Subadult bears also risk cannibalism if they are foolish enough to stand up to a big boar.

GRIZZLY BEAR

Ursus arctos ssp. *horribilis* (**ur**-sus: Roman name; **arc**-toce: Greek name; hor-**rib**-i-liss: horrible). 6–8' long (3" tail hard to see); shoulder height 4½'; brown (rarely black) ± grizzled with light tan, but color is not reliable for distinguishing from black bear; instead, look for hump over front shoulder, concave facial profile, and very long, pale claws (foreclaws typically 3" or longer). Canada; endangered in the Lower 48, but recently increasing in Yellowstone region and nw MT, n ID. Ursidae.

The grizzly is the largest terrestrial carnivore, and one of the most intimidating. It's big and fierce enough to take any hoofed mammal, but it isn't fast enough to hunt healthy deer regularly. So its diet is much like a black bear's, with roots, berries, insect grubs, pine nuts, spawning salmonids, small mammals, and dead large mammals. Our region's ability to support grizzly bears may decline sharply as whitebark pines decline due to blister rust and pine beetles; scientists are concerned about how well grizzlies will get by without them during fall. At Yellowstone, grizzly mortality correlates with poor whitebark cone crops more strongly than with any other factor. Even if pine-nut shortages were merely to drive the bears down to lower elevations, increasing their contacts with people, the bottom line would be the same: dead bears. Two other rich foods that are important to at least some Yellowstone bears are also threatened: cutthroat trout, displaced by introduced lake trout, and army cutworm moths, which migrate up from the plains, where they are poisoned as agricultural pests.

How does a huge bear gather and eat little pine nuts? By raiding squirrel middens and

Grizzly bear with cubs

Bear Safety

Don't go hiking in terror of bears; go in knowledgeable wariness of them. Consider all bears dangerous even though they normally withdraw from contact with people. Black bears encountered by hikers often fit one of the two "abnormal" types: sows with cubs, and "problem bears" familiar with campers and camper food. Never camp in an area with torn-up camp food strewn about, or long strips of overturned sod, or recent large-mammal remains. If a bear enters your camp, distance yourself from your food.

When in grizzly country, let them know you're around. The sound that best gets their attention is the human voice. Shout, talk, laugh, and sing. (If you are brave enough, and hope to see *other* kinds of wildlife, make noise only when you're in likely grizzly bear range and habitat, such as thickets, avalanche tracks, streamsides, lily meadows, and berry patches.) Bear bells are useless. Bear-bangers sound like a good idea, but grizzlies commonly ignore them. Bear spray, on the other hand, is likely to save some lives. Carry it and study the instructions, then save it in case you are charged.

If you meet a bear in the open, act calm. Form a group, look large, avoid eye contact, and retreat discreetly—never run, as you cannot outrun a bear. (Bears being slow is an optical illusion.) Sometimes you can slowly circle widely around a black bear, if it hasn't already run off; if you are forced to try that with a grizzly, allow it a quarter-mile buffer.

If a bear charges you, stand and act calm. If you have spray, spray now. Climb a tree if you have both a strong tree and time to climb. Most charges are "bluff charges," or are abandoned when the bear can see that you're a person. If attacked, lie prone and play dead with your legs spread and your hands and your pack protecting the back of your neck. Don't move until you're sure the bear is gone.

The Rockies have for years had bears that look for their dinner in campers' gear. And while human injuries from black bears are extremely rare historically, they will increase if naive campers leave more and more food around. Some backcountry campsites have bearproof vaults for your food, toothpaste, and cooking utensils. Elsewhere, bring your own bear canisters or hang your food well out of reach. Never discard food or "leave it for the chipmunks." When car-camping, store food and cookware in your car when not in use. (Black bears in the Sierra Nevada rip car doors off to obtain food, but fortunately, they have not yet passed that technological skill along to their Rocky Mountain cousins.) Detailed advice on food storage and other recommended behavior is ubiquitous in the national parks.

Dealing with food in bear country requires constant thoughtfulness, with two objectives: to protect your body and equipment from collateral damage by bears going after your food, and to save innocent bears from becoming problem bears that will end up euthanized. Don't tempt them.

trampling the heaps of cones to smash them. Its big tongue is remarkably efficient at separating out the nuts—and, for that matter, separating grubs from rotten wood or bee honey from combs, and gleaning swarms of ladybugs high in the Mission Range.

During the long fattening-up season, when a grown grizzly needs about 35,000 calories a day, 250,000 berries would not be too many. Flower bulbs and taproots are a mainstay. In the southern Canadian Rockies, the staple species is yellow sweetvetch, a legume;

in Yellowstone, it's biscuitroot, a carrot; in between, it tends to be glacier lilies. Patches to be dug up and dined on are selected for two main criteria: proximity to a forest edge and digability, which correlates with gravelly soil and with a history of grizzlies digging. The bears till big patches of meadow sod in order to nip off bulbs from the sod's underside. If they also turn up voles and insect larvae, so much the better. For reasons not clear to scientists nor, presumably, to grizzlies, tillage has the effect of increasing available nitrogen in the soil and hence the size of next year's lily bulbs (the ones the bears missed this year), as well as their nutritional content, the number of seeds they set, and the seeds' viability. The usual word for behavior that enhances one's food crop is "gardening," and scientists do use that word. Bears return to their gardens, with the big, rich bulbs and loosened soil, year after year. Any meadow thick with glacier lilies is likely to have been a grizzly garden at some point, and conversely, meadows not gardened tend to end up lily-poor.

Some 30 percent of Yellowstone-area grizzlies make a habit of overturning alpine talus stones in search of army cutworm moths to eat, and end up with notably shorter, blunter claws. It's worth it: these moths are a bear's richest food, at 75 percent fat content. They fly to high mountains to eat alpine plants, then congregate under talus to escape the midday heat.

Grizzlies diets vary enormously, both by region and over time. We know this from study of isotopes in hairs or bone, including museum specimens and hairs snagged in the woods. Isotope study reveals ratios of animal to vegetable foods, and of marine to nonmarine foods. This is more accurate than either direct observation or scat analysis in two key ways: it averages the bear's diet over several years, and it quantifies foods to the degree they were digested, whereas scats overemphasize the indigestible. From marine isotopes, we learn that salmon contributed 60 percent of the diet of three

nineteenth-century bears in central Idaho; grizzly recovery there today would require an entirely different diet. Glacier National Park grizzlies are 90 percent vegetarian! Coastal Alaska bears are 90 percent piscivorous or carnivorous, and Yellowstone bears 50/50.

Life cycle and hibernation are similar to those of the black bear except that grizzlies dig very serious dens on hillsides, commonly moving a ton of earth. Cubs weigh about a pound at birth, leave their mother after one and a half or two and a half years, and live fifteen to twenty-five years if they're lucky.

Subspecies *horribilis,* the grizzly bear of the Western states and provinces, was once treated as a full species. Other *U. arctos* subspecies include the even larger Kodiak brown bear and the much smaller brown bears of Eurasia. The polar bear, *U. maritimis,* is such a close relative that polar bears and grizzlies have produced fertile "pizzly" offspring together. Climate change could conceivably force the two bears into more frequent proximity, leading to extensive hybridization.

Grizzlies avoid contact with people, and do not normally view people as food; on the other hand, they show little fear of people in close encounters. Avoid those.

SKUNK

Mephitidae. Glossy black with white patterns.
Western Spotted Skunk
Spilogale gracilis (spil-**og**-a-lee: spot weasel; **grass**-il-iss: slender). Also *S. putorius.* 11" + 5" tail (kitten-sized); many ± lengthwise, intermittent white stripes; tail tip a rosette of long white hairs. Lowlands w of the CD, from Missoula s.
Striped Skunk
Mephitis mephitis (**mef**-it-iss: pestilential vapor). 18" + 11" tail (cat-sized); 2 broad white stripes diverging at nape to run down sides of back, plus thin white stripe on forehead. Widespread, more common in farmlands than in mtns, and rare at high elevs; A.

SKUNK COLORATION IS the opposite of camouflage; it's to a skunk's advantage to be conspicuous and recognized, since its defenses are so good. The rare animal that fails to stay clear may receive additional warnings such as forefoot stamping, tail raising, or (from a spotted skunk) a handstand with the tail displayed forward like a big white pom-pom. Spotteds can spray from the handstand, but usually return to all fours. As a last resort, the skunk sharply bends its spine and fires its notorious defensive weapon—up to six well-aimed rounds of a cocktail of thiols and thioacetates in a musky vehicle secreted just above the anus. This substance burns the eyes, chokes the throat, can instigate retching, and of course stinks like hell. It can be shot either in an atomizer-style mist or, more typically, in a water pistol–style stream fanned across a 30-to-45-degree arc for greater coverage. Range is well over 12 feet.

Folk antidotes to skunk spray include tomato juice, ammonia, gasoline, and incinerating your clothes; juice is the least unpleasant, fire the most effective. (In all seriousness, try a five-minute tub soak, avoiding your eyes, with hydrogen peroxide with a little baking soda and dish detergent mixed in; or, for your clothes, use bleach for things that you don't mind bleaching.)

Only great horned owls—with built-in protective "goggles" and very little sense of smell—seem to prey on skunks regularly. Many big owls smell skunky and have skunk-bitten feet. As far as the oddsmakers of natural selection are concerned, skunk defenses are superlative. But like porcupines, skunks seem to be as vulnerable to tiny parasites as they are well defended against big predators.

Skunk foods include insects and grubs, mice, shrews, ground-nesting birds and their eggs, berries, grain, carrion, and kitchen scraps. Our skunks remain active most winters. Farther north, they sleep torpidly for days or weeks during cold spells. Typical dens are burrows dug by other animals.

Long known as denizens of North America's farms more than of its wilderness,

Western spotted skunk

Striped skunk

skunks in this century are making their move on suburbs and cities.

AMERICAN MINK

Neogale vison (ne-**og**-a-lee: New World weasel; **vice**-un: archaic French name). Also *Mustela vison, Vison vison, Neovison vison*. 13–16" + 6–8" tail; long, narrow, and short-legged; dark glossy brown exc variable white patches on chin, chest, belly; ears inconspicuous. In and near streams, marshes, and sometimes lakes; A. Mustelidae.

The mink, a semiaquatic weasel, preys on fish, frogs, crayfish, ducks, water voles, and muskrats. Some populations become terrestrial for a while, subsisting on hares and voles; others, on the coast, prey on crabs. Too slow in the water to chase fish efficiently (unlike otters), they more often pounce on slow fish in shallow water. The muskrat may be a mink's most dangerous game—heavy enough to drown a mink by dragging it under. In deep water, muskrats even attack mink fearlessly. On the other hand, a duck that thinks it can shake a mink by taking to the air may be in for a fatal surprise: cases are on record of mink hanging on for the flight until the duck weakens and drops.

The mink version of delayed implantation (p. 347) allows the female to mate and conceive, and then a few weeks later ovulate and mate again, often producing a litter with multiple fathers.

Foul discharges from under the tail are characteristic of the weasel family, which used to include skunks. While skunks developed their marksmanship and range, optimizing the anal gland as a defensive weapon, mink evolved an even worse smell. They spray when angered, alarmed, or captured, when fighting each other (they are antisocial), and to mark territory or repel would-be raiders of their meat caches.

We don't know how many kinds of mammals can contract COVID-19 nor, of those, how many suffer severe illness, but we know that mink do. Denmark's entire population of farmed American mink died of either the virus or euthanasia in 2021. The bird flu H5N1 also wipes out mink farms. Respiratory viruses are highly contagious in artificially crowded populations. Twenty European nations have banned mink farming, and the rest of the world would be wise to follow suit.

WEASEL

Very fast, slinky, slender, short-legged, long-necked animals; in characteristic running gait, the back is arched; ears inconspicuous; rich medium brown above, pale beneath, incl feet and insides of legs and (long-tailed only) some of tail; most high-elev weasels turn pure white in winter, exc tip of tail always black. Ubiquitous; A. Mustelidae.

Long-tailed Weasel

Neogale frenata (ne-**og**-a-lee: New World weasel; fren-**ay**-ta: bridled). Also *Mustela frenata*. Deep cream to orange-yellow beneath (in summer); ♂ 11" + 5¾" tail; ♀ 9" + 5⅜" tail.

Short-tailed Weasel

Mustela richardsoni (mus-**tee**-la: Roman name; **richard**-so-nye: for Sir John Richardson, p. 447). Also American stoat, ermine; *Neogale* or *Mustela erminea*. White to light cream beneath (in summer); ♂ 8½" + 3" tail; ♀ 7" + 2⅝" tail.

American mink

Short-tailed weasel

NARROW, LINEAR SHAPES like the weasel's are rare among the smaller warm-blooded animals because they are inefficient to heat. When asleep or torpid, weasels can roll up into, at best, a lumpy sort of disk, which takes 50–100 percent more calories to maintain at a given temperature than it would if rolled up into a sphere, as rodents do. Weasels have to eat around 30 percent of their body weight daily, and more in winter. But there's no question of the shape being worth that price: when it needs to eat, the weasel can chase a small rodent down any hole or through any crevice, or find it hibernating there. A weasel is thinner, faster, and fiercer of tooth and claw than any animal anywhere near its own weight (2 to 10 ounces). Though mouse-sized prey are their staple, long-tailed weasels can also run down squirrels in trees, pikas in talus, and snowshoe hares on snow—prey as much as ten times their size. In general, the smallest members of predatory families go after the most abundant prey, and are the surest of catching it. The food chain is really an extremely broad pyramid: few individuals can fit at the top as large predators.

Weasels are ferociously aggressive. Reports of "killing sprees" in which they kill far more than they can eat are numerous and confirmed. Granted, human observation may have inhibited or overlooked the weasel's efforts to cache the leftovers for later use. There are also cases of weasel cannibalism, including juveniles eating littermates, seemingly carried away with the taste or smell of blood. They nest in burrows of chipmunks, ground squirrels, moles, and so on, often lining these with fur plucked from the late homeowner. Weasels may themselves fall prey to owls, foxes, bobcats, or snakes.

The name "ermine" is applied by naturalists to the short-tailed weasel only, but by furriers and the general public to the fur of both species interchangeably, so long as it's in the white, winter pelage. In fact, most ermine coats are made from long-tailed weasels because it takes fewer of them to make a coat. It still takes hundreds, though, so a single pelt commands a surprisingly low price. The black tail-tip on the otherwise-white winter fur may serve as a decoy: the weasel can usually escape hawk or owl talons that strike this one body part that's conspicuous against snow.

NORTHERN RIVER OTTER

Lontra canadensis (**lon**-tra: Italian name). Also river otter; *Lutra canadensis*. 27–29" + thick, tapering, muscular, 17–19" tail; dark brown with silvery belly, pale whiskers, very small ears, webbed feet. In or near water; A. Mustelidae.

Otters are among the unlucky animals people are belatedly adoring—only after reducing them to near rarity. In both Europe and America, they were trapped for fur or shot on sight as vermin, largely because anglers accused them of more predation on trout than they actually inflict. Ancient Chinese fishermen, in contrast, trained them to herd fish into nets. A few European hunters trained them to retrieve waterfowl.

Today, otters come up in arguments over the existence of nonhuman play. Reputable observers report them running up snowy hills again and again just to slide down, or bodysurfing in river rapids for no apparent reason. Others call these behaviors efficient transportation, or just youngsters honing their transportation skills. Otters need to spend a lot of time rolling around, to clean and align their fur to maintain its warmth when in water, and this extended rolling can

Northern river otter

look playful. Oddly, the pups seem afraid of the water and have to be taught to swim. Their most efficient and powerful stroke is a lengthwise undulation, using the massive tail but not the paws. They can also swim with their hind paws or with all four. On land, they look awkward, and would rather slide on their bellies than walk, especially on snow; nevertheless, they regularly travel many miles on land.

Otters have a low, mumbly "chuckle" while nuzzling or mating. Females mating may caterwaul. Otters are relatively social, for weasels. Two main nonfamily social patterns are known: a female raising kits may have an unrelated female helper; and unrelated males may form groups that stay together for a long time, groom each other, hunt together, and share food—but of course do not want their buddies' company when a female in estrus is around. Even where large, tight brotherhoods have been studied (in coastal Alaska), a percentage of males and most females are solitary.

Anglers in the Rockies see otters regularly. Look on riverbanks and lakeshores for otters' slides, tracks, rolling sites, or spraints. The latter are fecal scent-markers placed just above water on rocks, mud banks, or floating logs, often atop small debris piles and usually showing fish bones, scales, or crayfish-shell bits under a greenish, gelatinous (when fresh) coating that smells distinctive but not unpleasant. Otter staples are crayfish and slow-moving fish, which are much easier to catch than trout. Otters commonly occupy beaver lodges, even sharing them with the beavers, a species they hardly ever prey on.

PACIFIC MARTEN

Martes caurina (**mar**-teez: Roman name; cor-**eye**-na: northwestern). Also pine marten; *M. americana*. 14–18" + 8" tail; body narrow, legs short, tail fluffy, nose pointy; variably blackish brown, buff, or cinnamon-brown (like a small red fox on short legs). Remote wilderness with trees; A. Mustelidae.

Pacific marten

Martens have evolved a resemblance in form, color, and habits to their favorite prey, tree squirrels. They even eat conifer seeds sometimes, or a few berries, and raid birds' nests. Like weasels, they excel as predators because their prey is abundant and has only meager defenses against them. But unlike weasels, they have been reduced from infrequent to almost rare by trapping and by their aversion to civilization. The Bighorns and some other ranges have isolated "island" populations, since martens can't migrate across treeless steppe.

We rarely see them because they are usually up in the branches, where they are fast, camouflaged, and active mainly at dawn, dusk, or under heavy overcast. Still, curiosity and appetite sometimes lure one right into a hiker's camp. Winter forces them to the ground more, where they hunt voles and hares. They are light enough to move well on top of snow, and can also pursue prey in snow tunnels.

Their musk, so mild as to be almost undetectable to us, is used mainly to mark tree branches to ward off other martens, except of course briefly during summer when about 50 percent of other martens find the smell not repellent, but quite attractive. Like their weasel relatives, martens are solitary and aggressively territorial vis-à-vis members of the same sex. Males and females whose territories overlap ignore each other except when ready to mate.

A 2012 paper separated Pacific from American martens. The two can be distinguished only by using skull measurements or molecular analysis. The ones in eastern Montana are said to be *M. americana,* while many in northern Montana and Idaho are hybrids.

FISHER

Pekania pennanti (pek-**an**-ia: Abenaki name; **pen**-an-tie: after Thomas Pennant). Also wejack, pekan; *Martes pennanti.* 20–25" + 14" tail; long, thin, and short-legged; glossy black-brown, occasionally with small, white throat patch; ears slightly protruding; paw print avg 2½–3" wide, often showing 5 toes and claws. Rare; dense forest; n ID and w-most MT. Mustelidae.

Fishers don't actually fish. The name may derive from the Dutch *visse,* meaning "nasty." Fishers eat a variety of birds and mammals, snowshoe hares predominating. Fast enough to run down and kill martens, they can also rotate their hind feet almost 180 degrees to scamper down tree trunks, and use that technique to force porcupines to the ground. Their endgame for porcupines (who outweigh them about two to one) features darting, dodging attacks to the face, with both tooth and claw, repeated for maybe half an hour, until the porcupine is too weak to flail with its tail. Fishers end up with quill bits scattered throughout their organs and muscles like shrapnel, though the majority soften in the stomach and pass safely through. Scats containing bits of quill are generally either fisher or cougar.

The only animal that significantly threatens the fisher is people. Fisher pelts, resembling Siberian sable, usually rank as the highest-priced North American pelt. By 1930, trapping and habitat loss had very nearly eliminated fishers from the Lower 48. Foresters came to see their economic benefits after porcupine populations exploded in their absence. Montana and Idaho reintroduced fishers several times between 1959 and 1991; they are hanging on by a thread, and have failed to take up residence in Glacier National Park. After being reintroduced to western Washington, they are faring better in that region. But in the Rockies, their fate is uncertain since the US Fish and Wildlife Service took away their ESA Threatened status in the region in 2017, under the Trump administration.

Reproduction in the weasel and bear families usually involves delayed implantation: the fertilized ovum undergoes its first few cell divisions and then goes dormant for weeks or months before implanting in the uterus and resuming its growth in time for springtime birth. The delay is extra long in the fisher, producing a total gestation of up to 370 days, with only around 60 of these active. Thus the female often goes into heat just a few days before or after giving birth, and mates before weaning.

Fisher

WOLVERINE

Gulo gulo (**goo**-low: gullet or glutton). Also skunk-bear. 26–31" + 8" tail; somewhat like a small bear but with ± distinct gray-brown to yellowish striping across the brow and down the sides to the tail; fur thick and long; paw print 4–4½" wide; unlike lynx or puma, many prints will show 5 small, well-separated toes with claws; unlike black bear, the "heel" of the paw is small and often doesn't print, and rear edge of pad is usually very concave. Near timberline; rare; northerly. Mustelidae.

Wolverine

Badger

Wolverines' reputation as extreme fighters should come as no surprise, as they're the biggest terrestrial predators in the weasel family. Biologists in the Selkirks found a three-hundred-pound caribou brought down by a twenty-five-pound wolverine, which is surely at the extreme end of the size ratio of a lone mammal predator to its prey. Yet these awesome predators often scavenge carrion, cruising avalanche basins, sniffing and digging out goat and deer carcasses deep in the packed avalanche snow. Extra-strong jaws let wolverines pulverize frozen meat and bone. They sometimes raid trappers' cabins and caches, trashing them and stinking them up with truly execrable musk. The powerful scent repels other carnivores from the wolverines' caches, and is crucial when wolverines seek compatible wolverines during the summer-long mating season.

Wolverines have long been poorly studied because they scrupulously and easily avoid humans (as well as each other). Advances in radio-tagging have at last facilitated study, especially in Glacier National Park. One unsurprising finding is that they are, after all, lovable, at least to their researchers. They are not as solitary as was once thought: after dispersal from the birth den, young wolverines were seen consorting with littermates, or even with their dads (boot camp in pre-

dation and survival?). The paths they take from A to B make mountaineers' jaws drop: a male was GPS-tracked ascending 4,900 feet in ninety minutes, crossing Glacier's highest summit—in January.

Wolverines were almost certainly extirpated from the Lower 48 by 1930, via trapping for their fur and deliberate poisoning for predator control. Since then, they've expanded south from Canada to reoccupy much of their former range. One male wandered hundreds of miles from the Tetons, crossing a desert and a freeway, to recolonize Colorado. After a few fruitless years there, he went northeast to try North Dakota, where he was shot, dooming his lineage, as there were no female wolverines in either state.

Climate change will almost inevitably reverse the wolverine's expansion. One particular key metric has been identified: deep snowpack in mid-May, when wolverine kits demand the most milk from their mother. That demand, in time and in calories, keeps her from roaming her home range to find food, so she prepares well ahead of giving birth by burying meat to freeze in the snow near her den. She requires a home with a freezer, and she will move as far north as she has to to get it. Snow also insulates the kits in the den—some biologists think that this is its more critical benefit.

Climate projections find that the area in the Lower 48 that will have snowpack in mid-May will decrease in this century by between 50 and 90 percent. Such terrain is found only at the highest parts of the West's mountain ranges, like a far-flung chain of islands. If gene flow occurs only through a precarious lacework of corridors, populations may be threatened by inbreeding depression—in which too much mating is between close relatives, leading to mutations, birth defects, and poor health. According to generally accepted formulations of "minimum viable population," wolverines should have already died out from the margins of their present range. Their sheer sparseness is anomalous among mammals. They are true outliers.

BADGER

Taxidea taxus (tax-**eye**-dee-a **tax**-us: both from the Roman name). 25" + 5" tail; broad, low, flat animal with thick fur grizzled gray-brown, while ± yellowish, esp the tail; white stripe down face; forefeet heavily clawed for digging. Grasslands and sagelands; A. Mustelidae.

This animale burrows in the ground & feeds on Bugs and flesh . . . his head Mouth &c is like a Dog with its ears cut off, his hair and tale like that of a Ground hog.
—Capt. William Clark

So runs the first written description of an American badger. This squat, ungainly, but fantastically powerful burrowing creature lives mainly by digging ground squirrels, gophers, mice, and snakes out of their holes. It often digs a temporary den for resting

Paw Print ID Simplified

One big clue to identifying prints is the number of toes and size.

Five toes: Bear, raccoon, and weasel families. Distinguish among these by size.

LEFT: *cat family*; RIGHT: *dog family*

Four toes, with claw prints in front of most toes: Dog family. Wild canids indistinguishable from domestic dogs.

Four toes, mostly without claw prints: Cat family. Cougar prints are 3–3½ inches long and wide, bobcat about 1¾–2 inches. Claws are normally retracted but may extend for traction in soft mud.

Four toes, the larger two paws side by side, alternating with staggered forepaw prints: Hare or rabbit.

Since some cat prints show claws, it helps to learn additional differences between cat and dog prints. The pad (ball of foot) of a cougar or bobcat print is indented or scalloped once in front and twice in back; dog pads, meanwhile, are convex (not indented) in front, as are lynx pads—though the fur often obscures this feature in a lynx print. Scats and scratchings near the tracks also help. Skilled trackers learn the patterns of how forepaws and hindpaws land in various gaits.

Lynx with prey

Bobcat

Often thought of as a larger version of the bobcat, the lynx actually averages a bit lighter, but looks larger with its much longer fur and legs and bigger feet—all adaptations to deep snow and cold. It preys almost single-mindedly on the snowshoe hare. Lynx numbers rise and plummet roughly in tandem with hare population cycles, peaking typically one to three years after the hares do. When hares are scarce, lynx can turn to squirrels for food, but forgo reproducing until hares are back, and individuals migrate as far as 700 miles in search of happier hunting. Lynx of the Lower 48 are listed as threatened, but probably have always been marginal dispersers from population irruptions up north. Lynx maintain an evanescent but more or less full-time presence in the northern tiers of Montana, Idaho, Wyoming, and Washington. In Canada, they are declining but still numerous—and still lucrative for fur trappers. They are genuinely shy of people, staying far away, unlike bobcats.

Between 1999 and 2006, 218 lynx were released in the San Juan Mountains. They have reproduced and spread out a bit, maintaining numbers in that same ballpark.

through the heat of the day. It spends winter mostly torpid, in twenty-nine-hour cycles with brief active periods, rather than in prolonged hibernation. Burrow entrances are large (a foot across) and may have a latrine nearby—a heap of mixed bones, fur, and scat.

Badger-hair brushes are made from the very different European badger, genus *Meles*. "Badgering" is not something badgers do, but what is done to them by European hunting dogs, who wisely avoid fang-to-fang combat.

LYNX

Lynx canadensis (links: Greek name). Also *L. lynx, Felis canadensis*. 31" long + 4" tail; gray cat ± tawny-tinged, not clearly spotted or barred exc the black tip of stubby tail; long hairs tuft the ears and ruff the cheeks; to confirm a lynx sighting, you must measure several footprints well over 2" long, or see that tail-tip is black both above and below. Rare; in wilderness with deep winter snowpack; WA and MT, to CO (where reintroduced). Felidae.

BOBCAT

Lynx rufus (**roo**-fus: red). Also *Felis rufus*. 28" long + 6" tail; tawny to gray cat, generally with visible darker spots, and bars on outside of legs and top (only) of tail; ears may show tufts, and cheeks ruffs, but these tend to be shorter than on lynx. Widespread, favoring brushy, broken, or logged terrain; A. Felidae.

The bobcat is a lovely creature we see all too rarely even though it lives throughout our range, probably as abundantly as it ever did. You just might surprise one if you travel quietly, but generally they keep out of sight.

Wild cats all like to work out their claws and clawing muscles on tree trunks, just like house cats scratching furniture. Bobcat or lynx scratchings will be 2 to 5 feet up the trunk, and cougar scratchings 5 to 8 feet up. These gashes may be deep, but rarely take off much bark; tree-clawing that strips big patches of bark is more likely the work of bears. Wild cats also often scratch dirt or leaves to cover their scats, at least partly. These scratchings may be accurately aimed at the scats, unlike the random pawings of male dogs next to their fecal markers. Bobcat and cougar scats also tend to be more segmented than coyote scats.

To preserve their claws' sharpness for slashing or gripping prey, cats retract them most of the time, and they rarely show in cat tracks. One toe (the first, or "big toe") has been lost from the hind foot, but on the forefoot it has only moved a short way up the paw, enlarging the grip. The hind legs are powerful, for long, leaping pounces, but cats other than the cheetah aren't especially fast runners. The cat jaw is shorter and "lower-geared" than most carnivore jaws, and it has fewer teeth. (Mammals evolved from reptiles with many teeth. When teeth are fewer, they're also often more specialized.) Cats have relatively small and unimportant incisors, huge canines for gripping and tearing, and a quartet of enlarged, pointed molars called carnassials, which, rather than meeting, shear past each other like scissor blades for cutting up meat. Cat tongues are raspy, with tiny, recurved, horny papillae, which can clean meat from a bone or hair from a hide. The cat nose is short, suggesting less reliance on smell than in the dog family. As in owls, the eyes are large, far apart, and aimed strictly forward to maximize three-dimensionality. Their eyes, like an owl's, reflect fire or flashlight beams in the dark. A reflective layer right behind the receptor cells on the retina redoubles light intensity at night.

Except in cougars, which have round pupils, cat pupils narrow to vertical slits for maximum differentiation between night and day openness.

Hares and rodents predominate in bobcat diets. In winter, bobcats turn to deer a little more, occasionally hunting fawns but more often finding carrion.

COUGAR

Puma concolor (**poo**-ma: name for it in the Quechua language of Peru; **con**-color: all one color). Also mountain lion, puma; *Felis concolor.* 4' long + 2½' tail; our only cat with a long, thick tail; ours typically grayish sandy-yellow (the species ranges from nearly black or slate gray to reddish brown); kittens spotted until about 6 months old; varied purrs, chirps, and yowls, infrequent. Widespread but elusive; A. Felidae.

Cougars are solitary, with large home ranges, the males' overlapping those of females. Males rarely fight over territory; they scent-mark by scraping piles of dirt mixed with their urine or scats. Near one cougar sighting, I found scats on the trail in several piles a few feet apart, each with radiating claw marks in the dirt. Scent-markers help the sexes find each other when a female is in estrus, which may be at any time of year. A male roams with and sleeps near the female for about two weeks, but no longer—if she let

Cougar

him approach the kittens, he might eat them. She rears the young for well over a year, and breeds in alternate years. She has a loud, eerie mating scream that sounds strangely human.

Cougars may follow solitary hikers, unseen, even for days, but hardly ever let hikers catch a glimpse of them. Consider yourself lucky if you so much as find a clear set of tracks.

Cougars take diverse prey, from grasshoppers and mice on up through porcupines and coyotes to elk, but their staples are young deer and elk. A male (the larger sex) can eat about one deer every ten to fourteen days, up to twenty pounds of meat at a time, burying the remains to come back to later. Buried meat, which may assault your nose, is a sign of cougar. She locates deer by smell or sound, stalks it slowly—crouching and freezing for periods in a position a deer might mistake for a log—and then pounces the last 30 feet or so in a few bounding leaps. She bites the prey in the nape, and may either bite through the spinal cord or snap it by twisting the head back. If that fails, she tries to hang on until the prey suffocates. It's risky for cougars to hunt deer and elk, who outweigh them. They are sometimes trampled or thrown hard enough to kill them, and one starved to death pinned under an elk it had killed. After wolves returned to Yellowstone, cougars there had to increase their predation on elk because wolves often drive them from their kills. With their advantage in numbers, wolves even kill cougars occasionally.

Game managers value cougars as the major predator of deer. They keep deer and elk moving on their winter range, helping to prevent local overbrowsing. Cougars select young, old, or diseased herd members—the easiest and safest to attack—minimizing their impact on total deer and elk numbers, which are affected more by hunting.

Hunting of cougars tends to disrupt cougar territories and to increase the proportion of young adult males—trends that may increase the threat to human safety. Nevertheless, attacks on humans remain rare. Fatalities are far fewer than those due to fatal dog attacks, beestings, lightning, and so on. The US Rockies have had four in the past century. Most victims are children; pets also may be snatched from large campgrounds and even backyards.

Some safety rules (note that some are the opposite of safety rules for bears, p. 341): Don't leave small children or pets unattended in likely cougar habitat. If you see a cougar, pick children up immediately, stand tall and confident, and maintain eye contact. Move your arms and backpack in any way that will make you look bigger. Act unlike prey: do not run or turn your back, and retreat slowly. If the cougar still seems aggressive, throw things at it (without prolonged crouching to pick them up). Remain standing. If attacked, fight back aggressively with any weapon you can grab: the cougar may give up in favor of easier prey.

Cougars roam a vast range—most of North and South America—with remarkably little genetic variation. This is attributed to growing from a very small remnant population in South America after North America was reopened to them by the extinction of saber-toothed cats, part of the great megafauna extinction around ten thousand years ago.

Order Artiodactyla (Split-hoofed Mammals and Cetaceans)

We have three families in the order Artiodactyla: deer, cattle, and pronghorns. The cattle family includes sheep and goats, but not horses. Another family, the pigs, is native to North America (that would be peccaries) but not to our Rockies. The order was long called "cloven-hoofed mammals," but the American Society of Mammalogists now includes whales, porpoises, and orcas in it.

WHITE-TAILED DEER

Odocoileus virginianus (oh-doe-co-**ill**-ius: hollow teeth). 4¾' long + 12" tail; medium tawny brown or in winter grayish, with white patches on throat, inside of legs, and rump just under the tail; upper side of tail is brown with white fringe and tip, but tail is raised in flight, the pure white underside "flashing" and waving; antlers branch from 1 main tine; fawns are white-spotted; ears smaller (4") than mule deer's. Mainly lowland and riparian areas; A. Cervidae.

White-tailed deer

America has a lot more deer than it did two hundred years ago. Especially white-tails. In fact, there's more biomass of white-tails alive today than any other wild vertebrate on land. (That stat comes from a study with a horrifying bottom line: humans and our domesticated stock outweigh all wild vertebrates on land, globally, by a factor of fifty to one. Small consolation: fish and invertebrates are more massive still.)

Anglo civilization has been good to white-tails, thanks more to its war on forests than to its war on predators: deep forest supports relatively few deer, but brushy clearcuts are deer heaven. At the other extreme, the wide-open Great Plains once supported very few deer, but farmers opened that vast area to white-tails by planting trees for windbreaks. Though the clear pattern has been westward expansion, early explorers did report a few

Pheromones in Hooved Mammals

Deer are fine subjects for the study of pheromones—chemical messages conveyed among conspecifics. Mule deer are rather antisocial, using some pheromones apparently to repel each other—a function akin to territoriality, though mule deer are not territorial. A subordinate deer will sniff a dominant deer's tarsal patches, get the message, and retreat to a respectful distance. (Tarsal glands are buried in patches of dark hair on the inside of the ankle joints, midway up the rear legs.) To activate tarsal pheromones, deer of any age or sex urinate on these patches and rub them together. While most pheromones are secreted in sweat or sebum (skin oil), deer urine is itself pheromonal in that it reveals the animal's health and strength. Glands on the forehead are rubbed on shrub twigs to advertise the condition of a dominant buck, or to mark possession of a sleeping bed. Interdigitate glands, between the two toes, secrete an attractant pheromone, marking a deer's trail for other deer to follow. Metatarsal glands, on the outside of the lower hind leg, secrete a garlicky odor signaling fear or alarm.

Sensing pheromones looks like sniffing, or sometimes a caricature of sniffing (the "lip curl," demonstrated by the bighorn ram on p. 361). It is a distinct ("sixth") sense, using its own organ, the vomeronasal organ within the nose chamber. Whereas smelling detects chemical recipes made by combining six or seven basic odors, each vomeronasal nerve ending—one thousand to ten thousand times more sensitive than an olfactory nerve ending—is lit up by only one pheromone. Bingo!

white-tails in the Rockies. (We don't know their exact presettlement range.)

Our two deer tend to separate ecologically, white-tails preferring riparian areas and mule deer (below) the montane slopes. The two are similar in size, color, teeth, and other adaptable body parts, but white-tails are easy to tell when fleeing, because of their flashing, erected white tail. White-tails can bound or simply run very fast—a good strategy on prairies and in deciduous woods.

MULE DEER

Odocoileus hemionus (oh-doe-co-**ill**-ius: hollow teeth; hem-ee-**oh**-nus: half ass, i.e., mule). 4½' long + 6–8" tail; medium tawny brown or in winter grayish, with white patches on throat, inside of legs, and rump just under the tail; belly paler; first branch of antlers (♂ only) may itself fork; fawns are white-spotted; ears mulishly large, 8", rotate independently; adult tracks 2–3" long, the 2 toes each tapered their full length; scats ½–¾", ovoid pellets. A. Cervidae.

Browsing is a sophisticated, serious business. Spend a while sitting quietly and watching deer eat. Notice their odd way of clipping their browse, almost gumming it with lower incisors against an upper pad; they lack upper incisors. They may show an intense preference for an individual plant that has either less tannin or more of some

Mule deer

nutrient. They lap up springwater (no matter how muddy it has become from trampling hooves) containing mineral salts they crave. They chomp mushrooms, "nature's salt licks," for other hard-to-find minerals. They strip horsehair lichen from tree limbs; it has modest nutritive value on its own, but offers minerals that enhance their austere winter diet of twigs and evergreen needles. In recent burns, they find the particular species of grass that exhibits the sharpest post-burn spike in protein content. Like other cud chewers that can live on a high-roughage diet thanks to cellulose-digesting bacteria in their first stomach, deer have to browse for the nutritional demands of their bacteria. Insufficient protein can kill the bacteria, leaving the browser dangerously malnourished, even with its belly full.

In the absence of wolves, deer may increase beyond the carrying capacity of their habitat; then, during severe winters, they are malnourished, and fall prey to parasites and/or coyotes. (That's the conventional wisdom, anyway.) Degraded deer habitat isn't stripped bare; it's just short on the particular species that supply adequate fats, proteins, and trace minerals. Wolves and cougars, before we decimated them, were the chief nonhuman predators of our deer, but now rank behind domestic dogs, hunters, cars, and trains as deer-killers. Even dogs that wouldn't know what to do with a deer if they caught up with one can chase it to a death via barbed-wire entanglement or a broken leg. Deer are transfixed by a strong beam of light at night, making them frequent victims of cars and trains. (If you count car-crash fatalities, deer are by far the deadliest animal to Americans.) Coyotes, bears, and bobcats are infrequent predators of deer, taking mainly fawns or critically weakened adults.

When fleeing, mule deer break into a bouncy, high-bounding gait called "stotting." Though slower than a flat-out run, it allows them to react to rough terrain with abrupt,

Chronic Wasting Disease

An insidious, incurable disease called chronic wasting disease (CWD) threatens American deer and elk. The infectious agent is neither virus nor bacterium nor parasite: it's a misfolded protein called a prion. It spreads inside nervous-system tissue by inducing other prion proteins to refold to conform to itself. This works very slowly, but incurably, so far. It may take a few years for symptoms—weakness, drooping head, wasting away, salivating, depression, and eventually death—to appear.

In the wild, this disease has only infected the deer family. It was first identified in 1967 in a captive deer herd in Colorado, possibly having crossed over from sheep carrying a long-known prion disease called scrapie. It has reached thirty states, but remains most prevalent in Colorado, with over half the state's deer herds and a third of its elk herds showing signs of infection. It is rare or absent west of Montana.

Infective prions persist out in the world to a creepy degree, after they are left on the ground in saliva, urine, scats, or decomposing carcasses. They have been found in blades of grass and in ticks that fed on infected elk. Transmission is faster in herds (including captive herds), which suggests a couple of interesting proposals: One, end Wyoming's winter elk-feeding programs at Jackson Hole and elsewhere. Two, reintroduce wolves more widely, to keep the elk moving and to break up herds. State wildlife agencies advocate for increased hunting, for the same reasons. In some places, entire infected herds have been killed to stop the spread of disease.

Given the current state of knowledge, it's hard to envision anything averting a relentless, slow decline of our deer and elk. Currently, only dead animals can be tested, and that's months if not years after they get infected. Intensive testing and culling of animals as soon as they are infected could help stall the epidemic, once live-animal tests are developed. Tests are in the works, and more likely to arrive than futuristic proposed solutions such as vaccinating a large portion of the population, or disseminating genes from naturally disease-resistant individuals.

The urgent question on hunters' minds is whether they can catch CWD by eating venison. Not a single case has been seen in a human. On the other hand, more than two hundred people worldwide have acquired a fatal prion disease by eating infected beef (the cattle version is called mad cow disease), and experiments have come up with indirect evidence that transmission to people may be possible. The prion is unscathed by cooking. (Game should be cooked to kill other germs, especially *E. coli*.) Fortunately, wise hunters in the Rockies can get their kills tested for CWD before eating any. Get the latest information and regulations from state wildlife agencies.

unpredictable changes of direction—a hard act to follow for any predator giving chase.

Deer and elk do of course move away from intense wildfires, but not in a panic as depicted in *Bambi*. They are remarkably adept at finding refugia within fire perimeters. They often return to their old home range within hours of a fire sweeping through it, and remain surprisingly faithful to it for life, even at some cost to their health after severe fires. After the low-to-moderate-severity fires that are the long-term norm in much of the West, deer forage improves and deer numbers grow. Megafires with large, high-severity patches may be another story; the data are not in yet.

Steep south-facing slopes just above river bottoms are ideal winter range. The low elevation and insulating tree canopy offer warmer temperatures and shallower snow, while the south aspect lets in low-angle sunlight and tends to have more shrubs to nibble. In summer, our deer move upslope to meadows, clearcuts, and open woods, where they fatten up on herbaceous plants.

The only close social tie among deer is between a doe and her fawns and yearlings. Males are solitary or form loose, small groups, except during rutting when the sufficiently dominant ones follow single does in heat for a few days. A doe seeks seclusion even from her yearlings before and after giving birth; it's up to the yearlings to reunite with her afterward. For the new fawns' first few weeks, she hides them in separate nestlike depressions under brush. She browses in the vicinity, strives to repel other does, and comes back to nurse mainly at night. If a threatening, large animal like you approaches, she nonchalantly ignores the fawns. (She will meet a fox or bobcat, however, with a bold counterattack.) Occasionally, people come across a hidden fawn and not its foraging mom, and make the cruel mistake of rescuing the "abandoned" fawn. Unless you actually find the mother dead, assume a fawn is being properly cared for and leave it in peace, untainted by human scent.

ELK

Cervus canadensis (**sir**-vus: Roman name). Also wapiti; *C. elaphus.* 7–8' long + 5" tail; tan with a large, well-defined creamy pale patch on rump; extensive brown tinges on neck, face, legs, and belly; ♂ have antlers; fawns are white-spotted; adult tracks 3–4⅝" long, more rounded than deer; scats ⅝–1¼" ovoids in winter, patties in summer. A. Cervidae.

Today, elk inhabit coniferous forests and high mountains of the West. Before white settlement, they were common all the way to Vermont and South Carolina. Great herds of them on the plains were second only to buffalo in sheer biomass and as a food and material resource for humans. Like bison, they were shot in huge numbers. They were able to hold on in the mountains and deep forests until an alarmed public, rallied by a famous hunter named Teddy Roosevelt, got refuges and hunting restrictions enacted to allow elk to recover. The very idea of conserving species was novel to white culture; the object, in those early days, was to conserve the species for future generations to hunt.

Politicized controversies have raged over the "natural" population levels of elk. The best data show that they are native to most parts of the United States and Canada that supported either forest or grassland. Elk may have been reduced a few thousand years ago by Indigenous hunting, and rebounded a few hundred years ago when Native peoples were decimated by introduced diseases. Elk were savaged by market hunting in the late nineteenth century. They've done well in the twentieth with regulated sport hunting, and grew thick in some national parks with no hunting, no wolves, and reduced cougar numbers.

Yellowstone was one such park where elk browsing was preventing aspen and willow shoots from reaching tree stature over large areas. After wolves were reintroduced into the park in 1995, elk numbers dropped by more than half, aspen and willows returned to some but not all areas, and beavers returned thanks to having willows to eat again. With the smaller population, elk have not suffered severe winter-kill years; but they are still more dense at Yellowstone than in most of the Rockies, and their browsing still challenges aspen.

Most of the year, elk travel in segregated herds, the mature bulls in bands of ten or fifteen, and the females and young males in larger herds. In late summer, the bulls become mutually hostile, and the largest, most aggressive of them (primary bulls) divide out harems from the cow herds; they tolerate yearlings, but drive away two- and three-year-old males. The harem's movements, like those of the cow herd in winter, are subtly directed by a matriarch apparently respected for her maturity rather than her size or strength. The bull seems to tag along rather than to lead, and he may lose part or all of

Bull elk jousting

his harem if he gets distracted. Other bulls strive to distract him, hoping to take over the harem by overcoming him in a clash of antlers staged at dusk or dawn. This is the rutting season.

Elk have a remarkably extravagant courtship system. Bulls challenging each other or working up their sexual or combative frenzies behave curiously:

- **Thrashing.** These attacks on small trees and brush (common also among deer bucks) have been called "polishing the antlers" or "rubbing off the itchy velvet" but are actually visual challenges, warm-up or practice for sparring, or autoerotic stimulation. Profligate urination typically accompanies thrashing, and antlers are used to toss urine-soaked sod onto their backs.
- **Pit wallowing.** Shallow wallows are dug and trampled out and lined with urine and feces. Elk cows appear to get excited when exploring them. Wallows often become long-lasting pools.
- **Bugling.** This unique call includes a deep bellow and a farther-carrying whistle. Its clearest function is to herd cows. Elk cows occasionally bugle too, when calving in spring.

The reek of urine advertises the bull's physical condition. Others can smell how much he has been metabolizing fat as opposed to fresh food or, worse yet, muscle. A bull metabolizing only fat is one so well fed that he can devote all of his energy to the rut without weakening from hunger. Keeping track of a harem, defending it from other bulls, and reaping the sexual rewards are not only hard work, but so time-consuming that the bull can no longer eat or rest. Almost inevitably, primary bulls succumb in the same season to lesser but better-rested rivals, and often these secondary bulls yield in turn to tertiary bulls. Defeated dominants wander off alone, catch up on food and sleep, and show no further interest in sex that year. They may not last the winter. Weaker "opportunistic bulls" live longer. They spend the rutting season alone, but stand at least a chance of mating with a stray cow.

Rocky Mountain elk cows may bear young every year, starting at age two. Browsing techniques and care of the young among elk are much like those described for deer (p. 354), except that to an elk cow, humans are puny meddlers to be chased from the nursery. Biologists who tag elk calves learn to be quick tree-climbers.

In Europe, a relative, *C. elaphus*, is the red deer or stag that Robin Hood hunted. If you're thinking that the genus known to the English as a "deer" ought, in English, to be a deer, you

Horns and Antlers

Horns are sheaths of keratin, like fingernails, generally found on both sexes; they form from epidermal tissue at their bases, and slowly slide outward over bony cores. Cattle-family members (sheep, goats, and bison) grow just one unbranched pair for life, but pronghorns shed and regrow theirs.

Antlers, a defining character of the deer family, are solid bone, usually branched, and in most species confined to males. They take form inside living skin—complete with hair and blood vessels—and stretch the skin outward as they grow. This skin, or velvet, dies and sloughs off before the antlers come into use in fall. In late winter, the antlers weaken at their bases and are soon knocked off, and a new pair will begin to grow by early summer.

On the ground, rodents set to work with tiny incisors recycling shed antlers into mouse or squirrel bones. Increasingly, humans compete with the rodents for sheds, which can be crafted into decor or sold in the Asian medicinal trade.

Cumbersome and easily entangled in brush, antlers make poor survival tools. Hooves, not antlers, are a deer's defensive weapon. With the exception of small antlers grown by some female caribou, antlers exist solely to establish dominance among males. Like bright plumages on small male birds, they exemplify a sort of fitness via sexual selection. (The fittest do survive—but as a genetic lineage, not necessarily as long-lived individuals. Large-antlered elk bulls tend to die younger than weak ones who rarely fight over cows, but the latter, leaving few offspring, are by definition less fit.) Some scholars think that antlers' chief value is visual—attracting females or intimidating rival males. Like boxing gloves, antlers can regulate and extend combat, reducing risk to life and limb. No doubt they're less deadly than smaller, sharper headgear would be, but this case has been over-stated: even though bulls most often show restraint in sparring, combat injuries inflict a major portion of adult male mortality.

An elk or white-tail deer antler has one main beam from which all the other points branch, starting with a brow tine. On mule and blacktail deer, both branches off the first Y may again branch. The number of points increases with age, but also responds to nutrition, and thus advertises physical condition. Rich feeding in captivity has produced five-point yearlings, while meager range can limit even dominant bucks to two tines. Five points are typical on full-grown elk; older bulls may grow six, seven, or rarely eight. While mature mule deer typically have four or five, British Columbia has seen freakishly large antlers—forty-eight points on a mule deer, twenty-five on a white-tail.

have a point. A different genus exists that we ought to call "elk," as Europeans do and did for centuries before Americans dubbed it a "moose." The names were misplaced because American species of *Cervus* and *Alces* are larger than their European congeners. Colonists on these shores met white-tails about the size of European deer (*Cervus*) and

wapiti about the size of European elk (*Alces*) and named them accordingly. The misno-mers stuck, despite generations of efforts to replace "elk" with the Shawnee name, wapiti.

MOOSE

Alces alces (**al**-sees: Norse name that, in English, became "elk"). Also *A. americanus*. 9' long, 7' high at shoulder; ♀ smaller; tail negligible; dark gray-brown; long, horselike head with loose, over-hanging upper lip; antlers have broad, flat blades and many points; long dewlap ("bell," of no known function) hangs from throat; pro-nounced shoulder hump; tracks 5–6" long, more pointed than elk; winter scats oval, 1½–1⅞" long; sum-mer scats often loose like cow pies. Willow thickets with numerous bro-ken, chewed, and bent-over tops are a sign of moose; A exc NM. Cervidae.

Moose

The twentieth century saw a continent-wide moose-population explosion where logging had converted old forests to brushy second-growth, including willow twigs, a moose's winter staple. Almost unknown in Idaho and Washington before 1920, moose were common there by 1950. In the great boreal forests, forest fires are infrequent and vast, producing moose heavens that last a decade or two. Moose are well adapted to that pattern: they migrate into burns and then double their birth rate. Logging in the coniferous West gives them similar habitat bonanzas.

Moose abound today in the Yellowstone ecosystem, but the earliest report of them there dates from 1870; not even archeological digs have found earlier evidence. It's possible they hadn't been there since before the last ice age. Many of Yellowstone's moose died of malnutrition in the winter of 1988, after fires burned 36 percent of the park, because a post-fire increase in moose browse takes a season or two to grow. Moose herds have grown in Colorado since they were first brought in, in 1978.

Moose were hardly ever preyed upon during their 130 years of coexistence with grizzlies at Yellowstone. However, as wolves moved into Jackson Hole recently, they soon killed ten adult moose that, presumably, had not learned to be wary of them. Mothers who lost offspring to wolves learned quickly, soon responding hyperdefensively to wolf howls. (Joel Berger, analyzing data on these and other "naive moose," concluded that they support the "blitzkrieg hypothesis"— that the earliest humans in North America might have wiped out the ice age megafauna because even the largest species are easy to kill when they have been without predators for generations.)

For most of the year, bull moose are soli-tary, and cows are accompanied only by their yearlings. Moose form loose groups in fall, for mating, and become more vocal, with a sort of tremolo moan.

With massive bodies, and stomachs hard at work fermenting cellulose, moose run hot. Heat limits their range southward, and when they retreat into water in summer, it's to cool off as well as to dine on succulent pond vegetation. They act like they don't mind the biting flies and mosquitoes that swarm them. On the other hand, heavy infestation with winter ticks (forty thousand ticks per moose are common during bad tick seasons) can cause severe blood loss, hair loss, heat loss, and winter mortality. Dramatically pale, fuzzy "ghost moose" are individuals who have lost most of their guard hairs, leaving the paler underhair. Infested moose have been observed paying little attention to magpies pecking at their backs, leading some scientists to infer that a symbiotic "cleaner" relationship is in effect. Others claim that the magpies aren't just gleaning ticks but are eating scabs (and keeping wounds open) and that few cleaner symbioses offer the host a net benefit beyond a pleasurable back-scratching.

Another odd mortality factor is gum disease and molar wear. Moose reach peak foraging efficiency around ages five to seven. Males therefore grow the biggest racks then, and win the most mates of any period in their lives. Wear and tear from the sheer volume of woody stuff chewed takes its toll on teeth, and as browsing efficiency dwindles, the adults become wolf bait—or first parasite bait and then wolf bait. Moose in their prime are rarely preyed upon, and calves are well defended by their mothers. Moose fleeing from wolves are amazingly fleet—almost as fast as white-tail deer. They glide over logs and brush, taking advantage of their long legs to hold their center of gravity at constant height over a course that looks like high hurdles to a wolf. (In the minus column, long legs require moose to kneel, awkwardly, in order to reach food near ground level.) On soft snow, moose easily outrun wolves, which flounder; but wolves can run fast on top of crusted snow and catch moose, which break through.

Moose tolerate humans, up to a point. They're often described as docile, but "self-confident and relaxed" would be more accurate. A skittish, unpredictable streak combines with sheer mass to produce the surprising statistic that more people are killed by moose than by grizzly bears. While some deaths are in moose-car crashes, many others are from aggressive behavior. Lowered ears and raised hair on the neck and shoulders are a serious warning. Do not try to "shoo" a moose. Since cleared, hard ground or snow is an advantageous platform for a moose defending itself in nature, it may fiercely defend a spot on a road or a groomed ski trail. The heavy hooves are deadly weapons against wolves and humans, and effective deterrents even against grizzlies. They can put a serious dent in a car.

BIGHORN SHEEP

Ovis canadensis (**oh**-vis: Roman name). Also mountain sheep. 5–6' + 6" tail; 3⅓' to shoulder; ♀ smaller; brown with white rump patch; very ragged and ♀ cream-colored during May or June molt; rams have massive, curled horns, often broken-tipped; ewe horns smaller, merely curved; horns brown and ± blunt, vs. black and very sharp in mtn goats. Open habitats; A. Bovidae.

Bighorn sheep habitat has three characteristics: it's wide open, making it easy to see predators approaching; it's near cliffy terrain to escape to; and it has high-quality grasses or sedges to eat. In winter, both the forage and the escape terrain have to be kept nearly snow-free by wind or steepness. Some herds have winter and summer habitat a few hundred yards apart, whereas others migrate 10 miles or more. Bighorns are spottily distributed, due in part to their exacting habitat needs. Habitat is reduced when forest densifies due to fire suppression.

Highly social, bighorns are usually seen in herds of five to a hundred. Outside of the November–December rutting season, adult males herd together, often far from the herd of females and juveniles, who seek range with few predators. Males, being bigger and more heavily armed, are less vulnerable to cougars and wolves, and are not at all vulnerable to coyotes and bobcats, freeing them to focus on forage quality, which determines horn size and thus social rank—the year-round obsession of rams. Rank is usually settled by merely displaying, and perhaps kicking a little. Only when two rams' horns are equal do they butt heads, facing off at several paces, rearing on hind legs, charging forward with forelegs flailing. Slam. Stagger. Reel. Repeat, until one gives up, up to twenty-four hours later. Ram necks and skulls are massive, with spongy bone mass around the horn base to absorb impacts, which may be audible a mile off. The victor gets to tend estrous ewes, though subordinate rams do manage to sire about half the lambs after either dashing past

Bighorn sheep pair

sive, while those that live in national parks are brazenly bold. Never take advantage of their boldness by attempting to touch or feed them. As with other mammals, that can be hazardous to your health.

In Wyoming, a group known as the Mountain Shoshone specialized in hunting bighorns, building elaborate traps out of intertwined small trees and limbs.

BISON

Bos bison (boce: cattle in Latin; **bye**-sun: Greek name). Also buffalo—a misnomer; *Bison bison*. 10–12' long, 5–6' to shoulder; shaggy, dark-brown, cattlelike beast with upcurved horns; woolly mane and goatee, esp on older ♂; calves are red-brown. Low-elev prairies, grassy openings. Bovidae.

Saw large herds of buffalo on the plains of the Sandy river, grazing in every direction on the short and dry grass. Domestic cattle would certainly starve here, and yet the bison exists and even becomes fat.
—John Kirk Townsend in Wyoming, 1834

the tending dominant ram, or blocking a ewe, beginning early, so that when she goes into estrus she isn't anywhere near a dominant male.

The horns are effective only for that one style of combat. Mountain goats, with sharp horns and aggressive behavior, easily drive bighorns off whenever the two species compete for a resource such as a mineral lick. The arrival of the non-native goats in Colorado and Wyoming adds to the challenges bighorns face.

The arrival of domestic sheep has been a worse challenge, due to diseases they transmit to bighorns. Many local bighorn populations died out due to disease, hunting, or both, and many of today's herds descend from individuals brought from elsewhere by wildlife agencies. Introduced herds tend to fluctuate—multiplying for a while, then plummeting during an epidemic, then either recovering or not. Failure often follows contact with domestic sheep, or sometimes high cougar populations. When herds are doing well, agencies may open them to limited trophy hunting, which seems to pose no threat to numbers or to the herd's health. Herds that are hunted become very shy and elu-

Townsend's comment hits the nail on the head, and offers a kernel of the modern view that reconverting large areas of the High Plains into bison range (a "Buffalo Commons") would be the most economically efficient, as well as ecologically beneficial, way to produce food there. Grazing by bison benefits the shortgrasses that coevolved there with the bison.

The Great Plains as they were first described in writing were an unusual ecosystem in that they supported grazing animals of awesome total biomass, but oddly few species: mainly elk, pronghorns, bighorn sheep in a few locales, and millions upon millions of bison. Some scientists believe early Americans wiped out most large mammal species here between sixteen thousand and ten thousand years ago; bison may have survived because their huge herds could lose themselves in the vastness of the plains.

Bison

commanded the army in Sioux country.) Army-subsidized hunters with long-range rifles littered the plains with carcasses after shipping the hides east to be made into robes or drive belts for machinery. Later, picking up the bleached bones became a cottage industry for struggling settlers; bones were exported to England to make bone china. In Canada, market hunting proved equally effective without military backing.

The herds were so far apart, and moved so far and so fast, that hunters without horses couldn't keep up with them, and year-round sustenance was a severe challenge in many parts of the High Plains. People may not have lived there in large numbers until the 1700s, after they acquired horses in trade from southwestern groups. At that point, in flight from white men and their diseases in the East, groups that had been farming in permanent villages on the eastern plains moved west and created the famous equestrian Plains culture. Bison were that culture's heart and soul.

The downward spiral of the bison began in the 1600s with several causes. They were pushed out of the East by the destruction of prairie habitat and the decimation of Native cultures whose fires had produced and maintained prairies. In the West, it was more the arrival of horses. Wild horses competed directly for grass, and domesticated ones enabled Native groups to move onto the plains and make a good living hunting bison.

But those declines were nothing compared to the years 1868–82. Plains warriors were frustrating the US Army's efforts to make the West safe for white settlers, so the army decided that exterminating the bison would starve and demoralize Native peoples into submission. It worked. (General Sherman, whose notorious March to the Sea showed the effectiveness of scorched earth,

Today's bison descend from fewer than six hundred individuals that were spared. The majority descend from about one hundred in six private breeding operations, and the rest were in three wild herds that were protected just in time: in far northern Alberta, in Texas, and at Yellowstone. Poachers reduced the Yellowstone herd to twenty-three animals when the new park had no law enforcement and little local support. Ironically, some US Army units that oversaw bison extermination in the 1870s were reassigned to protecting the remnant herd. Again, it worked.

Species that come back from very small populations suffer from a lack of genetic diversity. Today's bison are prone to brucellosis and tuberculosis, which has led to the notorious slaughter of Yellowstone bison as they leave the park.

All that extra wool around a bison's head—females have the goatee but lack the forehead bush—has at least two benefits. First, it pads male heads when they butt, in rutting competition (and is progressively shorn by scissoring between opposing horns over the course of the butting season). To gore each other, they have to lower their heads extremely, since the horns point almost backward. Second, it insulates them while they clear snow for winter browsing by swinging their huge heads back and forth, rather than by kicking with puny, little hooves as deer and sheep do. Bison can clear snow depths of up to about 30 inches. The frequency of greater depths west of the Continental Divide helped keep their numbers low there. Cold itself is no problem: bison coats are densely haired and warm (hence the popularity of buffalo robes). In summer, bison pant, which cools them by rapidly evaporating water from the lungs. They also used to spend time surprisingly high in the mountains, escaping flies and cooling themselves on snow patches. Wallowing in dust may help cool them and discourage parasites, as well as serve rut-related purposes. As with elk, a few dominant bulls sire almost all the calves. A strong bull can tend a few dozen cows serially, identifying when each goes into estrus.

MOUNTAIN GOAT

Oreamnos americanus (or-ee-**am**-nos: mountain lamb). 5' long + 6" tail, 3' high at shoulder; all white, with "beard," shoulder "hump," and "pantaloons" formed of longer hair; black hooves; sharp, curved black horns. Alp/subalp; native to nw MT, introduced in the other states. Bovidae.

Cold wind may not penetrate mountain goats' insulation, but floundering through wet snow all day and night can be a problem, hence in the Rockies they seek bare, windswept ridges in winter. They eat enough huckleberry twigs, lichens, and fir needles to keep their stomachs busy, but extract so little nutrition from this winter fare that they live mainly on stored fat. They eat snow rather than seeking running water.

Apparently, mountain goats are not limited primarily by predation. Eagles have been known to take kids and yearlings, dive-bombing to knock them off ledges, and cougars occasionally take even adults. But, overall, goats are fairly predator-proof with their proverbial evasive skills on precipitous terrain, as well as hooves and

horns as defensive weapons. The hooves have strong, sharp outer edges and a hard, rubbery, corrugated sole for superlative grip. Forage and climate seem to be the limiting factors on goat populations. Winter starvation and disease are ranking causes of death, along with the inevitable attrition from falling.

Mountain goats have been decimated wherever they were freely hunted, but they lose their fear of people after a few generations in a national park. Be careful to suspend your gear and bestow your urine only on the rocks, as goats may be brazen enough to enter camps to eat urine-soaked earth, in the process demolishing precious alpine turf. (Mountain goats get too little sodium in their diets, so they crave the sodium in our urine, as well as in mineral springs.) This isn't entirely new, unnatural behavior: goats travel many miles just to lick earth with minerals they need. The herd on Mount Blue Sky lost two recent cohorts of kids to epidemics of *E. coli* after too much contact with humans.

Mountain goats are not native south of Glacier National Park. The introduced Rocky Mountain populations do not seem to threaten alpine vegetation extensively, as introduced goats did in Olympic National Park.

Mountain goat with kid

The paths mountain goats beat often lure hikers onto rock faces that require climbing technique or hardware. Tufts of white floss snagged on branch tips remind us these are not paths for mere boots.

Mountain goats are classified in a tribe with chamois, which lead a similar life in the Alps. Both differ from true goats and sheep in several ways.

Mountain-goat horns are sharp, dangerous defensive weapons. One grizzly-bear death by goring is recorded, as is at least one human death. During rutting competition, the horns and skulls are not massive enough to sustain bouts of butting. Male mountain goats are thickened at the other end instead: the skin of their rumps, where they are likely to be gored in their flank-to-flank style of fighting, has been known to reach ⅞ inch thick, and was used in Alaska for chest armor. Males also seem to be inhibited against real fighting or sparring—the occasional pair who get carried away and actually fight are typically both forced into retirement by broken horns if not severe wounds. The effective breeding males are those who manage to intimidate the others with visual and olfactory displays, without ever coming to blows. Pit wallowing is indulged in, as among elk, and males often spend the winter in filthy coats, looking rattier still in spring while these molt in big sheets, revealing immaculate new white.

Females and immature males, who aren't as inhibited from using their horns, commonly chase the males around. Females may viciously charge males who come too close, except during a brief sexually receptive period in fall when they allow the male a creeping, submissive-looking courtship. A single kid is normally born in May or June. Mothers are legendarily protective, walking against a kid's downslope flank to prevent a fall.

PRONGHORN

Antilocapra americana (an-til-o-**cap**-ra: antelope goat). Also antelope—a misleading term that scientists avoid. 4' long + 5" tail; tan with a white rump patch that can "flash" in alarm, tan tail, white lower jaw, white blazes on neck, white belly and inside of legs; adults of both sexes have black horns; ♂ horns up to 20", with one forward prong and a sharp rearward curve at tip, or tip frequently broken off from combat; ♀ horns 4", prongless; the slight mane (tan) is erectile. Open plains; A. Antilocapridae.

Pronghorns are evolutionary orphans. They have no living relatives—no other species in their family. Their horns are shed annually, like antlers, but are made of keratin, like horns, and are unique in growing around a permanent bony core.

Their great speed—sustainable at better than 45 miles per hour—is as curious as their horns. Other than pronghorns, you just don't find animals a lot faster than their predators: evolution seems to always come up with a predator fast enough to catch each prey species at least some of the time. The likely explanation is that pronghorns evolved their speed at a time when North America held a far greater variety of large mammals. Not one, but thirteen species of pronghorn lived during those glory eras. Predators included a cheetah, a hyena, and a bear each longer-limbed and faster than their modern relatives. The African cheetah, capable of speeds up to 70 miles per hour, is the only living predator much faster than a pronghorn, but a pronghorn can outpace a cheetah over any distance greater than a couple of hundred yards. It may have been packs of long-legged hyenas that drove pronghorn evolution to produce their unequaled endurance at high speed.

We would have to look to migratory birds to find anatomical adaptations for endurance that rival those of the pronghorn: a heart three times larger than that of a similar-sized goat, and lungs with five times the oxygen diffusion rate. To maximize breath intake, pronghorns run with their mouths open.

Their chief surviving predator is the coyote, which any adult pronghorn can not only outrun but also chase away. It's the fawns up

Pronghorn

for chasing a coyote, and sometimes chases deceptively in a fawnless area. She varies the length of time between nursing reunions, so it won't be worth the coyote's time to come back and follow her every four hours or so. After five days of age, the fawn's instinct to lie perfectly still when approached begins to switch to an instinct to hop up and run, emitting the bleat that any mother will respond to with a chase. At that age, they can outrun you or me, but not a coyote. Before long—growing more rewarding to eat and easier to spot—their mortality rate rises. In Montana in the fenced-in Confederated Salish and Klamath Tribes Bison Range, 90 percent of fawns die, suggesting that the hiding strategy works better where there are vast plains to spread out over.

To minimize the vulnerable period, pronghorn mothers invest lavishly in feeding their young. Compared to mammals of similar size, the total birth weight of the litter (usually twins) and their growth rate while nursing are extreme.

Pronghorns have adapted poorly, so far, to fences, paved roads, and the expanding population of wild horses. They hardly ever jump fences, even though they're perfectly capable of it, and they cross roads or step through wire fences reluctantly—possibly their severest human-related problem. Populations are fairly stable at around half a million, compared with perhaps 35 million in 1820, and only 20,000 in 1920. (Before 1920, Westerners slaughtered pronghorns, fearing competition with livestock.) Most are out on the Wyoming basins where people are few, but pronghorns are easily seen at Yellowstone, Grand Teton, and the Bison Range. They rank among the more avid consumers of sagebrush.

to forty-five days old that are vulnerable, and suffer predation rates of 40 to 80 percent in most herds. Pronghorns and coyotes play a fascinating shell game of hiding and searching strategies. After a bout of nursing, fawns trot off alone to pick featureless hiding spots in the grass, where they lie motionless, chin down, for several hours. The mother's interdigitate glands leave a scent trail a coyote can follow, but the fawn is odorless except for its urine and feces, which it holds in until the mother is there to eat them.

During hiding, the mother's job is to remember the fawns' exact locations without ever conspicuously looking at them, while grazing far enough away to not lead a coyote to them, but close enough to chase away any coyote that happens to approach. To not give anything away just by her location or her reactions, she varies her distance from the fawns and her sensitivity threshold

Birds

The surviving descendants of winged dinosaurs, birds are winged, warm-blooded vertebrates with feathers. According to DNA evidence, crocodilians are their closest living relatives—closer than crocodiles are to lizards, or birds to mammals. In one conspicuous way, birds and mammals (which diverged from reptiles much earlier) evolved in parallel: both bird and mammal lines produced keratinous skin growths to serve as insulation, enabling warm-bloodedness. While mammals got hair, birds got feathers, whose light weight is ideal for flying. In insulation value per unit of weight, down feathers still have no equal among resilient, deformable, durable materials.

In addition to insulation, feathers provide most of a bird's shape, size, and color: plucked, a duck, a crow, a hawk, and a gull would look surprisingly alike in form, and pathetically small. Body plumage as much as doubles a bird's girth. The "fat" jays you see while skiing aren't fat; they're fluffing out their plumage to maximize its insulation value. (Goosebumps on your skin are a vestige of a similar response.) Less fluffed-out during flight, body plumage serves the equally crucial function of streamlining. The long outer feathers of wings and tail, meanwhile, provide most of the bird's airfoil surface at very little cost in weight. They constitute 35 to 60 percent of a typical bird's wingspread and 10 to 40 percent of its length. Feathers make the bird.

Color in most plumages—nearly all female and juvenile plumages, and many fall and winter male plumages—emphasizes camouflage. Since pale colors make good camouflage against the sky, and darker colors against foliage or earth, birds tend to be paler underneath than on top. A mother and young in the nest especially need to be cryptic, since all they can do when predators pass overhead is sit tight. (The mother could fly, but she likely stays to protect her flightless young.) Males are freer from the need for camouflage than females, and much freer than earthbound mammals; in males (and in the females of a very few species), sexual selection went wild evolving colorful plumage.

A bird that looks drab to us, or two species that look just alike to us, or the two sexes of one species, may look very different to a bird. Many bird orders can see ultraviolet light, and have UV markings that our eyes do not register. These birds see their way around easily before dawn and in the dusk, in a landscape richly illumined with UV light.

For efficient pairing up, conspecifics need to recognize each other quickly. A showy plumage doesn't look equally good to all female birds—it looks best to females of the same species. The same goes for birdsongs. (As traditionally defined, songs differ from calls in being longer and more complex, and characteristic of male Passeriformes, often called songbirds.) Experienced birders—and at least one phone app—can identify many species by song alone; females and juveniles in many genera are nearly impossible to identify when no males or songs are present. But females also sing in some species. Indeed, scientists find ever more exceptions to almost any generalization about birdsongs.

Birds molt, or replace all of their feathers, at least once a year, and twice if they have different seasonal plumages, which serve to make the male showy for courtship and then camouflaged for the nonbreeding seasons. Molting has a different purpose among ptarmigan, whose females, juveniles, and males all may go through two or three seasonal plumages—to camouflage them against snow, or no snow, or patchy snow. With a few exceptions like the mallard and the dipper, the large feathers of the wing and tail molt just a few at a time, leaving the bird tattered but still flightworthy.

Much time is spent on the vital task of preening with the toes or beak, applying oil from a preen gland. Aligning and oiling plumage optimizes it as an airfoil, as insulation, and as waterproofing.

To fly, birds had to evolve not only wings and feathers, but radically larger, more efficient respiratory systems. The stamina for a single day of flying, let alone for migrating across oceans (p. 379), would be inconceivable in a mammal. It demands a lavish supply of oxygen to the blood, and of blood to the muscles. Cooling also is done through the breath. Breathing capacity is augmented by several air sacs and, in many birds, by hollow interiors of the large upper leg and wing bones, all interconnected with little air tubes. Each breath passing through the lungs to the sacs and bones and back out through the lungs is efficiently scoured of its oxygen.

Hollow bones, their interiors criss-crossed with tiny, strutlike bone fibers in accord with the best engineering principles, doubtless evolved to save weight, yet some birds lack them. Loons—birds that dive for a living and don't fly much—have solid bones. But some diving birds have hollow bones, and a few birds with solid bones manage to soar.

A bird's sternum, or breastbone, projects keel-like in front of the rib cage to provide a mechanically advantaged point of attachment for the flight muscles. Known on the dinner table as breast meat, these muscles are a flying bird's largest organs. They power the wings and also guide flight by controlling the orientation of each and every feather along the wing edge through a system of tendons like ropes and pulleys. There is scant muscle in the wing and none in the foot, which is moved via tendons from upper-leg muscles. A flying animal ideally has nearly all its weight in an aerodynamic fuselage close to its center of gravity, and little weight in gangly limbs.

The wing is analogous at once to both the wing and propeller of an airplane. This is no accident; pioneers of mechanical flight studied and tried to copy bird flight for centuries before Wilbur and Orville finally, albeit crudely, got it right by separating the propelling and lifting functions of the bird wing. Like a propeller, wings provide forward thrust by slicing vertically through the air while held at a diagonal—the rear edge angled upward on the downstroke and vice versa.

Once there is enough forward motion or headwind to provide strong airflow across the wings (bird or airplane), their shape provides vertical lift by creating a low-pressure pocket in the air curving over their convex upper surface, while the lower surface is effectively flat. This upperside-convex principle is common to all flying birds except perhaps hummingbirds, but wing outlines have diverged in many specializations. (Read about soaring wings on p. 386, speed wings on p. 398, little-used wings on p. 372, and hovering wings on p. 378.)

Bird names, both common and scientific, follow the American Ornithological Society (AOS) checklist, which is official throughout the continent. The size figure that begins each description is the length from tip of bill to tip of tail of an average adult male; most species' females are a little smaller, but bird-of-prey females are bigger. I use Golden Field Guide's measurements of "live birds hand-

Mallard male

Northern pintail males

Green-winged teal male

shrubs, low shrubs, or streamside thickets without a canopy over them. After nesting season, many birds migrate either south or downslope. Ones that don't migrate may shift to a different habitat as the weather, their caloric needs, and available foods change.

Order Anseriformes (Waterfowl)

The waterfowl and the chicken-like fowl are placed first and second because they evolved very early. Their ancestors survived the end-Cretaceous extinction, already distinct from dinosaurs. One curious ancient trait is that waterfowl have penises, which were lost from most other bird orders. Most waterfowl species (and all of the ones described here—even swans) are in the duck family, Anatidae.

DUCK

Anas spp. (**an**-us: Roman name). ♂ makes a short piping whistle; ♀ quacks and clucks. Marshy lakes, ± year-round.

Mallard

A. platyrhynchos (plat-i-**rink**-os: broad nose). 16", ws 36"; breeding (Sep–June) ♂ have iridescent dark-green head separated from red-brown breast and brown back by white neck-band; bill yellow on ♂, black and orange on ♀; both sexes have a band of bright blue with black-and-white trim on upper, rear edge of wing, and much white under wings; ♀, juveniles, and summer ♂ speckled drab.

Northern Pintail

A. acuta. 18½", ws 35"; breeding ♂ (Sep–June) has chocolate-brown head divided from gray body by a narrow white line reaching up from the white breast; extended, long black "pintail" feathers are distinctive; ♀ dull brown with some red-brown on wings.

Green-winged Teal

A. crecca (**krek**-a: Swedish name imitating its call). Also *A. carolinensis*. 10½", ws 24"—smaller than other dabbling ducks; breeding ♂ (Nov–July) has cinnamon-brown head with broad, iridescent-green crescent arching from eye to a short crest on nape; body gray with yellowish-

held in natural positions." These run shorter than those in other manuals that, following taxonomic tradition, are measurements of long-dead specimens or skins forcibly hand-stretched. For a mental yardstick, think of sparrow size (4–6 inches), robin size (8½ inches), jay size (9–12 inches), crow size (17 inches), and raven size (21 inches). But it can be hard to get much sense of a bird's size when you see it against the open sky.

Habitat and behavior are often useful ID clues. Many small birds are faithful to certain plant communities or to levels within a forest—upper canopy, lower canopy, tall

white rump patch and white side stripe when not in flight; ♀ gray-brown; both sexes have a black-and-bright-green patch on rear edge of wings. Mainly in winter in s CO and NM; farther n, winter in some valleys but mainly summer or spring/fall migration.

DABBLING DUCKS TAKE flight abruptly and steeply, unlike diving ducks (below), which splash along the surface. They feed near the water surface, upending themselves in shallow water and plucking food from the bottom. To meet high protein needs before and during egg-laying, they eat mollusks, insect larvae, and small fish. They eat mainly plants the rest of the year.

Unlike most birds, the Anatidae molt their flight feathers all at once, leaving themselves flightless. To escape predators, they hide out for the duration in large groups in marshes. In summer, we see mallard mothers with chicks trailing behind, but no fathers. Newly hatched ducklings can swim a mile, but they can't fly for almost two months. At that point (late summer), it's the mother's turn to molt. While flightless, they are "sitting ducks," but the rest of the time they can practice "sleeping with one eye open." If four mallards are asleep in a row, the two in the middle close both eyes while the two on the ends tend to keep the outside eye open for predators. Brain scans on mallards in this state found that the brain hemisphere connected to the open eye remained semi-wakeful and capable of rousing the bird upon sighting a predator, while the other hemisphere experienced normal sleep.

Ducks are the primary reservoir for bird-flu viruses. Climate change pushes birds to migrate to new places where they contact different species than before, increasing the rate of interspecies viral transmission, and ultimately increasing risks to humans.

These three species abound over much of the Northern Hemisphere, conspicuous except during duck-hunting season, when they make themselves scarce. Domestic ducks were bred largely from mallards, centuries ago, and breed with them when given the chance; city-park duck flocks often mix mallards and hybrids.

HARLEQUIN DUCK

Histrionicus histrionicus (hiss-tree-**ah**-nic-us: actor or jester). 12", ws 26"; breeding ♂ plumed in a clownlike patchwork of slaty blue-gray, rich brown on the flanks, and white splotches with black trim—but may appear merely dark from a distance; others dark brown with several small white patches on head, and whitish belly. Rough water; mainly Yellowstone and Glacier.

Whether in whitewater rivers or heavy surf, harlequins display phenomenal pluck and strength as swimmers. Many live on rocky seashores, where they dive for shellfish, and migrate inland to build their grass-lined nest among brush or boulders along mountain streams, where they feed largely on insects. But Wyoming's few harlequins may be full-year residents. Unlike most ducks, harlequins rarely mix with other species.

Harlequins, goldeneyes, and buffleheads are carnivorous diving ducks, as opposed to dabbling ducks who merely tip or dip, feeding at the surface for their mostly plant diet. Divers tend to be smaller and chunkier, with shorter bills. (These are informal groupings, not taxonomic ones.) Some witty people have compared harlequin plumage to unfinished paint-by-number art.

Harlequin duck male

Bufflehead male

Buffleheads are the only ducks small enough to move right into a flicker's old nest without alteration, and seem to rely on that convenience. They excel at diving, and mainly eat crustaceans, molluscs, and insect nymphs. A courting male flies over his target female and makes a splashy ski-landing, with white crest erect, just past her.

GOLDENEYE

Bucephala spp. Also whistler. 13"; ♂ has black head with green or purple iridescence, white spot below bright-yellow eye, white belly, row of white spots on black wings; ♀ brown with dark-brown head, pale-yellow eye; wings whistle distinctively in flight. Lakes, beaver ponds, slow rivers; A.

Barrow's Goldeneye

B. islandica (iss-**land**-ic-a: of Iceland). ♂ has white crescent below eye; ♀ bill usually yellow.

Common Goldeneye

B. clangula (**clang**-you-la: small racket). ♂ has round white spot; ♀ bill usually black with small yellow tip.

Common goldeneye pair

COLD-LOVING DUCKS, GOLDENEYES stay on their boreal-forest nesting grounds as long as the progress of fall freeze-up allows. Areas with the climate they use for breeding will shrink severely. Many individuals (especially *clangula*) see our range as the balmy south, merely wintering here. All three *Bucephala* species are seen throughout our range, but commonly breed only from Wyoming north. Goldeneye males have striking courtship displays, including a "head-throw-kick": first thrusting his head forward, then pressing it way back into his rump, bill pointing skyward, then forward again while kicking up a splash.

Common goldeneye males

BUFFLEHEAD

Bucephala albeola (bew-**sef**-a-la: buffalo head; al-be-**oh**-la: small white). 10", ws 24"; ♂ white with black back, black head with big erectile white patch crossing crown behind eye; ♀ brown with elongated white patch behind eye; short, thick bill. Lakes, slow rivers; A.

SWAN

Cygnus spp. (**sig**-nus: Greek name). Huge, white waterfowl with very long neck; black bill and skin reaching up from the bill to a point at each eye (see species distinctions); immatures light gray.

Trumpeter Swan
C. buccinator (**buc**-sin-ay-tor: trumpeter). 45",
ws 8'; black all the way from bill to eye. Call is 1
or 2 deep, resonant honks. Rare and few; lakes
of nw WY and nearby MT and ID.

Tundra Swan
C. columbianus. Also whistling swan. 39", ws 7';
most adults have yellow spot in front of eye; calls
include high yelps, long reedy whoops. Flocks
migrate very high (1,000–5,000' up) in V formation, stop over at RM marshes and lakes;
spring and fall; A.

Trumpeter swan

ONCE WIDESPREAD IN the United States and
Canada, the second-largest North American
bird—the trumpeter swan—was brought to
the brink of extinction by market hunting.
Red Rocks Lake National Wildlife Refuge,
Montana, nurtured its subsequent recovery.
We have breeding populations there and at
Yellowstone and Grand Teton. If you see
them, maintain a quiet and respectful distance, as they need undisturbed habitat for
nesting.

Tundra swan

The tundra swan, breeding only in the
remote Arctic, has kept its numbers above
a hundred thousand. City parks have mute
swans, *Cygnus olor,* introduced from Europe.

CANADA GOOSE

Branta canadensis (**bran**-ta: origin moot, perhaps from Anglo-Saxon "burnt"). 25"; ws 68";
head and long neck (held straight forward in
flight) are black exc for a white chin strap; body
brown, massive; tail black/white; wings brown;
belly white; deep "honk"; migrating flocks form
a V or skein, and honk continuously. On and
near water; A.

Canada goose

The best-established explanation for the
V formation of flocks in flight is that a goose
saves energy (75 percent in one calculation,
3.5 percent in another) by drafting off the
goose ahead of it. Different birds, usually
females, take turns in the leader spot, the
only position that doesn't save any energy.

Canada geese declined due to hunting and
draining of wetlands in many areas, but then
they found a lot of new habitat on lawns,
golf courses, and croplands. Populations
have increased perhaps tenfold over recent
decades, making geese equally unpopular
among urbanites and farmers. This fowl
feeds sometimes by dabbling, more often by
grazing on land, and is a serious crop pest in
a few locales.

COMMON MERGANSER

Mergus merganser (**mer**-gus: diver; mer-**gan**-ser:
diving goose). 18", ws 37"; breeding (Nov–July)
♂ white beneath with black back and head, and

Common merganser pair

Common merganser female and young

red bill, the head greenish-iridescent in strong light, becoming red-brown in Aug–Oct; ♀ gray with red-brown head sometimes showing slight crest on nape, and red bill, white throat; hoarse croaks and cackling. Lakes and streams; A.

Instead of a Donald Duck–like broad bill, mergansers have a long, narrow bill with a hooked tip and serrated edges for gripping slippery fish and other aquatic prey. Of our three merganser species, the common is the largest, most montane, and most common. Many winter here and fly north to breed, but others move up to our mountain lakes to breed, nesting in woodpecker holes or other cavities.

Order Galliformes (Chicken-Like Fowl)

The Galliformes feed and nest on the ground, and fly in short, infrequent bursts. They have undersized wings and pale breast meat, indicating mostly fast-twitch muscles suited for brief, intense use. After the fowl has been flushed a few times in quick succession, these muscles are too oxygen-poor to fly again until rested. Tirelessness in flight is more the norm among other kinds of birds, which, being infrequent walkers, have dark, iron-rich breast meat and pale and scanty leg meat.

WHITE-TAILED PTARMIGAN
Lagopus leucura (la-**go**-pus: hare foot; loo-**cue**-ra: white tail). Also snow grouse. 10"; underparts white; upper parts white in winter, mottled brown in summer, patchwork of brown and white in spring and fall; feet feathered;

bright-red eyebrow comb on ♂ in spring; soft clucks and hoots. Alp/subalp, a few locations in each state, esp CO. Phasianidae.

Little ptarmigan are the largest creatures that make our alpine zone their exclusive home. Dwarf willows are their staple in winter, crowberries in fall. Ptarmigan moms have been seen dropping a tidbit of one of these foods in front of their young, pointing at it and at the plant, and uttering a food call. Such clear teaching behavior has mainly been described in chimpanzees and a few other mammals. Like other grouse, ptarmigan rely on camouflage to protect them from predators. In winter, they switch to pure-white plumage and stay on pure-white snow as much as possible, digging into it for shelter and to reach willow buds. Once their summer plumage grows in, they stay off the snowfields.

The *p* in "ptarmigan" is not only silent but silly. It must have found its way into

White-tailed ptarmigan summer plumage

print long ago in the work of some pedant who imagined "ptarmigan" to be Greek. It's Gaelic.

DUSKY GROUSE

Dendragapus obscurus (den-**drag**-a-pus: tree-lover; ob-**skew**-rus: dark). Formerly blue grouse. Also hooter. 17"; adult ♂ ± mottled dark gray above, pale gray beneath, with yellow eyebrow comb; others mottled gray-brown; both sexes have blackish tail; ♂ courtship call a series of 5 or 6 low hoots; hen with chicks clucks. Fairly common and frequently seen; all elevs in montane forest; A. Phasianidae.

Dusky grouse spend the breeding season in relatively open habitats, enriching their diets with caterpillars, plant shoots, berries, and even mushrooms. For winter, dense conifers offer thermal cover, visual cover, and plenty of needles, which are most of what they'll find to eat then. In some populations, this pattern means they migrate upslope for winter. Conifer needles are a very poor, but plentiful, diet. To adapt, the grouse intestines lengthen by about 40 percent each fall; grouse pack their digestive tracts full each evening and spend all night digesting, and still some 75 percent of the fibrous matter passes through undigested, creating a thick buildup under roost trees. They lose muscle mass inexorably on this diet, restoring it in summer on insects and berries.

Both the courting "hoot" and the males' chief visual display—exposing bare yellow patches on the neck—are performed by inflating a pair of air sacs in the throat.

RUFFED GROUSE

Bonasa umbellus (bo-**nay**-sa: bull; um-**bel**-us: umbrella). Also drummer. 14"; mottled gray-buff; tail red-brown or gray (2 color phases) with heavy black band at tip and faint ones above; ♂ has slight crest; black neck "ruff" is erected only in courtship display; no vocal mating call; distinctive "drumming" is common in late spring, occasional in other seasons; an owl-like hoot is sometimes heard. Lower coniferous forests, usually near broadleaf trees; WA to UT. Phasianidae.

With sharp downstrokes of his wings while perched on a log, the male ruffed grouse makes a mysterious accelerating series of muffled thumps, known as drumming or booming. You can't tell which direction or distance it's coming from, which is characteristic of very low-pitched sounds. An attracted female must have a tantalizing search in store for her. At the end of it, she can watch his fan-tailed, ruff-necked dance, and then mate with him, but that's all he has

Dusky grouse male

Ruffed grouse male drumming

to offer her—she will incubate and raise the young by herself.

She nests on the ground, like other grouse, and trusts her excellent camouflage up until the last second, when you're about to step on her unaware. Then she flushes explosively, right under your nose. This may scare you away; if not, she may try her famous broken-wing routine: a predator would follow the seemingly easy prey until she gets far enough away from the chicks, and at that point she flies.

Order Podicipediformes (Grebes)

Like loons, grebes are a primitive order of birds poorly adapted for flying and worse for walking, but superlatively for diving. Grebe toes—fat, scaly lobes—paddle more efficiently than webbed feet; while underwater, grebes use a side-by-side stroke resembling our butterfly stroke but without any help from the forelimbs. (Diving ducks, in contrast, paddle their feet alternately, and also use their wings.) On the surface, grebes can troll with one foot while tucking the other. They build floating rafts to nest upon. Hatchlings take up submersible life before they can swim, clinging to a diving parent's back. Like loons, grebes are denser than ducks, and float lower on the water—a key trait for recognizing them. The order contains only one family, Podicipedidae.

WESTERN GREBE
Aechmophorus occidentalis (eek-**mof**-or-us: spear bearer; oxy-den-**tay**-lis: western). 18"; black-and-white bird with red eyes; needlelike yellow bill; long, graceful neck, held nearly straight, is white in front, black in back; head white from eyes down, black above; in flight, head and feet are lower than body. Lakes, marshes; May–Oct; A. Population declining. Very similar Clark's grebe, *A. clarkii*, occupies much the same range and was treated as part of this species from 1886 to 1985.

Courting western (and Clark's) grebes run on the water surface in pairs, side by side, necks arched. It's a remarkable display: the next heaviest animal that can run on water, a kind of lizard, weighs only a tenth as much. Grebes do it by slapping their three toes on the water up to twenty times per second, keeping the birds up for four to seven seconds.

Western grebes sometimes use their bills as spears, stabbing competitors from beneath.

PIED-BILLED GREBE
Podilymbus podiceps (pod-i-**lim**-bus: Greek name; **pod**-i-seps: rump feet). 9"; stocky; bill high, stout, with downcurved ridge, pale with a black band across it (on summer adults); both sexes drab-brown mottled white, white under tail; summer adults marked with black on face and throat; in flight (rarely seen), head is slightly

Western grebe

Pied-billed grebe

Mourning dove

lower than body, and feet trail behind the very short tail. Lakes, marshes; year-round; A.

By exhaling deeply to decrease their buoyancy, pied-billed grebes quietly sink and skulk with only head and nostrils above water. Their calls have been compared to those of a braying donkey or a squealing pig. Crayfish are a favorite food.

Order Columbiformes (Pigeons)

This order contains only one family, Columbidae, called either pigeons or doves with no clear distinction between them.

MOURNING DOVE

Zenaida macroura (zen-**eye**-da: for the namer's wife; mac-**roo**-ra: big tail). Also Carolina turtledove. 10½"; light gray-brown with pinkish cast on underside; ♂ neck ± iridescent; black-spotted wings; tail shows white edge in flight; wingbeats whinny during takeoff; unmistakeable cooing song jumps upward on 2nd syllable: "ooOOOO, oo, oo." Cottonwoods, ponderosa, aspen, brush; A.

It takes a mourning dove pair only around thirty days from laying eggs to seeing the fledglings off, and then they're ready to lay another clutch. They feed the nestlings "pigeon milk," a regurgitated liquid made from their all-seed diet. They are less common in the montane West than the rest of the country, where they are one of the most abundant birds, holding steady despite hunt-ers taking around 7 percent of the population annually. The worst threat from hunting is lead shot left scattered over the ground. Doves eat it, perhaps mistaking it for seeds, and suffer lead poisoning. Lead shot is banned for waterfowl hunting nationally, but only in some states for dove hunting.

Doves bob their heads as they walk, like their cousins the domesticated rock pigeon, *Columba livia,* which is native to dry canyon habitats in Eurasia and, after spreading across our continent since 1602, is sometimes seen on rocky canyon walls here.

Order Caprimulgiformes (Nightjars)

This very small order was recently separated from the swifts and hummingbirds.

COMMON NIGHTHAWK

Chordeiles minor (cor-**die**-leez: evening dance; **mi**-nor: lesser—a false name since this is now the larger species of nighthawk). 9", ws 23"; wings long, bent backward, pointed, falconlike; mottled brown/black, with white wing bar; ♂ also has white throat and a narrow white bar across tail. Marshes and open areas; June–Sep; A.

Nighthawks are often seen flying at dusk: no self-respecting nighthawk would be on

Common nighthawk

the ground then, when so many insects are on the wing. Like swifts and swallows, nighthawks prey on insects by flying around with their mouths open. Their flight is wild and erratic like a bat's, but swift like a falcon's—though they aren't really any kind of hawk at all. Males may interrupt their erratic feeding flights with long, steep dives that bottom out abruptly with a terrific raspy, farting noise of air rushing through the wing feathers. This is done relatively close to the nest. The vocal call is a softer, nasal beep repeated while feeding. Nighthawks have an odd style of perching, lying lengthwise along a branch. Like several unrelated ground-nesting birds, they employ the broken-wing act to decoy mammals like us away from the nest or the young. Common nighthawks winter in South America. (Of course, it isn't winter there then.) Their population has declined sharply along with that of their prey.

Vaux's swift

White-throated swift

Order Apodiformes (Swifts and Hummingbirds)

The name Apodiformes derives from "footless" and refers to the very short legs of both swifts and hummingbirds. These birds don't walk, basically; they're always either flying, roosting, or in the nest.

SWIFT

Family Apodidae. Small birds with relatively long, pointed, gently curved wings, short bodies, and very short legs.

Vaux's Swift

Chaetura vauxi (key-**too**-ra: hair-thin tail; **vawks**-eye: after William S. Vaux, p. 377). 4½"; slaty to brownish gray, somewhat paler beneath; wings long, pointed, gently curved; tail short, rounded; staccato chipping. ID and MT from Bitterroots n. Apodidae.

White-throated Swift

Aeronautes saxatilis (ar-row-**naught**-ies: flyer; sax-**at**-il-iss: rock-dweller). 6½"; black with black/white underside; tail long and forked;

Black swift

rapid twittering. Around cliffs; ID, WY, and sw MT (rare farther n).

Black Swift

Cypseloides niger (sip-sel-**oy**-deez: like *Cypselus*, the Old World swift; **nye**-jer: black). 7"; black all over; tail slightly forked; call rarely heard. Subalpine; uncommon; from Bitterroots n.

SWIFTS MAY NOT be the swiftest, but they are extreme flyers nevertheless, doing all their

A First Look at Shrinking Glaciers

William Sansom Vaux, curator of the Philadelphia Museum of Natural History, never saw the Rockies, but his niece and nephews Mary, George, and William Vaux turned their Western peak-bagging vacations into the first annual study of glacial retreat in North America. The idea began when Mary, on her second trip to British Columbia, was shocked by the visible shrinkage of the Illecillewaet Glacier. They photographed it every year from 1897 to 1913, developing their big glass plates in the lodge's wine cellar—a major contribution to the growing understanding of glaciated landscapes and glacial cycles.

When the Smithsonian published Mary's wildflower paintings in five volumes in 1925, reviews called her "the Audubon of American wildflowers."

hunting, eating, and drinking aloft. They even mate in free fall, disengaging after several seconds to resume flying. They often fly 600 miles in a day—and make it look playful, interspersing short glides between spurts of flapping, and creating their famous optical illusion of flapping left and right wings alternately. Too bad most of this goes on too high up for us to see well.

"Strainers of aerial plankton," swifts can only maintain their high metabolisms on a steady supply of flying insects. Few insects fly during cold weather, so when it's cold and gray, swifts go torpid (p. 318) or else fly off for a few days, as far as they need to to find sun. Nestlings remain torpid while Mom and Dad cavort in the sun. People have picked up torpid swifts and nighthawks, mistaking them for dead, and been startled when they fly off in perfect health as soon as they warm up.

Swifts' stiff, spine-tipped tail feathers help them perch on sheer surfaces. Black swifts nest on cliffs, often behind waterfalls. Vaux's swifts glue their nests to the insides of hollow trees or chimneys. Rare (and fetching rarefied prices) Asian "bird's nest soup" features filmy masses like boiled egg whites—the special saliva that swifts work up as nest glue.

Groups of Vaux's swifts numbering in the thousands congregate at a hollow tree or chimney over a period of a few weeks in fall, building up for massed migration. After sundown, they create a tornadolike vortex as they all fly into the roost. Their numbers are in decline as fewer big trees are allowed to remain standing after rot hollows them out.

HUMMINGBIRD

Family Trochilidae. Very small birds seen with wings a buzzing blur, hovering or flying; bill long and needle-thin, for sucking nectar from flowers.

Broad-tailed Hummingbird

Selasphorus platycercus (se-**lass**-for-us: light-bearer; plat-i-**sir**-cus: broad tail). ♂ 3¾"; ♀ larger; green back and crown, whitish breast and belly; ♂ have large, iridescent, red throat patch, some rust color on tail; to assert territory, either a vocal rattle or a distinctive penetrating trill added to flight hum. Southerly in summer; winters in Mexico. Trochilidae.

Calliope Hummingbird

S. calliope (ca-**lie**-o-pee: a Muse). Formerly *Stellula calliope*. 2¾"; bronze-green above, white below with reddish tinges; throat has long deep-purple streaks on adult ♂, small, dark speckles on others. Meadows, low shrublands, alpine in Aug; A.

Broad-tailed hummingbird male

Calliope hummingbird male

gram of flesh than any other warm-blooded creature, tending to make a clockwise loop, north along the Pacific, then inland in summer, and south along the Rockies to central Mexico. Hummingbird individuals strictly follow their route programming, which must reside in their genes.

The smallest of all birds, hummingbirds have frantic metabolisms in common with shrews, the smallest mammals. Ounce for ounce, a hummingbird flying has ten times the caloric requirement of a person running—and we don't have to spend all day running, while hummingbirds do spend most of their waking hours flying around in search of nectar for those calories. They consume up to half their body weight in sugar daily normally, and even more when preparing to migrate.

Their whirring flight, suggestive of a huge dragonfly, works more like insect flight than like that of other birds, and allows stationary and backward hovering, but no gliding. The wings beat many times faster than other birds', thanks to an extremely shortened wing with long feathers; there's very little mass to flap. Their hyperkinetic days are complemented by torpid nights, with body temperature and metabolism sharply lowered.

Hummingbirds are belligerent toward their own and larger species: calliopes have been seen dive-bombing red-tailed hawks. Small birds harassing larger birds is called "mobbing." There are records of predators actually being killed by mobs of smaller birds, but single hummingbirds aren't a threat. Still, studies have shown that small birds that mob—even when it's two or three little birds harassing a predator twenty times their size—are preyed upon less than those that don't. Possibly the predator conserves energy by saving its killer moves for less aggressive prey.

Hummingbird territoriality seems to concern food rather than sex. (Hummingbirds mate freely, without pairing up.) Both sexes

MY RED BOOTLACES alone draw several hummingbirds a day in the alpine country in July; they pause only an instant to confirm my boots are no bed of paintbrush. I can only hope the pleasure they bring me doesn't cost them much.

Most bright-red flowers that bear nectar at the base of tubes 1–2 inches long are adapted to pollination by hummingbirds. Columbine and skyrocket are stereotypical, and are certainly beloved of hummingbirds, but various pinkish to red flowers—currants, fireweed, paintbrush, lilies, fool's huckleberry—are also important nectar sources. When hummingbirds migrate north a little too early for nectar, they may compensate by gleaning aphids, hawking midges, and sucking sap from sapsucker wells. At Gothic, Colorado, where several decades of records exist, broad-tailed hummingbirds arrive in spring twelve days earlier than they did thirty years ago.

At one-tenth of an ounce, the calliope is the smallest US bird. It migrates farther per

stake out their patch of the nectar resource even during two-week refueling stops during migration, and defend it with the same tall, elliptical flights they use for courtship: a courting male swoops around and around in an ellipse hundreds of feet high, at the lowest point passing at high speed only inches from the demure object of his attentions and eliciting a shrieking noise from his wings. Males flash their gorgets (iridescent, erectile throat patches) both in courtship and in aggression.

These two species, as well as the rufous hummingbird, another Rockies fall migrant, are in decline and highly vulnerable to climate warming.

Order Gruiformes (Cranes)

Birds currently considered Gruiformes are generally wading birds (they like shallow water), but the order is not actually very cohesive, and may get split many ways in the future.

SANDHILL CRANE

Antigone canadensis (an-**tig**-a-nee: see below). Formerly *Grus canadensis*. 37"; ws 80"; gray (sometimes rusty) with red forehead; long black bill; long legs; long neck held straight out front in flight; when standing, tailfeather "bustle" droops over rump; loud, guttural "kraaaak"

Finding Their Way

Some birds migrate thousands of miles and find the exact same home territory they occupied the year before. Some do this over the ocean. Some do it at night. And each bird has to do it a first time, with no previous experience. "Trans-Pacific migrants . . . almost always behave as if they have a GPS on board, the nature of which represents one of the great mysteries in biology."

How they find the energy to do it, in some cases flapping their wings continuously for days without resting, sleeping, eating, or drinking, is an equal astonishment. Techniques include building up fat reserves equal to half of their body weight, supplying water by burning off proteins, and temporarily but severely shrinking the digestive tract to conserve weight, which rules out stopping to eat.

As for navigation, several methods have been demonstrated to at least contribute. No single one of them can explain all cases:
  · Celestial navigation, by the constellations
  · Polarity of sunlight, which they perceive as a band of darker sky somewhat like you might see through a camera's polarizing filter
  · The sun's direction, while knowing the time of day
  · Recognition of a "map" based on smells or visible landmarks
  · Magnetoreception—directly perceiving Earth's magnetic field, to detect not only which way is magnetic north, but also how far they are from the equator, based on the field's strength
Magnetoreception is likely the main method for the greatest diversity of animals, especially birds. There are two hypotheses for how magnetism is perceived; both may be correct. One involves tiny particles of magnetite, an iron mineral found in many animals that responds to magnetism. In birds, the useful particles apparently reside in the beak.

The second is harder to grasp, but is ascendant in recent science. It involves quantum mechanics—a magnetic effect on the spin of electrons within special proteins called cryptochromes, located in the retina. Birds would basically *see* Earth's magnetic field.

Sandhill crane dancing

Spotted sandpiper

esp when flying in flock or dancing. Lake shores, beaver ponds, willow thickets; A. Gruidae.

The song and dance of this huge bird are impressive, albeit indelicate. In the dance, pairs bounce like balls several feet into the air, with wings half-spread, legs dangling, and voices full-throated. This is associated with courtship and nesting, but also performed at other times of year—for fun?

Sandhill cranes eat all kinds of small animals and mass quantities of roots and tubers, from submerged to dry.

The genus name *Antigone,* long in disuse, was resurrected in 2010 to accommodate a genus split. In Greek myth, Antigone—not Sophocles's tragic heroine, but a Trojan princess—was turned into a stork for saying her hair was more beautiful than Hera's.

Order Charadriiformes (Shorebirds)

This order is ancient, likely having survived the Cretaceous extinction. It has a huge population, mainly in the form of seagulls.

SPOTTED SANDPIPER

Actitis macularia (ak-**tie**-tiss: shore-dweller; mac-you-**lair**-ia: spotted). 6¼"; light brown above with white eye stripe and wing bars, white below; in summer, adults dark-spotted white below, with ± yellowish legs and bill; dips and teeters constantly when on the ground; flies with wings stiffly down-curved; call a high, clear "peep-peep," usually in flight or when landing. Single or in pairs, along streams and lakes, esp subalpine; A. Scolapacidae.

Like the dipper, which also feeds in frothy streams, the spotted sandpiper dips from the knees; it tips forward and back, hence its nickname "teeter-tail." When threatened by a hawk, it can dive like a dipper as well. To feed, it plucks prey from shallow water or the bank, or sometimes from midair. Prey range in size up to trout fry, though insects predominate. No plant foods are eaten. Spotted sandpipers are widespread, breeding in almost any mountains north of Mexico and then wintering on seacoasts with mild weather.

In this and a few other sandpipers, the females are larger, more aggressive, and dominant; they migrate to the breeding grounds first, establish and defend territories, court the males, initiate sex, do less than their share of sitting on the eggs, and do very little raising of the young aside from serving as a sentinel. They have less prolactin—a hormone that promotes parental caregiving—than the males. Many females are monogamous, but the fittest are serially polyandrous: a pair courts, bonds, builds a nest, and provisions it with four fertile eggs, and then she leaves that family and goes off to do it all over again with a second, a third, and even a fourth male in one season. It's different and it works, but you have to wonder why sandpipers developed this way.

WILSON'S SNIPE

Gallinago delicata (gal-in-**ay**-go: chicken-like). Formerly in *G. gallinago*, common snipe. 9"; ♀ larger; plump bird with straight, 2½"-long bill, brown, longitudinally striped on head and streaked on back; belly ± white; grating "ski-ape" call when flushed; rarely seen until flushed. Marshy ground; A. Scolapacidae.

Male snipes (and sometimes females) have an audible flight display called "winnowing" or "whinnying." Rushing air vibrates the spread tail feathers as the snipe plummets from 100 feet up, making a unique, hollow, quavering "whoooo." This may go on all night during courtship (late May), but is also heard during other seasons and times of day. The slender, somewhat pliable bill probes for larvae, earthworms, and crustaceans in mud or shallow water.

KILLDEER

Charadrius vociferus (ka-**rad**-ree-us: Greek name). 8"; brown back, rusty rump; white belly, breast, and neck, with a heavy black neck ring and a second black ring across breast; another black band from black bill across eye and nape; red eye ring; tan legs; "kil-dee," "kil-dee-dee,"

Wilson's snipe

Killdeer

Sora

and other sharp calls; often heard at night. Low to mid-elev gravelly shores and fields; A. Charadriidae.

A gleaner of insects, the killdeer feeds and nests on the ground. A parent may employ the broken-wing ruse to distract predators from the nest, or chase off trampling cattle by flying at their faces.

SORA

Porzana carolina (por-**zah**-na: Italian word for "crake," a relative). Also sora rail. 6¾"; thick, stubby yellow bill; short, upturned tail; looks plump from the side but is laterally compressed; gray-brown with white flecks lengthwise on back, crosswise on belly; black face, gray throat; calls include a squeaky chirp, a rising "sor-UH," and a haunting nocturnal whinny—a rapid, then slowing, musical descent of notes. Widespread; rarely seen until flushed; marshy ground; nests among cattails; A. Rallidae.

Despite their clumsy-looking flight, soras migrate as far as South America.

CALIFORNIA GULL AND RING-BILLED GULL

Larus californicus and *L. delawarensis* (**lair**-us: Roman name). 17", ws 52" (ring-billed is slightly smaller); white exc top of wings, where black at tip and otherwise gray; yellow bill; California has greenish legs; immatures are speckly brownish with black tails. Apr–Nov; mtn lakes, town dumps, croplands; A. Laridae.

The California gull is the Utah state bird, revered there for supposedly consuming a plague of locusts that threatened the Mormon pioneers' crops in 1848. Wintering at the coast but coming inland to breed colonially at intermountain lakes and reservoirs, it owes much of its increase in numbers to the Bureau of Reclamation creating many large lakes. From these, individuals make long, foraging day trips, and are often seen around alpine lakes. Gull species are mixed in the large colonies, and these shockingly opportunistic feeders commit a fair amount of nest predation upon their neighbors. Even the smaller ring-bill can eat amazingly large prey, such as young ground squirrels.

The Great Salt Lake hosts the largest breeding colony of California gulls, but that could collapse any year now. As the lake shrinks due to overuse of water, it gets saltier. It's close to the point where brine shrimp and brine flies—the base of the food chain—can't survive in it.

Order Gaviiformes (Loons)

Like ducks, loons use wings as well as webbed feet to swim underwater. Diving either headfirst or submerging like submarines, they can go deeper than any other birds (300 feet down!) thanks to being almost too heavy to float. (They float lower in the water than ducks—a trait you can learn to spot easily.) Loons eat fish, plus some frogs, reptiles, leeches, insects, and aquatic plants.

COMMON LOON

Gavia immer (**gay**-vi-a: Roman name; **im**-er: sooty). 24", ws 58" (variable size, larger than ducks); bill heavy, tapered; in breeding season, both sexes have iridescent-green/black head, white collar, black/white-checked back, white belly; winter plumage dark gray-brown above, white below; in flight, head is held lower than body, and feet trail behind tail. Lakes; uncommon; northerly for breeding; A in other seasons. Gaviidae.

The loud nocturnal "laughs" and "yodels" of loons have been called beautiful, horrible, hair-raising, bloodcurdling, magical, and maniacal. Hearing them is a memory to cherish.

Loons' heavy bones are a primitive trait that remains advantageous for diving. It is, however, disadvantageous for flying: though loons can fly fast and far, they take off only

Ring-billed gull

Common loon

Great blue heron

bluish-gray with some white, black, and dark-red markings; bill, neck, and legs extremely long; neck held "goosenecked" in flight; huge birds seen in low, slow-flapping flight over rivers and lakes are generally this species; various loud, guttural croaks. A. Ardeidae.

The heron stands perfectly still in shallow water until some oblivious frog or small fish happens by, and then plucks or spears it with a quick thrust of the beak. Prey see little of the heron but its legs and shadow, and perhaps mistake it for an odd reed or cattail. Thanks to excellent night vision, a heron can nail prey even at night.

with great effort and splashing, and plop down gracelessly. They can get trapped for days or weeks on forest-lined lakes too small for their low-angle takeoff pattern—waiting to take off into a headwind. On their feet, they're inept and cumbersome—the extreme rear placement of their legs is great for swimming but awful for walking. They go ashore only to breed, on an island, and nest in soggy plant debris at the water's edge. She and he take turns on the eggs.

After wintering near the coast, some loons move to mountain lakes, arriving soon after the ice breaks up. Do not approach loons closely during the summer nesting season, when they are very sensitive.

Order Pelecaniformes (Herons)

Herons, egrets, and bitterns form the Ardeidae, treated as the largest family in the Pelecaniformes. Whether they truly belong with the pelicans is debated; the order was originally defined as all birds with all four toes webbed, but that trait is likely to have evolved convergently in barely related birds.

GREAT BLUE HERON

Ardea herodias (**ar**-di-a her-**oh**-di-as: Roman and Greek names for heron). 38", ws 70";

Order Cathartiformes (New World Vultures)

After getting juggled around in the bird family tree for over a century (see p. 384), New World vultures are now an order of their own, with only seven species alive today. They are highly adapted to feeding on carrion. Old World vultures somewhat resemble our vultures, but are in the hawk family.

TURKEY VULTURE

Cathartes aura (cath-**ar**-teez: purifier; **aur**-a: breeze). Also turkey buzzard. 25", ws 72"; plumage black exc whitish rear ½ of wing underside; head naked, wrinkled, pink (exc black when young); soars with wings in a shallow V, often tipping left or right, rarely flapping. Open country; late Feb–Oct; A. Cathartidae.

> *bird of rebirth*
> *buzzard*
> *meat is rotten meat made sweet again and*
> *lean*
> —Lew Welch

Linnaeus savvily named this creature "purifier" to counter its unsavory reputation. Its digestive tract is immune to disturbance by the meat-rotting organisms that would sicken the rest of us. Its beak and talons are too weak to tear up freshly dead mammals,

Turkey vulture

let alone to kill live ones—so it must feast on carrion. (Predatory birds, in contrast, are equally ready to either kill or to eat carrion.) Vultures go long periods without food, and when they find it, they gorge themselves. This may account for their lethargy on foot and difficulty in taking flight. (If hungry enough, they'll fill their stomachs with plant foods.) Once on the wing, they are the best soarers of all land birds, rarely needing to flap.

A congregation of vultures wheeling usually means carrion below. Vultures slowly gather from miles around. Several may crowd a carcass, but they eat one at a time, in order of dominance yielding occasionally to desperation. Dominant individuals have

Raptors Reconfigured

Once, long ago, the hawks, eagles, falcons, owls, and vultures comprised a single order, Raptores—from the Latin for "snatcher"—embracing nearly all birds that prey on vertebrates. Later, owls were given an order of their own, but "raptor" persevered as a casual term that usually includes owls and vultures. Starting in 1873, our New World vultures were said by some to belong among the storks, and to have evolved some traits in common with hawks convergently. The AOS checklist briefly adopted that position in 1998. In 2010, it switched to the view that vultures are hawks after all, and that the raptors most distantly related to the others are the falcons, which are almost songbirds; order Accipitriformes was adopted for the hawks and vultures. Then, in 2016, the AOS decided that vultures, having branched off almost as early as owls—56 million years ago—deserve an order to themselves, Cathartiformes. Some taxonomists disagree.

Raptors of all four orders still offer a great case of convergent evolution over geologic time. Shared traits include heavy, hooked bills; talons with strong muscles; and females larger than males. (Female northern harriers outweigh males by as much as three to two, and in golden eagles the ratio can reach two to one. In other bird orders as well as mammals, males tend to be bigger than females; invertebrates tend to have bigger females.)

The usefulness of beaks and talons in predation is clear, but correlating female size with predation had ornithologists wracking their brains for a century. It turns out to be a case of sexual selection, the same force that makes males grow ever bigger in some species! Female birds are bigger than their mates in bird species where females choose mates largely by appraising acrobatic courtship flights, with quick turns, twists, and rolls. As we know, small size helps gymnasts—even avian ones. So while acrobatic skill might be useful when hunting, this trait must instead result from sexual selection; if it were from hunting success, it would shrink both sexes. (Vultures, in contrast, are not stunt fliers, and their males are as big as their females.)

redder heads, and seem to flush even redder as a dominance display. Vultures locate their food by both sight and smell: they almost uniquely have a sharp sense at least for a few carrion odors, in contrast to the weak sense of smell in many birds. Fossil-gas companies found that by perfuming their gas with ethyl mercaptan, they could get vultures to lead them to pipeline leaks.

"Buzzard," an old word tracing from the Roman *buteo,* is used in Britain to include hawks (genus *Buteo*) as well as Old World vultures, which are in the hawk family. American vultures and condors resemble Old World vultures due to convergent evolution, but are not closely related.

Osprey

Order Accipitriformes (Hawks and Eagles)

The archetypal birds of prey, Accipitriformes carry to perfection traits like the large, hooked beak; large, sharp talons; and acute powers of vision. Compared to falcons, they have broader wings adapted for soaring over land, though the style and degree of soaring differs between genera within this order (see p. 386).

OSPREY

Pandion haliaetus (pan-**die**-un: a mythic king; hal-ee-**ay**-et-us: sea eagle, Greek name). Also fish hawk. ♂ 22", ws 54", ♀ larger; blackish above, exc white crown; white beneath, with black markings most concentrated at wing tips and "elbows"; the "elbow" break is sharper both rearward and downward than on similar birds, suggesting a shallow M shape while soaring; frequent calls include loud whistles and squeals. Near rivers and lakes; A. Pandionidae.

In a hunting technique that wins respect if not utter astonishment, the osprey dives into water from 50 to 100 feet up, plucks a fish from a depth of 1 to 3 feet, and bursts immediately back into flight, gripping a squirming fish sometimes as heavy as itself.

Osprey feet bear minute barbules that help grasp wet fish.

The osprey is a nearly worldwide species comprising a genus and a family by itself.

Ospreys are among several birds of prey that suffered heavy losses in some areas from the pesticide DDT before it was banned here. Toxins from prey concentrate in the tissues of consumers of poisoned prey; birds are especially sensitive. But in the West, so far, ospreys coexist with civilization. Dams have created a lot of new osprey habitat.

GOLDEN EAGLE

Aquila chrysaetos (ak-**will**-a: Roman name; cris-**ay**-et-os: gold eagle). ♂ 32", ws 78", ♀ much larger; adults dark brown all over; juveniles have white bases to their main wing and tail feathers

Golden eagle

Soaring Wings

Vultures, eagles, and *Buteo* hawks share a broad, spread-tipped wing and tail optimal for soaring over land. This contrasts with the short wings and long tails of *Accipiter* hawks, the narrower wings of falcons, and the clean, linear wings of gulls and albatrosses, which soar superlatively on the very different air currents over the ocean.

The overland soarers stay aloft largely by seeking out thermals, or columnar upwellings of warm air—the daytime convection pattern of low, warm air rising to trade places with higher, cooler air. Since thermal air is warmed by land absorbing and reradiating sunlight, soarers rarely cross large lakes, or travel at night or in the early morning. Dry, sparsely vegetated land does it best, so steppes and prairies are popular with soaring birds. They also like mountains—which deflect prevailing winds upward—and shift to using mountain updrafts more during windier weather. To travel long distances, they ride one thermal up in a spiral, then glide obliquely downward to the next thermal, then spiral again. To migrate, they often wait for a low-pressure weather trough, or follow the updrafts of long north-south mountain ridges.

(but never an all-white tail, as in bald eagle, nor a white rear edge of wing, as in turkey vulture); call (rarely used) is rapid chipping. Open country. Accipitridae.

Golden-eagle range encircles the Northern Hemisphere. In North America, they inhabit any terrain that has a lot of vertical relief, few trees, and populations of hares and large diurnal rodents like marmots or ground squirrels. The hares provide winter fare when the rodents hibernate. Winter may impel an eagle to attempt larger prey such as a fox, or rarely a coyote or deer, by means of a plummeting, falconlike dive. Momentum multiplies the eagle's ten pounds into enough force to overpower heavier prey. More often, golden eagles hunt from a low, fast-soaring cruise, using angular topography both for visual cover and for updrafts. They often rob hawks and falcons of their kills—hawks rarely venture, let alone nest, within half a mile of an eagle's aerie. The latter is a stick structure 4 to 6 feet in diameter, and growing from 1 foot deep to as much as 5 feet with many years of reuse—yet it remains somehow hard to spot against its cliff.

Superlative soarers, golden eagles think nothing of climbing 6,000 feet just to take greatest advantage of wind currents aloft. Some elsewhere migrate long distances, but ours mostly stay for the Rocky Mountain winter.

Eagle feathers and talons for ceremonial use are crucial to Indigenous groups, who receive permits to take a number of eagles, and also receive dead eagles obtained by government agencies. Humans inflict at least half of golden eagle mortality, mainly by illegal shooting and poisoning, ingestion of toxic lead fragments in hunters' kills, collisions with vehicles and structures, and electrocution on power lines; yet their numbers seem to be declining only modestly in the West. Wind turbine collisions are far down on the mortality rankings currently, but with rapid development, turbines could become a serious threat, as could climate change itself.

BALD EAGLE

Haliaeetus leucocephalus (hal-ee-**ee**-et-us: sea eagle; lew-co-**sef**-a-lus: white head). ♂ 32", ws 75", ♀ larger; adults blackish with entire head and tail white; immatures (3+ years) brown;

wings held flat while soaring; various weak chips and squawks, or louder, gull-like shrieks. Near large lakes and rivers; year-round, but with peaks when migrating in spring and late fall; A. Accipitridae.

The US national symbol, the bald eagle was for most of America's history considered vermin to be shot on sight. Shooting them was even rewarded with

Bald eagle

bounties in Alaska until 1952, twelve years after it became a crime in the Lower 48. Some Westerners still fear and hate eagles enough to break the law.

Despite a public image as a fearsome predator, the bald eagle far more often scavenges carrion or robs other birds—ospreys, gulls, kingfishers, and smaller bald eagles—of their prey. Most salmon it eats are spawned-out carcasses. At lambing time on sheep ranches, bald eagles eat afterbirths and stillborn lambs. Grossly exaggerated fears of lamb and salmon predation fueled America's animosity toward eagles for the last two centuries.

Our bald eagles breed in spring. A fondness for fish and waterfowl tends to keep bald eagles near water. They gather in large numbers at spawning runs, including those of freshwater species like kokanee. They used to do that along the North Fork of the Flathead, but those kokanee runs collapsed after non-native opossum shrimp invaded the watershed.

Bald eagles' spectacular courtship "dance" is well known: the male dive-bombs the female in midair, she rolls over to meet him, and then they lock talons and plummet earthward, breaking out of their death-taunting embrace at the last possible instant. Juvenile bald eagles engage in every kind of courtship behavior short of mating, which waits for the fifth or sixth winter.

Indigenous people associated the bald eagle with sickness, death, and healing, and hence viewed it as an ally or guardian of shamans. Eagle feathers were used in healing rituals, and it was variously said that an eagle would fly over a sick person, eat a dead person, or scream to a person soon to be killed by an arrow.

NORTHERN HARRIER

Circus hudsonius (**cir**-cus: Greek for some circling hawk; hud-**so**-ni-us: of Hudson's Bay). Formerly marsh hawk; *C. cyaneus.* ♂ 16¼", ws 42", ♀ much larger and colored differently; conspicuous white rump patch on both sexes; ♂ ash-gray above, whitish underneath with black wing tips and tail bars; ♀ speckled reddish brown, paler beneath; tail long and narrow; wings held well

Northern harrier female

Northern harrier male

above horizontal while gliding. Grasslands and marshes; A. Accipitridae.

The harrier nests among tall grasses, and spends hours a day cruising just inches above the tops of reeds, grasses, or low shrubs. When a vole takes off running underneath it, the agile hawk may harry it through many dashes and turns before dropping on it. Unlike most hawks, it hunts primarily by sound; in experiments, harriers precisely locate and attack tiny tape players concealed in meadow grass peeping and rustling like voles. In this respect, harriers evolved convergently with owls, developing an owl-like facial ruff of feathers to focus sound.

In years of abundance, some males take multiple mates and manage to feed two or three nests of about five young each. The mothers often leave the nest briefly to take midair delivery of morsels from their harried breadwinners. A female, having paired and begun a nest, may dump that mate if he turns out to be a poor provider, and join the harem of a better one. A male courts by means of a "sky dance" in a grand U pattern, landing on the nest site he proposes. Agile, vigorous dancers get more mates.

ACCIPITER

Accipiter spp. (ak-**sip**-it-er: Roman name for a fast flier). Tell these from other hawks by behavior, and by proportionately short, broad wings and long tails; tail often narrow (not fanned out), broadly barred (± black/white) its full length, with a pure-white tuft under its base; 3 species hard to tell from one another—size ranges overlap, ♀ being about as big as the next bigger species' ♂; both sexes slaty-gray above, or ± brownish; yearlings brown above, pale beneath with red-brown streaking; calls cackling, infrequent. Accipitridae.

Sharp-shinned Hawk

A. striatus (stry-**ay**-tus: striped). ♂ 10½", ws 21" (jay-sized); tail square-cornered, dark-tipped; underside pale with fine brownish barring. Often at forest/clearing edges, or alpine; A.

Cooper's Hawk

A. cooperii (**coop**-er-ee-eye: after William Cooper). ♂ 15½", ws 28" (crow-sized); tail rounded, with slight white tip; underside pale with red-brown barring. Forest understory or edges; streamside brush; A.

Northern Goshawk

A. gentilis (jen-**tie**-lis: aristocratic). ♂ 21", ws 42" (red-tailed hawk–sized); tail rounded, with slight white tip; underside pale, with fine blue-gray barring. Deep wilderness forests; A.

Sharp-shinned hawk juvenile

Cooper's hawk male

THESE ARE FOREST hawks. Their long tails lend maneuverability, while the short wings avoid branches. They typically burst from a concealed perch, or less often fly around and chase prey they happen across.

Each day while young are in the nest, a sharp-shinned father brings home about two sparrow-sized birds apiece for his family of seven or so. That's a lot of hunting. He seems to stay in shape with playful harassment of larger birds, even ravens. Accipiter males select a limb not far from the nest and take prey there to pluck it before eating it or delivering it to the nest. An indiscriminate scattering of thousands of small feathers may be a sign of one such pluckery in the branches above. Either parent may make a nasty fuss over an animal, such as yourself, happening near the nest.

The larger accipiters prey on larger birds and also mammals. Thanks to their speed and strong talons, they can take larger prey, relative to their own size, than *Buteo* hawks. Goshawks take many rabbits, and crow-sized Cooper's hawks take many crows, which are dangerous prey for them.

Cooper's and sharpies have adjusted to urban life, as long as there are trees, and even goshawks are starting to do so in Europe, but in the US, goshawks still seem to insist on remote forest.

Accipiters are often described as nonsoarers, with a flight rhythm of five flaps and a short glide—sometimes a useful identifying trait. But in the strong, sustained updrafts of rugged mountains, they appear capable of soaring till hell freezes over.

RED-TAILED HAWK

Buteo jamaicensis (**bew**-tee-o: Roman name; ja-may-ik-**en**-sis: of Jamaica). ♂ 18", ws 48", ♀ larger; brown above, highly variable beneath, from dark brown to (most often) white with delicate red-brown patterning; tail (adults) often red-brown above, pink below, broadly fanned out; call a short, hoarse, descending scream. Widespread; A. Accipitridae.

Red-tailed hawk

Comfortable around freeways or perched on fenceposts, the red-tailed hawk may be the bird of prey we see most often. It has been on the increase in nearly all regions, perhaps thanks to increases in handy perches like utility poles, wire fences, and juniper trees, and to handy prey like voles and ground squirrels. Its preferred natural habitat is grassland with some trees. It also may be the bird of prey humans *hear* most often, thanks to Hollywood. Sound editors have a sample that used to be labeled "red-tail hawk scream (for Westerns)." Today they insert it anywhere from the Alps to Antarctica whenever the director says, "Dial up some of that wild-and-free lonesomeness."

Adaptability, opportunism, and economy are key to the red-tail's success. Lacking the speed that makes falcons and accipiters perfect specialists, it employs patience, craft, and acute vision to hunt, often from a perch. Hares and ground squirrels may get an aerial

Swainson's hawk

teetering, low cruises like a harrier's. In fall, it gathers in huge flocks and migrates 7,000 miles south, making much use of thermals.

Up until around 1940, Swainson's hawk was commonly shot as a varmint—despite being in fact an efficient predator of crop pests—and though that practice has largely ceased, the species is still in decline. Habitat is lost as agriculture intensifies and fewer trees for nesting remain.

Order Strigiformes (Owls)

With their ghostly voices and silent, nocturnal predatory flight, owls have universally evoked human dread, superstition, and tall tales. Various parts of their anatomies found their way into talismans and potions both medical and magical. Owl pellets seem almost ready-made talismans, and are also great data sources on owls. They are strikingly neat, oblong bundles coughed up by owls to rid them of indigestible parts of the small prey they swallow. Hard, angular bone fragments end up smoothly encased in fur. A spot with many pellets suggests an owl's roost on a limb above. (Hawks make similar, but smaller and fewer, pellets.)

Owls' sensory adaptations are the stuff of legends. They have the broadest skulls of any bird's, and their wide-set eyes and ears widely optimize three-dimensional vision and directional hearing. Their ears pinpoint prey along the vertical axis as well as the horizontal, thanks to asymmetrical skulls. By hearing alone, great gray owls locate voles under snow as deep as 18 inches, hovering above them and then plunging through snow to snatch them. The facial ruff of feathers, plus ear flaps hidden under its outer rim, funnel sound to the highly developed inner ears. (The "ear tufts" or "horns" on top of some species' heads are expressive and decorative, unrelated to hearing.)

Owls' eyes are the most frontally directed of any bird's; this narrows the field of vision but makes nearly all of it three-dimensional.

swoop after a stealthy approach. Or the hawk may brazenly land between a rodent or lizard and its refuge, forcing the prey to attempt an end run. Red-tails are also pirates, robbing hawks and even eagles or great horned owls of their kills.

Courting red-tails perform impressive aerial duets, often initiated by the male with his sky dance, a series of steep dives that brake abruptly and reverse into steep ascents.

SWAINSON'S HAWK

Buteo swainsoni (**swain**-son-eye: for William John Swainson). ♂ 18", ws 49", ♀ larger; tail broadly fanned out, finely banded gray with wider black outer band; brown above, highly variable beneath: front ½ of wing commonly white, but flight feathers are dark; wings more pointed than other big soarers; call a squeaky, descending shriek. Prairies and sparse woodland; Apr–Sept; A. Accipitridae.

In May and June (breeding season) this grand hawk preys on mammals up to the size of rabbits and large ground squirrels; in arid country, snakes and lizards join the diet. In other seasons, it snatches dozens of large insects daily—dragonflies in mid-air, grasshoppers in the fields—hunted with

The squashed-down bill also expands the three-dimensional area. (Force your eyes far left, then far right, to see the profile of your nose on either side; only that portion—about a third—of your field of view lying between the "two noses" is seen in three dimensions, by both eyes.) The owl's adaptive tradeoff—narrower field of view, but all of it three-dimensional—favors zeroing in on prey, not watching out for predators. To look around, an owl (like other birds) can twist its neck in a split second to anywhere within a 270-degree arc. The eyeballs, which don't rotate in their sockets (the neck does it all), have the optimal light-gathering shape: somewhat conical, with a thick, powerful lens. The retina has a reflective backing (the kind that makes nocturnal mammals' eyes gleam in your headlights) behind the photoreceptor cells. These are almost all rods (high-sensitivity vision) and few cones (color vision). Owls' light-perception threshold is between a thirty-fifth and a hundredth of ours. They have much sharper acuity for detail than we do, even by day, and a modicum of color perception as well.

Most owls practice utterly silent flight, a magical thing to witness. Their feathers are literally muffled, or damped, with a velvety surface and soft-fringed edges, incurring a tradeoff in efficiency and speed. Their flapping is slow and easy, thanks to low body weight per wing area. (Fluffy body plumage means owls are far slimmer and lighter than they appear.) Silent flight enables owls to hear the movements of small rodents, while it keeps sharp-eared prey in the dark over someone coming for dinner. Perhaps mice live in ignorance of owls except as invisible agents of disappearances from the family.

Great horned owl with prey

GREAT HORNED OWL

Bubo virginianus (**bew**-bo: Roman name; vir-jin-ee-**ay**-nus: of Virginia). ♂ 20", ws 55", ♀ larger; large "ear" tufts or "horns," yellow eyes, and reddish-tan facial ruff; white throat patch, otherwise finely barred and mottled gray-brown; long, low hoots (4–6 in series) year-round, but esp in Jan–Feb breeding season; pairs sing duets; nocturnal. Common in forests; A.

The nocturnal great horned owl can hunt mammals much heavier than itself, such as porcupines and large skunks, as well as almost any sort of creature down to beetles and worms. In places where cottontails are common, they are the owl's chief prey. Great horned owlets may attempt flight from the nest before they are ready; landing on the forest floor, they must reside there until fully fledged. The parents continue to feed and guard them aggressively. If you find an adorable, fluffy owlet on the ground, back off.

GREAT GRAY OWL

Strix nebulosa (strix: as in "strident," Greek name imitating owl; neb-you-**low**-sa: cloudy). ♂ 22", ws 60", ♀ larger; gray with vertically streaked breast; large facial ruff of many concentric

Great gray owl

Barred owl

Burrowing owl

BARRED OWL

Strix varia (**vair**-ia: variegated). ♂ 17", ws 44", ♀ larger; dull brown, white-barred above and vertically flecked beneath; no "ear" tufts; dark eyes; yellow bill; strident, reedy hoots (6–9) like a deep-voiced rooster; nocturnal. Northerly.

Barred owls were eastern birds unable to penetrate the treeless Great Plains and unknown in the US Rockies until the 1980s. It may have been global warming that helped them expand their range northward into the boreal forests, enabling a Saskatchewan end run around the plains.

BURROWING OWL

Athene cunicularia (a-**thee**-nee: for Athena; cue-nic-you-**lair**-ia: burrower). Both sexes 8", ws 22"; light brown with white; relatively long legs that bob frequently; hoot is 2 notes, the second louder, longer, slightly higher; day or night, esp crepuscular. In grassland—famously in prairie dog towns where they move into existing borrows; Apr–Sep, plus a few overwinter; A.

Burrowing owls are listed as threatened in Colorado and of concern in the other states.

rings; yellow eyes; no tufts; low hoots (4–6) evenly spaced; nocturnal. Rarely seen; in conifers; northerly.

They declined in the late twentieth century apparently due to habitat loss (grasslands plowed up), and widespread poisoning of their prey (ground squirrels) and their home builders (prairie dogs).

NORTHERN SAW-WHET OWL

Aegolius acadicus (ee-**go**-lius: Greek name; a-**kay**-di-cus: of eastern Canada). ♂ 7", ws 17", ♀ larger; reddish-brown above, white- and brown-smeared (adults) or golden (juveniles) beneath; blackish bill; yellow eyes; V-shaped white eyebrows; short toots repeated with machinelike monotony (like a truck backing up); nocturnal. Sites with both deciduous trees and conifers; A.

Northern saw-whet owl

The small *Aegolius* owls are very hard to spot in their forests, but studies using mist nets catch plenty of them. If you do spot one, it may let you approach quite close.

BOREAL OWL

Boreal owl

A. funereus (few-**near**-i-us: deathly). ♂ 10", ws 24", ♀ 70 percent heavier; brown with white-spotted wings, white-streaked breast; obvious facial disk whitish-centered, blackish-rimmed; yellow bill and eyes; from 8 to 15 rapid, staccato hoots on 1 note; primarily nocturnal. Mature spruce-fir forests; uncommon; A.

Boreal owls turn out to live in most mature spruce-fir forests in our range, even though old range maps show none south of the 49th parallel. They may be as dependent on old growth in the Northern Rockies as spotted owls are in the coastal forest. Old forests provide abundant voles, combined with enough maneuvering room for owls to hunt them.

NORTHERN PYGMY OWL

Glaucidium gnoma (glaw-**sid**-ium: small owl, from "gleaming"; **no**-ma: gnome). ♂ 6", ws 15"; ♀ larger (barely robin-sized, but owl-shaped); brown with slight pale barring, 2 dark eyelike

Northern pygmy owl with vole prey

Belted kingfisher female

The Greeks had a myth that the halcyon, a kind of bird we presume was a kingfisher, floated its nest on the waves of the sea while incubating and hatching the young. "Halcyon days" are a respite from the storms of life.

Our kingfishers, in real life, raise their young amid a heap of regurgitated fish bones at the end of a hole in a mud bank. Is "nest" too sweet a term for such debris? They look for their prey—fish, crayfish, waterbugs, and larvae—from a perch over a stream or occasionally a lake. (They can also hunt from a hover where branches are in short supply.) After diving and catching a fish, they often return to thrash it against their branch before swallowing it headfirst. Anglers have long resented this bird's success rate, but statistically kingfishers are unlikely to reduce trout numbers significantly. The kingfisher population has plummeted during the advance of civilization; there was once a pair of kingfishers on virtually every creek in the United States.

spots on nape; yellow eyes; longish tail often held cocked; flight swift, darting, audible, with rapid beats—atypical of owl flight; steady hoots like whistling, a bit faster than 1 per sec; primarily diurnal. ± Open forest; A, breeding northerly.

While all owls are formidable hunters, evolution has specialized them for a broad spectrum of habitats and roles. The fierce, little pygmy owl darts about catching insects and also birds up to and greater than its own size. Like some 40 percent of owl species, it hunts mainly by day.

Order Coraciiformes (Kingfishers)

Kingfishers and their relatives form a large order, with many species tropical and bright: motmots, rollers, and bee-eaters. Relatively few inhabit the Americas. They typically subdue prey by thrashing them against objects.

BELTED KINGFISHER

Megaceryle alcyon (me-ga-**ser**-i-lee **al**-see-on: big, followed by 2 Greek names). 12"; head looks oversized because of large bill and extensive crest of feathers; blue-gray, with white neck and underparts, and on ♀ a reddish breast band; call a long, peculiar rattle. Along streams, year-round; A. Alcedinidae.

Order Piciformes (Woodpeckers)

Woodpeckers resemble songbirds (Passeriformes), but differ in several specialized traits. Most have two front and two rear toes, rather than three and one. (Black-backed and three-toed woodpeckers have a two-and-one arrangement.) The order includes toucans and several other tropical families, but ours are all in family Picidae.

Family members with "woodpecker" in their name forage for insects by jackhammering on trees. Females are generally thinner-billed than males, and pry bark perhaps more than they jackhammer it. Flickers mostly pry, and sapsuckers have their own foraging specialty.

Woodpecking also serves to chop large, squarish holes for nests. Padded with wood chips, such nests are dry and easy to defend, allowing prolonged rearing—advantages that have led several other species to rely on abandoned woodpecker holes for their nests. Lastly, woodpeckers peck on resonant trees to communicate: the classic woodpecker drum roll asserts territory or points out a nest site to a prospective mate.

Hairy woodpecker

Strong, sharp claws on their feet, together with short, stiff tail feathers, give woodpeckers the grip and the bracing they need for hammering the full force of their bodies into a tree. They also have adamantine, chisel-like beaks, and thick, shock-absorbing skulls to prevent boxer's dementia. Before diving into work on a tree, they listen for the minute rustlings of their insect prey boring around under the bark. Insect larvae and adults provide the bulk of their diets, with seeds and berries providing the rest. For snatching grubs out of their tunnels, a woodpecker has a barb-tipped tongue longer than its head. The tongue shoots out and then pulls back into a tiny, tubular cavity that loops around the circumference of the skull.

Black-backed woodpecker

HAIRY WOODPECKER

Dryobates villosus (dry-a-**bay**-teez: tree walker; vil-**oh**-sus: woolly). Formerly *Picoides villosus*. Also *Leuconotopicus villosus*. 7½"; black and white exc (♂ only) a small red patch on peak of head; wings barred. Wooded habitats, esp with aspen; A. Picidae.

DOWNY WOODPECKER

D. pubescens (pew-**bes**-enz: fuzzy). Formerly *Picoides pubescens*. 5¾"; like hairy woodpecker only smaller. Wooded habitats; A.

BLACK-BACKED WOODPECKER

Picoides arcticus (pic-**oy**-deez: like a *Pica*— Roman name for both magpies and woodpeck-

ers). 8"; almost entirely black above and white below, with modest flecking on wings and sides; ♀ has small yellow crown. In and near burns; northerly. Picidae.

The black-backed, specializing in insects of freshly fire-killed trees, is our top avian fire-follower. It declined during the decades of successful fire suppression, and is listed as sensitive in several states. Increasing fire may be alleviating that problem.

American three-toed woodpecker

Pileated woodpecker

Red-naped sapsucker

and from bill; ♂ has yellowish crown. Conifer forests with bark beetles; moves in en masse a few weeks after forest fire, nipping at the heels of bark beetles; A.

Uncommon and restricted to high elevations, the three-toed resembles its black-backed kin in liking burns and eating many bark beetles.

PILEATED WOODPECKER

Dryocopus pileatus (dry-**oc**-o-pus: tree sword; pie-lee-**ay**-tus: crested). 15"; all black exc bright-red, large, pointed crest, black/white-streaked head with red moustache on ♂; white underwing markings visible in flight; drumming very loud, slow, irregular; call a loud, rattling shriek with a slight initial rise in pitch. Deep forest with many standing dead trees or snags; from nw WY n and w. Picidae.

Our largest woodpecker, the pileated has a strident call like a maniacal laugh, which it seems to use when you or I come around. When foraging, it excavates large, square holes, but for a nest it turns lazy, carving an entrance into an already-hollow, rotten

AMERICAN THREE-TOED WOODPECKER

P. dorsalis. Formerly *P. tridactylus.* 7½"; black/white barred back and sides, black wings; white belly, face black with white streaks from eye

Northern red-shafted flicker

RED-NAPED SAPSUCKER

Sphyrapicus nuchalis (sfie-ra-**pie**-cus: hammer woodpecker; noo-**kal**-iss: nape—). 7¾"; sexes alike; head has 3 red patches—forehead, throat, and back of crown—and is otherwise black/white-striped; back and wings black, with white bars and rump; belly pale, dull yellowish; taps in a syncopated rhythm, but does not jack-hammer, on trees; calls are catlike mews and "cherrrrs." Forest, favoring aspen; A. Picidae.

Neat, horizontal rings of ¼-inch holes drilled through tree bark are the work of sapsuckers, who drill these "wells," leave, and come back another day. Time allows sap not only to flow but also sometimes to ferment, and, most importantly, to attract the sapsucker's main course: insects. Butterflies and moths in experiments prefer their foods fermented, and inebriated behavior has been observed in butterflies and sapsuckers alike. Other birds also dine at these sap wells; they are crucial for hummingbirds that inadvertently migrate north too soon for adequate nectar supplies.

Damage to the bark is insufficient to kill the tree by girdling it, but any breaching of the bark can help fungal diseases invade. On balance, sapsuckers are good for trees, as they are top predators of spruce budworm moths, a major pest. They nab insects in midair, glean ants from bark crevices, and vary their diets with berries.

Order Falconiformes (Falcons)

Distinguish falcons in flight from other hawks by their pointed, swept-back wings, and straight or slightly tapering tails when in flapping flight. (When soaring, they fan their tails out as other hawks do.)

PEREGRINE FALCON

Falco peregrinus (**fahl**-co: Roman name; pair-eg-**rye**-nus: wandering). ♂ 15", ws 40", ♀ larger; wings pointed, tail narrow; color variable, ours mostly slate blue to (immatures) dark brown above; white beneath with dark mottling and

tree. It eats ants primarily, and may chop many inches deep into rotten wood to dig out carpenter-ant nests. Each bird requires a lot of dead trees, so this species dwindled as old-growth forests vanished from most of the continent. Pileateds were also hunted. More recently, they have recovered, and are quite willing to brave urban woods if the woods have enough snags. A few have become pests, chopping away at house and phone-pole timbers.

NORTHERN RED-SHAFTED FLICKER

Colaptes auratus ssp. *cafer* (co-**lap**-teez: pecker; aw-**ray**-tus: golden; **caf**-er: a mistaken name). 11"; gray-brown with black-spotted, paler belly, black-barred back, black "bib"; ♂ has red-to-orange crown, moustache, and underside of wings and tail; flies in a roller-coaster path; varied calls: one a flat-pitched rattle. Semi-open and edge habitats; A. Picidae.

Flickers grab ants and preen with them, using ants' formic acid to kill parasites. They peck for insects in the soil or catch them in midair, or pry bark up to get them. After the breeding season, seeds and berries are major foods.

Peregrine falcon

± reddish wash except white throat and upper breast; conspicuous rounded dark bar descends across and below eye. Widespread, often near rivers or cliffs; A. Falconidae.

Peregrines certainly earn their name. Some that breed in Alaska migrate to south-ern Chile, a longer migration than any other raptor. One was GPS-tracked migrating 3,953 miles at a rate of 179 miles a day. Other individuals don't migrate—paradoxical peregrines.

Peregrine falcons hunt birds almost exclusively, often by stooping and striking the prey a stunning blow to the neck, with fist outstretched.

Peregrines came back after their near-extirpation from our range in the 1970s led to US and Canadian bans on the pesticide DDT and on the capture of peregrines for falconry. Pesticides get more concentrated as they move up the food chain, and DDT was causing eggshells to be so thin they broke under the mother's weight. Captive-breeding-and-release programs completed this famous success story, but it came slowly, hindered by poaching and by insecticides still in use. Now peregrines have moved into many cities, where two of their favorite things abound: high ledges to perch and nest on, and pigeons to eat. The Audubon Society

Speed Wings

Swept back, narrow, and pointed must be the optimal wing design for sustained speed, since so many fast flyers—falcons, swifts, swallows, and nighthawks—evolved it independently.

The fastest of all falcons in level flight is said to be the largest falcon, the gyrfalcon, *Falco rusticolus,* of the Arctic (rarely reported here). In ancient falconry, only kings were entitled to fly gyrfalcons. The slightly smaller peregrine edges out the gyrfalcon as the world's fastest animal, however, based on stooping—not exactly flying but plummeting at an airborne target. To minimize drag, the wings are cupped close to the sides. Most hawks and eagles stoop occasionally, when an opportunity presents, but the peregrine specializes in midair prey and has perfected the stoop.

The figure 180 miles per hour started making the rounds after 1930, when a small-plane pilot reported a stooping peregrine passing him while he was diving at 175 miles per hour. He was disbelieved by some, who maintained that the record belongs to the swifts. The ornithologist Vance Tucker retorts that swifts are probably no swifter than pigeons, whereas his instruments optically tracked a peregrine stooping at 157 miles per hour. The pilot Ken Franklin trained his peregrine, Frightful, to skydive with him. Using multiple electronic altimeters over several stoops, he clocked her at 242 miles per hour.

American kestrel

Prairie falcon

a field or roadside. From this vantage, it can drop and strike prey quickly, as a larger, broader-winged hawk might do from a low soar. Grasshoppers and other big insects are staples; mice are also taken. If all these are scarce, the kestrel may chase down sparrow-sized birds.

PRAIRIE FALCON

Falco mexicanus. ♂ 16", ws 40", ♀ larger; wings pointed, tail narrow; dusty- to (less often) slaty-brown above, strongly mottled cream-white beneath, with blackish under-wing patches; may have a vertical streak below eye, narrower and less distinct than on peregrine. In steppe canyons year-round, or timberline meadows in late summer; A.

A grassland bird that nests in steppe canyons, the prairie falcon may visit alpine meadows after its nesting season is over. It's been known to prey on Colorado pikas, but ground squirrels are by far the most frequent prey, followed by horned larks and meadowlarks in winter when ground squirrels hibernate. It stoops at a shallow angle, or overtakes small birds in fast, low flight, or by repeatedly diving through a flock of swallows. A parent may defend its nest by dive-bombing intruders' heads, veering away at the last instant—providing a lucky hiker with a great adrenalin-pumping wildlife encounter.

predicts that warming may actually enlarge their suitable range.

AMERICAN KESTREL

Falco sparverius (spar-**ver**-i-us: sparrow). Formerly sparrow hawk. ♂ 9", ws 21", ♀ slightly larger; red-brown above, exc wings blue-gray on adult ♂ only, and tail tipped with heavy black band and slight white fringe; brown-flecked white beneath (exc dark tail); call a sharp, fast "killy-killy-killy." Open country; A.

The kestrel is one of the smallest, most successful, least shy, and most frequently seen raptors. When not perched on a limb or telephone wire, it often hovers in place, wings fluttering, body tipped about 45 degrees, facing upwind, 10 to 20 feet above

Order Passeriformes (Perching Birds, a.k.a. Songbirds)

Perching birds are the most recently evolved avian order, with the greatest number of species—nearly 60 percent of all bird species—which indicates success and rapid evolution.

OLIVE-SIDED FLYCATCHER

Contopus cooperi (**cont**-o-pus: short foot; **coop**-er-eye: after James or William Cooper). 6¼";

Willow flycatcher

olive-gray with pearly white smear down throat and breast; white, downy tufts on lower back visible in flight. Forest edges; A. Tyrannidae.

The olive-sided's song, usually asserted from some conifer pinnacle, has been transcribed as "quick! . . . FREEbeer" or "tuck . . . THREE bears." These flycatchers summer throughout our forests, clearcuts, and timberlines. They eat winged insects up to bee sized, spotting individuals from their perch and darting out to snatch them.

WILLOW FLYCATCHER

Empidonax traillii (em-**pid**-o-nax: gnat king; **trail**-ee-eye: after Thomas Traill). 4¾"; olive-brown above with dull wing bars; pale yellowish olive beneath; 2-toned bill; no conspicuous eye ring. Summer; brush, incl streamsides and wetlands; A. Tyrannidae.

It's next to impossible to identify *Empidonax* birds in the bush, camouflaged among foliage. This one has a recognizable song, a sneezelike "fitz-bew."

Warbling vireo

WARBLING VIREO

Vireo gilvus (**veer**-ee-oh: green; **jil**-vus: pale yellow). 4¾"; greenish gray above; heavy white eyebrow, no eye ring; solid gray wings; white belly, ± yellowish flanks; slow-moving; song a 2-to-4-second twitter with final upbeat accented. Forest, esp riparian; May–Sep; A. Vireonidae.

Female and male vireos take turns on the eggs. The nest dangles from a fork in branches high in a tree, and consists of a small cup of grasses lined with lichens, moss, or feathers, often decorated outside with bark, petals, or catkins held on with repurposed spider webbing.

LOGGERHEAD SHRIKE

Lanius ludovicianus (**lan**-ius: butcher; loo-do-vis-i-**ay**-nus: of Louisiana). 7"; gray above, white beneath, separated by black wings and tail and wide black eye stripe; head large for a songbird; very rapid wingbeats. Juniper, grassland, farms;

Olive-sided flycatcher

thickets for nesting; mainly Mar–Sep, sometimes overwintering; A. Laridae.

Shrikes, uniquely, are songbirds of prey. They converged with raptors in evolving a hooked bill to sever the neck vertebrae of small mammals. Lacking large, powerful talons, they came up with a workaround: they immobilize vertebrate prey by impaling it on a thorn. (Insects, a bigger part of their diet, don't need impaling.) "Loggerhead" describes the disproportionately large head and neck (another convergence with raptors), while the word "shrike" denotes their shrieking call. They have declined for decades; the clearest culprit is the decline in insects, fueled by insecticides.

Loggerhead shrike

COMMON RAVEN

Common raven

Corvus corax (**cor**-vus: Roman name, from **cor**-ax: Greek name mimicking a raven). 21", ws 48"; black with purple-green iridescence, and shaggy grayish ruff at throat; bill heavy; tail long, flared then tapered; alternates periods of flapping and flat-winged soaring; calls include a throaty croak and a deep "tock"-like rapping on a big wooden block. Forest and alpine; A. Corvidae.

You can tell a raven from its little sibling the crow by the raven's greater size, heavier bill, ruffed throat, more prolonged soaring, and voice, an outrageously hoarse, guttural croak rather than a nasal "caw." High in the mountains, you are more likely to see ravens.

Their spring courtship flights are a sight to remember: barrel rolls or chaotic tumbles while plummeting, then they swoop, hang motionless, all the while flaunting their vocabularies. The ensuing family group— four or five young are typically flying by June—often stays together through the fol-

lowing winter, and may flock with other families. The nest is high, solid, and cozy, perhaps lined with scavenged deer hair, but filthy and often flea-ridden.

Ravens were once abundant throughout the West; flocks followed bison herds to eat carrion. Then the bison resource was erased, and countless ravens were killed with poisoned bait left out for wolves and coyotes, or simply shot. Perceived threats to agriculture, though mostly imaginary, were deeply ingrained in European culture. No dummies, ravens became shy, retreating to remote deserts and mountains and becoming scarce on the Great Plains and farther east. In recent decades, however, they are thriving, losing their shyness, and even moving into cities.

Ravens were valued members of Native villages. Their thievery of food was tolerated; killing them was taboo. Raven was at once a trickster and a powerful, aggressive, chiefly figure—much like Coyote. He was a creator in origin myths. His croaks were prophetic—a person who could interpret them would become a great seer. To inculcate prophetic powers in a chosen newborn, the Kwakiutl fed the infant's afterbirth to the ravens. On other continents, a mythic view of ravens and crows as powerful, knowing, and a bit sinister is almost universal. French peasants once thought that bad priests became ravens, and bad nuns, crows.

The evolution of seed plants was accompanied by the specialization of seed-eaters—songbirds and rodents, mainly—and a wave of predators that eat seed-eaters. The raven rides the crest of both waves: it eats rodents, nestlings, and eggs, as well as seeds. It is the largest of all Passeriformes, and might be the most advanced bird.

Bernd Heinrich, who devoted years to the close observation of ravens, casts doubt on one well-known anecdote said to prove raven intelligence: their teaming up to heist meat from coyotes. He does, however, believe (as do many Indigenous people) that ravens will lead a wolf or a human hunter to prey, on the odds that they will end up with scraps from the kill. Devising ingenious tests of raven intelligence, Heinrich showed that ravens remember where they cache seeds under snow, rather than smelling them out; moreover, they keep a corner of their eye open for where other ravens cache seeds, and try to rob those caches first.

A 1911 paper on animal longevity reported a captive raven living sixty-nine years, which, if true, would easily be the record for a bird. A raven in the wild would be lucky to reach fifteen.

AMERICAN CROW

Corvus brachyrhynchos (**cor**-vus: raven; brak-i-**rink**-os: short beak). 17"; black with blue-purple-green iridescence; tail squared-off; rarely glides more than 2–3 seconds at a time. Woodland, forest openings; A. Corvidae.

Crows are fond of farmland, and are most often seen here along lowland streams and meadows, or in partly deciduous forest. In our mountain wilderness, they are less common than ravens, and seem to avoid potential habitat that has a lot of ravens. They scavenge carrion, garbage, fruit, snails, grubs, insects, frogs, and eggs and nestlings from birds' nests. Species as large as ducks, gulls, and falcons may lose nestlings, and some songbirds count crows among their major enemies.

Crow populations in North America grew from 1966 until 1999, when the mosquito-vectored bird virus called West Nile arrived and hit crows hard in Eastern cities. Crow numbers in the West were not strongly affected, and nationally the disease peaked by 2005; crows recovered somewhat to stabilize near their pre-1966 level. Now it looks like the virus caused significant, long-lasting declines in just a small number of bird species—but not crows. In 2013, it caused a huge die-off of eared grebes, a scarce species, at the Great Salt Lake.

CLARK'S NUTCRACKER

Nucifraga columbiana (new-**sif**-ra-ga: nut-breaker; co-lum-be-**ay**-na: of the Columbia River). 11"; pale gray (incl crown and back), with white-marked black wings and tail (white outer tail feathers and rear wing patch); long, thin, grayish bill; call a grating "kraaa, kraaa," or various rasps and rattles; may run through them for minutes at a time. Subalpine habitats with whitebark, limber, or lodgepole pines; A. Corvidae.

Clark's nutcracker

Pinyon jay

Flashy black-and-white wings and tail distinguish Clark's nutcracker from the Canada jay. Both have been nicknamed "camp robber."

In a classic case of coevolution (described on p. 58), the nutcracker looks to whitebark pines for food, while whitebark pines depend on nutcrackers to disseminate their seeds. The bird pries the cones open while they're still on the tree, selects high-quality pine nuts (the most nutritious and most viable), eats some on the spot, and buries the rest, a few at a time, for retrieval later. Though canny enough to remember cache locations for nine months, the birds are so industrious that they store two or three times what they can consume—well upward of thirty thousand pine nuts per year. The uneaten ones have thus been planted, often many miles from their source—a great advantage to the pines, especially in recolonizing large, severe burns. The strong-flying nutcrackers commonly fly up to 20 miles carrying as many as one hundred whitebark pine seeds in a sublingual pouch that bulges conspicuously at the throat.

Poor pine-cone-crop years drive nutcrackers far and wide in search of substitute fare on other conifers. The decline in whitebark pines has led to a decline in nutcrackers.

PINYON JAY

Gymnorhinus cyanocephalus (jim-no-**rye**-nus: naked nose; cy-a-no-**sef**-a-lus: blue head). 9"; entirely dull blue, sometimes grayer beneath; long, thin black bill; no crest on head. Piñon-juniper woodland; from c MT s. Corvidae. (The AOS checklist of bird names insists on Anglicizing "piñon.")

Pinyon jays and piñon pines have evolved in a mutualism much like that of Clark's nutcrackers and whitebark pines, but less extremely specialized. The jays do almost everything in flocks: pine-nut caching is a big party. Over past millennia, piñons showed a greater ability than most trees to move around in response to climate, thanks to jay-assisted long-distance seeding. For at least fifty years now, pinyon jays have been in decline following declines in piñon pines— first due to silver mining, then clearing for cattle, then megadrought and consequent bark beetles and fires.

Pinyon jays are the Bobby McFerrins of the usually tone-deaf crow family. *Birds of*

Canada jay

brownish-gray wings, tail, rump, and (variably) nape and crown; short, ± black bill; juveniles all dark gray exc pale cheek streak; varied calls include a whistled "whee-oh"; flight quite silent. Deep forest and timberline, often with spruces; A. Corvidae.

Like their close relatives the crows, jays seem coarse and vulgar, but are in fact intelligent, versatile, and successful. Their voices are harsh and noisy, yet capable of amazing variety and accurate mimicry of other birds. Versatile in feeding, ours eat mainly conifer seeds, berries, and insects, but also relish meat when they can scrounge some up or kill a small bird or rodent. Canada jays, or "camp robbers," were known for thronging around logging-camp mess halls; a trapper would sometimes find them nipping at a carcass while he was still skinning it. You will find them just as interested in your lunch, and bold enough to snatch proffered food from your fingertips.

Jays' nonchalance reminds skiers that winter here is perfectly livable for the well adapted. They're so comfortable in the cold that they nest and incubate their young in the depth of winter. We can only speculate as to how that is adaptive; no nearby birds do it except Clark's nutcracker, which is more variable in timing. Both species store huge amounts of food for winter.

the World reports that "varied sounds of the subsong are not jaylike, and contain beeps, toots, musical whistles, bell-like sounds, notes reminiscent of a woodwind instrument, clicks, chuckles, and grating sounds. Once [we] recorded > 20 min of continuous subsong given by a single bird. No two subsong bouts are alike, and each session seems to be spontaneously composed."

CANADA JAY

Perisoreus canadensis (pair-i-**sor**-ee-us: "I-heap-up"). Also gray jay, camp robber, whiskey-jack. 10"; fluffy, pale gray with dark-

A Naturalist Marooned

Georg Wilhelm Steller was the first European naturalist-explorer on Alaskan soil—though only on an island and only for one day. A German, he crossed Siberia to accompany the Danish captain Vitus Bering on a Russian ship built and launched from Kamchatka in 1741. They were just east of Cordova, Alaska, when they turned around late that year, and they didn't quite make it back to Kamchatka. Marooned for a terrible scurvy-ridden winter on small, rocky Bering Island, the captain and many of the crew died. Steller owed his survival to eating vitamin-rich native plants, which the non-scientists scorned. But Steller took to drink and died in Siberia without seeing Europe again. Though Steller's jay thrives today, Steller's sea eagle, Steller's sea lion, and Steller's eider are not doing well, and Steller's sea cow went extinct.

Canada jays have a special trick for storing food—saliva that coagulates in contact with air. This enables them to glue little seed bundles in bark crevices or foliage, up where they won't get buried in snow. Food storing as a winter strategy is common in the crow and chickadee families, but apparently in no other birds. It entails a phenomenal ability to remember exact locations—feats of memory that humans can't match. Unlike pinyon jays, Canada jays store perishable foods like mushrooms, berries, and meat. Recent winters have failed to keep it all frozen, causing spoilage and sharp drops in jay numbers locally.

STELLER'S JAY

Cyanocitta stelleri (sigh-an-o-**sit**-a: blue jay; **stell**-er-eye: after Georg Steller, p. 404). 11"; deep ultramarine blue with ± black shoulders and dramatically crested head. Among conifers. Corvidae.

Birds of North America describes the following repertoire of vocalizations documented by students of the Steller's jay: creak, squawk, growl, rattle, song, "ut," "ow," "wah," "wek," "aap," "tjar," "tee-ar," guttural notes, and mimicry. Mimicry includes an uncanny imitation of a red-tailed hawk's scream, which presumably deceives other birds into clearing out while the jay feeds, but perhaps it sometimes warns of an actual hawk. Jays and crows are the birds most often seen harassing or "mobbing" birds of prey—a defense of smaller birds against larger ones. This jay is omnivorous, smart, and aggressive—traits that run in the crow family. It robs food cached by its cousin, Clark's nutcracker.

"Blue jay" in Rocky Mountain parlance refers to Steller's. The smaller, paler blue jay proper, *C. cristata,* is a rare visitor here from the East.

BLACK-BILLED MAGPIE

Pica hudsonia (**pie**-ca: Roman name). Formerly *P. pica.* 16"; black with white belly; white wing patches, and wing undersides flash in flight; wings and very long tail show blue or green iridescence; sexes similar, ♂ bigger; harsh calls include a whining "mack," a chattering babble at larger beasts, and a soft song between pairs. Deciduous or open woods, streamsides, farmland; ± year-round. Corvidae.

More beautiful than its jay and crow cousins, the magpie, like them, is omnivorous, adaptable, smart, aggressive, noisy, and

Steller's jay

Black-billed magpie on a moose

unloved by farmers. In days of old, magpies flocked around bison herds, picking parasites from woolly backs. To Lewis and Clark, they were relentless camp robbers, but after decades of bison genocide and magpie genocide, magpies are fewer and warier.

As for foraging on big beasts, we no longer assume that such "cleaner symbioses" are always mutually beneficial. Magpies pick maggots from sores on livestock, but also pick at the sores. They follow elk and moose and pick winter ticks from them, but also cache live ticks on soil, where they lay eggs and multiply. Carrion is as important as ticks as a winter food for magpies, so it could be in their interest for ticks to kill more moose.

Mobbing behavior reaches a peak of audacity in the magpie. One magpie pecks at the tail of an eagle holding prey, while others stand by to snatch the prey when the distracted eagle drops it; or the eagle may end the harassment by flying off with just part of his meal. Young male magpies form gangs, and, thanks to strength in numbers, are socially dominant over adult males (able to chase them from carrion). The males are family men, building a huge stick nest in

Recycling Bird Brains

To the long list of weight-saving adaptations in birds, add their ability to grow new brain cells. You may have heard that you can lose brain cells throughout life (when you over-imbibe, perhaps) but never grow new ones. Not true. The discovery of newly formed neurons in chickadees led the way to a similar find in humans. In both species, the new cells are in the hippocampus, the brain region involved in learning and memory.

Chickadees grow new neurons in a burst each fall, when they cache seeds in thousands of crevices they will have to find again if they are to survive the winter. An even bigger hippocampal growth spurt benefits dispersing juveniles as they learn the geography they will inhabit for the rest of their lives. But the total hippocampal neuron count does not increase over the chickadee life span. One hypothesis holds that each neuron stores one memory, and is resorbed and replaced at some point after that memory is no longer needed. In a tiny flying animal, a brain big enough for a lifetime of memories might never get off the ground.

Very small birds actually have a far higher percentage of their body mass in the form of brain than we do. The fact that natural selection produced big chickadee brains shows that brainpower is critical for food-caching birds, as brains are energy hogs compared to other kinds of tissue. A phenomenal ability to remember precise locations evolved separately in the chickadees and the jays that cache food for winter, and in many migrating species. Some of these species have nonmigratory, non-caching relatives whose powers of recall amount to diddly-squat.

Another kind of memory that must be worth holding on to is a male's memory of conspecific males' songs. As long as each singer remembers his neighbor's song from the year before, and stays on his own territory, both are spared a fight. Warblers remember songs from year to year as they return from Central America to reclaim their old haunts. In male white-crowned sparrows, neurons in a part of the brain used for song skyrocket in number from around 100,000 to around 170,000, and then quickly drop after mating, when the birds lose interest in singing.

Black-capped chickadee

Horned lark

riparian brush or trees, and bringing home much of the bacon; but they aren't particularly faithful, and are generally "randy" (the word used by Craig Birkhead, who wrote the book on magpies). Magpies are good at burying food and remembering the location; utilizing their sense of smell, they (especially females) often rob each other's caches.

Magpies are susceptible to West Nile virus. Their lowland Colorado numbers apparently declined after 2003 when the virus arrived, but subsequently rebounded.

In a "magpie funeral," a magpie finds a magpie corpse, calls loudly, and within minutes dozens of magpies congregate raucously, perching in trees and flying down two or three at a time to walk around the corpse. They'll even do this around a few magpie feathers.

CHICKADEE

Poecile spp. (**pee**-sil-ee: many-colored). Formerly *Parus* spp. Paridae.

Mountain Chickadee
P. gambeli (**gam**-bel-eye: for William Gambel). 4¼"; black crown, eye streak, and throat; white eyebrow and cheeks, grayish belly, gray-brown upperparts; typical song a lazy, descending "fee-bee" or "fee-bee-bee-bee." Mid-elev forest; A.

Black-capped Chickadee
P. atricapilla (ay-tri-ca-**pill**-a: black hair). 4½"; black crown and throat, white cheeks; grayish

belly, gray-brown upperparts; call "chick-a-dee-dee-dee." Lower thickets, streamsides; A.

CHICKADEES SEEM TO epitomize the chipper dispositions people want to see in songbirds, and they let us see it, being less wary than most. The various chickadees—gleaners of caterpillars and other insects from the branches—reside here year-round, nesting in fur-lined holes dug rather low in tree trunks either by woodpeckers or by themselves, in soft, punky wood. For winter, they store both insects and seeds in thousands of bark crevices (see sidebar p. 406), speedily wedging them into place or gluing them with saliva, and often hiding them with lichens or bark bits. Some mountain chickadees move upslope and winter near timberline, while others move down to the plains.

"Chick-a-dee-dee-dee" is an alarm call, or a call to arms, gathering allies to help mob a predator. For higher danger levels, the birds speed up the "chick-a-dee" and add more "dee-dees." Researchers once counted twenty-three "dees" for a pygmy owl—more dangerous than a big owl because being close to chickadee size makes it agile enough to catch them.

HORNED LARK

Eremophila alpestris (air-em-**ah**-fill-a: loving lonely places; al-**pes**-tris: alpine). 6½"; pinkish-brown back; ± white underparts; sharp, black

Violet-green swallow

Tree swallow

Cliff swallow

breastband and cheek smear; tiny "horns" (feather tufts) above eyes; white outer tail feathers show in flight; song a long, tinkling twitter prefaced by a few ascending notes, mainly before sunrise. Nests and forages on ground, walking (not hopping); tundra in summer, year-round on lowland steppe, grassland, overgrazed rangeland; A. Alaudidae.

In his famous courtship display, the male spirals steeply to several hundred feet up, circles around and around singing, then plummets with wings held to his sides. One song type, the "recitative," lasts several minutes.

The horned lark is the only New World member of the lark family, and is widespread across North America, with many subspecies specializing in diverse habitats that all feature sparse low vegetation. For safety from predators, it simply ducks into slight depressions and, I would imagine, prays.

SWALLOW
Family Hirundinidae.
Violet-green Swallow
Tachycineta thalassina (tack-y-sin-**ee**-ta: fast-moving; tha-**lass**-in-a: sea-green). 4¾"; dark with green/violet iridescence above; white below, extending around eyes and up sides of rump to show as 2 white spots when seen from above; wings backswept and pointed, tail shallowly forked; various high tweets. Open areas, in flocks; Apr–Sep; A.
Tree Swallow
T. bicolor (**by**-col-or: 2-colored). 5"; iridescent blue-black above, white below (not extending above eye or rump); otherwise like violet-green swallow. Open lower elevs; Mar–Aug; A.
Cliff Swallow
Petrochelidon pyrrhonota (pet-ro-**kel**-i-don: rock swallow; per-o-**no**-ta: red back). Formerly *Hirundo pyrrhonota*. 5"; blackish wings and tail, whitish below, with cinnamon-red cheeks and paler rust-red rump and forehead; wings backswept and pointed, tail square or barely notched. River cliffs, bridges; Mar–Aug; A.

A SWALLOW'S FLIGHT is graceful, slick, and fast (though less swift than a swift's). A swallow's swallow, or rather its gape, is striking. The wide, weak jaws are held open almost 180 degrees while the swallow knifes back and forth through the air intercepting insects. Lightning-quick visual and neck-muscle reflexes, and ultraviolet vision that may make insect wings glitter, aid in gathering flying insects; a single mouthful can hold as many as fifty of them. One expert swears he has seen swallows jerking their heads this way and that to catch raindrops.

Swallows nest in large colonies. Cliff swallows build striking, gourd-shaped mud nests on cliffs or under bridges or eaves. Tree and violet-green swallows use ready-made crevices and tree holes.

KINGLET
Family Regulidae.
Golden-crowned Kinglet
Regulus satrapa (**reg**-you-lus: little king; **sat**-ra-pa: ruler). 3½"; gray-green above, whitish below, with 2 white wing bars; central yellow ♀ or orange ♂ stripe on head is flanked by black and then white stripes at eyebrow; very high, lisping "chee, chee" call. High in conifers year-round; A.

Extra-Pair Goings-On

The DNA police have been looking into avian paternity, and they've demolished the faithful chirping-couples stereotype that inspired centuries of bad verse. Move over, Cole Porter—reggae duo Sly and Robbie accompanied by Bootsy Collins had it right when they sang, "Birds don't do it, bees don't do it, / WE are the only ones that fall in love."

In many bird species, a male may help build the nest, sit on the eggs, defend the territory, and bring food to the young, but that species's social monogamy rating bears little correlation with that male's likelihood of being the sire of all the young in that nest.

Patterns of extra-pair copulations (EPCs) vary widely among closely related species, or even populations of the same species. For example, a study of cliff swallows found that 2 percent of fertilizations were extra-pair, whereas the study of tree swallows found 44 percent. The EPC picture in swallow colonies is complicated by plenty of intraspecific brood parasitism, in which a female waits for a neighboring nest to be momentarily unattended, and then slips in and lays an egg; her unsuspecting victims will feed her young along with their own. This gets worse: a male slips into an unattended nest and rolls an egg out, to its doom, perhaps to make room for his mate to lay an egg there, or perhaps to keep his female neighbor receptive to his advances. Either way, he's promoting his own genes.

Given the many reproductive strategies, there have been academic imbroglios—replete with anthropomorphic terms like "divorce," "harem," and "cuckold"—over how to interpret them. There are male strategies and female strategies; the latter are usually the key, since females apparently control the fertilization success of copulations. One scientist who watched black-capped chickadees in a small area over a twenty-year period witnessed thirteen extra-pair trysts; in each case, the female sought out a male of higher social rank than her own mate. A later study using DNA corroborated that conclusion. A study in one warbler species found that males with larger song repertoires seduced more females, and that the female got what she sought: fitter genes, as measured by the breeding success of her offspring.

Golden-crowned kinglet

Ruby-crowned kinglet

Ruby-crowned Kinglet

Corthylio calendula (cor-**thil**-i-o: Greek name; ca-**lend**-you-la: larklet). Formerly *R. calendula*. 3¾"; gray-green above, whitish below, with white eye ring and 2 white wing bars; rarely visible scarlet spot on crown is displayed only by excited ♂; scolding "jit-it" calls, and long, variable song of chatters, warbles, and rising triplets. Conifer forest; May–Oct; A.

CONSTANT MOVEMENT—WINGS TWITCHING even when perched—characterizes kinglets. They catch insects, sometimes in flight but mostly on foliage. The birds' tiny size and ability to hover enable them to forage on twigs too weak to support other birds. Insects are a less concentrated and less predictable food than seeds, and so kinglets supplement with some seeds. Golden-crowned kinglets, chickadees, and woodpeckers commonly flock together in winter; this may help them locate insect populations.

A kinglet weighs less than three pennies (1.5 ounces total). Among birds and mammals, only a few kinds of hummingbirds and shrews are smaller, and not by much—and hummingbirds and shrews are rarely exposed to such cold air as kinglets. In winter, they spend nearly every waking minute foraging, raising their body fat daily from around 5 to 11 percent, enough to survive the night if they are lucky. When it gets too dark for foraging to yield a net gain, kinglets plunge into the first insulative spot they see, such as soft snow under thick brush, twist their necks to wedge their eyes and bills in among their back feathers, and go to sleep. Occasionally, several may huddle together overnight. It's a harsh existence. Mortality in northerly kinglets is estimated to be 87 percent per year. They sustain their numbers by producing eighteen to twenty eggs per pair, in two broods per summer.

WAXWING

Bombycilla spp. (bom-bi-**sil**-a: silky tail). Sexes alike; largely gray; yellow/black tail tip;

Bohemian waxwing

Cedar waxwing

White-breasted nuthatch

swept-back crest on rusty-brown head with black mask, throat, and beak; in flocks, making constant high hisses. Forest and woodland; A. Bombycillidae.

Bohemian Waxwing
B. garrulus (garrulous: the European jay). 6¼"; rusty under tail; wings have yellow and red bits. Year-round in n MT; Nov–Apr farther s.

Cedar Waxwing
B. cedrorum. 5¾"; white under tail; yellowish belly; wings have yellow bits. Mainly May–Oct, but a few overwinter at Grand Teton National Park.

THOUGH THEY LIKE insects—the Bohemian is big and fast enough to hawk dragonflies— these avian sugar-freaks mainly seek out sweet berries. Staples in the Rockies would be mountain-ash, chokecherry, and junipers (hence "cedar" waxwing). They like their berries fermented, and don't seem to know their limits: mass deaths from fermented fruit are on record. The need for abundant ripe fruit makes them unfaithful to any particular locale for either wintering or nesting—they were said to come and go like

"Bohemians"—and leads cedar waxwings to delay laying eggs until full summer, later than any other North American bird. Waxwings do not sing, are not territorial, and are not aggressive with each other, perhaps because they're better off cooperating than com- peting in seeking out flushes of ripe fruit. Mature waxwings develop waxlike deep-red or yellow nubs on the tips of wing feathers; these apparently mark them as high-value mates.

NUTHATCH
Sitta spp. (**sit**-a: Greek name). Blue-gray above, paler beneath. Sittidae.

Red-breasted Nuthatch
S. canadensis. 4"; pale reddish below, with white throat and eyebrow, black ♂ or dark-gray crown and eye streak. Conifer forest; A.

White-breasted Nuthatch
S. carolinensis (car-ol-in-**en**-sis: of the Carolinas). 5"; white below (may have a little red under tail), with black ♂ or gray crown, but all-white face. Deciduous or pine woodland; A.

Pygmy Nuthatch
S. pygmaea (pig-**me**-a). 3½"; white below, with brownish-gray crown, no eyebrow stripe; gleans from branches more than trunk. Ponderosa pine woodland; A.

NUTHATCHES WALK HEADFIRST down tree trunks, apparently finding that way just as right-side-up as the other. They glean insects from the bark, and eat seeds in fall through

Brown creeper

Pacific wren

This well-camouflaged, full-time bark-dweller gleans its insect prey from bark crevices, and nests behind loose bark. In contrast to nuthatches, which usually walk down tree trunks, the creeper spirals up them, propping itself with stiff tail feathers like a woodpecker's. Crevices approached from above versus below would reveal different types of prey, putting creepers and nuthatches in different ecological niches.

The Treecreeper family has only nine species, and only this one in the Americas.

PACIFIC WREN

Troglodytes pacificus (tra-**glod**-i-teez: cave goer, a misleading name that happens to also be that of the chimpanzee). Formerly winter wren; *T. troglodytes*. 3¼"; finely barred reddish-brown all over; tail rounded, very short, often (as in all wrens) held upturned at 90° to line of back. Near the ground in forest with dense herb layer; year-round; mainly from c MT w. Troglodytidae.

Though we hear it often, the Pacific wren's song never ceases to amaze with its sheer virtuosity. The notes go by too fast—ten per second—for us to catch them individually, but the singer knows his repertoire perfectly, and would be as defiant as the fictional Amadeus if criticized for "too many notes." Each song, of up to three hundred notes, is a permutation of short building blocks from a vocabulary of perhaps eighty different blocks. The wren can repeat the song exactly, and so can his neighbor; they learn each other's songs. He typically repeats it several times, with slight variations, before moving on to the next permutation, and he takes many days to go through his repertoire. (An astonishing recording is available of a Pacific Wren song slowed down to one-sixth speed.) Pacific wrens were formerly included in *T. troglodytes*, the winter wren of the East; the two visually indistinguishable species sing just differently enough to isolate them reproductively in regions where they overlap.

The wren moves in a darting, mouselike manner, eats insects, maintains a low profile

spring. They nest in dead snags, smearing pitch around the nest hole. Even in deep wilderness they aren't shy, and draw our attention with their penetrating little call, a tinny "nyank." They move to different elevations or subregions when cone crops are poor.

BROWN CREEPER

Certhia americana (**serth**-ia: Greek name). ¾"; mottled brown above, white beneath; long, downcurved bill; call a single, very high, soft, sibilant note. On tree trunks in older forest, year-round; A. Certhiidae.

You probably won't see a brown creeper unless you happen to catch its faint, high call and then patiently let your eyes scour nearby bark. (Many of us are literally deaf to its call, if our hearing is high-frequency-challenged—a defect correlated with the male sex and perhaps with rock and roll.)

Rock wren   American dipper

among the brush, and goes to great lengths to keep its nest a secret. Wrens will build several extra nests just as decoys, while the occupied nest has a decoy entrance, larger than the real entrance but coming to a dead end.

ROCK WREN

Salpinctes obsoletus (sal-**pink**-tease: trumpeter; ob-so-**lee**-tus: dull). 4¾"; gray-brown, sparsely speckled with pale ocher-tinged rump; white breast with fine streaking; belly white with yellow flanks; tail much longer than winter wren; dips body frequently; song is variations, sung at strikingly regular intervals, on a single, short note ornamented complexly. Canyons; Apr–Sep; A. Troglodytidae.

Before building their nest within a rock crevice, rock wrens arrange small stones on its floor and as a "sidewalk" to its entrance. Stones placed right in the opening ensure that it's tight enough to exclude anyone bigger than a rock wren.

AMERICAN DIPPER

Cinclus mexicanus (**sink**-lus: Greek name). Also water ouzel. 5¾"; slate-gray all over, scarcely paler beneath, often with white eye ring; tail short; feet yellow. In or near cold mtn streams; A. Cinclidae.

It's no wonder that dippers impress campers throughout America's western mountains, considering how much trouble campers have

keeping warm. Winter and summer, snow, rain, or shine, dippers spend most of their time plunging into and out of frigid, frothing torrents, plucking out aquatic insects such as dragonfly and caddis fly larvae, and sometimes tiny fish. Somehow they walk on the river bottom, gripping with their big feet. They can also swim with their wings, quickly reverting to flight if they get swept out of control downstream. They can dive deep into mountain lakes, and occasionally forage on snowfields. They often nest behind waterfalls. In August, they have a flightless molt period when swimming becomes their only escape from predators. Fledglings swim on their first day out of the nest, at around twenty-five days old. Dippers never get soaked to the skin, thanks to extremely dense body plumage and extra glands to keep it well oiled.

Dippers are named for their unexplained, jerky knee-bending, performed as often as once a second while standing, accompanied by blinking of their flashy white eyelids. Their call, "dipit dipit," is forceful enough to carry over the din of the creek. Even in midwinter, they occasionally break into song. Both sexes are virtuosi, with long, loud, lyrical, bell-like, and extremely varied songs.

VARIED THRUSH

Ixoreus naevius (ix-**or**-i-us: mistletoe mountain, its food and habitat; **neev**-i-us: spotted).

Varied thrush

Swainson's thrush

Hermit thrush

forest; Mar–Oct; from nw MT w, plus rare in A. Turdidae.

The varied thrush sings a single note with odd, rough overtones; it's actually two slightly dissonant notes at once. After several seconds' rest, he sings another tone, similar but higher or lower by some irrational interval. Birds have two sets of vocal chords, one in each bronchus. Many songbirds alternate them rapidly to warble or burble euphoniously. Several thrush species use them for simultaneous note pairs, but only the varied thrush performs just this one vocal trick in the starkest possible way. Prolonged early or late in the day, in deep forest and fog, this minimal music acquires powers of enchantment.

AMERICAN ROBIN

Turdus migratorius (**tur**-dus: Roman name). 8½"; like varied thrush (above) but no breast band, eye streak, or orange on wings, and belly is brighter red, posture more steeply angled; song brash, overly cheerful, aggravates crepuscular insomnia; eats worms. Anywhere with partial cover; Apr–Oct; A. Turdidae.

SWAINSON'S THRUSH AND HERMIT THRUSH

Catharus spp. (**cath**-a-rus: pure). 6¼"; gray-brown above; pale eye ring; belly white, breast spotted. Turdidae.

Swainson's Thrush

C. ustulatus (ust-you-**lay**-tus: singed). Back and head ± reddish; tail less so. Nests in riparian thickets or aspens, wanders in forest; May–Sep; A.

Hermit Thrush

C. guttatus (ga-**tay**-tus: spotted). Tail, but not back, is rusty red; tail is "nervously" raised and lowered every few seconds, while wings may twitch. Shady forest; Apr–Nov; A.

8"; breast, throat, eyebrows, and wing bars rich rusty-orange ♂ or yellow buff ♀, contrasting with slate-gray breast band, cheeks, crown, back, etc.; whitish belly. Deep conifer

BOTH THE THRUSH and the hawk bearing the name Swainson migrate as far south as Argentina, putting each near the top of the distance charts for its genus. Cold-hardy

Townsend's solitaire   *Mountain bluebird*

hermit thrushes, in contrast, arrive early and migrate modestly: some overwinter in New Mexico lowlands, and few go farther than Mexico.

Catharus thrushes sing lyrically, virtuosically. The hermit prefaces a fast phrase with a single long, clear note, then performs variations at different pitches. Swainson's begins with a slow phrase of a few notes, then spirals upward, flutelike, and often repeats at different pitches. Both forage on the ground for earthworms, insects, and berries.

TOWNSEND'S SOLITAIRE

Myadestes townsendi (my-a-**des**-teez: fly eater; **town**-send-eye: after J. K. Townsend, p. 312). 6¾"; gray with white eye ring; dark tail has white feathers on sides (as does the abundant junco; solitaire's tail longer, slenderer, more upright, more arboreal); dark wings have buff patches, visible underneath in flight; song a long, melodious warble heard at any season; call a repeated high "eep." Forest; year-round, wintering lower; A. Turdidae.

The solitaire returns to high elevations early, searching the first snow-free areas for a nesting cavity in a stump or rotting log. After the breeding season, it may gather in large flocks, belying its name. Most seek out juniper woodlands for winter, subsisting on juniper berries. In summer, they eat insects.

MOUNTAIN BLUEBIRD

Sialia currucoides (sigh-**ay**-lia: Greek name; cue-roo-**coy**-deez: warblerlike). 6"; summer ♂ intense sky-blue above, shading to paler blue or gray beneath, or grayish where not sunlit; ♀ and winter ♂ gray-brown with variable touches of blue on tail, rump, and wings; song a long series of chips and descending warbles; "phew" call. Grassy meadows, sparse forest, burns, clearcuts; Mar–Oct, longer in s RMs. Turdidae.

These hardy birds arrive early, with snow still on the ground, and some nest in rock crevices in alpine tundra. They mainly nest in woodpecker holes, and thrive in burns where woodpeckers have been perforating snags. They drop on insect prey from a low hover or perch, or forage on the ground. Their brilliant-blue color comes from the reflective structure of the feathers, rather than a pigment, and nearly disappears when shaded. Mountain bluebirds are sustaining their populations, at least within our mountains, while the western bluebird, *S. mexicana*, has declined severely.

AMERICAN PIPIT

Anthus rubescens (**anth**-us: Greek name for some bird; roo-**bes**-enz: reddish). Formerly water pipit; *A spinoletta*. 5½"; sexes alike: gray-brown above, buff below, with white outer tail feathers and dark legs; bill slenderer; flight-call note suggests "tsipit." Alp/subalp; Apr–Oct; A. Motacillidae.

American pipit

Evening grosbeak

chattering, social whistle calls. Conifers, riparian woodland; winters in UT, CO, NM lowlands; A in summer. Fringillidae.

Though highly capable of cracking seeds with its massive bill, an evening grosbeak will switch its diet to 80 percent insect larvae when presented with an outbreak of western spruce budworm—one of our worst forest pest insects. In one small study, the local grosbeaks were eating between three and nine million caterpillars in a season.

ROSY-FINCH

Leucosticte spp. (lew-co-**stick**-tee: white patch). Species formerly combined as rosy finch, *L. arctoa.* 4¼"; brown with blackish face, gray crown or entire head, and (♂ esp) reddish tinges on rump, shoulders, belly; either hoarse or high chips and chipping chatters. Arctic fellfields, snow, etc. Fringillidae.

Gray-crowned Rosy-Finch
L. tephrocotis (tef-**roc**-o-tiss: ashen head). From Bitterroots n and w.

Black Rosy-Finch
L. atrata (a-**tray**-ta: black). Darker. nw WY, sw MT, c ID.

Brown-capped Rosy-Finch
L. australis (aus-**trah**-lis: southern). Gray-brown crown. CO, NM.

THE FINCH-FAMILY NAME, Fringillidae, comes from the same Latin root as "frigid." Rosy-finches are the most alpine birds of the West, at least in summer. They nest in high rock crevices, reportedly even in glacier crevasses, and often forage on glaciers, utilizing a resource all mountaineers have noticed—insects that collapse on the snow, numbed by cold after being carried astray by upvalley winds. The bulk of their diet is vegetable, including seeds, white heather flowers, and succulent alpine-saxifrage leaves. In winter, they form flocks in valleys or on the plains, seeking conveniences like window feeding boxes, grain-elevator yards, and spilled grain on railroad beds—but they commonly commute up to bare tundra patches for the

Distinguish the pipit from our other common species with white outer tail feathers (junco and solitaire) by its habit of regularly jerking its tail down as it walks along foraging for invertebrate prey, sometimes in shallow water or on snow. To attract a mate and, once he has one, to assert territory, the male flies straight up and then drifts down on spread wings, singing a long series of similar thin, high cheeps.

EVENING GROSBEAK

Coccothraustes vespertinus (cock-uth-**roust**-eez: kernel cracker; ves-per-**tie**-nus: of dusk, a misleading name). 7¼"; ♂ yellow grading to dark-gray head with bright-yellow eyebrow stripe, black/white wings; ♀ dull yellowish with gray head and back; very thick, heavy bill; loud,

Brown-capped rosy-finch

Cassin's finch

Red crossbill

day, even at 30°F below zero. Their climatic niche is shrinking rapidly, dimming their future prospects.

For unknown reasons there are six times more male than female rosy-finches. Instead of a fixed territory, mated males guard the space around their mate, wherever she goes.

CASSIN'S FINCH

Haemorhous cassinii (**he**-mo-roos; blood-red—; cas-**in**-ee-eye: for John Cassin). Formerly *Carpodacus cassinii*. 4"; older ♂ has crimson crown, pale red breast; ♀ (and ♂ up to breeding yearlings) sparrowlike gray-brown with streaked white breast and belly; undertail feathers dark-streaked white; song a varied warble full of breaks and squeaks, often with a series of imitations of other birds; call "tidilit." Open forest; A. Fringillidae.

Flocks of these mountain-loving birds subsist through winter eating conifer buds, then switch to emerging male catkins of aspen in time to fuel their nesting efforts. They nest in the conifer canopy, often in colonies of several hundred. Locally abundant some years and scarce in others, they're also erratic about migration. Some stay near their nests in our mountains, while greater numbers migrate downslope or modestly southward.

The Audubon climate model indicates they may be rare and local by 2080.

RED CROSSBILL

Loxia curvirostra (**lox**-ia: oblique; cur-vi-**ros**-tra: curved bill). 5½"; older ♂ red; ♀ yellowish beneath, greenish-gray above; young mature ♂ often orange while grading from yellowish to red; wings solid dark gray; "chip, chip" call; warbling song. In flocks, high in conifers; A. Fringillidae.

A small shower of conifer seed coats and seed wings often means a crossbill flock is above. You have to be close to see the crossed bill: the lower mandible hooks upward almost as much as the upper one hooks downward. These odd bills can move sideways to efficiently pry cone scales apart. Evolution has specialized crossbills to clean up conifer seed crops when the remaining seeds are too sparse for larger seed-eaters like squirrels and woodpeckers, and too tightly encased in the cone for siskins and finches.

Pine siskin   American goldfinch   White-crowned sparrow

Red crossbills clean up leftovers from the fall crop well through winter and usually spring, depending on tree species; in summer, their jaws open wide to glean insects.

Red crossbills come with different sizes and styles of crossed bill, each specialized for the cones of one conifer species. To pass their bill size on to their offspring, the birds must find mates of similar bill size. Slight differences in their call notes enable them to do so, and thus the different-billed populations minimize interbreeding even while they mix within most flocks. It was proposed that the red crossbill is several different species, distinguishable by bill size and call note but not by plumage. In fact, their degree of genetic separation is substantial but not decisive. Most types readily feed on whichever conifer species currently has plentiful seeds, and migrate wherever necessary to find them—including the East Coast.

One new call-note species appears on the AOS checklist so far: *L. sinesciuris,* named for the absence of squirrels from two small mountain ranges in southernmost Idaho, which leaves the bird with first dibs on lodgepole pine seeds there. Three years after it was named, its doom was predicted. It had declined by 60 percent in five years, because the lodgepoles there are doing poorly in the warming climate.

PINE SISKIN

Spinus pinus (**spy**-nus: Greek name; **pie**-nus: pine). Formerly *Carduelis pinus*. 4¼"; gray-brown with subtle lengthwise streaking; yellow in wings and tail may show in flight; various distinctive, scratchy twitters and sucking wheezes. In large flocks in treetops year-round; A. Fringillidae.

Siskins' narrow, sharp bills (in contrast to the heavy, conical ones typical of their family) limit them to lighter seeds, like thistle and birch, along with insects and buds. Siskins hang upside down from catkins while extracting the seeds. The populations move around sweepingly, and tend to be locally abundant in alternate years.

Pine siskins look nondescript, but call and fly quite distinctively. A flock's twitterings seem to match its breathtaking undulations and pulselike contractions and expansions in flight. I once saw one of these aerial ballets on the same day as a performance by the Blue Angel fighter jets, and thought, "Why would I watch those noisy, clutzy contraptions?"

AMERICAN GOLDFINCH

Spinus tristis (**tris**-tis: sad). Formerly *Carduelis tristis*. 4¼"; breeding ♂ bright yellow with black cap, wings, and tail; ♀ dull yellow with white-barred black wings; both sexes much duller in winter; flies with roller-coaster ups and downs. Woodland, ranches, roadsides; in flocks;

year-round; A. Ours are ssp. *pallidus*, somewhat paler than the other spp. Fringillidae.

Goldfinch chicks are fed a low-protein, all-seed diet, and manage to thrive. The payoff is that cowbirds rarely parasitize goldfinch nests, because the cowbird chicks would die of malnutrition. Roadside weeds including thistles are prized. The nests are in thick brush.

Chipping sparrow

SPARROW

4¾–5¾", brown to gray, often flecked, sometimes with a contrasting crown; bill rather thick. Passerellidae.

White-crowned Sparrow

Zonotrichia leucophrys (zo-no-**trick**-ia: striped head; loo-**coaf**-riss: white brow). 5¾"; striking black/white-striped crown; mostly brown above, gray beneath; golden bill; song 2 seconds long, starts pure and ends buzzy. Riparian willows, subalpine; abundant; A.

Chipping Sparrow

Spizella passerina (spy-**zel**-a: finchlet; pas-ser-**eye**-na: sparrow—). 4¾"; brown back with dark streaking, gray nape and underside; red-brown crown; whitish eyebrow streak above dark eye streak; song is 3–4 seconds of very rapid chips. Mar–Oct; edge (and urban) habitats; A.

Song Sparrow

Melospiza melodia (mel-o-**spy**-za: song finch). 5½"; brown with blackish streaking above, white below with brown streaking convergent at throat, above a mid-breast brown spot; pumps its tail in flight; 4-second song starts with 2–3 emphatic notes, then varied series of buzzes and warbles. Streamsides, cattails; year-round; A.

Gray-headed form of dark-eyed junco

Song sparrow

These three sparrows are widespread across the continent. Chipping sparrows nest in conifers, forage in brush, and consequently spend their time where open forest meets brush—that is, edge habitats. They multiplied in number in the nineteenth century, as farmers clearing forests created millions of new edges. They declined in the twentieth century, partly due to competition from introduced house sparrows and to nest parasitism: brown-headed cowbirds lay eggs in smaller birds' nests. When host parents try to feed all the ensuing chicks, the bigger cowbird chicks get most of the food and are soon strong enough to push the true offspring out of the nest.

DARK-EYED JUNCO

Junco hyemalis (**junk**-oh: rush, the plant, for no clear reason; hi-em-**ay**-lis: of winter). 5¼"; tail dark gray-brown exc for white side feathers;

Red-winged blackbird female

Yellow-headed blackbird

Bullock's oriole

young. After the young fledge, juncos travel in loose flocks until the next summer. Ours mostly migrate short distances downslope for winter, or sometimes a few hundred miles south.

RED-WINGED BLACKBIRD
Agelaius phoeniceus (ad-gel-**eye**-us: flocking: fee-**niss**-i-us: reddish-purple). 7¼"; ♂ all black exc for erectile red shoulder-patches with yellow edge; ♀ gray-brown with a touch of red on shoulder; song "BONK-ra-leee." Nest in wet habitat; at other times wander widely, e.g., in grain fields; often in large flocks; A. Icteridae.

Red-winged males sing their gurgling-brook songs and flash their red epaulets to defend territory. Conversely, trespassing or submissive males cover their epaulets, and in an experiment, blackening their shoulders left them with no chutzpah whatsoever. While males pay attention to epaulets in divying prime territory, females pay attention mainly to territory. The male with the best territory will find that a harem of ten or fifteen has moved in with him—but he will not sire all the young. The very best marshy habitat with cattails or bulrushes acquires blackbird densities high enough to describe as a colony.

YELLOW-HEADED BLACKBIRD
Xanthocephalus xanthocephalus (zanth-o-**sef**-a-lus: yellow head). 8½"; ♂ black with bright-yellow head, neck, and breast; white wing patches shown in flight; ♀ gray-brown with dull-yellow face and breast; raspy notes. Wetlands; Apr–Oct; A. Icteridae.

Yellow-headed and red-winged blackbirds share some marshes, partitioning the resources. This one builds nests on cattails; it feeds on aquatic insects in summer and switches to seeds (many of them from crops) the rest of the year. Female and male flocks migrate separately.

BULLOCK'S ORIOLE
Icterus bullockii (**ick**-ter-us: Greek name; **bull**-oc-ee-eye: for William Bullock, Sr. and Jr.). 7";

otherwise, subspecies vary: "gray-headed" all medium gray (paler beneath) exc rust patch across upper back; "pink-sided" medium-gray head distinct from pale rust-colored body; "Montana" has dark-gray head distinct from whitish belly, rust flanks, and red-brown back; in all, ♀ tends to be a little duller or paler than ♂; simple, hard trill. Nests and forages on ground; A. Passerellidae.

Juncos are seed-eaters who turn to insects in summer, feeding insects and larvae to their

Western meadowlark

Yellow-rumped warbler

Yellow warbler

♂ bright orange to yellow exc for black cap, chin, back, and wings (with white bars); ♀ gray-brown suffused with dull yellow around head and tail tip; charming song of whistles and slides; sharp chip calls. Cottonwoods, other broadleaf trees; May–Sep; A. Icteridae.

This flashy songbird seems to tolerate heat well, and may even expand its summer range in a warming climate. It eats sizable insects such as grasshoppers and moths, and larvae including tent caterpillars. It also has a sweet tooth for nectar and fruit.

WESTERN MEADOWLARK

Sturnella neglecta (stir-**nel**-a: little starling; neg-**lec**-ta: overlooked). 8½"; streaky light-brown back; yellow belly and throat separated by heavy black chest V; long, pointed bill; striking song: pure flute notes, then a watery burble. On ground or low perches; grasslands up to alpine meadows; A. Icteridae.

Eastern and western meadowlarks look alike, and were not officially separated as species until 1908. Where they overlap, a male can't tell the species of another male until females arrive and males begin to sing. Their songs attract conspecific females only, and that is solely responsible for the reproductive isolation (and the existence) of two species.

Males have an average repertoire of seven songs, and can learn new ones in adulthood. A male with a large repertoire typically gets two mates. Females build the nests in and of grass, with elaboration ranging up to arched roofs and long, roofed entrance runways.

Charmed by its song, humans proclaimed the western meadowlark the state bird of which state: Kansas? Oregon? Wyoming? North Dakota? Montana? Nebraska? All six.

WARBLER
4–4¾"; slender bill. Parulidae.
Yellow-rumped Warbler
Setophaga coronata (see-toe-**fay**-ga: moth eater; cor-o-**nay**-ta: crowned). Formerly Audubon's and myrtle warblers; *Dendroica coronata*. 4¾"; yellow in small patches (mere tinges on ♀) on crown, rump, throat, and sides; mostly gray/black (breeding ♂) to soft gray-brown (others), with white eye ring and 1 vague wing bar; song is 2 seconds of "chwich-wich," rising. Conifers, aspen; abundant, Apr–Oct; A.
Yellow Warbler
S. petechia (pet-**eek**-ia: red-dotted). Formerly *Dendroica petechia*. 4"; yellow ± overall; wings and tail gray with some yellow; ♂ breast

Townsend's warbler

red-streaked; song a rapid string of "sweets." Willows, forest edges; May–Sep; A.

Townsend's Warbler

S. townsendi (**town**-send-eye: after J. K. Townsend, p. 312). Formerly *Dendroica townsendi.* 4¼"; whitish beneath, greenish-gray above, with 2 white wing bars; crown black; sides of face (exc dark cheek patch) and breast bright yellow; song 3 "sweezes," then 2 higher, louder ones. Mid-elev to high forest; breeds OR to nw MT; migrates through all states.

Wilson's Warbler

Cardellina pusilla (car-del-**ee**-na: Italian name for a small goldfinch; pew-**sil**-a: very small). Formerly *Wilsonia pusilla.* 4¼"; yellow beneath, olive above; ♂ has shiny, black crown; tail entirely dark; song a burst of around 7 "cheeps." Brushy riparian areas; May–Oct; A.

MacGillvray's Warbler

Oporornis tolmiei (op-or-**or**-nis: autumn bird; **tole**-me-eye: for William Fraser Tolmie). 4½"; yellow beneath, grayish-olive above; solid gray "hood" (head and throat) with broken white eye ring; short song of about 3 rising and 2 falling notes. Thickets, burns, clearcuts; May–Sep; A.

Common Yellowthroat

Geothlypis trichas (gee-oth-**lip**-iss: ground finch; **try**-cus: thrush—both names misleading). 4¼"; olive-brown with yellow throat and breast, ♂ has black mask; duller ♀ has gray cheeks; song "waCHEEry-waCHEEry." Brush, esp wetland; May–early Oct; A.

Wilson's warbler

Common yellowthroat

WARBLERS ARE KNOWN for long winter migrations to the tropics. Birders concentrate on learning the songs, since there are so many kinds of warblers and they all tend to keep themselves inconspicuous among foliage. Most have "chip-chip" calls, which often serve as alarms, or to warn another male that aggressive defense of territory will come next. They glean insects and may also hawk at larger insects, and some vary their diets with fruits and seeds. Wilson's warbler is a "leapfrog" migrator: individuals that breed farther north winter farther south, in Central America. This was demonstrated by

Lazuli bunting

Western tanager

a latitude-correlated ratio of hydrogen ions in their feathers.

LAZULI BUNTING

Passerina amoena (pas-ser-**eye**-na: sparrow—; a-**me**-na: delightful). 4½"; breeding ♂ turquoise with russet breast, white belly, white bars on blackish wings; blue parts speckled with brown during offseason; ♀ unstreaked brown above, brown-streaked white beneath; various sweet, twittering songs lasting 2–4 secs. Brushlands incl burns, sagebrush, juniper; late May–Aug. Cardinalidae.

A yearling male lazuli bunting assembles his song out of fragments he hears from nearby males. After a few days of playing around with it, his song crystallizes and he sings that one song for the rest of his life. It's unique to him, but the neighborhood develops kind of a dialect because males there share most of the same syllables, among the 122 different syllables that have been described for the species.

WESTERN TANAGER

Piranga ludoviciana (pir-**ang**-a: Tupi name from Brazil; loo-do-vis-i-**ay**-na: of Louisiana). 6¼"; summer ♂ have bright-red to orange head, yellow breast, belly, and rump; black back band, tail, and wings; white wing bars; others yellowish- to greenish-gray above, yellow beneath; short, robinlike song. Open forest; late May–Sep; A. Cardinalidae.

Lewis and Clark described many new plant species, but only four new birds: Lewis's woodpecker, Clark's nutcracker, the tundra swan, and the western tanager, whose scientific name refers to the Louisiana Purchase, the tract they explored. "Tanager" and "Piranga" are native words from deep in the Amazon Rainforest, where some tanagers winter. This particular species travels only as far as Central America. Here, in breeding plumage, males look like gaudy jungle birds, but in the jungle they wear drab winter coats.

Tanagers switch from an insectivorous diet to one of ripe berries in late summer. Their beaks are intermediate between the insect-picking thin type and the heavy, seed-crushing beaks of some finches.

12

Reptiles

We may say "reptiles and amphibians" in one breath, or lump them together as "herps," but they are very different—and probably aren't each other's closest relatives. Early amphibians pioneered onto land with the use of two innovations: air-breathing organs (lungs or skin) and legs. Certain lunged amphibian ancestors later developed two further innovations—watertight skin and sturdy eggshells—and could exploit terrain farther from water. That evolutionary step eventually led to birds, mammals, and the groups traditionally called reptiles.

Evolution turned egg-layers into live-bearers in dozens of separate events over the eons. Boas, though they are one of the more primitive snake families, bear live young. Meanwhile *Sceloporus* lizards include both egg layers and live-bearers; the southerly ones (ours) lay eggs. Eggs predominate in the tropics, and births in cooler climates, suggesting that carrying embryos around in her body helps a mother keep them warm.

The traditional class Reptilia comprises tortoises, crocodilians, tuataras, snakes, and lizards. Taxonomists have agreed for decades that those groups originated separately and are not as closely related to each other as crocodiles are to birds. So Reptilia ought to be either enlarged to include birds, or broken up. One solution is to keep only snakes and lizards in class Reptilia, and put turtles and crocodilians each in their own new class. Colorado, having named the painted turtle its state reptile, would have to revise that to stay taxonomically correct. Much of the profession, so far, chooses to side with Colorado, and against taxonomic perfectionism.

Suborder Sauria (Lizards)

Lizards are generally well adapted to hot, dry climates. Our mountains are therefore lizard-rich and amphibian-poor. Nevertheless, some studies find that lizard populations are in sharp decline due to warming. They share a climate adaptation with pikas and marmots: they avoid the heat of the day by retreating underground, but when the heat of the day gets longer, the time available for foraging gets shorter, and they are underfed. This doesn't kill lizards, but they may fail to reproduce, and then die out locally. Because arid lands are relatively unproductive, warm-blooded thermoregulation is something of a handicap for mammals there, demanding frequent feeding. Cold-blooded lizards (and snakes), in contrast, can go a very long time between meals, burning few calories. They can wait two years or more and then reproduce when a richer season comes along.

In cold weather, our reptiles retreat into crevices. Herpetologists call this hibernating, though it doesn't entail a sharp drop in metabolic rate, as hibernation does in mammals. They can easily crawl out on warm days in any month of the year.

SAGEBRUSH LIZARD

Sceloporus graciosus (skel-**op**-or-us: leg pores; grass-ee-**oh**-sus: agreeable). 5–6¼" long; belly relatively wide; very long toes on hind legs; scales abrasive to the touch; back gray-brown to greenish, striped lengthwise; belly pale; ♂ develop long blue patches on each side of belly, and blue mottling under chin; ♀ in Apr–May may develop red-orange on flanks. Juniper or shrubland; A, limited to w slope in CO and NM, but to e slope in MT. Iguanidae.

Sagebrush lizard

For courtship, territory assertion, and other communication, Iguanid species each have a precise gestural language including pushups, head-bobbing, teeth-baring, throat-puffing, and side-flattening. Contact with you may alarm one enough to provoke these gestures, or a quick scurry under a bush. Sagebrush lizards also climb shrubs or rocks as an escape. They seek out crevices or rodent burrows for keeping cool. At Yellowstone, they seek out hydrothermally heated areas, apparently occupying higher-elevation habitat than they would be able to without the heat.

Plateau fence lizard

PLATEAU FENCE LIZARD

Sceloporus tristichus (skel-**op**-or-us: leg pores; **tris**-tic-us: 3 rows). Also fence swift, spiny lizard, eastern fence lizard; *S. ustulatus.* 5–6¼" long; similar to sagebrush lizard except spiny scales on back are more pronounced (prickly) and the blue on throat is 2 distinct side patches rather than mottling. Canyons; UT, NM, and CO w or s of the Front Range. Iguanidae. Similar western fence lizard, *S. occidentalis,* is in e OR and w ID.

Collared lizard

A lizard like this in the Laramie Range or the Colorado Front Range would be a prairie lizard, *S. consobrinus.* The two species were both included in *S. ustulatus,* the eastern fence lizard, until a 2002 molecular study divided that species four ways without being able to describe a single visible characteristic to separate the new species. The genetic lines of descent did not correlate with the several morphology-based subspecies that had been recognized for decades. This meant that all the records of where different subspecies live were as useless as the physical descriptions that enabled telling the subspecies apart, so defining the new species distributions will take a lot of time and lab work.

COLLARED LIZARD

Crotaphytus collaris (cro-ta-**fie**-tus: on the temples). Variable but often very colorful lizard up to 14" long; 2 black collars around back of neck; tan to pale olive overall, but sometimes extensively turquoise, green, or strong blue toward

Plateau striped whiptail

Western terrestrial garter snake

the tail; head sometimes yellow. Rocky places in UT, CO, NM. Iguanidae.

Watch out—this showy lizard bites! Without venom, fortunately. When fleeing, it may run on its hind legs like a tiny *Tyrannosaurus.*

PLATEAU STRIPED WHIPTAIL

Aspidoscelis velox (as-pid-a-**skel**-iss: wearing shin guards; **vee**-lox: swift). Also *Cnemidophorus velox*. 8–10¾" long; dark brown to black with 6 or 7 full-length cream-yellow stripes; tail grades to blue, brightest on juveniles; scales tiny and smooth on the body, raised and prickly on the tail. Low elevs in NM, se UT, and w CO. Teiidae.

This species of quick, active, beautiful lizard has no males. Hatchlings, in August, are identical triplets to quintuplets, all clones of their mother. The species originated from a single mother, but over time small mutations divided it into at least six genetic lines. The genus includes other all-female species as well as species with both sexes, such as the tiger whiptail, *A. tigris,* which ranges from eastern Oregon to a corner of New Mexico.

Suborder Serpentes (Snakes)

Reptilian scales overlap rearward, so that a snake easily slides forward, but not at all easily backward, thus enabling slithering: the entire body slides forward, following the serpentine path that a still photo captures. This lateral-undulation mode is the norm for snakes on land; they continuously push off against bumps on the ground, and also

briefly lift portions of their bellies that are leaping forward.

Undulation takes just as much energy as a lizard takes to walk on four legs, so we can rule out energy savings as the reason snakes lost their legs in evolution. (It remains a mystery.) A completely different locomotion mode, sidewinding, is more efficient, and yet only one kind of snake (a rattler called a sidewinder) specializes in it; several other species can do it, but save it for crossing surfaces too smooth for lateral undulation. Then there's rectilinear slithering, in which snakes creep straight forward by alternately flexing and relaxing tiny muscles just under their scales, in waves that travel along the body. Very slow, and seven times costlier in energy used per inch traveled, it is used only for sneaking up on prey, and only by large snakes (including our largest, the gopher snake). There are still more modes of slithering for swimming, for climbing trees, and for passing through narrow spaces.

GARTER SNAKES

Thamnophis spp. (tham-**no**-fis: shrub snake). 16–46" long; variable in color, but generally with 3 yellow-buff stripes down back, and ± distinctly spotted between the stripes. Often in or near marshes; also meadows, forest. Colubridae.

Blackneck garter snake

Northern rubber boa

Western Terrestrial Garter Snake

T. elegans. Background color most often gray-brown or (esp in nw) black; spots ± dull, often indistinct. A.

Blackneck Garter Snake

T. cyrtopsis (sir-**top**-sis: round—). Darker and more dramatically striped; top of head gray; top of neck black, split by a single yellow stripe that extends the length of the back. NM and s CO.

Common Garter Snake

T. sirtalis (sir-**tay**-lis: garter). Flanks above the lateral yellow stripes regularly spotted or checkered with red. Northerly, plus foothill elevs southerly.

GARTER SNAKES LIKE to bask in the sun, sometimes intertwined in groups. They give birth in summer to three to fifteen live, worm-sized snakelets. In our region, they generally hibernate. They are partially aquatic, even the so-called western terrestrial.

While snakes of this family are often called nonvenomous, many of them, including the familiar garter, actually employ neurotoxins in their saliva to help subdue prey. They're rarely dangerous to large mammals like us because the toxins are pretty weak and are not injected through hollow fangs. However, that does not mean you can pick up a garter snake and get off scot-free—you'll have stinky hands coated with snake musk and feces.

NORTHERN RUBBER BOA

Charina bottae (care-**eye**-na: graceful; **bo**-tee: for Paolo Botta). 18–27"; unpatterned dark greenish brown above, grading to ± yellowish belly; scales small, smooth, shiny; skin rather loose; tail not tapered, almost as broadly rounded as the head. Forests; WA to UT and nw WY. Boidae.

This smaller cousin of tropical boa constrictors is quite an odd fish in a Northern Rockies context. Our other reptiles prefer sunny habitats where they can bask, get good and hot, and then be active. The boa avoids hot sun like the plague. It lives only where there's ample shade along with a little sun, and tends to be active at dawn, dusk, night, and on cloudy days. It tolerates higher elevations—up to 8,600 feet at Yellowstone. But calling it "active" is a stretch. This is one sluggish snake. Somehow it manages to catch small mammals, mainly the young in their nests. It kills by coiling and squeezing, like its relatives the tropical constrictors. This works by stopping the heart, not the breath, and is a quick death. With one part of its length breathing while another part squeezes, a boa's own breath is uninterrupted: its left lung extends about a fourth of its body length. Lacking a diaphragm, snakes power their breath with groups of individual ribs.

The rubber boa's only defenses are a foul smell and the use of its blunt tail as a fake-head decoy, or as a club to fend off a mother mouse while devouring her young. You can pick a boa up and handle it safely—just expect to wash afterward.

RACER

Coluber constrictor (**col**-ub-er: Roman term for some snake). Our ssp. is the western yellowbellied racer. 20–70" long; brown to olive above, yellowish beneath; adults plain, but juveniles (up to 2 or 3 years old) patterned with brown

Racer

Milksnake

Smooth green snake

MILKSNAKE

Lampropeltis gentilis (lam-pro-**pel**-tiss: shiny shields; gen-**tie**-liss: aristocratic). Also *L. triangulum*. 18–48" long; entire upper side brightly banded in the sequence red-black-white-black, with the red bands widest, except sometimes substituting orange for red, or pale yellow for white. Foothills, and up to 9,000' in NM; NM n to e foothills of MT Rockies. Colubridae.

The beautiful color patterns on milksnakes are much-debated as an example of Batesian mimicry, which posits that mimics get a wide berth from other animals when they resemble a dangerous species—in this case the highly venomous coral snake. Various mnemonic rhymes offer to help you distinguish milk from coral, but we don't need them here since the nearest records of coral snakes are from semideserts southwest of Albuquerque. In theory, Batesian mimics should live only in places where their models live, but milksnakes in the Rockies are well outside coral snake range.

A study using rubber replicas of snakes found that predators unfamiliar with coral snakes attacked rubber milksnakes more than they attacked dull, inconspicuous rubber snakes. But would that hold true for real snakes? Another hypothesis, called flicker-fusion, holds that milksnakes make themselves disappear by sprinting forward so fast that the bands blur—at least to a hawk's eyes in dusky light.

SMOOTH GREEN SNAKE

Opheodrys vernalis (o-**fee**-a-driss: tree snake; ver-**nay**-liss: of spring). Up to 26" long, slender, bright green with a narrow cream belly. Marshes, moist grassland or grassy woods; southerly. Colubridae.

This attractive, modest-sized snake can be approached without risk (other than bad smells). It hibernates in large groups, and lives on insect prey, exploring by flicking its black-tipped red tongue around to catch scents.

for all or part of their length. Foothills; A. Colubridae.

This agile, slender snake is true to its name "racer," but not to its name *constrictor* since it kills its prey—mostly insects and some small rodents and reptiles—with its mouth, not with its coils. Though nonvenomous, it may bite and thrash when captured. Clutches of four to ten eggs hatch in late summer.

Great Basin gopher snake

Prairie rattlesnake

GOPHER SNAKE AND BULLSNAKE

Pituophis catenifer (pit-you-**oh**-fis: pine snake; ca-**ten**-if-er: chain-bearer, referring to the pattern). 3–6' long, heavy-bodied, relatively small-headed; tan with blackish blotches in a checkerlike pattern, largest ones along the back. Colubridae.

Great Basin Gopher Snake

P. catenifer ssp. *deserticola* (dez-er-**tic**-a-la: desert dweller). WA, OR, ID, UT, Snake River basin in WY, w CO.

Bullsnake

P. catenifer ssp. *sayi* (**say**-eye: for Thomas Say, p. 95). Nose looks upturned and pointed due to 1 enlarged scale; the tan background color may be tinged reddish to yellowish. MT, and e of CD s.

THE ONLY SNAKE you would likely mistake for a rattler around here is the gopher snake. The resemblance may be another case of mimicry for fright value; it even includes a threatening vibration of the tail that, when it rustles dry leaves, can sound much like a rattler. This snake is nonvenomous, killing its prey, including gophers, by constriction. It can also climb trees and prey on nestling birds. Constricting characterizes one small branch of the huge family Colubridae, which contains some 68 percent of all snakes.

The young are already many inches long when they hatch from eggs in moist, often communal burrows.

RATTLESNAKE

Crotalus spp. (**crot**-a-lus: rattle). Adults usually 16–36" long, or rarely up to 5', heavily built, with large, triangular head; tail terminates in rattle segments; notoriously variable brown, black, gray, and tan colors in a pattern of regularly spaced, large, ± geometric dark blotches against paler crossbars. Dry, rocky low to mid elevs. Viperidae.

Prairie Rattlesnake

Crotalus viridis (**veer**-id-iss: green, a misleading name for this species, which ranges at most to a greenish-yellow background shade). MT, WY, CO, NM, ec ID, se UT.

Pacific Rattlesnake

C. oreganus. OR, WA, ID, c UT.

RATTLESNAKES HUNT AT dusk, dawn, and night. A brisk, dry buzz of the tail rattle, sounding a bit like a cicada, is their typical warning to large intruders. Far from being aggressive toward humans, they will usually

Pacific rattlesnake

Painted turtle

to a subfamily, the pit vipers, that share this sixth sense.

Two other rattlers may be encountered near the hot fringes of our range. The western diamondback rattlesnake, *C. atrox,* lives in New Mexico and runs larger than the prairie rattler. The midget faded rattlesnake, *C. oreganus concolor,* lives on sun-drenched rocks of the Green River basin in Utah, western Colorado, and southwest Wyoming. True to its name, it is pale and on the small side, but its venom is more potent than that of its relatives.

Order Testudines (Turtles)

A bony shell—developed originally from ribs—distinguishes all turtles, tortoises, and terrapins. (Those are not taxonomic groups; they're just vernacular words with no clear distinction.) The entire order Testudines holds fewer than four hundred species.

PAINTED TURTLE

Chrysemys picta (**cris**-em-iss: golden turtle; **pic**-ta: painted). 4–10"; head, neck, and legs dark green with bright-yellow stripes; underside, side flesh (between shell halves), and rim of upper shell marked with bright red, sometimes fading to orange. Slow-moving water at elevs usually below 3,500'; A. Emydidae.

The herpetologist of the Hayden Survey of 1871 reported collecting a painted turtle at Yellowstone Lake, but there may have been a cataloguing error. In all the years since, no breeding populations of turtles have been found anywhere near that high, anywhere in our range. (Abandoned turtle pets are reported from time to time.)

The Southern Rockies have several native turtles, but the Northern Rockies have only this one, an aquatic omnivore. It spends most of its life in the water, even while asleep, but goes ashore to lay eggs and perhaps to explore for a new body of water.

flee if given the chance. This species can inject a dose of venom lethal to small children and many mammals, though rarely lethal to adult humans—just excruciating. (A solitary victim might possibly be incapacitated long enough to die of hypothermia if warm clothes and food are far away.)

The strike itself is stunningly quick: starting from any of several positions (not necessarily coiled), the rattler takes half a second to strike and pull back. The main advantage of this speed is that rodents are hardly ever quick enough to bite the snake's head. There's no lack of precision, but there are disadvantages: the quick strike limits the amount of venom delivered, sometimes allowing prey to get away, and it risks damaging the jaw or fangs if the impact is too hard. High-speed films show the snake taking care to have its head already decelerating before impact, while its "neck" is still accelerating, all within that quarter-second time frame.

Reptilian eyesight is generally sharp, but poor in low light. Rattlesnakes, however, can hunt warm-blooded prey in pitch darkness thanks to a pair of pits near their eyes, sensory organs exquisitely sensitive to infrared radiation. Rattlesnakes belong

13

Amphibians

Though often listed among the terrestrial vertebrates, amphibians barely qualify: they lack an effective moisture barrier in either their skins or their eggs, so to avoid lethal drying they must return to water frequently, venturing from it mainly at night or in the shade, and rarely far. Many of them hatch in water as gilled, water-breathing, legless, swimming larvae (e.g., tadpoles); later metamorphose into terrestrial adults; and return to water to breed—hence the word *amphibious,* from "life on both sides." Some individuals fail to make the move onto land, but reach full size and sexual maturity while still in aquatic larval form. One entire family of salamanders has neither gills nor lungs, but breathes instead through their very thin skin; they may be the most truly terrestrial amphibians.

Among vertebrate animals, mammals and birds are "warm-blooded," whereas reptiles, amphibians, and fish are "cold-blooded." This doesn't mean they're self-refrigerating, but they're never a lot warmer than their environment. Ambient heat definitely helps them be active, yet they can sustain activity in astonishing cold—long-toed salamanders in our high country typically breed in sub-40°F water with winter's ice still on it (after hibernating through most of the freezing season). At the other extreme, amphibians rarely survive heat over 100°F. Ironically, their intolerance of heat won salamanders and newts a superstitious reputation as fireproof. They do know how to survive a ground

fire: they take refuge in a familiar wet crevice or burrow, just as they do from the midday sun. Most populations hold up quite well after forest fires. When Mount St. Helens blew up, it was mid-May; amphibians that were underwater at that season survived and carried on in the blowdown zone, while terrestrial amphibians died and were slow to return.

GLOBAL AMPHIBIAN DECLINE

Amphibians are in trouble worldwide. A majority of amphibian populations are thought to have fallen severely, and several species have gone extinct—way more than their share, compared to other taxonomic groups. There's a tangled swarm of interacting issues, varying by region and species.

Why amphibians? Their permeable skin gives them little protection from pollutants, pathogens, and ultraviolet radiation. And they seem slow to adapt.

And why now? Many small bodies of water that amphibians love are drying up in the warming climate. But climate may be less of a factor than other anthropogenic processes, including habitat loss such as draining of wetlands, introduction of predatory fish to high lakes, water pollution, increased ultraviolet radiation due to ozone depletion, and the transport around the world of pathogenic amphibian chytrid fungi.

The chytrid we have so far is *Batrachochytrium dendrobatidis,* or Bd for short. Starting in the 1970s, it attacked our spotted frogs and toads, especially in the south, extirpating the western toad from New Mexico and leaving it endangered in Colorado. Our salamanders remain almost unscathed, but there's a second *Batrachochytrium* species especially for salamanders. Bsal for short, it decimated many European salamander species in the early 2000s after reaching Europe from Asia via the exotic salamander trade. The same path seems likely to bring it to our shores, with unpredictable results. Lab studies find that our salamander species produce skin secretions that make them resistant or tolerant to either Bd or Bsal, varying from one salamander species to the next.

Bd and Bsal are very odd: they're in a phylum with only one other species. We don't know whether they are a new disease, or one that was endemic in our amphibians for a long time but to which they abruptly became more susceptible. Several studies find that climate warming makes amphibians more susceptible to them. On the hopeful side, while Bd has driven several amphibian species to extinction, others have declined but survived and even increased, acquiring resistance to the disease. Some toads in the Rockies survive and then, if reinfected, show lighter symptoms.

Order Caudata (Salamanders)

In high mountain lakes that did not naturally acquire fish after the ice ages, salamanders were the top predators until they were usurped by stocked trout. Finding themselves prey for the first time, the salamanders became scarce and secretive, but generally survived. They reoccupy the lakes quickly if the fish are removed. To restore the natural ecology, some national parks do in fact kill off fish from some high lakes.

WESTERN TIGER SALAMANDER

Ambystoma mavortium (am-**bis**-ta-ma: to cram into the mouth; ma-**vor**-tee-um: like the god of

Long-toed salamander

Western tiger salamander

war). Also *A. tigrinum*. Very wide, blunt head; dirty yellow to olive, marbled or barred with black (photo features the barred species); c CO and e; markings on the blotched (ID, MT, WY, n CO), our largest salamander, are smaller and less dramatic; the clouded (UT, w CO) is dark and murky. A. Ambystomatidae.

Adults occur in four morphs, permutations of gilled versus terrestrial and "typical" versus cannibalistic. The latter have broader heads and an extra row of teeth, and, sure enough, they count tiger salamander larvae as a major part of their diet. Individuals can eat larvae nearly their own size.

LONG-TOED SALAMANDER

Ambystoma macrodactylum (macro-**dac**-til-um: big toes). 4–6" long; wide, blunt head; long legs; dark gray-brown with an irregular, often blotchy, full-length back stripe bright yellow to tan. ID, w MT.

Like other amphibians, this one stays close to water, especially in arid terrain. In the high country, it breeds even before the ice is gone, to make the most of its brief, frigid active season, and still the larvae need two summers before metamorphosing.

Ambystoma and Plethodon belong to two families with elaborate mating rituals ending, in many species, with a procession: he walks along dropping gelatinous sperm cases, while she follows, picking them up with her cloacal lips for internal fertilization. A long-toed male may literally interlope: he slips in between the romantic duo, mimics a female walk to escape notice, and places his sperm cases atop the first male's, assuring himself of paternity.

COEUR D'ALENE SALAMANDER

Plethodon idahoensis (**pleth**-o-don: full of teeth). Also *P. vandykei*. 3–4½" long, slender; blackish except for a nearly full-length narrow back stripe that may be yellow, red, orange, or green. n ID, Cabinet Mtns of MT; rare and little studied. Plethodontidae.

Most salamanders have a larval stage with intricate, plumelike, red (blood-filled) structures on the sides of their heads. These are external gills that "breathe" or absorb oxygen suspended in water. But not in this family, the Lungless Salamanders: they have neither lungs nor gills, and breathe through their skins exclusively. Most hatch as miniature adults, having breezed through the larval stage within the egg; they are completely terrestrial. This species spends much time underground—for example, under logs or stones in wet talus—venturing out mainly at night and when rain or a creek splash zone assure it of staying wet. It gleans insect larvae from wet rocks.

Order Anura (Frogs and Toads)

Toads are distinguished from frogs by their warty skin, toothless upper jaw, sluggish movement (generally walking rather than hopping), and parotoid glands behind the eyes. These bulbous protrusions exude a thick, white poison related to digitalin, effec-

Pacific treefrog

Boreal chorus frog

tively deterring predators. (It may nauseate and burn, but does not cause warts.)

CHORUS FROG

Pseudacris spp. (sue-**day**-cris: false cricket-frog). Small frogs with variable coloring; ♂ throat inflates when calling, and at other times looks like a dark, loose skin flap. Hylidae.

Pacific Treefrog

P. regilla (ra-**jil**-a: queenlet). Also *Hyla regilla*. 1–2¾" long; skin bumpy; toes bulbous-tipped; color extremely variable, and may include green or red; in n RM, usually gray-brown with large, irregular black splotches; ♂ throats gray. n ID, w MT.

Boreal Chorus Frog

P. maculata (mac-you-**lay**-ta: spotted). Also *P. triseriata, Hyla triseriata*. ¾–1½" long; gray-brown to green, with generally 3 dark stripes down back, though these may be broken or reduced to spots; pale lip line; toes scarcely bulbous. Mainly near and e of CD.

Western toad

Columbia spotted frog

Western Toad

A. boreas (**bor**-ius: northern). Also boreal toad; *Bufo boreas.* The 2 glands bulging behind the eyes are oval; voice weak, peeping. Lakes, streams, and forests; now northerly, as its numbers have collapsed in NM, CO, and s WY. Bufonidae.

Woodhouse's Toad

A. woodhousii (for Samuel Woodhouse). Also *Bufo woodhousii.* Pale, somewhat yellowish; the 2 glands bulging behind the eyes are elongated; croak is loud, nasal—like a sheep with a cold, bleating. Largely nocturnal, in and along waterways in arid country; southerly.

TOADS' SLOW PACE limits them to slow, creeping invertebrate prey.

Toads resist drying; they can get away from water better than most amphibians. The western toad inhabits animal burrows and rock crevices, and is often seen in mountain meadows and woodlands. It is less strictly nocturnal than most toads, especially at elevations where nights are too cold for much toad activity.

COLUMBIA SPOTTED FROG

Rana luteiventris (**ray**-na: Roman name; loo-tee-ih-**ven**-tris: yellow belly). Also *R. pretiosa.* 2½–3½" long; brown with poorly defined blackish spots above; ridges on back ± same color; belly yellow to salmon (or red in some outlying populations); toes fully webbed. Near, or most often in, cold lakes or streams; n and w from Tetons. Ranidae.

Spotted frogs range higher in the Rockies than other frogs, but are considerably slowed down (compared to conspecifics at sea level) at higher elevations, taking four to six years to mature, and then laying eggs only at two-to-three-year intervals. The male's deep croaks, in a series of six to nine, are modest in volume.

NORTHERN LEOPARD FROG

Lithobates pipiens (lith-a-**bay**-teez: rock walker; **pip**-ee-enz: peeping). Also *Rana pipiens.* 2¼–4"

VOICED FROGS EMPLOY a variety of calls, including alarm, warning, territorial, and male and female release calls. Pond-frog choruses include mating-call duets and male trios. The male treefrog amplifies his rather musical, high-pitched call, heard both day and night, with a resonating throat sac he blows up to three times the size of his head. Boreal chorus frogs call at night in raspy, toneless voices, beginning in early spring.

Treefrogs are distinguished by their bulbous toe pads, which offer amazing grip on vertical surfaces such as trees. Our species probably spends more time in water and on the ground than on shrubs and trees. It has a sticky tongue for catching insects.

TOAD

Anaxyrus spp. (annex-**eye**-rus: king—) Also *Bufo* spp. 2–5" long; thick; sluggish; skin has large bumps, the largest being 2 glands behind the eyes; olive to grayish, with a narrow, pale stripe down back, blotches on belly. Bufonidae.

Northern leopard frog

Rocky Mountain tailed frog

long; green or brown with light-haloed dark spots; back has conspicous pale ridge from each eye to groin, and also along jawline; toes fully webbed. In and near water; A. Ranidae.

Leopard frogs hunt with a quick dart of the tongue. Their croak is feeble, rough, and prolonged. Some call it a "snore" followed by "clucking" or "moaning." Gravid females become huge, and lay several thousand eggs.

They are in serious decline. Predation by introduced bullfrogs (*R. catesbeiana*) is a big problem for them, along with stream pollution and habitat alteration.

ROCKY MOUNTAIN TAILED FROG

Ascaphus montanus (**ask**-a-fus: lacking a spade—a body part). 1–2" long; skin has sparse, small warts; olive to dark brown with large, irregular black splotches and black eye stripe; voiceless; nocturnal. In undisturbed mtn streams; OR to nw MT. Ascaphidae (a family of just this 1 genus).

Tailed frogs spend most of their time in fast, cold creeks, where their little-seen but sizable populations are perfectly able to withstand trout, but not logging. They attach their eggs like strings of beads to the downstream side of rocks. The tadpoles suck firmly onto rocks, or perhaps to your leg or boot when you wade a creek. But don't worry; they aren't bloodsuckers. The adults don't have real tails; those soft protuberances are male cloacas, and they fertilize the females internally. You might think a penis prototype to be an advanced item on an amphibian, but it's a primitive trait that many other amphibians once had but lost. Tailed frogs may have kept it to enable mating in fast streams without the semen washing away. Their closest relatives are in New Zealand.

14

Fishes

When Euro-Americans arrived and began throwing new fish species into watersheds, the existing species distribution was still in flux from the last ice age. Where glaciers advanced, aquatic life retreated; when glaciers retreated, fish returned only as far as they could swim. Salmon may be phenomenal waterfall leapers, but still, there are limits. Most streams would, in nature, be fishless above some impassable waterfall. There are the odd exceptions. At Two Ocean Pass in the Absarokas, a stream splits, with one branch flowing down each side of the Continental Divide. Atlantic Creek has no high waterfalls, so fish were able to ascend it all the way to the pass, where cutthroat trout and dace—but not sculpins, for some reason—swam over the Divide into Pacific Creek and occupied Yellowstone Lake and its drainage, which until then was fishless above Yellowstone Falls.

Above the critical waterfall live aquatic communities whose animals got there on foot or on the wing—invertebrates, small mammals, and amphibians. Salamanders would have ruled many lake food chains back in the day. But most of these aquatic communities were augmented—since 1872 in Colorado—with trout arriving in saddlebags or buckets, or skydiving from airplanes. Nongame fish were sometimes stocked for the game fish to eat, or accidentally as escaped bait. Sometimes baitfish have multiplied so much that they threaten to outcompete trout; managers have poisoned entire lakes to prepare them for restocking. Amphibians almost disappear from high lakes when fish are added, and effects cascade down through the food chain.

Another fish-limiting factor in some high lakes, strange as it may sound, is high temperature. Trout may suffer or die of heat in water that could chill and "freeze" a person in short order. Since fish are "cold-blooded," their body temperatures drop with that of the water, but their health isn't at risk; only their activity and growth rates are reduced. Under ice for nine months, high country trout spend a very slow winter without danger or discomfort, but also without growing. In the smallest, highest bodies of water that can support trout, they never grow very big, maturing and spawning at three to six years of age while only 3–5 inches long and still displaying the parr marks typical of juvenile trout elsewhere.

Some lakes are too clean to make good fish habitat. The aquatic food pyramid rests upon algae, which in turn depend on minerals not present in rain or snow; water has to pick these up in its passage over or through the earth. Small drainage basins, high snowfall, barren or impermeable terrain, and rapid turnover of lake water sometimes combine to severely limit nutrients. In those lakes, stocked trout may die out unless replenished often.

Check a lakeshore near an inlet or outlet stream, and follow the stream to its first waterfall; if you find a shallow, gravelly spot in early summer, you may see a spawner busy swimming back and forth over it. A logjam at the outlet stream, or slabby shallows nearby, may be good places to spot fish at any time of year. Polarizing sunglasses (or a camera polarizer) can help you see fish. Trout don't feed all day—only when the insects are most active. When the lake is first ice-free, feeding may go on from midmorning to midday. By October, they may feed all afternoon. If luck brings you a calm feeding period after an extended blow, look for frenzied feeding where floating insects are concentrated against the downwind shore.

Most fishes in this chapter are omnivores, generally taking the biggest foods luck brings them that will fit down their throats. They eat plankton throughout life, straining it with their gill rakers. That's all they can handle as tiny "fry," but later they move up to insects and larvae, snails, worms, isopods, freshwater shrimp, amphibian larvae, fish eggs and fry, and finally adult fish or even birds. Any fish species is fair game, even their own.

Civilization Versus Fish

Why are so many salmonids threatened? Answers start with overfishing and the ubiquitous introduction of non-native fish. Beyond that, almost everything people have done that makes the West less natural makes life harder for fish.

- Silt from logging, construction, and livestock operations ruins the clean stream gravel salmonids need for spawning.
- Loss of shade makes small streams too warm.
- Logs that slow streams down and create pools and meanders, essential fish habitat, were removed.
- Seasonal fluctuation is aggravated by logging and by severe fires, because rain and snowmelt run off much faster, leading to floods in winter and spring and reduced, warmer streamflow in summer.
- Pollution comes from mines, factories, farms, sewage, and lawn and pavement runoff.
- Draining of marshes and beaver ponds eliminates habitat.
- Irrigation of farms diverts enough water to reduce some creeks to warm, muddy trickles in summer.
- Dams may be too high to have fish ladders: Grand Coulee Dam shut the door on an Idaho-sized patch of habitat. Electric turbines cut up smolts on the way down. Smolts die of bubbles that form in their blood due to pressure changes during the drop. Big reservoirs warm up in the sun, harbor non-native warm-water predators, and lack enough current to move smolts downstream on schedule.
- Hatchery programs began with high hopes of mitigating harm from overfishing and dams, but turned out to be problematic. Hatchery fish interbreed with wild fish and dilute the latter's finely tuned local genetic adaptations. (Montana does not stock its rivers with hatchery trout.)
- Though anglers are limited in how many fish they can remove, even catch-and-release fishing can kill fish when handling leaves them prone to skin lesions.
- And last but not least: warming.

CUTTHROAT TROUT

Oncorhynchus clarkii (onk-o-**rink**-us: swollen snout; **clark**-ee-eye: for William Clark, p. 241) Also *Salmo clarki*. Lower jaw bears a long, heavy red line (occasionally pink or orange); jaw longer than other trout, opening to well behind the eye; dorsal fin has 9–11 rays, pelvic fin usually 9, anal fin 9 (range 8–12); adults dark-spotted, with spots densest toward tail and above middle; various degrees of pink and red on sides and belly; gold to brown tones predominate, brightest in the Colorado subspecies and duller in the northerly ones. Salmonidae.

The seven surviving subspecies here are native to distinct drainages:

- **Colorado River cutthroat:** Colorado, San Juan, and Green river basin
- **Greenback cutthroat:** east of the Continental Divide in Colorado, now rare, and the subject of much taxonomic confusion
- **Rio Grande cutthroat:** Rio Grande and Pecos basins
- **Yellowstone cutthroat:** Yellowstone and Upper Snake basins
- **Snake River cutthroat:** small range on Upper Snake, including Jackson Hole
- **Westslope cutthroat:** Columbia and Missouri drainages in Montana, north and central Idaho, bits of Washington and Oregon
- **Bonneville cutthroat:** Utah closed basins of Sevier and Great Salt lakes

The cutthroat is the archetypal Rocky Mountain fish, the natural top fish of crystal-clear, icy-cold bodies of water in our range. Colorado, Wyoming, Montana, Utah, Idaho, and New Mexico each name it as their state fish. (Five of the six name a particular subspecies.) They do so under threat: its populations are diminished and precarious. Overfishing began early: until 1909, Yellowstone Lake had a commercial fishery for cutthroats to supply the park's hotels. Stocking of trout in the Rockies—beginning with individuals bringing brook trout by 1872—often favored introduced fish over cutthroat, partly because cutthroat was too easy to catch and considered not much of a fighter. But few ever complain about their beauty, flavor, or size—they can get larger than rainbows, browns, goldens, or brookies.

The introduced fish often outcompete cutthroats. Though they may be poor competitors against challenges they didn't evolve with, cutthroats are formidable predators, with big mouths and extra teeth on their tongues. Fish use teeth for gripping prey prior to swallowing it, not for chewing. (Teeth can be a useful ID character, but only for fish on their way to the frypan. Never grope around in a catch-and-release fish's mouth; it can be lethal to the fish.)

The greatest single drop in cutthroat numbers took the form of hybridization with rainbow trout. Hybrids occur in nearly all waters where stocked rainbows put the two species together. A few river systems had both species as natives, and in those rivers cutthroats kept themselves genetically separate by choosing colder water to spawn in. However, in rivers that never had rainbows before, they interbreed. The "cutbows" can look like either parent species; having one key character of the wrong species is a giveaway—tongue teeth or a red jawline on a rainbow, or white-bordered anal and pelvic fins, or a dark-spotted head, on a cutthroat.

Fishery agencies' efforts to rebuild cut-

Colorado cutthroat trout

The Lateral Line

Fish hear with their skin. Sensory neuromast cells are scattered over their bodies near the surface, but these are so sensitive that the noise of rushing water can overwhelm them. Therefore, while slow-water fish have a lot of these to hear with, fast-water fish like trout and sculpins have just a few distributed across their skin, and use them to detect and orient to the current.

For hearing, fast-water fish have a highly developed lateral line on each flank, a visible, full-body-length set of neuromasts buried in a sort of canal. These screen out the white noise, yet remain sensitive to non-constant splashes and ripples, like when an insect—or an angler's dry fly—hits riffled water.

throat numbers have accelerated. Most involve selecting headwater basins with barriers such as waterfalls to separate them from all downstream fish, then poisoning these basins to kill all their fish, and finally—after thoroughly checking that all the fish are gone—restocking with genetically pure hatchery cutthroats. (Ironically, that same barrier commonly meant that the basin above it was fishless two hundred years ago, so the cutthroats are not exactly being restored to former habitat.) The usual poison, rotenone, is put into the streams at all their sources, while being neutralized with a second chemical down near the barrier. This laborious and harsh method has succeeded in many cases, and there are no known preferable methods for sustaining cutthroats as a pure species, yet, unsurprisingly, groups from many points on the political spectrum have fired up opposition.

For spawning (and here I describe a generalized salmon-family behavior), the female chooses a gravelly spot, usually in a riffle, and begins digging a trough, or "redd," by turning on her side and beating with her tail. A dominant male moves in to attend to her, and becomes aggressive to outsiders, attacking any who come too close. He may nudge her repeatedly. When she signals her readiness by lowering a fin into the redd, he swims up alongside her and they open their jaws, arch their backs, quiver, and simultaneously drop her eggs and his milt. Other males hitherto kept at bay now dash in and try to eject some of their own milt onto the eggs. Hundreds to thousands of orange-red eggs (a thousand, on average, from a 14-inch cutthroat) come out tiny, then quickly swell, absorbing water along with the fertilizing milt. The female's last maternal act is to cover them with gravel as she extends her redd, digging the next trench immediately upstream from the last. Pacific salmon die soon after spawning, as do a majority of steelhead.

The eggs take several weeks to hatch; the colder the water, the longer they take. The newly hatched alevin remains in the gravel a few weeks more, still nutritionally dependent on the yolk sac suspended from its belly. As it depletes the yolk, it adapts to a diet of zooplankton and emerges from the gravel as a fry. In a year or two, they grow to parr size, displaying the parr marks that make them easier to identify.

On the coast there are anadromous runs of cutthroats, but none swim far enough upstream to reach our range.

The red "cuts" on a cutthroat's throat serve to emphasize gestures and displays. One reason for the wide variability in color is that trout alter their camouflage in response to colors they see around them—like chameleons, only much slower. Experimentally blinded trout eventually contrast sharply with their associates.

RAINBOW AND STEELHEAD TROUT

Oncorhyncus mykiss (onk-o-**rink**-us: swollen snout; **me**-kiss: Kamchatkan Indigenous name). Also redband trout and many other names for local races; *Salmo gairdneri*. Coloring extremely variable; flesh bright red to white; spawning adults usually have a red to pink streak (the "rainbow") full length on each side, much deeper on ♂; returning sea-run fish (steelhead) are silvery all over with guanine (a protein coating on all salmonids fresh from the sea), which obscures coloring underneath; dorsal-fin rays typically 11–12, pelvic-fin rays 10, anal-fin rays 10 (range 8–12); juveniles have a distinct row of 5–10 small, dark spots along the back straight

Whirling Disease

In the early 1990s, an introduced parasite of trout arrived and soon nearly wiped out the famous rainbow-trout fisheries of Montana's Madison River and several Colorado rivers.

Whirling disease can infect almost all of our salmonid species. However, different species, different rivers, different spawning areas, and different years show wide variation in mortality levels.

The beast in question, *Myxobolus cerebralis* (mix-**ob**-o-lus sar-ah-**bra**-liss: "slimy lump on the brain") is a myxozoan ("slime animal") with a complex, multi-stage, two-host life cycle. In trout, it infects cartilage and compresses spinal nerves, causing sluggishness, contorted tails, and blackened heads. Some victims swim in circles (whirl) after spinal damage curls their tails to one side. Many fish die before they reach that point, disabled from evading predators or from foraging effectively.

The decomposing victim releases myriad parasites in a resting-spore form that can survive for decades in the mud. These come back to life if eaten by a tiny, primitive worm named *Tubifex tubifex*. After growing in the worm's gut, they take form as larger triactinomyxons ("TAMs" for short). These were named as a species for years before they were identified as a life stage of *M. cerebralis*. They swim, attach to trout fry like grappling hooks, and fire shotgun-shell loads of sixty-four small spores each into their victims' bodies.

The worms abound in fertile, sediment-laden streams, which often result from streambank damage or agricultural runoff. Temperature is another key factor: fry that hatch in our coldest streams rarely get severe cases, and once they grow up and their cartilage turns to bone, they can live anywhere. Whirling disease has hit our rainbows hardest; cutthroats often choose those very cold streams to spawn in, and they're somewhat more resistant genetically.

Whirling disease may kill off nearly all the fish stocks most susceptible to it within susceptible areas, thus selecting for resistant stocks, which could eventually reoccupy most trout range. Some basins have already recovered significantly. A highly resistant strain of rainbow turned up in a trout farm in Germany, of all places, and was then hatchery-crossbred with Colorado River rainbows and stocked into rivers here. In the first few years, the results looked dicey, until a few generations of wild interbreeding and natural selection on the Upper Gunnison River refined the fish into a new strain that appears able to thrive in the wild here.

Rainbow trout

in front of the dorsal fin, plus 8–13 oval parr marks along the lateral line. Common in high lakes and streams. Salmonidae.

The West's best-known sport fish is native up and down the Pacific Slope from Mexico to the Alaska Peninsula and inland through most of the Columbia River system, though apparently not on the Snake above Hells Canyon. Today, the rainbow trout lives almost anywhere in the world with a temperate climate, both in fish farms and stocked in streams. Anglers love its leaping ability. In hatcheries, it grows faster and suffers less mortality than other trout, and in streams it thrives in diverse conditions. Climate-based projections predict it will continue to thrive in the Northern Rockies while other trout decline.

Yellowstone provides an extreme example of the latter: stocked rainbows in the Firehole River tolerate water temperatures up to the high eighties for brief periods, and this is hot-spring water heavily laden with minerals. Under these conditions, trout that can survive actually grow much more quickly than normal, but have shorter life expectancies.

Subspecies *aguabonita,* the golden trout, is native only to a small area in the southern Sierra Nevada, in California, but is so sought after by anglers that it gets stocked widely in the Rockies, especially in remote, high lakes without competing fish. The Wind Rivers are a prime destination.

After 150 years of stocking, much of it ungoverned and unrecorded, the genetics of rainbow trout in the West are irretrievably muddled. Not only are the different stocks of rainbow interblended, but rainbow-cutthroat hybrids are widespread.

The sea-run form of this species, known as steelhead, reaches our region in the Wallowas and central Idaho, though its numbers there are getting precariously low. These are summer-run steelhead; they spend fall and winter in streams, and spawn in early spring. Most swim to sea when two years old, feeding there for one to four salt years before their first spawning run. Some individuals return to sea and spawn again the following year. At sea, steelhead grow faster and reach larger sizes than their freshwater cousins. Steelhead do eat while running upstream, so they're more inclined to bite than salmon, and their flesh is in good condition. Big and strong, steelhead are notoriously hard to locate and hard to land. A favorite riffle for steelhead is a fiercely guarded secret.

BULL TROUT

Salvelinus confluentus (sal-vel-**eye**-nus: German name Latinized; con-flu-**en**-tus: of rivers). Formerly included in Dolly Varden trout, *S. malma.* Also bull char. Adult head and midsection typically as wide as they are tall; olive-greenish back and sides regularly pink- to yellow-dotted; juvenile parr marks wider than the light spaces between; dorsal fin unmarked, with 10 or 11 rays, anal fin usually with 9. Cold, clean lakes and streams; northerly. Salmonidae.

Commonly called trout, fish of this genus are more properly char, or charr. Mature charr generally have light spots against a darker background, whereas trout have the reverse. Charr are praised as crafty in Washington Department of Fish & Wildlife literature: "Log jams, cascades and falls that are barriers to the chinook's brute strength and the steelhead's acrobatic abilities may be only minor obstacles to the cunning and guile of [charr]. . . . Some go as far as to stick their heads out of the water to peek and find the easiest route."

The bull trout, once an abundant western trout with huge catch limits and even bounties (because it eats salmon fry), is now extinct in many watersheds, endangered in others, and listed as threatened in the United States. It requires even colder, cleaner streams for spawning than other salmonids. When logging in a watershed muddies the gravel and warms the water by reducing shade, it may take decades before the stream is again good enough for a bull. Climate change dims its prognosis even in pristine streams.

Two introduced charr species (below) also threaten bulls. Brook trout have interbred with bull trout and produced both sterile and fertile hybrids, reducing bull trout reproduction. (Brookies were, after rainbows, the second most stocked trout; they are no longer stocked in watersheds with bull trout.) A third charr species, lake trout, seems to occupy pretty much the same ecological niche as bull trout once it's in the same lake, and almost always wins out competitively. In Flathead Lake, Montana, the two species coexisted for a while, until the introduction of opossum shrimp, intended as a food base for kokanee, tipped the scales in favor of lake trout. In response, the Confederated Salish and Kootenai Tribes will incentivize net fishing for lake trout until the bull trout recover.

BROOK TROUT

Salvelinus fontinalis (sal-vel-**eye**-nus: German name Latinized; fon-tin-**ay**-lis: of springs). Dark-green back and sides with "wormy" patterns and red spots encircled by blue haloes; dorsal fin dark-spotted, with 8–10 rays, anal fin with 7–9; juveniles' parr marks dotted with lighter red and yellow; tail hardly forked. Introduced. Salmonidae.

This eastern North American native was the most popular trout for stocking in the early years of western sport fishing, and became very widespread. Some consider it the tastiest and most beautiful. It requires cold water, and has succeeded in some small alpine lakes where rainbows did not, apparently because brookies will spawn on a beach, whereas rainbows require an inlet stream. They sometimes thrive until they crowd out other species and stunt their own growth through overcrowding. In Yellowstone, they are exceptionally tolerant of both acidic and alkaline water; they grow extra large and abundant in the alkaline water of limestone terrain.

Today, agencies work to get rid of brookies because they are non-native and tend to outcompete native trout. A new approach is to stock cutthroat or bull trout streams with brook trout altered (not by genetic modification but by putting hormones in their parents' feed) to have two Y chromosomes. Eggs fertilized by these hypermales will all be

Bull trout

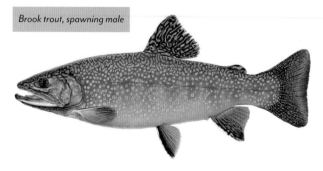

Brook trout, spawning male

males. Skewing the population's sex ratio enough would suppress reproduction overall, tipping the competitive scales in favor of natives—if it works.

Lake trout

LAKE TROUT

Salvelinus namaycush (sal-vel-**eye**-nus: German name Latinized; nam-**ee**-kush: Cree name). Also mackinaw, gray trout. Bluish- or greenish-gray to brown, with pale spots and whitish belly; proportionally narrow, as seen from side; tail deeply forked. Introduced in deep, cold lakes. Salmonidae.

Widespread in northeastern North America, lake trout are native here only as ice age relics in a few lakes high in Montana's Jefferson River and Hudson Bay watersheds.

Until rather recently, anglers and fishery managers alike assumed that the object of management is to provide the most popular catches possible wherever they can be induced to live. If anyone made a generous, little donation of a few live fish into waters that lacked that species, they did so freely, without fear of dissent. In recent decades, it has been illegal to introduce non-natives, but there probably are still a few scofflaws who think that doing so is their inalienable right. The illegal introduction, in the 1980s, of lake trout into Yellowstone Lake provides a textbook case of a highly esteemed gamefish becoming an ecological catastrophe.

The sole native trout in the lake is the smaller Yellowstone cutthroat, whose size disadvantage is exacerbated because it inhabits a lake's deeper waters in its youth, whereas lake trout live there throughout life: the bigger trout both prey on cutthroats and outcompete them for smaller prey. In most western lakes where they have been introduced, lake trout have either almost or entirely eliminated the native trout, especially bull trout. They will do the same to Yellowstone Lake's cutthroats, if allowed to. But the loss of a sport fishery worth $36 mil-

lion a year isn't the worst of it. The worst is that grizzly bears, ospreys, otters, mink, pelicans, dippers, and kingfishers all can only catch shallow fish, which would be the vanishing cutthroats, not the lakers. If the lake loses its cutthroats, it will lose its ospreys, and the park will support fewer grizzlies.

There is no way to eliminate the lake trout, but park scientists calculate that for about $300,000 a year they can gillnet enough of them to maintain a strong cutthroat population, following careful study of the right mesh sizes, locations, and best times to avoid netting cutthroats. Needless to say, the program welcomes lake trout anglers who prefer catch-and-eat over catch-and-release.

Ironically, lake trout were themselves decimated in the Great Lakes by overfishing and the arrival of sea lampreys.

Lakers reach their great sizes slowly, and are more likely to reach age twenty than any of our mammals. Some record lakers were 42 pounds in Yellowstone, well over 60 pounds in British Columbia, and 102 pounds in Lake Athabasca. In Yellowstone's Lewis and Shoshone lakes, where they were introduced in 1890, lake trout will probably continue to provide some of the West's biggest catches. Those lakes had no native fish, so the lakers' presence seems less dire there, but their ecological impact via salamanders and other prey has still been harsh.

BROWN TROUT

Salmo trutta (Roman names). The only trout with both black and red spots; the red spots are fewer—they're near and below the lateral line,

Brown trout

and have pale (cream to pink or bluish) haloes; yellowish to olive-brown background; dorsal and anal fins each 10-to-11-rayed. Widely introduced. Salmonidae.

Angling as a sport was nurtured in Scotland more than anywhere else, and browns are the favored trout there, so Americans were always eager to have them. They were first introduced in the East in 1882, and in Yellowstone just eight years later. They are warier of fishhooks than any other trout, perhaps due to survival of the wariest over a thousand years of angling in Europe. They have declined sharply since 2015 in some Montana basins.

KOKANEE AND SOCKEYE SALMON

Oncorhyncus nerka (onk-o-**rink**-us: swollen snout; **ner**-ka: Kamchatkan Indigenous name). No dark spots on back or tail fin; 28–40" long; thin, rough gill rakers in 1st gill arch; 13+ rays in the anal fin; silvery overall, finely black-spotted blue-greenish above lateral line; spawning adult has greenish dark head, crimson body; ♂ is slightly humpbacked; 4" juveniles have small, oval parr marks almost entirely above lateral line. Kokanee are widespread in bigger lakes and their tributaries; anadromous sockeye runs hang on in the Grande Ronde and Salmon river systems. Salmonidae.

Pacific salmon range the Pacific from California to Korea, and the Arctic from the Mackenzie to the Lena in Siberia. First described to science by Georg Steller (p. 404) in Siberia, they were given species names from Russian and Kamchatkan languages.

Most salmon are anadromous: they migrate up from the sea to spawn. They require clear, cold, well-aerated, gravelly creeks to nurse them in infancy, and larger bodies of water richer in animal life to rear them to maturity. It may be no surprise that the ocean is ideal for the latter phase; what's amazing is that salmon are perfectly adapted, as water-breathing creatures, to handle the chemical shock between salt and fresh water.

The ability to home back to their natal stream is a wonder of animal navigation. Once they get close, salmon zero in on the mineral "recipe" of the precise tributary of their birth by means of smell, a sense located in fish in two shallow nostrils. Some go only a few miles upriver, but some chinook salmon used to go 2,000 miles up the Columbia to its source, Columbia Lake, British Columbia. Today, dams stop them from reaching our region except for a few that make it to the Grande Ronde, Clearwater, and Salmon drainages.

Sockeyes can swim to still higher elevations. Ascending 6,996 feet to the foot of Idaho's Sawtooths, Pettit Lake sockeyes hold the record. Nearby Redfish Lake was named for them back when they were abundant. They became a cause célèbre when Redfish Lake numbers dwindled to four, one, and zero in some years in the 1990s. Those few survivors were captured and bred in hatcheries in an effort to show that, with enough money, we can sustain even a fish that migrates through eight dams. During extra-hot summers like 2015 and 2021, managers improve their odds of survival by catching most of them at Lower Granite Dam and trucking them the rest of the way. The odds of the run surviving the next twenty years are poor, and might be improved by dismantling the four Snake

River dams, which raise river temperatures as the dammed water sits in the sun.

Sockeye fry, unlike other salmon, spend a year or two in a lake before migrating to sea. Sockeyes that were landlocked in several lakes when the glaciers retreated (and in more lakes when dams were built) became a distinct freshwater form. Called kokanee, they stay pretty small on a diet of micro-animals including the crustaceans that turn their flesh pink. Salmon that get big in fresh water can eat waterfowl chicks, adult amphibians, and water shrews. Kokanee are stocked in many Colorado lakes; they reproduce on their own in none of them.

On returning to fresh water to spawn, most salmon eat little or nothing, instead metabolizing stored fat and muscle tissue. Sockeyes do feed en route to spawning during a last summer vacation in a lake where they typically turn deep red both inside and out. By spawning time, salmon are at death's door—and look it. With pale, fungus-ravaged flesh, they make poor eating. Bears and eagles don't complain.

The decomposition of spawners is far more than a spooky detail. A central glory of salmon—one whose loss, with their decline, is an ecocatastrophe in progress—it evolved because the parents' dead bodies become their offspring's major food source. Salmon used to move about eight million tons of nutrients per year from the sea far into the interior of the greater Pacific Northwest. Bears that ate them spread the nutrients around, dropping them in urine and feces. This has been confirmed and measured in Alaska: the isotopic signature of marine-based nitrogen is found in plant leaves up to a mile from salmon streams, but not in similar leaves near streams that lack salmon.

Since nitrogen is a main limiting factor on plant growth, the decimation of salmon literally stunts tree growth, as well as populations of bears, of course, and bald eagles, mink, and salmon themselves, which shrank in both size and number.

We always knew dead fish are good fertilizer; we knew the rivers used to shimmer crimson with spawners; the people who lived here told us in many ways. The truth was always staring us in the face, but no one got it and no one ran the numbers on salmon as fertilizer until the 1990s. Too few people in the public policy sphere get it even now.

MOUNTAIN WHITEFISH

Prosopium williamsoni (pro-**soap**-ium: mask; for Lieutenant Robert S. Williamson, p. 217). Pale silver fish without spots; mouth small, without teeth; scales relatively large, bumpy; avg 12" fully grown; 10–13 rays in anal fin. Widespread from n CO and n UT n. Salmonidae.

Whitefish mostly use their little mouths to search the bottom for insect larvae, but occasionally they will rise to a surface hatch. They spawn in late fall and winter, in water as cold as 35°F. The eggs have a sticky coating that adheres them to rocks, so no redd needs to be dug. Some heavily fished streams lost

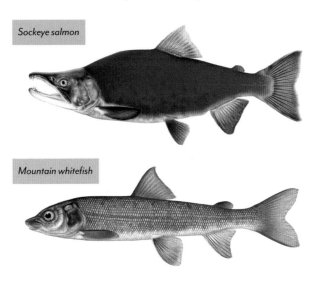

Sockeye salmon

Mountain whitefish

all their wild-salmonid stocks except white-fish, so I suspect whitefish are pretty wily. In autumn, their fry and eggs are important foods for game fish.

ARCTIC GRAYLING

Thymallus arcticus (thigh-**mal**-us: Roman name, ref to thyme aroma). Body shape simple, streamlined, laterally compressed; generally silver-gray, but in various lights and seasons this gray can flash a gorgeous range of tints and hues; a few small black spots near front end and on the often-turquoise dorsal fin, which is large, with 16–21 rays, and can extend and look sail-like. A few high lakes and streams. Any grayling caught must be released immediately. Salmonidae.

ARCTIC GRAYLING RANGE encircles the Arctic Ocean. The ice ages evicted grayling from the Canadian Rockies and pushed them down into the Lower 48. The retreating ice left two small disjunct populations, but the one in Michigan died out in the 1930s. Montana grayling thrive where stocked in small, high lakes, but efforts to restock them in their former river habitats have failed. In the nineteenth century, they shared some rivers with cutthroats and whitefish (an unusual instance of resource partitioning by three salmonid species); today, they share the Big Hole River and some lakes with cutthroats and rainbows. Brown trout, on the other hand, seem to make life all but impossible for grayling, and we can't get browns out of any major stretch of river.

SMALLMOUTH BASS

Micropterus dolomieu (my-**crop**-ter-us: small wing; **doh**-lo-mew: for Déodat de Dolomieu). Golden olive, with vertical blotching overall but strong horizontal lines across face and gill covers; lover lip protrudes; body taller than trout; pectoral fins small; 2 dorsal fins, front one coarse, low, and stiff, rear one finer and soft. Introduced; A. Centrarchidae.

Beloved by some anglers thanks to their outstanding feistiness on the hook, these Eastern fish are now widely scattered in the Rockies, where thay have become a problem, preying on young trout. Most live in large, clear reservoirs. They tolerate warmer water than native trout, and go into a torpid state when the water gets cold.

LONGNOSE SUCKER

Catastomus catastomus (cat-a-**sto**-mus: mouth underneath). Mouth set back on underside of head, with fleshy lips; no adipose fin; anal fin as long as or longer than pelvic and pectoral fins;

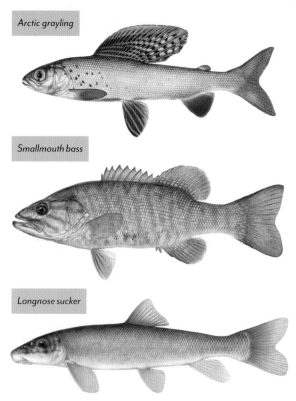

Arctic grayling

Smallmouth bass

Longnose sucker

generally dark greenish-gray, with pale belly, but breeding fish (esp ♂) develop reddish streaks and tinges; some subspecies occasionally reach 24" at ages up to the upper 20s. Widespread. Catastomidae.

Suckers move around slowly, using their mouth to scrape algae, bacteria, and some tiny animals off the rocks.

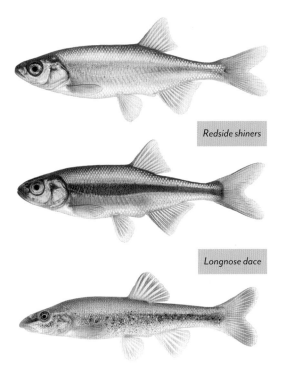

Redside shiners

Longnose dace

MINNOW

These spp. rarely longer than 5". Cyprinidae.

Redside Shiner

Richardonius balteatus (for John Richardson, see below; bal-tee-**ay**-tus: girdled). Mouth at tip of snout; flanks pink to red; dorsal fin entirely behind midlength point; no adipose fin; 8 rays in anal fin. Native only northerly, but now found in A.

Longnose Dace

Rhinichthys cataractae (rye-**nick**-thiss: snout fish). Mouth just behind (and below) tip of snout; generally drab and dark above, grading paler toward belly; breeding ♂ may get reddish lips, gills, belly, etc.; no adipose fin. A.

DACE SEEM EQUALLY at home in either slow or very fast water. They even live at the "cool" 90°F bottom of one hot-spring creek, just inches below its 147°F surface.

Scientists in the Deadly North

The Scottish naturalist **Sir John Richardson** was on John Franklin's first and second expeditions into Arctic Canada. The first was beset by starvation and cannibalism. Cartographer **Robert Hood** wrote the first explanation of the aurora borealis as an electrical and magnetic phenomenon, and then was murdered. Richardson shot the murderer and rescued Franklin, and they escaped with their lives—but without Richardson's specimens. They were game to try again three years later, and this second expedition avoided catastrophe. Soon both were knighted. Richardson wrote a fauna of boreal North America, and stayed home from Franklin's tragic third expedition, which ended with all members starving, freezing, or getting eaten. He then commanded a ship in one of the many futile Franklin party rescue attempts.

Franklin's ships, HMS *Terror* and *Erebus*, were found in deep water 167 years later, in 2014. Thanks to climate change, tourists may soon cruise the Northwest Passage that Franklin found so lethally impenetrable.

The minnow family includes goldfish and common carp, both invasive in many parts of the United States.

Sculpin

SCULPIN

Cottus spp. (**cot**-us: Greek name). Scaleless, sometimes ± prickly, minnow-sized fishes with wide mouths, thick lips, depressed foreheads, very large pectoral fins, a ¾-length dorsal fin in 2 parts, and a rounded, unforked tail fin. Widespread in streams, much of the time under rocks, logs, etc. Cottidae.

These funny-looking little fish are adapted to life on the bottom, with wide, depressed mouths for bottom-feeding; motley, drab colors for camouflage against the bottom; eyes directed upward (the only direction one can look from the bottom); and huge pectoral fins to reduce the energy cost of anchoring them in strong current. They actually swim with those fins, in a darting motion, rather than by wiggling their whole bodies like most fish. The eggs, laid in spring, adhere to the underside of stones, and are guarded by the father.

Sculpins eat some trout eggs and fry, and compete with young trout for aquatic insect prey, their main food. Larger salmon and trout, in return, are fond of sculpins, inspiring a trout fly pattern called the "muddler minnow." (Colloquially, sculpins are called muddler minnows, bullheads, or blobs.) In all likelihood, most trout eggs that end up in sculpin bellies were ones not adequately buried in gravel by their mothers, and would have perished anyway.

The most common sculpins here were known as mottled sculpins, *C. bairdii,* and later *C. bondi,* but are now split into three species: *C. punctulatus* in the Colorado River basin, *C. semiscaber* in the Columbia and Great basins, and one in Montana that awaits a name.

Sculpins are found above impassable waterfalls on some rivers, presumably getting there via periglacial lakes that crossed present-day drainage divides briefly while the ice sheet was retreating.

15

Insects

Insects—six-legged animals with jointed external skeletons made of chitin—are far and away the most diverse and successful type of animal life on Earth. Over half a million species have been named, and at least as many remain to be discovered. Only a tiny sample of our insects can be discussed in this space: some of the ever-charming butterflies and dragonflies, the invaluable pollinators—bumblebees and hover flies—and a rogues' gallery of unloved insects.

A majority of insect life cycles entail metamorphosis from a wingless larva, which does most or all of the eating and growing, to an enclosed and stationary pupa, and finally to a sexual adult, often winged. Butterfly and moth larvae are called caterpillars ("furry cats" in Latin). The larva typically molts several times while growing. Each stage, from one molt to the next, is an instar (think of it as an instance of the larva). For butterflies, the larval food plant is a good indicator of where you'll see the adults, since they are born on or very near that plant and complete their life cycle by leaving eggs on that plant. At least the females do, and for the males to complete their life cycle they need to be with the females. Some species migrate, temporarily taking them far from the larval plant. Insect migration is a rapidly growing field of study, thanks to new technology.

The pupa—far from being the "resting stage" it was once called—"hosts the greatest magical trick in the natural world: construction of a butterfly from caterpillar soup . . .

taking as little as 5–7 days" (David James and David Nunnallee in *Life Histories of Cascadia Butterflies*). It's chunky soup, full of items that were inert in the larva but are predetermined to develop within the pupa into each of the required adult body parts.

Metamorphosis in dragonflies is "incomplete" in that there's no pupal stage. Their larvae go through ten to fifteen aquatic instars of incrementally increasing complexity and mobility before molting into a winged adult. The late instars have proper legs, and the final one shows the (nonfunctional) beginnings of wings. Aphids and mayflies demonstrate other versions of incomplete metamorphosis.

Order Diptera (Flies)

Most insects have four wings, but Diptera are true to their name, which means "two wing."

MOSQUITO
Family Culicidae.

A female mosquito has a most elaborate mouth. What we see as a mere proboscis is a tiny set of surgical tools—six "stylets" wrapped in the groove of a heavier "labium" flanked by two "palps." The operation begins with the two palps exploring your skin for a weakness or pore. There the labium sets down. Delicately and precisely, two pairs of stylets—one for piercing, one for slicing—set to work. They quickly locate and pierce a capillary, then bend and travel a short way

Aedes communis *female engorged with blood*

point of blood stimulation, to which you may be allergic, you probably feel nothing.

Unlike beestings, whose function is to inflict pain or injury, a mosquito bite raises its itchy bump—an allergic reaction to mosquito saliva—only incidentally. The sensitivity is acquired. The first few times a given species of mosquito bites you, there's no pain or welt. As you develop sensitivity to that species, your response speeds up from a day or so, at first, to one or two minutes. Eventually, you may become again desensitized to that species. We may fail to notice this cycle because we rarely distinguish among the species.

within it. The remaining two stylets are tubes; one sucks blood out while the other pumps saliva in, stimulating blood flow to the vicinity and inhibiting coagulation. Up to the

The females need at least one big blood meal (greater than their own weight) for nourishment to lay eggs. It takes human-feeding species about two minutes to draw

Insects in Decline

Most of us who have paid attention to the West's butterflies over several decades have noticed an ongoing drop in their abundance. Beyond butterflies, insect populations are in an alarming decline worldwide. Insect numbers are likely dropping faster than those of birds, mammals, and reptiles. It's a difficult thing to pin down or to attribute because, as you would expect, insect populations are hard to count.

Of course, there are different causes for the declines of different kinds of insects, and in different regions. But from what scientists have been able to tell, the decline has not relented in recent decades, and the taxonomic and regional breadth has been global, so some scientists turn to the one factor that matches that scale: climate.

Additional issues are clearly involved:

- Intensification of farming practices—for example, heavier use of insecticides, and plowing and cropping to the edges, rather than neglecting some strips and corners where wildflowers can grow
- New insecticides that directly target everything that eats leaves, pollen, or nectar
- Changes in land use, such as urbanization, ever more paving, and, yes, solar farms
- Novel pathogens and competing species arriving from other continents
- Development fragmenting the landscape
- Light pollution (the nearly ubiquitous spread of artificial light at night)

As for climate, we have only a few partial answers to the obvious first question: are insects migrating poleward (or uphill) as fast as their suitable climate migrates? The few studies suggest they are not. And do they need to? No, not strictly. They are constrained less by climate itself than by the other species they need for food and perhaps shelter. So

enough. Aside from that meal, they eat honeydew, sap, and nectar; some plants may depend on them as pollinators. Male mosquitoes eat nectar only, but contribute to our discomfort all the same: they hover around warm bodies in the reasonable expectation that that's where the girls are. If we wanted, we could get males to fly down our throats by singing the pitch that female wings hum. The plumey male antennae vibrate sympathetically with conspecific female wingbeats, and the males home in on any steady source of this pitch.

Each mosquito species has its own wingbeat frequencies, one for males and a slightly lower one for females; young adults speed up their wingbeats as they mature. Wingbeats range up to six hundred per second in mosquitoes, and peak at over one thousand in their relatives, the midges. Anything over fifty beats per second is too fast to be triggered by individual nerve impulses. Instead, the thoracic muscles are in two groups: each is stretched by the contraction of the other, then contracts in a twitchlike reflex. The thorax shell snaps back and forth between two stable shapes, like the lid of a shoe-polish tin, one shape holding the wings up and the other down. The nervous system supplies only a slow, unsteady pulse of signals to keep this vibration going. (Flies, bees, and many other insects fly similarly.)

Most mosquitoes are active only a short period each day, with the typical peaks coming around sunset and just before dawn. They need an air temperature of at least 40°F to

how well are they timing their northward (or upward) range extension to match those of other species? We don't have a lot of answers yet.

Insects need (and often fail) to match other species in their phenology, or seasonal timing. Somehow plants and animals determine the right time in spring to bloom or to migrate, and in fall to go dormant or to migrate. Many kinds of mismatches are possible. For example, insects suffer if they count on multiple plant species at one time and place for nectar, and if those species diverge in their phenological response to warming. Some organisms respond more to temperature, which helps them adapt to earlier springs; others respond more to day length, and may get left in the dust by climate change.

Insects may fail to survive winters that have increasing numbers of days with freezing temperatures but no snow on the ground. That trend seems to be happening, especially at mid-latitudes (like here). Parnassians, a classic high-elevation butterfly, were found less likely to survive when November temperatures reached either cold or warm extremes (each trending upward).

And then there's an unexpected trend that's just beginning to get studied; ominously, it affects everything that eats plants, including us. It's well known that plants tend to grow faster in an atmosphere richer in CO_2. Now it appears that they do so at the cost of diluting their nutrient content. In Kansas, the ecologist Ellen Welti blamed a 30 percent decline in grasshopper numbers partly on grasses getting less nutritious.

A study of butterfly numbers in the West found a 1.6 percent average annual rate of decline over forty-two years, with the greatest declines in arid regions. Meanwhile, the Southeast may have actually improved due to increased rain, and in the West the few areas that saw an increase in precipitation saw increased numbers of insects. Where precipitation does not increase, you have hotter droughts—a challenge for insects to survive.

fly, and their bodies are too small to retain a temperature much above that of the air.

Host choice is species-specific. Some varieties of the familiar mosquito *Culex pipiens* never bite people, but others rank among the peskiest. Most of our mountain mosquitoes are of genus *Aedes* (ay-**ee**-deez: repugnant). Unlike other genera that set their eggs afloat on still water, most *Aedes* species lay eggs in the fall on spots of bare ground likely to be

How to Not Get Bit

Clothing yourself from crown to toe may be safest. Where ticks are the concern, consider wearing long pants tucked into tall socks.

Clothes and gear impregnated with permethrin offer some of the best protection. Permethrin is toxic, but your exposure to it is minimal with factory impregnation, and modest even with do-it-yourself spraying—if you're careful. Wash these clothes separately.

DEET repellents are effective and turn out to be fairly innocuous. (The health risks of bites are worse than those of DEET.) Skip products with concentrations higher than 30 percent, or 10 percent for children and 0 percent for babies. DEET stinks, and dissolves some plastics.

Picaridin-based repellents are preferable: less toxic, not stinky, and not damaging to plastics. This synthetic analog of a compound in black pepper is effective against mosquitoes and ticks, but possibly less so against flies. That said, we all know the biters are sometimes fierce enough to thumb their little noses at lab-tested "efficacy."

Repellents employing 20 percent IR3535 are almost as effective as the preceding, and don't stink, but do damage plastics. They may be a good choice when ticks are your main concern, as they remain effective against ticks for twice as many hours.

PMD (related to and often combined with lemon eucalyptus) is the most durably effective repellent that's described as a botanical compound, though the products are factory-refined.

Most botanical oils are allergenic to some people, and a few even appear to be carcinogenic. Citronella, lemon eucalyptus, and some other pure plant oils are fairly effective if you like the smell and aren't allergic, but they have to be reapplied so often that they become impractical. "Encapsulated citronella oil nano-emulsion" may address that problem in the future. Catnip oil lasts longer but doesn't work against ticks.

Try looking for nootkatone, a promising compound in Alaska yellow cedar and in grapefruit skin. It received FDA approval but is not exactly on the market as of early 2023. (My internet search yielded two "sold out" repellent products, Mozzi Magic and Coco Noot, plus an essential oil blend that is 2.5 percent nootkatone.)

Forget wrist bands, dermal patches, vitamin B1, and especially sound-emitting devices, which actually increased the number of bites in one study. (That study liked clip-on diffusers of botanical aromatic oils, but the Environmental Working Group warns of possible health dangers from inhaling those aerosols.) Also, it's best to apply your sunblocks and repellents separately, because sunblocks need to be reapplied much more often.

There are humans whom mosquitos avoid. Scientists analyzed their skin secretions and identified the natural repellents as ketones smelling like nail polish remover and very ripe fruit. They hope someday to sell those compounds to those of us who are not so blessed.

briefly submerged in the spring, when the larvae take off swimming. These "wrigglers" feed by filtering algae and bacteria out of water. Two subtropical species of *Aedes* carry Zika virus, Chikungunya virus, and dengue fever, and are spreading northward in the US, but are not projected to thrive in the Rockies even in a scenario that is 10°F warmer.

Females zero in on the CO_2 we exhale, together with lactic acid and other skin and sweat aromas. Gruesome concentrations of mosquitoes typify the Far North, where musk oxen and nesting ducks make ideal hosts. One stoical researcher counted 189 bites on a forearm exposed for one minute in Manitoba. He extrapolated 9,000 bites per minute for one entire naked person, who could lose one-fourth of their blood in an hour.

Mosquitoes may be the best studied of all insect families because they are the deadliest to humans. In 1803, yellow fever killed nine-tenths of a French Army legion sent to conquer Haiti and the Mississippi Valley, leaving Napoleon in a mood to sell "Louisiana" to Thomas Jefferson at a price Congress couldn't refuse. That led to our region being part of the United States.

Malaria is increasing in the tropics; both the pathogen and the mosquito excel at evolving resistance to cures, repellents, and insecticides. Malaria and dengue fever are expected to spread north with climate change, though the arid climate should limit their threat in the Rockies. In the meantime, armies of researchers work full time testing compounds that repel or kill mosquitoes.

A mosquito-borne disease of recent concern is West Nile virus, which has been found in sixty-six mosquito species in the United States alone. (Also in three hundred bird species: several birds species suffered sharp declines at the time of peak infection in humans.) The virus can cause a debilitating fever or, in fewer than 1 percent of infections, life-threatening illness such as meningitis. In recent years, Arizona has been a US hot spot for West Nile, with Colorado and New

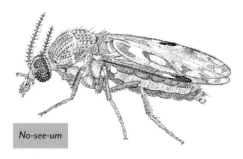

No-see-um

Mexico not far behind. Still, overall incidence remains far below the peak year, 2003, when it first reached the Rockies.

NO-SEE-UM

Culicoides spp. (cue-lic-**oy**-deez: gnatlike). Also punky, biting midge.

The common name suffices to identify these pests. At up to ⅛ inch long, they are just big enough to see, but small enough to invade screened cabins. They are hard to make out in the waning light of dusk, when they do most of their biting. Their stealth and our defenselessness irritate us more than the bites really hurt. No-see-ums are so localized that we can usually escape by walking 50 feet. Breeding grounds include puddles and humus. The chief victims of this bloodsucking family are other insects, ranging from their own size on up to dragonflies. In some species, the females prey on the males, and one pirate species sucks mammal blood from mosquito abdomens.

BLACK FLY

Simulium spp. (sim-**you**-lium: simulator). Also buffalo gnat. Stocky, dark fly scarcely ¼" long, appearing humpbacked because thorax is

Black fly

Deer fly

Horse fly

much bigger and higher than head; on landing on you, quick to bite, tilting forward.

These vicious biters raise a welt way out of proportion to their size, sometimes drawing blood, and have reportedly caused bovine and human deaths when biting en masse. They render some boreal North American vacation areas almost uninhabitable for the month of June. Fortunately, they tend to dissipate by midsummer.

Only the females bite. Once fed, they dive in and out of cold, fast streams, attaching eggs singly to submerged stones. The larvae stay underwater, straining plankton, moving around, and then re-anchoring themselves with a suction disk. After emerging from pupation, the adults burst up through the water in a bubble.

DEER FLY

Chrysops spp. (**cry**-sops: appearing golden). Fly the size of a smallish house fly but slower and softer; from dull gray-brown to nearly black; eyes spotted.

Deer flies are just too easy to kill—it's an irresistible exercise in utter futility. Where there's one deer fly, there are a thousand. Travel sometimes helps, since most deer flies stay within half a mile of the marsh or pond where they overwintered as larvae.

HORSE FLY

Tabanus spp. and *Hybomitra* spp. (ta-**bay**-nus: Roman name; high-bo-**my**-tra: humped cap). Fly like a very large house fly; black or gray-brown, and many have striking iridescent

stripes across their big eyes. Common in large-mammal habitat.

With larvae that overwinter twice, horse flies are long-lived, for flies. We see the nectar- and pollen-eating males less often than the bloodsucking females.

These are our biggest, fastest, strongest biting flies, so we're lucky they're sparse enough to view as individuals: when you manage to swat one, you may be ahead of the game for a while. Their obsession with the tops of our heads often diverts them from our more vulnerable parts. Even so, they're hard to catch up with.

Horse flies and their relatives, deer flies, are also known as gad flies. Dictionaries may spell them "horsefly" and "gadfly," but entomologists prefer to keep "fly" separate in the names of true flies (Diptera) as opposed to non-flies such as dragonflies and sawflies.

HEDGEHOG FLY

Adejeania vexatrix (ay-da-**zhon**-ia: not for Pierre François Marie Auguste Dejean, Napoleon's aide-de-camp who busied himself collecting beetles on the battlefield at Waterloo; **vex**-a-trix: a female who vexes). Round-bodied

Hedgehog fly

orange fly with black bristles in tidy rows across its abdomen; proboscis sticks out from its chin.

This easily recognized fly, often seen nectaring on flowers, is a significant pollinator in the mountains. Like other flies in the large Tachinid family, its larvae live by consuming various caterpillars from the inside out.

HOVER FLY

Family Syrphidae. Also flower fly.

You feel an insect on your arm. You look down, see a yellowjacket—hold that swat! Don't let a case of mistaken identity turn deadly.

Yellowjackets—a kind of wasp—crave your salami or your protein bar, but rarely land on your skin except to sting you. If this one hasn't stung, trust it. Chances are it is a hover fly, a stingless insect that mimics bees or wasps in order to ward off predators. Look more closely: are the two antennae stubby and short, and much smaller than the eyes? It's a hover fly. Or are they long, slender, and turned down? It's a bee or wasp. Is there a short, footlike, dark proboscis stepping around on your skin? Hover fly tongue, lapping up the salty sweat that attracted it to you. As for those long, downturned yellowjacket antennae, some hover flies heighten their mimicry with a "dance" behavior, waving their front legs to make them look like wasp antennae.

Do hover flies do more to earn our friendship than just slurp on us, look pretty, and decline to pack heat? Absolutely. They rank high among insects beneficial to plants, and hence to farmers. The larvae of many hover flies are the leading predators of aphids. In adulthood, hover flies are second only to bees as pollinators globally; in our high mountain meadows, flies may do more pollinating than bees. Their pollinating style has one advantage over that of bees (carrying pollen farther) and one disadvantage (being less faithful to a plant species on any given day). Whereas bees use their long tongues to draw nectar out of deep flowers, flies have short tongues, visit shallower flowers, and eat more pollen than nectar. Both males and females visit flowers, the females gorging on pollen proteins they require to develop their eggs, the males snacking while patrolling territory in hopes of a chance at a receptive female.

Since they seek sweat and resemble bees, hover flies are sometimes conflated with sweat bees, a family of actual bees. Sweat bees are typically black or green all over and just ¼–⅜ inch long. Our hover flies tend to measure ½–⅝ inch. A few thousand syrphid species have black-and-yellow-banded abdomens to mimic this or that well-armed bee or wasp. On several common ones, the thorax shines like polished brass and the huge eyes are maroon. Male eyes are plastered to each other on the top of the head; a narrow but distinct forehead separates female eyes.

Few insects can compete with a hover fly at hovering. Bumblebees sashay from side to side as they descend upon a blossom, whereas male hover flies spend minutes at a time hovering perfectly. They are the insects you see maintaining a fixed position in a beam of light above a forest trail. Watch for the fly to zoom off to the side, then retake the same spot it held before. This is a male hover fly defending aerial territory where a female may show up. When he darted off, he was bouncing a rival. (Or was that the rival

Black-margined hover fly, S. opinator

that won the confrontation and usurped the midair post? If you can tell which one of them came back, you've got quicker eyes than me.)

A position in a beam of light can be held longer than one in the shade because the fly absorbs solar heat, supplanting calories he would otherwise burn to maintain optimum body heat for efficient hovering. This allows him more minutes hovering before he has to go find the next snack. Cost-benefit analysis confirms that the sunlit fly comes out ahead.

Vertical-looking radar technology has improved such that it can identify and count thousands of migrating hover flies. In England, which has radar networks monitoring migrations year-round, scientists concluded that around four billion hover flies fly from the Continent to England each year, bringing with them significant volumes of pollen and high-nutrient biomass. The same scientists rushed out to California on hearing reports of migrating hover flies and, though forced to use old-school instrumentation (cameras), concluded that yes, some American hover flies migrate en masse as well.

Order Hemiptera (True Bugs)

Entomologists call this order the True Bugs, and insist that we should never call other insects "bugs."

COOLEY SPRUCE GALL ADELGID

Adelges cooleyi (a-**del**-jeez: unseen; **coo**-lee-eye: after Robert Alan Cooley). Soft, hemispherical bodies 0.04" long, covered at most stages with waxy, cottony white fluff; wings (if present) folded rooflike over body; on Douglas-fir needles (related spp. also on true firs, pines); more conspicuous are their conelike galls on branch tips. Widespread on and near spruces.

Many spruces seem to have an odd, spiky sort of cone in addition to their larger, papery-scaled ones. Looking more closely, we can see that these aren't really cones because they are at branch tips rather than

several inches back, and because they are fused wholes, not a set of wiggleable scales. The spikes turn out to be spruce needles with a hard brownish or greenish skin drawn tight like shrinkwrap. This is a gall, material secreted by the tree in response to chemical stimulation by an insect. Other galls include bright-red marginal swellings on shrub leaves, or lightweight tan orbs on oak limbs.

Though each spruce gall ends one branchlet's growth, galls themselves are rarely a serious drain on host plants. Adelgids and related aphids sometimes are, sucking plant juices through minute piercing tubes nearly as long as their own bodies. They suck in far more plant sugar than they consume, and then pass copious, sticky honeydew excretions. Some aphids are "herded" by ants or other insects that feed on aphid honeydew, but on our fir trees honeydew is more likely to end up consumed by a dreadful-looking black smut fungus. Overall, neither the feeding nor the housing of gall adelgids places them among our top forest pests.

Spruce gall adelgids actually feed mostly on Douglas-firs. Their life cycle includes no larvae, pupae, or males as those terms are normally defined. Instead, it is divided into five forms or castes, each egg-laying. One wingless form overwinters on spruce, then lays the eggs of the gall-making form, which emerges from the gall in late summer

Cooley spruce galls

and flies to a fir to lay eggs. These eggs produce fir-overwintering forms: wingless and winged. The first stays on firs, the second flies to spruces to beget either the spruce-overwintering form or a short-lived sexual generation. But in the Rockies, sexuals are more or less unknown and the various female forms perpetuate their clone parthenogenetically (without fertilization).

BALSAM WOOLLY ADELGID

Adelges piceae (a-**del**-jeez: unseen; pie-**see**-ee: of spruces, a misleading name). Soft, hemispherical bodies ½–5" long, covered at most stages with waxy white wool; rarely found in winged stage. Sporadically epidemic on true firs.

The balsam woolly adelgid coevolved in an unthreatening relationship with European true firs, but when accidentally introduced to North America, it proved deadly. Subalpine firs have suffered severe mortality in the northwestern Rockies. The tree's leader may droop and break off early in an attack, then all foliage may turn red-brown from the top downward while the trunk hemorrhages resin and twigs swell up in gouty knobs. Dispersing mainly on wind currents, this bug rarely grows wings. Males and galls are unknown.

With their complex (often asexual) life cycle, adelgids are poorly understood. Hemlock woolly adelgids, *A. tsugae,* are an existential threat to eastern hemlocks, which they first encountered in 1951, but they are unthreateningly present on the seven other hemlock species, including our two. Apparently, *A. tsugae* is native here and coevolved with our hemlocks; the Japanese strain attacking the East may get a new species name in due time.

WATER STRIDER

Aquarius remigis group (rem-**ee**-jiss: oarsman). Also Jesus bug; *Gerris remigis*. Body ⅜" long; head and thorax have fine gold hairs; forewings gold-flecked; skates rapidly on tips of

Water strider

4 very long legs. On still water; A. We have several smaller water strider spp.

BUG WALKS ON WATER—SCAVENGES FLOATING CORPSELETS could be the headline of a story about water striders. This insect's tarsi (feet) have greasy hairs and end in a claw that retracts into a hollow; somehow this arrangement rests on the surface tension of still water. Each foot dimples the surface tension; the dimple acts as a lens, making round shadows on the bottom in shallow water. Tiny insects falling on the water make rippling shadows on the bottom too, and striders watch for these, and then give chase. Once striders were able to read those shadows, it was a short step for them to learn to read the shadows of other striders, and then to use their ripples as a mode of communication.

Striders have a pair of shorter front legs for grasping prey, while floating on the other four. Metamorphosis is gradual, with around five nymphal stages. Some adults grow full-sized wings, and migrate; others have useless reduced wings. Genetics, day length, and ample food supply all help to determine who can take wing.

Order Hymenoptera (Bees and Wasps)

Hymenoptera have four wings, which logically ought to help you distinguish their two-winged mimics like hover flies; but no, the very narrow hind wings of bees and wasps are translucent and hard to see in a live situation.

BUMBLEBEE

Bombus spp. (**bom**-bus: buzz). Large, rotund, furry yellow-and-black bees; queens in most species are ⅝–⅞" long, and fly in the spring; workers ⅜–⅝" long, appearing late spring to fall. Ubiquitous.

Much "common knowledge" about bees is actually about honeybees, which aren't even native to the Americas. The European honeybee, *Apis mellifera,* invaded America along with European culture, but is not abundant in our wilderness. A honeybee worker can sting only once, but other bees have repeat-use stingers. Honeybees have an elaborate social order and communication rituals, but most native bees are "solitary bees"—they may live gregariously, but they don't organize societies with a division of labor. Honeybees are the top commercial species, but bumblebees are also raised and transported to provide pollination services for crops, such as blueberries, that demand buzz pollination. (Blueberry stamens don't release much pollen until they get vibrated by rapid wingbeats.)

Our most visible bees, bumblebees are in the honeybee family, and do have a strong social order, though their colonies are relatively small and don't overwinter. Each colony begins with a queen who hibernates, often among dead leaves, after mating in the fall. The first bumblebees every spring are extra large because they are queens. Typically taking a mouse or vole burrow for her nest, she secretes beeswax to make "pots." In some she lays eggs on a liner of pollen and nectar, sealing them over with more wax. Others she fills with honey (nectar concentrated within her body, by evaporation), which she sips for energy while working her muscles and pressing her abdomen onto her eggs, incubating them with body heat. The first brood is workers—small, nonmating females. Maturing in three or four weeks, they take over the nectar- and pollen-gathering chores while the queen retires to the nest to incubate eggs and feed larvae.

Protein-rich pollen nourishes the growing larvae, while pure carbohydrate nectar and honey are what the energetic adults require. Bumblebee honey is as tasty as honeybee honey but can't be exploited by people or bears because bumblebees don't store much for the future.

Late-season broods include increasing numbers of sexuals—male drones from unfertilized eggs, and new queens produced by more-generous feeding of a few female larvae—but only to the extent that the food supply allows. In one recent study in the Colorado alpine, only a minority of colonies were able to produce a queen; the majority died off at the end of the season. Only queens eat well enough to survive winter, after the drones and queens have mated. By feeding only a few individuals for winter, bumblebees conserve nectar and pollen resources.

Periods of torpor in insects are properly called diapause and can occur at any time of year, unlike true hibernation in mammals.

Generally, bumblebees are the most valuable plant pollinators at high latitudes and altitudes—cold places. Key adaptations include the use of ready-made insulated nests; the relatively large, furry bodies; and the skill of maintaining a thorax temperature of at least 86°F for flying in all weather conditions. Arctic bumblebees have been seen in flight in a snowstorm at 6°F below freezing. Bees are not cold-blooded animals.

For seeking the best flowers for nectar, bees utilize a superhuman toolkit of senses: they detect ultraviolet coloration, iridescence, microscopic surface texture, the polarity of light, subtle fragrances, and electrical charges. An insect arrives at a flower positively charged, while the flower is negatively charged; the opposing charges draw pollen from flower to bee. The flower's charge is strongest on its petal edges, providing electric guidelines that help the bee find the nectar quickly. Further, some of the bee's charge is transferred to the flower, neutralizing the latter's charge. The next bee that comes along detects the flower's neutral charge and moves on, since the nectar was just drained. The bee detects a charge without even touching the flower, via motion of the hairs on its thorax, like your hair being drawn toward a static-charged balloon. (Hover flies—and likely other insects—do this as well.)

Most bumblebee species have multiple color patterns; in a given area, several species may converge on similar patterns. As a result, identifying them to species may require capture, expertise, and a magnifying glass. *Bombus morrisoni* is distinctive, wearing a nearly full-length yellow cape, and no orange. *B. appositus* has a black back band separating its pale-yellow shoulders from a dull (sometimes banded) yellow-orange abdomen. *B. huntii,* abundant at low elevations, has a rich-red abdomen section framed by both yellow and black at each end.

Some bumblebee species are brood parasites, or "cuckoo" bumblebees. Their queens murder queens of other bumblebee species and usurp their colonies. They produce no workers, and their legs have no pollen sacs. Worker bees on guard duty sometimes get enough stingers through an invader's armor to kill her, but a false queen can pacify workers with pheromones. Then the workers bring home the bacon to the mixed family. Queens of non-parasitic species also sometimes usurp other queen's colonies. It is not unusual to find several dead queens in a nest, or to find colonies of mixed species.

Both honeybee and bumblebee populations are threatened by a host of modern problems. Some formerly abundant bumblebee species are now hard to find, while a few have expanded their ranges. Prolonged extreme heat waves cause bumblebee die-offs. Two top problems are habitat loss and neonicotinoid insecticides. Gardeners, please avoid any pesticide sold as "systemic," meaning that you put it on the soil and it goes

into every part of every plant in that soil, persisting for weeks. That may sound like a great way to prevent infestations, but only if you think that all insects are bad. It kills insects that eat pollen and nectar—the very insects your plants need the most.

(Europe has banned neonicotinoids since 2013, but manufacturers somehow convince US regulators that they are acceptable. Numerous studies show serious harm as well as ubiquity in the environment. They are plausibly linked to overall declines of birds and non-target insects, and to skyrocketing numbers of potentially life-threatening deformities in Montana deer. Neonics should be sharply restricted, if not banned outright.)

For bees and plants to thrive, it will help if they respond to climate change at similar rates—both their timing in spring and their migration north and upslope. It's already clear that mismatches cause declines in many species; competitors may benefit in some cases. So far, studies in the Colorado alpine find few of our bee species keeping up with either the flowers or the climate. Two Rocky Mountain *Bombus* species are evolving shorter tongues, which helps them adapt by being less specialized for particular flow-

Hunt's bumblebee, B. huntii

Elm sawfly | Convergent ladybug eating an aphid

ers; the losers may be those plants that specialize in getting pollinated by long-tongued bees. Subalpine bee species are already moving into the alpine zone, where the overly specialized alpine bee species may get outcompeted and have nowhere else to go.

To survive periods when floral food is scarce, many kinds of bees sometimes eat sugary honeydew excreted by scale insects onto plant stems. They somehow locate this resource even though it lacks attention-drawing petals. The best guess is that they keep a sharp eye out to notice the occasional lucky bee (of any species or family) enjoying a honeydew lunch. This phenomenon has gone almost unnoticed; fir krummholz and small willows in tundra may be good places to look for it, as both are noted as honeydew bars in other regions.

ELM SAWFLY

Cimbex americanus. ¾–1" adult, all glossy black with a yellow-white back band; feet and club-shaped antennae dull yellow; ♀ abdomen sometimes white-banded; caterpillars to 2" long, orange to greenish white, with a full-length black stripe. Around broadleaf trees; A.

These beasts look intimidating, but do not sting. They can defoliate birches, willows, and maples, but in our elmless region they aren't a major economic pest. Males find a high perch from which to watch for females and rival males.

Order Coleoptera (Beetles)

Coleoptera means "sheath wings." A beetle's forewings are modified into a hard sheath that encloses and protects the two hind wings when they are folded up, enabling it to bore into hard materials without jeopardizing its wings. This ability to both fly and bore may be a key to beetles' success—not that all beetles bore. They comprise by far the largest and most diverse animal order. One of every four animal species is a beetle.

CONVERGENT LADYBUG

Hippodamia convergens (hip-po-**day**-mi-a: horse biting?). Shiny, ¼", oval beetle, aging from orange to deep crimson, with (often 8) black spots.

Ladybugs, both as larvae and as adults, are terrific predators of aphids, scale insects, and other small herbivores. They are collected commercially and sold to gardeners as a benign form of pesticide. Don't fall for it: they will promptly fly away rather than sticking around your garden. Nearly all the ladybugs in commerce are this species, which is easy to collect because it converges—that is, gathers in massive heaps, typically on mountaintops in the fall, for overwintering. These masses are a valuable food resource for grizzly bears in Montana's Mission Mountains.

In June of 2018, Doppler weather radar in Southern California picked up an apparent rainstorm in the middle of dry weather. It turned out to be a county-sized migrating swarm of ladybugs flying a mile up in the sky.

Ladybugs are beetles, not members of the actual insect order known as True Bugs.

GOLDEN BUPRESTID

Buprestis aurulenta (bew-**pres**-tis: cow-swelling; or-you-**len**-ta: golden). Also *Cypryacis aurulenta*. ½–¾" beetle; metallic emerald-green, coppery iridescent down center, edges, and underneath; back ridged lengthwise. Widespread but shy, adults may be found feeding on foliage, esp Douglas-fir. The green rose chafer, *Dichelonyx backi*, is also bright metal-green, but rarely reaches ½" long.

This beetle beauty rivals our most glamorous butterflies. Foresters count it among our pests. Attacking trees in relatively small numbers, it rarely does more than cosmetic damage, but that can add up to a lot of money. The larvae bore deep into seasoned heartwood, and may continue to do so as long as fifty years after the wood is milled and built into houses. Such records make the buprestid a contender for the title of longest-lived insect, though the larvae mature in less than a decade in natural habitats.

TEN-LINED JUNE BEETLE

Polyphylla decemlineata (po-ly-**fil**-a: many leaves; des-em-lin-ee-**ay**-ta: 10-lined). ⅞–1¼" long, brown, with broad white stripes lengthwise on back; hairy underneath; ♂ have huge, wide, rakelike antennae; adults feed on conifer foliage, fly into lights at night.

The white, usually C-curved larvae (typical of this family, the Scarabs) feed on plant roots, becoming pests in crops and nurseries. When handled, the handsome adults make hissing or squealing sounds, which apparently suffice to scare off some threats.

Golden buprestid

Ten-lined June beetle

Ponderous borer

PONDEROUS BORER

Trichocnemis spiculatus (trick-uc-**nee**-mis: hairy leg; spic-you-**lay**-tus: with little spikes). Also pine sawyer, spined wood borer; *Ergates spiculatus*. Large, 1¾–2½" beetle; back minutely pebbled, dark reddish-brown to black; thorax ("shoulder" area) may bear many small spines; antennae long, jointed, curved outward.

Attracted to light, these clumsy nocturnal giants startle us when they come crashing into camp. The equally ponderous larvae take several years to grow to a mature size of up to 3 inches, chewing 1-to-2-inch holes through pine or Douglas-fir heartwood. Loggers call the larvae "timber worms." One logger was inspired by their mandibles in designing the modern saw chain.

BARK BEETLE

Dendroctonus spp. (den-**droc**-ton-us: tree murder). Small (⅛–¼") black to pale-brown or red beetle with tiny, elbowed, club-tipped antennae; adults and larvae both live in inner bark layer, hence are little seen, but their excavation patterns in the bark and cortex are distinctive. Ubiquitous.

Western Pine Beetle
D. brevicomis (brev-ic-**oh**-mis: short hair). On ponderosa pines.

Mountain Pine Beetle
D. ponderosae (pon-der-**oh**-see: of ponderosa pine, misleadingly). On most pines, but esp lodgepole and whitebark.

Douglas-fir Beetle
D. pseudotsugae (soo-doe-**tsoo**-ghee: of Douglas-fir). On Douglas-fir and larch.

Spruce Beetle
D. rufipennis (roo-fip-**en**-iss: red wing). On spruce.

BARK BEETLES ARE among the most devastating insect killers of western trees. The first signs of their attack are small, round entrance holes exuding pitch and/or boring dust. The pitch is the tree's counterattack, an attempt to incapacitate the beetles. The

Pitch extruding from where a mountain pine beetle bored into a lodgepole pine

many exit holes, a generation later, look as if they were made by a blast of buckshot.

After the bark falls away, you can see branched engravings underneath. The beetles, though less than ⅛ inch wide, chew much wider egg galleries through the tender inner bark layer, just barely cutting into the sapwood. The larvae set off at right angles to the gallery, widening as they grow, and then pupating at the ends of their tunnels. From there, they bore straight out through the bark upon emerging as adults. The tunnels are left packed with excreted wood dust, called frass. In most species, the egg gallery runs 6–40 inches straight up and down the wood grain, while the larval tunnels are horizontal, fanning out somewhat. Galleries of the western pine beetle curl and crisscross like spaghetti, and their larvae leave little impression on the inner bark, preferring the outer bark. Engraver beetles identify themselves with further variations on the theme.

These beetles are all native here. In typical years, they benefit the forest by culling the damaged, diseased, or slower-growing trees.

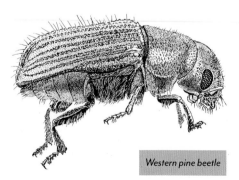

Western pine beetle

Vigorously growing trees are less palatable and better able to mount a rapid defense using pitch; the beetles must overpower the tree's health, or it will overpower them. They do so by attacking en masse and by bringing wood-eating fungi along in small pits on their bodies. Single females scout, then release scented pheromones to attract others to a vulnerable tree. Each female is followed into her entrance hole by a male, and they work as a pair.

Mountain and western pine beetles have occasional population outbreaks, when their economic equation shifts, offering advantages to them if they attack big, healthy trees. The quarter-century starting in 1990 produced sweeping epidemics of multiple beetle species: notably spruce beetles in Alaska, Yukon, and later Utah and Colorado; pine engravers (*Ips* spp.) in the Southwest; mountain pine beetles throughout the Northern Rockies and interior British Columbia; and the whole suite of beetles in California.

At least three known mechanisms correlate these epidemics to climate change:

- Bark beetles are periodically beaten back from the coldest parts of their range when severe cold snaps kill all their larvae. They expanded their range into central British Columbia and higher elevations in Montana and Wyoming after several decades without any snaps that cold.
- Warmer summers enable mountain pine beetles to produce two generations in one year, helping their populations to grow much faster.
- Warming has a drying effect on vegetation. Beetle epidemics tend to follow droughts, when most of a region's trees are stressed to some degree, weakening their defenses. That may go a long way toward explaining the shift from benign to destructive beetle ecology. Unfortunately, drought (relative to past normals) will increasingly be the condition of much of the West.

Thinning pine forests apparently helps somewhat to keep them vigorous in a drier climate, and can put trees out of range of beetle pheromonal communication—about 20 feet between trees. The forest management goal should be greater forest heterogeneity across the landscape—in species, in age and size, and in density.

Looking at a lodgepole forest devastated by beetles, many people intuitively call it a "tinderbox." Whether beetle kill actually leads to more or worse fires is an interesting question that has received intense study. The answer? It varies. Lodgepole forests in a dry summer are pretty damn flammable even when healthy. During the newly killed red stage, lasting a year or two, the trees are full of red-brown needles and are slightly more flammable than when green. Soon, in the gray stage, with needles fallen, the dead trees are too far apart to carry a crown fire easily, and fuels on the ground may be no worse than in a green forest, so fires are *less* likely and *less* intense than when the lodgepoles are green. Still later, if countless killed trees are fallen and jackstrawed across the landscape, fire intensity could get really scary; we started to see this in California's 2020 Creek Fire. Firefighters consistently report that fire behavior in heavy beetle mortality is more dangerous.

FIR ENGRAVER

Scolytus ventralis (sco-**lie**-tus: truncated; ven-**tray**-lis: on the belly). ⅛" long, shiny, blackish beetle; abdomen appearing "sawed off" a bit shorter than the end of the wing covers; known mainly by their egg galleries in the inner bark of dead and dying trees. Ubiquitous.

The fir engraver, like the western pine beetle, kills a lot of trees during occasional outbreaks, including the 2014–16 beetle epidemic in California. Its victims there were white firs, as it attacks true firs exclusively.

Its main gallery runs sideways across the grain, while the larval tunnels run with the grain.

Dozens of species in the subfamily Scolytinae take on specialized roles in recycling woody waste, a process crucial to the forest's vitality. Many of them have mycangia for holding and disseminating fungal spores that, along with bacteria, are partners in this great symbiosis. If not for its attackers, a tree would be virtually immortal—just as long as it could keep growing—since trees do not get old in the sense that animals do. Without beetles and fungi, would the forest stop growing short of impenetrability? It's hard to imagine such a world.

A majority of the Scolytinae rarely attack healthy branches, but this one's a killer. There has long been a forestry view of fir engravers as a natural rebalancing tool in the ponderosa pine region: white firs or grand firs may gradually encroach on the pines, especially in the absence of low-severity fire, but they aren't really drought-tolerant enough to persevere. When a few drier years come along, the firs will suffer and the fir engraver will finish them off.

STRIPED AMBROSIA BEETLE

Trypodendron lineatum (try-po-**den**-dron: bore tree; lin-ee-**ay**-tum: lined). ⅛" long, shiny black to brown beetle with faint, paler stripes lengthwise on wing covers; head hidden from above by thorax, so body appears to have only 2 sections; antennae shaped like very fat clubs; known mainly by their heavily black-stained pinholes deep in wood. In conifers.

Ambrosia beetles feed mainly on "ambrosia" fungi, which they cultivate, inoculating them into holes they bore in downed or dying

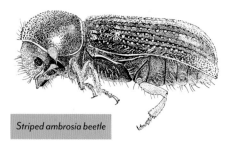

Striped ambrosia beetle

White-spotted pine sawyer

trees. If conditions are perfect, the fungi will grow just fast enough to feed the beetles and their larval brood without smothering them.

Since ambrosia beetles' fungal partner can break down the hardest of heartwood, they aren't confined to the tender layers under the bark, but burrow into the sapwood and sometimes even the heartwood. The fungus stains the wood around each borehole black, wreaking economic havoc on lumber felled in autumn and left out through winter.

Large fallen trees take a century or longer to decompose, and would take longer still if the decomposers (fungi and bacteria) didn't have borers to make a rapid initial penetration of the protective bark. To make both space and nutrients available for new plants, the forest ecosystem needs the logs to decompose.

WHITE-SPOTTED PINE SAWYER

Monochamus scutellatus (mo-no-**kay**-mus: 1 rein; scoot-el-**ay**-tus: with a small shield). ⅝–1" long, plus (on ♂) black/white antennae 1× to 3× body length; dark gray to black, often with white speckles like spatter paint. Among dead or declining conifers; A.

Pine sawyer larvae eat the wood of recently killed or attacked conifers. When a lot of them are munching in a quiet stand, it's quite audible. They often follow either fire or pine beetles, apparently detecting pine

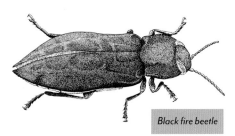
Black fire beetle

Snakefly

beetle pheromones. They eat pine beetle larvae, probably helping to end a pine beetle epidemic. On the other hand, they do attack and doom some trees that might otherwise have recovered after a fire. They also reduce the value of the timber by boring deeply into trees.

BLACK FIRE BEETLE

Melanophila acuminata (mel-on-**off**-il-a: loves black; a-cue-min-**ay**-ta: point-tipped). Also fire-bug. ¼" pitch-black beetle with minutely jointed antennae. Around burns. Patterns of 6–10 pale blotches on the back identify two very similar fire beetle spp.

Evolution gave fire beetles no way to deal with pitch or other tree defenses. What they got instead was a phenomenal ability to get there first when a lot of trees burn. Tiny infrared sensors in pits on the thorax can detect heat from as far as 70 miles away; that's far greater infrared sensitivity than any human-made device. They chase fires so efficiently that they get there in time to nip the firefighters. Laying eggs under the bark of dying pines, still smoldering, they facilitate recycling by providing entry points for decomposers.

Miscellaneous Orders

These insects all metamorphose through nymph instars rather than through a pupal stage. Snakeflies (Raphidioptera) are a small and unusual order, much reduced from when they were important during the age of dinosaurs. Orthoptera and Ephemeroptera are also ancient orders, but not so diminished.

Northern spur-throat grasshopper

SNAKEFLY

Agulla bicolor (ag-**you**-la: big throat?). Reddish, ½" fly with striking, long, flexible neck; black head ringed with red; clear wings with obvious veins, folded rooflike over back; ♀ has long, upcurved tail-like ovipositor. Order Raphidioptera.

Orchardists value snakeflies as fierce predators of aphids. Their larvae are also beneficial, preying on wood-boring insects.

GRASSHOPPER

Insect over 1" long; stiff, long forewings rest on back, making it look very straight; fan-shaped hind wings ± translucent, emerge only for flight, which is typically low and brief; forelegs very large, for jumping. Mountain meadows; A. Order Orthoptera.

Northern Spur-throat Grasshopper

Melanoplus borealis (bor-ee-**ay**-lis: of northern forests). Brown with showy green patterns; youngest nymphs green.

Alpine Grasshopper

M. alpinus (mel-on-**op**-lus: black tool). Like the preceding; spp. ID requires microscope.

Wrangler grasshopper

Bruner's Spur-throat Grasshopper

M. bruneri (**brew**-ner-eye: for Lawrence Bruner). Like the preceding; spp. ID requires microscope.

Wrangler Grasshopper

Circotettix rabula (circ-o-**tet**-ix: round grasshopper; **rab**-you-la: somewhat dark). Finely patterned drab brown. Southerly.

THE ORTHOPTERA ("STRAIGHT wings") are named for their long, leathery forewings. These protect the thin, collapsible hind wings of grasshoppers, used for flying in bursts. The order also includes flightless crickets and katydids like the Mormon cricket. Nearly all Orthoptera have huge hind legs, and can hop like crazy.

Meadows have shifting assemblages of several kinds of grasshoppers occupying different seasonal and spatial niches. Some crawl up plants and chew leaves off. These three *Melanoplus* species walk on the ground and eat vegetation harvested by other grasshoppers, along with seeds and small carrion.

"Locusts" are simply hoppers in flocks numbering in the multimillions. Their notorious population irruptions devastate crops. Though not quite qualifying as locusts, Bruner's and northern grasshoppers do irrupt, and sometimes overgraze mountain meadows, leaving little for the hoofed grazers. But since they would far rather eat lupine, locoweed, or dandelions than grasses, they more often benefit grasses and grazers alike. The alpine grasshopper does not irrupt, and rarely gets blown onto glaciers. There were three (and are now two) Grasshopper Glaciers in Wyoming and Montana, all named for their copious debris of frozen grasshoppers, especially *M. spretus,* a locust that plagued pioneers on the plains but was last seen alive in 1902—"the only documented extinction of a pest insect," as Jeffrey Lockwood points out.

Grasshoppers of our cold montane climate have a long diapause (period of dormancy) in the egg stage: they don't hatch until they've been in the soil for two or three years. After that, they develop quickly, requiring just four to six weeks before taking wing as adults in June. The eggs hatch into nymphs that resemble adults but lack wings. A series of five nymphal instars leads to adulthood without pupating.

A grasshopper's daily cycle in the Rockies begins with crawling out from a crevice and basking for two hours, with body held sideways to the sun and the sunny-side leg lowered to better warm the abdomen. A few hours of midday activity are followed by a late-afternoon bask before wedging down for the night.

These *Melanoplus* species fly briefly, low, and silently, but other genera such as *Trimerotropis,* with bright-yellow hind wings, and *Circotettix* rattle loudly in flight by rapidly snapping their hind wings.

Grasshoppers and crickets "stridulate," or make loud, raspy noises by rubbing the toothed edges of specialized body parts together. In crickets, it's two forewings rubbing together. Grasshoppers make a deeper pitch with big hind legs rubbing sections of the forewing. Some species also flash their hind legs in visual signals. Sections of the abdominal flanks are the hearing organs. These songs cover the usual set of essential communications—alarm, asserting territory, locating conspecific mates, jilting unwanted suitors. They are disproportionately loud for a small creature for the same reason that antlers are disproportionately bulky: sexual selection. Females go for the loudest males, inducing them to splurge their resources on racket in order for their genes to carry on.

MAYFLY

Order *Ephemeroptera*. Aquatic nymph around ⅓", cylindrical, with 3 "tails"; adults ⅜" + 2 or 3 very long, thin cerci (tails; in many spp. it's 3 tails on subimago, then 2 on imago); transparent forewings held vertical when at rest; hind wings negligible. In and near water; A.

Named *ephemeron* by Aristotle, mayflies have served since the *Epic of Gilgamesh* as symbols of life's brevity. Their adult lives are indeed short, averaging a day for many males and as brief as five minutes for males of one species. Just find a mate, copulate, die, perhaps first flying a few miles upstream to keep the population from drifting downhill. Females may hide for a week or two after mating, just letting their eggs mature.

The symbolism is misleading, though. Mayflies live for years—as nymphs in the water, going through several instars. They serve as exemplars of the generalization that when it comes to insects, the larva is the beast, and the adult is merely the reproductive organ, more often than not—an ephemeral bridge from one larval generation to the next. Adult mayflies, along with adults of some other orders, lack functional digestive systems; they can't eat. They can only procreate.

Mayfly nymphs "hatch," or metamorphose into adults, all on the same day for a species in any one locale, producing bug clouds massive enough to show up on weather radar. Mass hatches—like large flocks of birds, or mast years of conifer cones—overwhelm the capacity of predators to get too many individuals. They also make finding a mate quick and easy.

The larvae are major consumers of algae and detritus on stream bottoms, and are a dietary staple of many fish. Sensitive to water quality, mayflies are in decline.

Mayflies are the oldest insect order that survives today. They are unique in several traits: they have side-by-side sex organs— two penis counterparts and two female openings, and they are the only insects to molt once more after growing wings. The two instars with wings are called subimago and then imago—or, to fly fishers, duns and then spinners. Trout flies imitating mayflies have been popular since the Roman empire.

Order Odonata (Dragonflies and Damselflies)

Most Odonata are either dragonflies (suborder Anisoptera) or damselflies (suborder Zygoptera). Dragonflies rest their wings out flat; damselflies fold them over the body or spread them at an angle. Also, damselfly eyes bulge out left and right like giant earmuffs, without meeting in the middle. Darners, emeralds, meadowhawks, and whitefaces fly with the dragons; dancers and bluets with the damsels.

Dragonfly and damselfly nymphs, as fiercely predacious as the adults, rank among the chief predators of mosquito larvae. The largest, at 2 inches long, are big enough to take small tadpoles and fish. Dragonflies do without a pupal stage; instead, they

Female subimago of Calibaetis ferrugineus mayfly

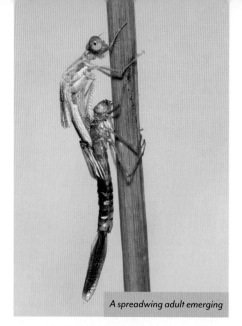

A spreadwing adult emerging

Hudsonian whiteface

adults' huge, round, compound eyes, comprising as many as thirty thousand simple eyes, see in all directions at once. Adults use their legs to grasp prey, not to walk. They have four similar wings that gyrate in independent loops, enabling hovering and turning on a dime as well as straight-line speed. Males hover territorially while watching for females, or while guarding a mate until she lays eggs.

MOUNTAIN EMERALD

Somatochlora semicircularis (so-mat-o-**clor**-a: body green). Body 2" long; eyes brilliant green; thorax dark with ocher spots; abdomen greenish black; wings clear. Sedge marshes and pond edges; A.

HUDSONIAN WHITEFACE

Leucorrhinia hudsonica (loo-co-**rye**-nia: white nose; hud-**so**-nic-a: of Hudson's Bay). Body 1⅜" long; black with many deep-red patches on thorax and all along abdomen (patches yellow on some ♀); white face below eyes; black dot near tip of black-veined wings; perches with wings below horizontal. Mid-to-high-elev ponds and wet meadows; A.

Some of the females are much duller-colored than males; others look just like males. This is puzzling (and also true of many darners, dancers, and bluets). There might be advantages in attracting either more or fewer males by looking just like one. However, to most observers, the males appear eager to mate with anything that moves on four long, clear wings, regardless of either gender or species.

DARNER

Aeshna and *Rhionaeschna* spp. (**eesh**-na: spear). Body 2¾" long; background color is dark brown; each flank of thorax has 2 pale diagonal stripes. **Paddle-tailed Darner**
A. palmata (pahl-**may**-ta: fanned-out). Thorax stripes and face yellow; dots on abdomen blue or (on some ♀) yellow. Forest lakes, ponds, and fens, often with sedges; A.

metamorphose gradually through ten to fifteen larval instars called naiads. Some either burrow and wait for prey or crawl sluggishly on their six sturdy legs. Others swim, and can use a kind of rectal jet propulsion for a burst of speed.

They crawl up out of the water on an emergent stem when ready to molt and release their inner adult. It takes the new adult about one very vulnerable hour to straighten and stiffen its wings (by pumping blood in through the veins) for flight. The nymphal skin, split down the back, is left behind, grasping the sedge stem (shown in photo).

Both larvae and adults hunt by eyesight, lacking much sense of hearing or smell. The

Paddle-tailed darner

Blue-eyed darner female placing eggs in water

Sedge Darner
A. juncea (**jun**-see-a: rush). Stripes and dots whitish; tip of abdomen bluish. Forest lakes, ponds, and fens, often with sedges; A.

Blue-eyed Darner
R. multicolor. Bright-blue eyes; stripes and dots all sky blue or (on some ♀) all yellow. Near ponds, often with pond lilies; sometimes seen well away from water; mainly lower elevs; A.

DARNERS, SOME OF the biggest, fastest dragonflies, may reach 33 miles per hour intercepting prey. (This is measured using radio-transmitters light enough, at 0.23 grams, to attach to a dragonfly without slowing it down much.) Dragonflies live in the fast lane, ceaselessly on the wing, feeding their high metabolisms with large numbers of unsuspecting mosquitoes and flies, quickly crushed and consumed. Their predation success rates, flight speed, maneuverability, and brain's executive function are all anomalously high, especially for their small brain size.

In darners, large size and frigid habitat tend to prolong development; our species spend between two and five years as naiads.

Though they live most of their lives in water, darners climb slopes to seek mates—hilltopping behavior, as seen in many butterflies. Before seeking a mate, a male curls his abdomen down and attaches a sperm packet just behind his waist. To mate, he uses his tail-tip claspers to grasp a female

by the head; to retrieve the packet, she then curls her abdomen way down, forward, and up. Together, they form the "wheel position." Female darners often lay their eggs surreptitiously to escape the harassment of repeated attempts to mate, since most species' males don't guard their mates as some dragonflies do. In the wheel position, they can both fly perfectly well, facing forward—one advantage of a long, skinny abdomen.

Another advantage is the long counterweight, for flight stability, and a third is thermoregulation: the abdomen's high surface-to-volume ratio makes it a good radiator for excess flying-muscle heat pumped out of the thorax. Giant dragonflies with 28-inch wingspans, back in the Carboniferous Period, may have been among the first thermoregulators. (Insect gigantism was rife 300 million years ago, possibly thanks to a richer atmosphere. Plants had recently evolved to large sizes, but with dinosaurs still 50 million years off, a lot of oxygen was being produced and not as much consumed.) The abdomen also absorbs radiant heat: either basking or in flight, it gets positioned—east to west, north to south, up and down—to maximize or minimize heat absorption, as needed.

CHERRY-FACED MEADOWHAWK
Sympetrum internum (sim-**pee**-trum: on rock, presumably to bask). 1½" long; ♂ deep red except black triangles running the length of abdomen; veins often orange-tinged, clear

Cherry-faced meadowhawk pair

wing area amber-tinted in some ♂; ♀ reddish to greenish-tan. Seasonally wet/dry marshes; A. Similar white-faced meadowhawk, *S. obtrusum*, has a large, pale blotch across face.

Our common, smallish red dragonflies are meadowhawks of several species. Couples fly in tandem for hours, sometimes over meadows. They drop the eggs while still in tandem, usually tapping them onto water, but sometimes where seasonally dried up; eggs need to be inundated to hatch, in spring. They perch horizontally, but when it gets too hot, they "obelisk" with abdomens pointed toward the sun to minimize exposure.

EMERALD SPREADWING

Lestes dryas (**les**-tease: robber; **dry**-us: wood nymph). Also stocky lestes. Body 1⅜" long; metallic green with bluish iridescence, blue eyes, blue tip, yellowish flanks and underside; wings clear with black spot near tips; held at angles at rest. Sedge marshes, fens; A.

Emerald spreadwing

Many damselflies become dull and dark when they cool off, helping them absorb heat more efficiently if the sun comes out again. They recover their brilliant colors when body temperature exceeds 68°F. Their color also shifts with maturation.

DANCER

Argia spp. (**ar**-gi-a: leisure). 1⅜" long; wings folded when at rest; bobs and weaves ("dances") in low flight. Naiads live in streams, not ponds or bogs.

Vivid Dancer

A. vivida. ♂ brilliant blue (to blue-purple) with black bands; ♀ range from blue through purple-brown to brown. Springs, small streams, irrigation ditches; A.

Vivid dancer

Emma's Dancer

A. emma. ♂ mostly lavender, except last 2 segments bluer; ♀ tan or pale blue-green. Rocky riverbanks; A–.

Emma's dancer

MALES OF MANY damselflies have developed an organ that, as they commence mating, can pluck out and discard a sperm packet left by a

Nobo bluet pair

NOBO BLUET

Northern bluet, *Enallagma annexum*, and boreal bluet, *E. boreale* (en-a-**lag**-ma: crosswise; an-**ex**-um: bound together). 1¼" long; ♂ abdomen brilliant blue with black bands; thorax has blue/black stripes running its full length; ♀ tan. Marshes and sedgey lake edges; also seen in tandem away from water; A.

Even experts can't tell these two species apart without putting them under a microscope, so in the field they just call them "nobo" or "borthern" bluets. Sometimes they're numerous enough to cast a blue haze over a lake. A boreal bluet typically enjoys just four days of life as an adult.

previous male, and replace it with their own sperm packet. (See darners, p. 469, for dragonfly sex basics.) In response, dancer males came up with what may be the surest way to guard a mate: keep her in tow. The pair fly around attached in tandem for several hours from morning mating to midday egg-laying. Using a pair of claspers at his abdomen tip to grasp her neck, he lifts her out of the water when she's done dipping her ovipositor to insert eggs onto a submerged leaf.

Order Lepidoptera (Butterflies and Moths)

This order is named for "scaly wings," and characterized by fine scales all over, which rub off easily, enabling Lepidoptera to escape sticky spiderwebs, and also playing the lead role in their coloration. We think of them as either moths or butterflies depending on the following traits, though there are exceptions to all of them except antenna shape:

MOTHS	BUTTERFLIES
• are mostly nocturnal	• are diurnal
• have slender-tipped, or else fernlike, antennae	• have club-tipped antennae
• have bigger bodies for their wing size	• have smaller bodies relative to wings
• may pupate in a chrysalis with an outer cocoon of silk	• pupate in a naked, more or less rigid shell, or chrysalis
• raise their body temperatures before flying by shivering their wings	• raise their body temperatures before flying by basking, or less often by shivering
• perch with wings spread to the sides, either flat or angled rooflike, forewings covering hind wings	• perch with left and right wings pressed together vertically, except (in some families) when basking
• in some genera do all their eating in the larval stages	• generally sip some nutrients as adults, not just as larvae

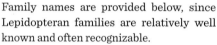
Aspen leaf miner

Western tent caterpillar moth

Family names are provided below, since Lepidopteran families are relatively well known and often recognizable.

ASPEN LEAF MINER

Phyllocnistis populiella (fill-oc-**nis**-tiss: leaf rasp; pop-you-lee-**el**-a: poplar—). Larva inside aspen leaf, carving a pale, mazelike, gradually enlarging path, consuming a green epidermal layer 1 cell thick; adult a small, drab moth. Gracillariidae.

Larvae that mine leaves evolved separately in moths, beetles, flies, and wasps. In each case, the larva is flattened to fit between the leaf's two cuticles. This one infests and weakens aspens, occasionally reaching pest levels. From an egg laid near a leaf tip in May, a new larva makes a beeline for the leaf base, right alongside the midvein, but is so small that this part of its path is hard to see. Growing, it mows back and forth on one side of the midvein—an obstacle it will cross only where small, near the leaf tip. It may consume much of the second half of the leaf before spinning its silken cocoon and rolling the leaf edge around itself. If two mothers lay eggs on the same leaf, each larva typically set-tles for one side of the midvein; it sounds like nice resource partitioning, but they won't grow as big on half a leaf. More than two eggs per leaf often lead to cannibalism: a caterpillar overtaken from behind gets eaten.

WESTERN TENT CATERPILLAR MOTH

Malacosoma californica (ma-la-co-**so**-ma: soft body). ¼–1½" ws; variably brown, forewing (the one in view when wings are folded) divided in thirds by 2 parallel fine lines; larvae (caterpillars) 2" when fully grown, bristling with tufted long hairs, generally dark brown with blue, orange, and reddish markings; egg masses plastered against twigs, esp at crotches, covered with a gray to brown foam that hardens to a nearly waterproof coating. On broadleaf trees and shrubs; occasionally on conifers; A. Lasiocampidae.

During tent caterpillar outbreaks, many trees and shrubs are defoliated and achieve little growth for a year or two, though few are killed.

The "tent" is a big, erratic silk web that affords caterpillar groups protection and insulation during resting periods between feeding sprees. Like their silkworm relatives, tent caterpillars pupate within silken cocoons coated with a skin-irritating dust, under curled leaves. Adult tent caterpillar moths have no working mouthparts; they survive purely on what they ate as caterpillars, living just a few days to mate and lay eggs. Tiny larvae are already formed within the egg cases by winter, ready to chew their way out when the leaf buds burst in spring. Larvae do all the eating and growing as well as much of the traveling—they have long bristles to help the wind carry them. Lacking wind, they crawl en masse. Tent building, mass migration, group basking, and other social activities rank them among the most social of caterpillars.

Trees are far from defenseless against foliage grazers. They need only slow the caterpillars' growth by a small percentage to multiply the number taken by predators and parasites. They can do this by loading their leaves with tannin and other hard-to-assimilate chemicals. However, that requires an investment of energy that the trees do well to avoid, so they may load only some of their leaves with tannin, forcing the caterpillars to expose themselves to predatory birds while searching for palatable leaves. Or the trees may wait until attacked, and only then step up tannin production. Remarkably, trees may increase their tannin in response to an insect attack on a nearby tree of the same species, sensing chemicals (alarm pheromones?) cast upon the breeze by the attacked tree. This is among the clearest evidence to date of communication among plants. In the words of one enthusiastic entomologist, "Plants are really just very slow animals."

POLICE CAR MOTH

Gnophaela vermiculata (noff-**ee**-la: darkness—; ver-mic-you-**lay**-ta: wormy-patterned). Also bluebell tiger moth. 2" ws; wide black margins surround white patches divided by black veins; yellow-orange fur on cheeks; antennae plume-like; flight erratic. Bristly black-yellow larvae eat bluebells; adults often on asters, other composites; June–July; A. Erebidae.

BLACK-VEINED FORESTER

Androloma maccullochii (an-dro-**lo**-ma: males fringed?). 1⅝" ws; distinctive hairy, yellowish shoulder pads on thorax; wide black margins surround cream patches divided by black veins; white fringe on trailing edge of wings; orange fur on upper legs; antennae threadlike, black-and-white-ringed on lower ½; greenish-yellow

Police car moth

Black-veined forester pair

Pandora pine moth

White-lined sphinx moth with tongue extended into nectary

larvae eat fireweed, sometimes stripping large patches. May–August. Noctuidae.

This and the police car moth—day-flying moths that rival butterflies in beauty—look similar except for their very different antennae, which show that they are not closely related.

PANDORA PINE MOTH

Coloradia pandora (pan-**dor**-a: all-gifted, a mythic original and perfect woman who unleashed trouble). 3–4½" ws; antennae plumelike, large, yellow-orange; hind wing nicely patterned above, white with gray and varying amounts of pink; forewing patterned in drab grays; body thick and furry. Around pines, the larval food; southerly, plus OR and WA. Saturniidae.

Pandora caterpillars take two years to develop, growing huge on a diet of pine needles, before pupating for up to five years. The largest instars are an important traditional food of Paiute people. After roasting in sand to remove the hairs, the caterpillars can be stored for a long time, then eaten as a snack or boiled to make a mushroom-flavored soup.

Populations irrupt occasionally, stunting growth of host pines for four to eight years, but killing fewer than 2 percent.

Lacking mouthparts for eating, the adults live only long enough to mate and lay eggs, and the females avoid flying until they have mated. At times, an outbreak may look like a bounty of huge, dead moths.

WHITE-LINED SPHINX MOTH

Hyles lineata (**hi**-leez: of woods; lin-ee-**ay**-ta: lined). 2½–3½" ws; large, heavy-bodied moth with rapid, buzzing wingbeats suggestive of hummingbirds; forewings dark olive-brown with a broad pale stripe to the tip, and white-lined veins; hind wings pink and black, much smaller; thorax white-lined; abdomen cross-banded. Wide range of host plants; A. Sphingidae. A very similar species without fine, parallel white lines on wings or thorax would be a bedstraw sphinx, *H. gallii*.

Moths in this family feed on nectar by day, night, or dawn and dusk. With a deep-pitched, buzzing flight style that includes hovering, they resemble other nectar feeders—

Eyed sphinx moth

bumblebees or, in the case of this genus, hummingbirds—almost more than they do other moths. Some have hairy bodies and clear wings, like bees. White-lined sphinx moths are the top pollinators of Colorado blue columbines, whose color variations reflect strategies for how best to get themselves pollinated in different areas.

EYED SPHINX MOTH

Smerinthus astarte (smer-**inth**-us: cord—; a-**star**-tee: a goddess of love and fertility). Also *S. cerisyi, S. ophthalmica*. 2⅜–3⅜" ws; variable; generally a gray to sepia camouflage pattern, exc each hind wing has a black-pupiled, split, blue eyespot in a pink/golden triangle. Streamsides; feeds on willow, aspen, cottonwood; A. Sphingidae.

SWALLOWTAIL

Papilio spp. (pa-**pil**-ee-o: butterfly). Also *Pterourus* spp. Large butterfly with "tail" lobe trailing each hind wing, dramatically black-patterned, often with blue or orange spots near tail. May–Aug. Papilionidae.

Western Tiger Swallowtail

P. rutulus (**root**-you-lus: shining, red-orange, or gilded). 2⅝–3⅝" ws; one of the several tiger stripes on forewing continues across hind wing; marginal crescents on hind wing are yellow. A.

Pale Tiger Swallowtail

P. eurymedon (you-**rim**-e-don: Greek charioteer in *The Iliad*). 2¾–3¾" ws; like the preceding but pale part is white to very pale yellow, and black border is wider. Larvae eat snowbrush; A.

Two-tailed Tiger Swallowtail

P. multicaudata (mul-ti-cow-**day**-ta: many-tailed). 3½–5" ws (largest in the West); black stripes narrower than on the preceding 2; each hind wing has 2 tails; ♀ has long rows of marginal blue spots. Prefers lower elevs; eats chokecherry, thistle, and (in towns) green ash; A.

Anise Swallowtail

P. zelicaon (zel-ic-**ay**-on: a Greek hero). 2¼–3" ws; forewing striped yellow and black, but fine black veins—not a stripe—divide yellow area of hind wing. Larvae eat parsley-family plants; A.

Western tiger swallowtail

Pale tiger swallowtail

Two-tailed tiger swallowtail

Black Swallowtail

P. polyxenes (pa-**lix**-en-eez: Trojan princess in *The Iliad*). Primarily black, with 2 long rows of

Anise swallowtail

yellow (to white) spots and a hind wing row of 5 or 6 blue spots that are big and prominent on ♀ only. Larvae eat parsley-family plants; NM, CO, s-most WY.

ANISE AND PALE swallowtails are often seen hilltopping—using landmarks (trees, buildings, and even TV towers where hills are lacking) to help locate mates. The males stake out sections of ridgeline where they await females. Similarly, western tiger swallowtails use riverbanks, canyons, or other lineations of trees as their singles bars. Many butterfly species will hilltop when their population is low, but when they are abundant, they seek mates by flying around near the larval host plants. Mate-seeking is the chief activity of butterflies. Females often succeed on their first day, as the males are all out and hot to trot, having emerged as adults a few days earlier. Males spend longer on the hunt, since most females are already mated and unreceptive.

Black swallowtail pair

The caterpillars flourish by late summer, and pupate over winter. The final instar of the black swallowtail has striking black, yellow, and white bands, deterring predators by mimicking the toxic caterpillars of monarch butterflies. Late instars of the tigers have eyespots; these faux eyes function defensively, as do the antlerlike scent glands that they stick out when disturbed. Early instars are camouflaged as bird droppings.

Papilio, the Latin word for butterfly, was the genus in which Linnaeus originally placed *all* butterflies. Let's just say he was more of a plant guy.

PARNASSIAN

Mountain parnassian

Parnassius spp. (par-**nas**-ius: of Mount Parnassus, home of the gods). 2–2¾" ws; white with black and translucent gray markings and small, red, often-black-rimmed eyespots. Papilionidae.

Clodius parnassian | Pine white

Mountain Parnassian

P. smintheus (**sminth**-i-us: Mouse God, a name of Apollo). Also *P. phoebus*. Fore- and hind wings usually with several red spots; antennae ringed black/white. High elevs, typically on or near stonecrop, the larval food; June–Aug; A.

Clodius Parnassian

P. clodius (**cloh**-dius: a Roman politician). Eyespots (usually 4) on hind wings only, antennae black. Widespread near forest edges; larvae feed on bleeding-heart; adults nectar on buckwheats; WA to UT and w WY; May–Aug.

The gray patches on parnassian wings are just windows through the minute scales that cover butterfly wings. Scales give butterfly and moth wings their opacity, color, and softness; without scales, they're translucent like the wings of bees or flies. If you attempt to grasp a butterfly, hundreds of scales will dust your fingertips.

Female parnassians are usually seen with a waxy object on the rear end. It hardens from a liquid extruded by the male during mating, and blocks any subsequent mating. It is largest on *clodius*, whose mating style some scientists describe as rape.

Early snowmelt may extirpate mountain parnassians by forcing them to begin feeding at a time of year when stonecrops are toxic to larvae. While ecologists have written papers assuming that many species will disappear from an area when it no longer has its current climate, few studies identify mechanisms that can actually cause that. This could be one such mechanism.

PINE WHITE

Neophasia menapia (ne-o-**fay**-zia: shines like new; men-**ah**-pia: moon bee?). 1¾–2" ws; white with darker veins that are very fine on ♂ and broad on ♀; forewing has blackish blotch pattern near tip and a short, blackish bar extending inward from front edge; a bit of orange around margins, mainly on ♀. Green larvae eat pine and other conifer needles; June–Oct. Pieridae.

Pine-white caterpillars irrupted in parts of central Idaho between 2010 and 2012, killing many ponderosa pines. Bark beetles are drawn to the weakened pines, and finish them off, while wasps that parasitize pine-white pupae multiply, and reduce the butterflies to normal numbers. The eggs are attached to pine needles in neat rows.

MARGINED WHITE

Pieris marginalis (**pie**-er-iss: a Greek Muse). Also veined white, mustard white; *P. napi, Artogeia marginalis*. 1⅜–1¾" ws, white, dark-stained near body; sometimes yellowish; early-spring individuals have darkened veins. Widespread on moist, cool sites; bright-green larvae feed on

Margined white

Western white

Orange sulphur

white, *P. rapae,* with forewing usually dark-spotted and dark-tipped, is a pest in vegetable gardens.

WESTERN WHITE

Pontia occidentalis (**pon**-sha: a name of Aphrodite; oxy-den-**tay**-lis: western). Also *P. callidice.* 1⅜–1¾" ws; white with broad gray-green vein lines, esp beneath hind wing, and variable gray pattern near margins of fore- or all wings; underside has only 2 colors— white and gray-green; black-stained near body. Alp/subalp; occasionally lower; larvae feed on mustard family; Apr–Sep. Pieridae. Similar Becker's white, of sagebrush country, has more green than white under hind wing, and black spot on leading forewing edge.

Pairs of butterflies, especially whites and sulphurs, follow each other upward in spirals. It may please us to think of these flights as prenuptial stairways to heaven, but it's more likely a sort of chase: a male trying to get close enough to determine the other's sex and receptivity, while the other flies upward to avoid contact. Males don't literally fight, being fragile and lacking weapons of any kind.

SULPHUR

Colias spp. (**coh**-lius: a name of Aphrodite). 1½–2⅜" ws; yellow to light orange above, ♂ often with a heavy black or gray border or corner; light green to yellow under hind wing (or some ♀ white overall); usually with a dark dot in center of forewing and a pale dot on hind wing. Larvae feed on pea family (lupine, sweet-vetch, milkvetch, sweet-clover, etc.). Pieridae.

Clouded Sulphur

C. philodice (fil-**odd**-iss-ee: a naiad of Argos). No orange tint; a close look usually reveals minute pink fringes, and not 1 but 2 small, pale, pink-rimmed spots on hind wing. Fields, roadsides; May–Oct; A.

Orange Sulphur

C. eurytheme (you-**rith**-im-ee: another Greek nymph?). Also alfalfa butterfly. Like clouded,

mustard family; Mar–Sep (has 2 broods each year). Pieridae.

This native feeds on native plants and is rarely seen in cities. The introduced cabbage

Eyespots and Tails

Dots framed in concentric rings decorate many butterfly wings, offering protection from predators. Birds are fooled into pecking at these fake eyes, thinking them vital targets, though butterflies have little trouble flying with tattered wings. "Tails" on hind wings are similar diversionary ornaments.

Very large eyespots may do more, actually startling and repelling predators. The butterfly responds to an exploratory first peck by abruptly opening its folded wings; between the two "eyes" suddenly revealed, the bird imagines a larger face than it had meant to take on. Since moths rest with wings spread, eyed sphinx moths (p. 475) keep the eyes on the hind wing hidden behind the forewing while at rest, camouflaged against bark; to suddenly reveal the eyespots, they quickly slide the forewings aside.

Newly emerged butterflies go through a few hours of flightlessness, their wings soft. Wings first take form all wadded up in the pupa, then when the adult emerges it pumps fluids in through the veins to extend the wings to their proper shape. Achieving rigidity takes several hours more. During those vulnerable hours, eyespots save lives.

but orange-blushed above; often shows a row of faint brown spots parallel to margin. July–Oct; A.

Queen Alexandra Sulphur
C. alexandra. Wings fringes greenish to pale yellow; pale spot on hind wing is ± rimless. Meadows, prairies, tundra; May–Sep; A.

ALEXANDRA WAS A mere subspecies of *C. occidentalis* until their wings were studied under ultraviolet light, and looked very different. Butterflies use the patterns of ultraviolet-reflective scales to recognize conspecifics. Females get better at it with experience; newly emerged ones are often fooled by arduous suitors from the wrong species. Dark scales that stain part of the wings absorb extra heat close to the body, making basking more efficient. Sulphurs bask with wings folded, tipping to adjust how much heat they're gaining.

The Old English origins of the word "butterfly" likely trace to a butter-colored sulphur. (Sorry, it didn't originate as "flutter by.") The spelling "sulphur" started out as a phony effort to make a Latin (or Arabic) word look Greek. Spellings with ph and with f then

Orange-tip

contended in European languages for millennia. By 1900, the British went with sulphur, Americans with sulfur, and Canadians straddled the phence. By 2000, chemists from all over agreed that the sixteenthth element shall henceforth be sulfur—but lepidopterists and geologists thumb their noses at them.

ORANGE-TIP

Anthocharis stella and *A. julia* (an-**thoc**-a-ris: flower grace). Also *A. sara.* 1–1¾" ws, white to (♀ only) ± yellow with deep-orange forewing tips, gray-green marbling under hind wing. Sunny

Common ringlet

Dark wood nymph

Common wood nymph

brown beneath; 2 black eyespots on each forewing; spots ± yellow-rimmed beneath and sometimes above; up to 6 tiny eyespots (often indistinct) on hind wings. Larvae eat grasses; adults sip nectar, sap, carrion. Nymphalidae.

Common Wood Nymph
C. pegala (**peg**-a-la: a mythic name). 1¾–2¾" ws, much paler beneath (esp ♀). Roadsides, meadows; June–Sep; A.

Dark Wood Nymph
C. oetus (**ee**-tus: doom, i.e., blackness). Also small wood nymph. 1½" ws, almost as dark beneath as above; 2nd forewing eyespot small and often inconspicuous. Dry grasslands, brush, open forest; June–Sep; A.

CHRYXUS ARCTIC

Oeneis chryxus (ee-**nee**-iss: a mythic king; **crix**-us: gold). 1¾–2" ws; orange- or gray-brown exc silvery gray-brown under hind wing (for camouflage against bark or soil); 1 to 3 smallish black eyespots; forewing long and angular. High ridges and meadows; June–Aug; A. Nymphalidae.

Arctics, like anise swallowtails, are hill-toppers with a powerful drive to travel uphill. Ridgelines acquire evenly spaced populations of hilltopping males, with scattered cameo appearances by eligible maidens foraying up from the more benign meadow habitat. A male will fly up to check out any

habitats; larvae feed on small mustard-family plants; an early butterfly: Apr (at low elevs) to July (high); A, with *stella* northerly and *julia* southerly. Pieridae.

COMMON RINGLET

Coenonympha tullia (see-no-**nim**-fa: common nymph; **too**-lia: a daughter of Cicero). Also *C. california, C. ampelos.* 1⅛–1½" ws, variable: ours usually creamy to bright yellow-orange exc under hind wing, where ± gray; underside usually has median white bands, and may have black eyespots near margins. Around grasses, the larval food; late May–Aug; A. Nymphalidae.

WOOD NYMPH

Cercyonis spp. (ser-**sigh**-o-nis: a legendary thief?). Drab black-brown above, barklike

lepidopteran, and if it's a conspecific, he engages it in a spiral flight while he tries to learn the visitor's sex and availability.

Arctics are among the minority of basking butterflies that bask with their wings folded and tipped aside. A butterfly basking in the morning is sluggish, offering you a good chance at a close view or a photo. Arctics spend two years as caterpillars, feeding on grasses and sedges.

Painted lady

COMMON ALPINE

Erebia epipsodea (er-**ee**-bi-a: underworld, ref to dark shade; ep-ip-**so**-dia: like *E. psodea*). 1½–1⅞" ws, dark brown with marginal rows of orange-haloed, yellow-pupiled black eyespots on all 4 wings; the forewing haloes coalesce into an orange band. Meadows, prairies, or tundra; May–Aug; A. Nymphalidae.

PAINTED LADY

Vanessa cardui (**car**-dew-eye: of thistles). 2–2¼" ws, salmon to orange above with short white bar and black mottling near forewing tip, row of 4–5 tiny eyespots near hind wing margin; larvae are spiny, lilac with yellow lines and black dots, and often spin nets around thistle tops. All elevs; Apr–Sep; A. Nymphalidae. Similar West Coast lady, *V. annabella*, has a row of 4 small, blue-filled spots above each hind wing.

Painted ladies are the most nearly worldwide of all insects, and are champion migrators. Regional populations explode occasionally—for example, in 2017 and 2019 here. Abundance seems to follow relatively wet winters in the southwestern deserts, which enable copious reproduction among the painted ladies wintering there. An irruption may take several generations, with individuals typically migrating a few hundred miles, the populations much farther. Under optimal conditions,

Chryxus Arctic

Common alpine

Lorquin's admiral

Weidemeyer's admiral

some individuals migrated 3,000 miles from northern Scandinavia to Morocco. To navigate, they use the direction of the sun together with an innate sense of time. Painted ladies are rather scarce here in between irruption years—unless you happen across survivors from a farm-bred flock released at a wedding. (Lepidopterists are not happy about these commercial releases.)

Feminine names are rife among butterfly scientific names. In most cases, like Vanessa, we have no record that a particular girl or woman was the honoree, but, with *V. annabella,* we do know that Annabella is the namer's daughter.

ADMIRAL

Limenitis spp. (lim-en-**eye**-tiss: harbor goddess). Also *Basilarchia* spp. 2–3" ws; mainly black and white above, the white in a heavy band (broken by black veins) across both wings parallel to margin; hind wing has a subtle row of orange spots; underside has same white band; sexes alike, but ♂ smaller; in flight, twitchlike beats alternate with gliding; larvae look like bird droppings, eat tree and shrub leaves. Nymphalidae.

Lorquin's Admiral
L. lorquini (lor-**kwin**-eye: after Pierre Joseph Michel Lorquin, who took his butterfly net with him on the California Gold Rush). Red-orange tips on forewings; underside dark parts mainly red-orange. Canyons, streamsides; MT, ID, WA, OR; May–Sep, with 2 generations per summer.

Weidemeyer's Admiral
L. weidemeyerii (wide-**my**-er-ee-eye: after John Weidemeyer). No orange tips above; underside variable, sometimes patterned with white, orange, black, and bluish. Streamsides, aspen groves, towns, tundra; June–July; A–.

THE HANDSOME ADMIRALS are scavengers of dung and dead things; they are also nectar feeders. The two species above hybridize where they meet in the Bitterroots and central Idaho.

Mourning cloak

MOURNING CLOAK

Nymphalis antiopa (nim-**fay**-lis: a minor nature deity; an-**tie**-o-pa: a leader of Amazons). 2–3¼" ws; dark red-brown above with cream-yellow border lined with blue spots; drab beneath, a barklike gray-brown with dirty-white border; margins ragged. Ubiquitous near broadleaf trees or shrubs; Feb–Oct; A. Nymphalidae.

The mourning cloak is familiar throughout the Northern Hemisphere. The adults hibernate—even in Alaska—occasionally coming out to fly on sunny winter days, and sometimes estivate through late summer. On dehibernating in spring, they look very ragged. The caterpillars—black with black spines, white speckles, and a chain of red spots—feed on broadleaf trees and shrubs. Colonies of them may be tied together with silk threads, or may choreograph unison dances to scare off predators.

Brush-footed butterflies—the large and colorful family Nymphalidae—perch on four legs. The brushlike forelegs alongside their faces are too short to walk on, and instead have sense organs for taste-testing food.

California tortoiseshell

Fire-rim tortoiseshell

CALIFORNIA TORTOISESHELL

Nymphalis californica. 1¾–2¼" ws; blotchy yellow-orange above; several large, dark blotches on leading edge, and a continuous dark margin; barklike mottled gray-brown beneath—much like a big comma but less ragged in outline, and not much like its *Nymphalis* kin. Open forest or brush; larvae eat *Ceanothus;* Feb–Nov; A. Nymphalidae.

Tortoiseshell and comma adults hibernate, often in tree hollows. Emerging in spring, they look worn, and paler than fresh adults. California tortoiseshells breed two or three generations a year. Their populations are volatile, with wild boom-and-bust cycles. During boom years, thousands smash into cars as they stream through mountain passes, and the larvae defoliate snowbrush far and wide.

FIRE-RIM TORTOISESHELL

Aglais milberti (a-**glay**-iss: Brilliance, one of the Graces; mil-**bear**-tee: for Jacques-Gérard Milbert). Also *Nymphalis milberti.* 1½–2" ws, dark brown above exc for a huge, firelike band parallel to margin, plus 2 small, less distinct orange blotches on each leading edge; gray-brown beneath, with a broad, paler band parallel to margin. All elevs; larvae eat nettles; adults often visit daisies; Apr–Oct; A. Nymphalidae.

Clusters of twenty to nine hundred pale-green eggs on nettles—or bristly, small, black, web-spinning caterpillars on nettles—are likely those of fire-rim or

Hoary comma

Gillette's checkerspot

Sagebrush checkerspot

Milbert's tortoiseshell. Adults migrate up to tundra in summer just to estivate; they descend to reproduce and to hibernate.

HOARY COMMA

Polygonia gracilis (pol-y-**go**-nia: many angles; **grass**-il-iss: slender). Also anglewings. 1¾–2¼" ws, wing margins lobed fancifully, as if tattered; burnt-orange above with dark margins and scattered black blotches; silvery gray-brown beneath, with a pale "comma." Ubiquitous; larvae eat currants and nettles; adults may eat sap, carrion, or scats. Brush, streamsides; Mar–Sep; A. Nymphalidae.

Commas are amazingly well camouflaged against bark: one moment they flit about almost too fast to follow, and in the next they disappear by abruptly alighting with wings folded. A tiny but distinct pale mark centered beneath the hind wing is variously C-, V-, boomerang-, or comma-shaped on different *Polygonia* species.

CHECKERSPOT

Euphydryas (you-fi-**dry**-us: shapely nymph) and *Chlosyne* spp. (clo-**sigh**-nee: a nymph?). Also *Charidryas*, *Occidryas*. 1⅛–2" ws, upperside black in fine, checkered patterns with red-orange, pale yellow, and/or cream; underside red-orange with yellow to white checks, black veins. Sunny, dry habitats; June–Aug. Nymphalidae.

Gillette's Checkerspot

E. gillettii (jil-**et**-ee-eye: after the namer's wife, Charlotte Gillette). Broad red-orange band across both wings near rear edge; otherwise mostly dark above, cream beneath. Larvae eat twinberry; northerly.

Sagebrush Checkerspot

C. acastus. Entirely checkered orange/black above. Larvae eat rabbitbrush and asters; A.

Like many subalpine butterflies, Gillette's checkerspot develops slowly, with two different larval instars hibernating in two consecutive winters. The younger caterpillars forage communally in a silken nest.

Field crescent

Mormon fritillary

Identifying the notoriously variable checkerspots often requires studying their genitalia under a microscope, and even that isn't final, since experts disagree on the number and scope of species.

FIELD CRESCENT
Phyciodes pulchella (fis-**eye**-o-deez: seaweed-red; pool-**kel**-a: beautiful). Also *P. campestris, P. pratensis*. 1¼" ws; antennae tips orange; largely black-brown above, with rows of red-orange and yellow-orange spots—the outermost spots crescent-shaped; low-contrast pale orange-yellow-brown patterns beneath, forewing usually black-edged. Common in sun, low to subalpine; larvae eat asters; May–Sep; A. Nymphalidae.

GREATER FRITILLARY
Argynnis spp. (ar-**jin**-iss: a boy loved by Agamemnon). Also *Speyeria* spp. 1½–3½" ws; orange with ± checkerlike black markings above; hind wing gray to golden tan beneath, with many round silvery-white to cream spots in both sexes; forewing golden brown beneath, with dark-brown checks. Black larvae (sometimes showing orange spots or spine bases) eat violets; May–Sep. Nymphalidae.

Mormon Fritillary
A. mormonia. 1½–2¼" ws; underside often greenish/brownish; spots usually silvered. Subalpine; varied larval hosts include bistort, dwarf willow, others; A.

Hydaspe fritillary

Great spangled fritillary

Hydaspe Fritillary
A. hydaspe (hi-**das**-pee: a river, or else a king). 2–2½" ws; sexes similar; underside has marginal white spots, hind wing redder rusty-brown,

Arctic fritillary

cream hind wing spots large and rarely silvered. Mid-elevs; nw CO to WA.

Great Spangled Fritillary

A. cybele (**sib**-el-ee: Sibyl, an earth-mother deity). Also *S. leto.* 2½–3½" ws; ♀ blackish to chestnut-brown with broad, pale-yellow, patterned margin; pale spots under hind wing are silvered. Aspen or oak woodland with ferns; A.

THE MANY *ARGYNNIS* species are easy on the eyes yet hard to tell apart. The great spangled is largest, and the most sexually dimorphic—the black or brown females look utterly different from the colorful males. In summer-dry mountains, mated females estivate in summer, then reawaken in fall, presumably the most favorable time to lay eggs. The larvae hatch in fall and then hibernate until spring. The Mormon fritillary depends, in adulthood, on a single nectar source, the aspen fleabane, and has declined in central Colorado as the fleabane has suffered climate-related decline.

Purplish copper

ARCTIC FRITILLARY

Boloria chariclea (bo-**lor**-ia: net—; car-ic-**lee**-ah: see below). Also purplish fritillary; *B. titania* ssp. *chariclea.* 1½" ws; tawny yellow-orange above with broad, fine patterns in brown (compared to checkerspots, the orange is paler, yellower, and there's much more of it than of brown); hind wing rust and cream beneath, with a central row of spots that may show as lavender. Dark, bristly larvae eat diverse subalpine plants; subalpine; June–Sep; A. Nymphalidae

Lilac-bordered copper

The name fritillary (both a lily and a butterfly) refers to a Roman dicebox with checkered markings. In the *Aethiopica* (one of the five ancient Greek romantic novels), Heliodorus tells the story of Chariclea, daughter of King Hydaspes and Queen Persina. Both parents are Black, and she fair, landing her in a world of trouble from day one. She triumphs. I suspect naming this butterfly for her was meant to imply a smaller, paler close relative of *B. hydaspe.*

A Guy Worth His Salt

Butterflies congregating on moist earth are "puddling"—sipping water and the nutrients that come with it. A behavior common in blues and swallowtails, it's also seen in a variety of insects. The vast majority of butterflies you see puddling are males, offering a clue that this isn't primarily driven by thirst. Close observation finds some puddlers "pumping"—siphoning many times their body weight per hour in one end and out the other, extracting sodium from the water and adding potassium salts to the water they excrete. Plant leaves evolved potassium-sodium imbalance as a defense, and all kinds of herbivores suffer the effects and are driven to seek out sodium salt sources such as your urine or your pack straps. Puddlers like the mud around mineral-spring seeps, or where a horse urinated on a trail.

Males gather nutrients and water to transfer them to a female when they mate. One spermatophore, or shot of ejaculate, commonly equals 15 percent of male body weight, and takes an hour to deliver. Zoologists call this a nuptial gift—something nutritious conveyed to a mate, which will nourish the progeny. For a bird, the appropriate gift might be a berry or a bug, but for a butterfly it's a big, salty spermatophore. Both parents benefit when the female doesn't waste time puddling, and has more time to spend searching for the perfect spot to lay eggs.

Additional useful gifts in the spermatophores of various species include water (conveyed as a gel), proteins, or nitrogen, the building block of proteins. That's why butterflies sip from protein-rich stuff like carrion and carnivore scats. To feed her eggs protein, a female who is undernourished in nitrogen may break down her own flight muscles. By saving her from doing that, the nuptial gift lengthens her life and raises her number of egg clutches.

Males of some whites spike their gift with methyl salicylate, whose odor repels other males. This benefits the female for a few days, protecting her from male harassment while she lays eggs; however, after a point she is ready to mate again and would be better off not reeking of antiaphrodisiac. Females are generally unready to mate again until they have consumed the spermatophore, so bigger is always better from the male point of view.

COPPER

Lycaena spp. (lie-**see**-na: she-wolf). Also *Tharsalea* spp., *Epidemia* spp. 1" ws, ♂ dark brown above with purple iridescence; ♀ yellow-orange above with brown speckles and borders. Lycaenidae.

Purplish Copper

L. helloides (hel-**oy**-deez: like species *helle* of Europe). Also *L. florus.* Both sexes have prominent orange crescents in a row (zigzag) along hind wing margin; forewing ochre beneath with spots, hind wing ± pinkish ochre. Roadsides, wet meadows, streamsides; larvae eat buckwheat family; May–Sep.

Lilac-bordered Copper

L. nivalis (niv-**ay**-lis: of snow). Upperside like the preceding; hind wing underside yellow with outer portion pale lilac; forewing underside yellow with dark spots; gorgeously colorful fresh, soon dulling. Meadows, canyons, sage steppe; larvae eat *Polygonum;* all but NM; May–Aug.

Mariposa Copper

L. mariposa (mar-i-**poh**-sa: "butterfly" in Spanish). Little or no orange marginal zigzag;

Mariposa copper

Lupine blue male

Arctic blue female

hind wing underside ashy gray with barklike pattern, white chevrons; forewings dull orange with spots. Mid-to-high-elev openings, bogs; larvae eat low blueberry species; northerly; July–Sep.

IRIDESCENCE IS COMMON among tropical butterflies. Here we find it in coppers—small, brown butterflies going, in Robert Pyle's words, "from penny-brown to neon purple in the flash of a sunbeam." Iridescent butterfly scales may have microscopic ridges that reflect light prismatically, or they may be of two types on one butterfly, regularly interspersed on the wing surface and held at different angles—not unlike the way fabric designers combine contrasting warp and weft threads to weave an iridescent satin.

The purplish copper hedges its overwintering strategy: some eggs overwinter before hatching, while others hatch into larvae that will endure winter.

BLUE

Plebejus spp. (pleb-**ee**-us: ordinary). Also *Icaricia* spp. Small butterfly; ♂ blue above, ♀ brown; both sexes dirty-white to gray beneath, with rows of spots darker on forewing than on hind wing. Lycaenidae.

Lupine Blue

P. icarioides (ic-air-y-**oy**-deez: like Icarus, who flew too near the sun). Also Boisduval's blue. ⅞–1¼" ws—large for a blue; white spots underneath with or without dark centers. Larvae eat lupine leaves, flowers, and seedpods; May–Aug; A.

Northern Blue

P. idas (**eye**-duss: one of Jason's Argonauts). ¾–1" ws; undersides of ♂ and both sides of ♀ usually show a bit of a line of orange crescents. Higher elevs; larvae eat heathers and legumes; June–Sep; A–.

Arctic Blue

P. glandon (**glan**-don: a pass in the Alps). Also *Agriades glandon, A. franklinii.* 1" ws; ♂ gray-blue above with black border; underside spots are large and white on hind wing, black with white rims on forewing. Larvae eat small flowers like rock-jasmine and saxifrage; June–Aug; A–.

Scarce and drab, the Arctic blue is none-theless conspicuous from time to time as the only butterfly flying in gray weather on high, windswept ridges. It roams Arctic Circle lowlands, and ventures south along Cordilleran summits.

ROCKY MOUNTAIN DOTTED BLUE

Euphilotes ancilla (you-fill-**oh**-teez: after genus *Philotes;* an-**sill**-a: maid). Also *E. enoptes, Philotes enoptes.* ⅝–1" ws, ♂ blue above; ♀ brown, usually with an orange flare near hind wing margin; fringes white, ± checkered with black; bluish white beneath, with scattered dark spots and orange hind wing flare. Sagebrush, open areas up to mid-elevs; near sulphur buckwheat, the larval food; May–Aug; A. Lycaenidae.

Rocky Mountain dotted blue

Caterpillars of many blues are attended by ants who eat "honeydew" secreted by the caterpillars. In this mutualistic symbiosis, the "stock-herding" ants sometimes attack and repulse insects that parasitize caterpillars.

Regarding species of *Euphilotes,* the trend is to recognize a distinct species for each species of *Eriogonum* buckwheat as larval host. With some 250 species of *Eriogonum,* this could keep lepidopterists busy for quite a while.

ELFIN

Callophrys spp. (cal-**oh**-fris: beautiful eyebrow). Also *Incisalia* spp. ¾–1" ws, brown above; no tail; outlined, often with a fine white line. Rocky places with stonecrop, the larval food; all but NM and UT; Mar–May. Lycaenidae.

Western Pine Elfin

C. eryphon (**air**-if-un: a Greek). Black/white-dashed marginal fringe usually visible above; underside often rose, brightly patterned all over with zigzags and often a fine white dividing line; older larvae pine-green, with paired, whitish, lengthwise stripes. Near pines, the larval food; Apr–July; A.

Western pine elfin

Moss's Elfin

C. mossii. Black/white-dashed marginal fringe ± visible; 2-toned beneath, the darker inner section sharply, jaggedly outlined, often with a fine white line. Rocky places with stonecrop, the larval food; all but NM and UT; Mar–May.

Brown elfin

Sheridan's green hairstreak

Sagebrush sooty hairstreak

Colorado hairstreak

Brown Elfin

C. augustinus (au-gust-**eye**-nus: for a helpful Inuit man nicknamed Augustus by Sir John Richardson, p. 447). No white fringe above; rosy brown beneath, with vaguely defined dark inner portion of hind wings. Larvae eat kinnickinnick, other shrubs; Mar–June; A.

BUTTERFLIES OF THIS genus overwinter as pupae, enabling them to emerge early, sometimes while snow still covers much of the ground. Any other butterflies flying then (anglewings or tortoiseshells, most likely) have almost surely hibernated as adults.

SHERIDAN'S GREEN HAIRSTREAK

Callophrys sheridanii cal-**oh**-fris: beautiful eyebrow; **share**-id-en-ee-eye: after General Philip Sheridan). ⅞" ws; unmistakable: both sexes apple-green beneath, with a white line crossing hind or both wings; gray above. Open slopes up to mid-elevs; more common s; larvae eat sulphur-flower and other buckwheats; Mar–May; A. The WY state butterfly.

SAGEBRUSH SOOTY HAIRSTREAK

Satyrium semiluna (sa-**tee**-ri-um: satyr—; sem-i-**loo**-na: half moon). Also sooty gossamer wing; S.

fuliginosum. 1" ws; uniformly dark gray above, ashy-gray beneath with faintly lighter and darker speckles. Sagebrush or meadows with lupine, the larval food plant; nw CO to WA; June–Aug. Lycaenidae.

COLORADO HAIRSTREAK

Hypaurotis crysalus (high-paw-**ro**-tis: gold eyes below; **cris**-a-lus: gold—). 1⅜" ws, with slender tails—the "hair" in hairstreak; upper side brilliant blue-purple with black margin interrupted by a few orange spots; underside brownish light gray with fine white lines and a few marginal orange flares. Near Gambel oak, its food; southerly. Lycaenidae.

The Colorado state insect explodes from drab to spectacular in an instant, simply by taking wing. As a butterfly, it eats no nectar, subsisting instead on sips from mud and oak-derived liquids, and avoids dehydration by preferring cloudy or rainy weather for most activity. That this deep-purple color is so rare in our butterflies surprises me, as it is highly visible and attractive in the ultraviolet range that insects see so well.

Other Creatures

Yellowstone's hot springs get their stunning "grand prismatic" hues mainly from biofilm organisms that tolerate—no, require—hot water. These thermophiles ("heat-lovers") differ radically from familiar life-forms in the elements they use in their metabolisms, and, inevitably, in the chemistry they produce in the process. That makes them a gold mine for both basic and applied (mainly pharmaceutical) research. Yellowstone provided science with its first good window into the world of thermophiles.

Thermophiles excite scientists for several reasons. First, that old favorite: profits. The bacterium *Thermus aquaticus,* discovered in Yellowstone, produces a heat-stable enzyme that enables polymerase chain reactions (PCRs), a pivotal technique in molecular biology. (The use of PCRs for applications like tests for COVID, the Human Genome Project, and Ancestry.com has created a total global market of over $8 billion in 2022.)

Second, sheer numbers. The heat limit for life on Earth is now placed at 250°F or hotter—still below the water-boiling point in high-pressure habitats in the deep sea and in deep rocks. Lithotrophic ("rock-eating") organisms live in pore spaces in rock way below the surface of the Earth, gleaning their energy by stripping electrons from minerals. Their metabolisms require neither sunlight nor air, not even indirectly. Clearly, the land surface cannot offer nearly as much habitat as the 2 or 3 miles of rock beneath it, nor as

much as the oceans, so we may well be outweighed by both of those dark biospheres.

Third, antiquity. Visits to "black smoker" hot springs at deep seafloor spreading centers led to great advances in our knowledge of extremophiles, a broader category combining thermophiles with organisms that live in acids or alkaline brines. Yellowstone has both acidic and alkaline springs, each with their respective extremophiles. Most are classed in the domain Archaea (ar-**kee**-a: ancients), formerly considered a type of bacteria.

Both archaea and bacteria lack cell nuclei. Both are traditionally considered one-celled, but they form gelatinous biofilms in which they communicate, aggregate in thick structures, and collectively defend themselves against antibiotics—in short, they behave rather like multicelled organisms. The free-floating single-cell phase differs from the biofilm phase chemically, and may be little more than a dispersal phase in between biofilms.

From about 3.8 to 2.5 billion years ago (the Archean Eon), the metabolic alchemy of primitive archaea and bacteria created the biosphere on Earth, sorting the elements, giving us our beneficent atmosphere, fixing nitrogen for plants, and even concentrating gold deposits.

Life on Earth most likely originated as extremophiles, either at deep-sea hot springs, in deep rocks, or in rocks crashing

Microbes are the source of the spectacular hues in Grand Prismatic Spring in Yellowstone National Park.

to Earth from space. Jack Farmer studied bacteria in Mammoth Hot Springs for clues to where NASA should look for life on Mars and Europa, a moon of Jupiter. He concluded that many extraterrestrial bodies have geothermal water systems that "could have provided cradles for the emergence of life in other planetary systems within our galaxy, and beyond."

If primitive life-forms are repeatedly synthesized in sulfide-rich, pitch-dark 250°F water, as many scientists believe, then perhaps we who have adapted to life on the frigid skin of the world, frying in all kinds of damaging radiation, are the true extremophiles.

WATERMELON SNOW

Chlamydomonas nivalis (clam-id-o-**mo**-nus: mantled unit; ni-**vay**-lis: of snow).

Pinkness in late-lying snowfields consists of pigments in living algae. These algae are the producers in an entire food cycle in snow. Ice worms, protozoans, spiders, and insects graze on the algae, and are in turn eaten by predators like the rosy finch. Droppings from the predators, bodies of producers and grazers, and pollen and spiders blown from downslope provide food for decomposers (bacteria and fungi) that complete the cycle.

The algae live on decomposition products and on minerals that blow in as dust.

Most of the pinkness is in energy-storing oils within resting spores that sit out the winter wherever they happen to end up in the fall. In spring, under many feet of new snow, they respond to increasing light and moisture by releasing four daughter cells that swim up to the surface. The year's crop becomes visibly pink only where concentrated on the snowpack surface.

Some say watermelon snow smells or tastes like watermelon; others warn of diarrhea.

C. nivalis is our most abundant among one hundred or more named species of

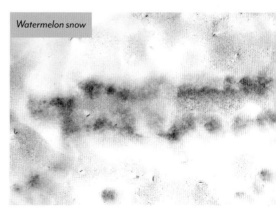

Watermelon snow

snow algae. Most are red, but some are yellow, green, or purple. With plants they share chlorophyll and a celluloselike cell wall, but their swimming ability and well-developed eyespot are animal-like. The Whittaker five-kingdom system took green algae out of the plant kingdom and into a kingdom Protista, but some systematists now put them back in with the plants.

GIARDIA

Giardia intestinalis (**jar**-dia: for Alfred Giard). Also *G. lamblia, G. duodenalis.* Higher classification moot.

In a paper titled "My Favorite Cell: Giardia," two Australian parasitologists rhapsodized over giardia—"the only microorganism we know to smile." Their affection is inspired less by the smile than by giardia's unusual traits bridging the realm of protozoa with that of bacteria. (The Tree of Life web project places them in a group called Fornicata, which just means "arched." No sexual processes have been seen in giardia.)

More to the point for hikers, giardia is a leading culprit worldwide in persistent diarrheal infections. Giardiasis is a most unpleasant experience; the drugs prescribed for it have unpleasant side effects, and don't always work on the first try; and, for some people, symptoms can linger long after the parasite is eliminated.

That said, the specter of giardia lurking in mountain streams has been kicked around for decades without ever being scientifically

Giardia intestinalis

demonstrated. Scientists who looked for it in streams in a popular part of the Sierra Nevada didn't find enough to worry about. A meta-study concluded that "education efforts aimed at outdoor recreationists should place more emphasis on handwashing than on water purification." In other words, the (rather weak) association between giardiasis and camping isn't about the water up there; it's about the hot water and soap that *aren't* up there next to the latrines.

So, decide for yourself. To avoid all risk of giardiasis, follow all the rules about purifying water and about hygiene before sharing food. As for me, I'm going to continue prostrating myself to the mountain stream goddess and drinking deeply—at least at selected streams descending from trail-free drainages. I'll purify water in drainages that are heavily trodden by humans or livestock.

(I'll also avoid being downstream from old mining claims. Especially in Colorado, some streams of clear, cold, beautiful water are contaminated with acid mine drainage, making them risky to drink and in some cases even to bathe in.)

What about beavers, water voles, and other giardia carriers? Yes, a great many vertebrate animals carry *Giardia,* but many of those giardia species don't infect humans. The often-heard claim that beavers are responsible for outbreaks in humans has, again, never been proven, and scientists are skeptical of it. The clear correlations with giardiasis are things like community swimming pools, sharing a household with a person or pet with giardia, and other old standbys of what epidemiologists call the "fecal-oral route."

For every person reporting symptoms, there are somewhere between ten and one thousand people carrying *Giardia* without getting symptoms. From that I infer, first, that there are a lot of people out there to get it from; second, even if you get it, you probably won't experience any symptoms; and, third, many of us are carriers and may not know it. That's why it behooves us to be totally scrupulous about burying our solid wastes or packing them out, and about disinfecting our hands. Gel hand sanitizers apparently work well.

If you go trekking in the developing world, where the water is riskier than in our mountains, take purification very seriously. You may face viruses or bacteria that require more stringent purification than giardia does.

Rocky Mountain wood tick in a typical waiting-for-host pose

ROCKY MOUNTAIN WOOD TICK

Dermacentor andersoni (der-ma-**sen**-ter: skin pricker). Adults about ⅛"; reddish-brown ♂ back has pale patterns across it; ♀ back has a pale, oval plate right behind the head; ♀ grows much larger and paler, resembling a dried bean, when engorged with victim's blood. Class Arachnida (Spiders), Order Acarina (mites and ticks). Dog tick, *D. variabilis,* is very similar.

Ticks, mites, and spiders differ from insects in having eight legs and two main body sections, rather than six and three. Ticks live on blood they suck from warm-blooded animals. They climb to the tips of grasses or shrubs and drop off when the stem shakes following a jump in CO_2 levels—a telltale sign of mammalian breath.

Ticks carry some serious diseases, but this is one of the better areas on the continent for not catching them. That's ironic: Rocky Mountain spotted fever got its name when it was a big problem in Montana a hundred years ago. It's rare here now. It can be fatal, so it must be recognized and treated with antibiotics. It starts two to twelve days after a tick bite, with high fever, severe headaches, and variable flulike symptoms. A distinctive rash near the hands and feet may appear later, but don't wait to see it—treatment needs to start as soon as possible

Colorado tick fever—uncommon outside of Colorado, fairly common inside—is neither life-threatening nor treatable. Fever and other flulike symptoms typically occur in two bouts of two or three days each, separated by a few days' respite, then go away on their own.

The ticks that transmit Lyme disease are absent from most parts of our range. (They do occur in Utah and in the Wallowas, but the disease hasn't been seen in them there.) The related disease borrelliosis is also serious, but quite rare here. It is transmitted by ticks other than *Dermacentor,* typically to people sleeping in rustic, rodent-infested cabins.

If you find a tick on you and can brush it off easily, do so at once; you haven't been bitten yet, so don't worry. If it has taken hold, pluck it carefully: using tweezers, grab the tick by the head, not the abdomen, and tug gently and repeatedly until it releases its grasp. Avoid either crushing it or leaving its mouthparts imbedded in your skin; both juices and mouthparts can transmit infection. Immediately wash and sterilize the site (e,g., with alcohol). If you can't get to a proper pair of tweezers, pull it with your fingers (technique as above) after dousing it with insect repellent to encourage it to back out. Ticks are a good reason to use insect repellent.

17

Geology

The Rockies offer as gorgeous and varied an assortment of rocks as you could ask for. That's no surprise, given the number of ages, causes, compositions, and stories they display. Before describing the kinds of rocks, this chapter will offer a history of the Rocky Mountains as a sequence of processes and their driving forces. Be warned: it's a saga, not a short story, and it has more loose ends than a bad murder mystery.

A hundred years ago, ancient strata in dramatic, varicolored bands earned the Rockies (Glacier National Park in particular) high rank as a geologists' playground. The strata beautifully illustrate the geosynclinal theory of mountain-building, which one eminent geologist complained was "a theory for the origin of mountains with the origin of mountains left out." After plate tectonics took over as the prevailing theory, geologists worked anew to explain why mountains rise. The Rockies resist simple explanation. No geologist doubts that plates exist and subduct, but all agree that plate tectonics don't explain mid-continents as tidily as they do oceans and continental margins.

PLATE TECTONICS

Earth's surface is divided into large plates that move. They grind along at speeds of a few inches per year. There are seven huge plates and many small ones. Convergence between them produces mountains.

New plate material is generated out of fresh basalt lava at mid-ocean ridges, where two plates move apart; where plates collide, one usually overrides the other, which plunges back into the depths, or subducts. New plate material is oceanic crust, a sheet of basalt rock averaging 4 miles thick. Under an ocean for millions of years, the sheet accumulates thick beds of marine sediments on top, which gradually become sedimentary rocks. All seven big plates carry some oceanic crust, and many also carry continental crust—thicker but less dense crust of many kinds of rock, with granitic rocks dominating.

You can think of crust as floating. Due to its thickness and low density, continental crust rides high upon its semifluid underpinnings; not much of it is covered with seawater, at least not in this day and age. Conversely, most oceanic crust does lie under an ocean.

What the crust floats on is the mantle layer, which extends down more than halfway to the center of the earth. Just to confuse us, the "floating" layer that moves as a plate is not just crust; it's thicker. Called the lithosphere, this layer includes all of the crust plus a top layer of mantle. Whereas crust is distinguished from mantle by chemistry and density, the lithosphere is distinguished from the mantle layer below it (asthenosphere) by its motion, as a plate.

The lithosphere cools as it ages—moving away from the mid-ocean ridge—and gets denser (less buoyant) as it cools. Though its crust portion remains buoyant, the deeper (mantle) portion becomes dense enough to pull the plate under at a subduction zone.

There, the oceanic plate descends as a slab under the overriding plate's edge. Textbook cross-sections of a subducting plate show it with a 20-to-60-degree slope, often with an arrow suggesting the slab slides into a slot. Those 20-to-60-degree slopes are well established, thanks to plotting the epicenters of a lot of deep earthquakes over time. However, the slab's slope is not necessarily the direction it moves in. Gravity is pulling it straight down. The mantle convection cell beneath it *might* support it enough to for it to slide along that slope, but alternatively its hinge line might roll back while the overriding plate advances.

Slabs may subduct as big sheets initially, until they eventually break up into pieces that sink into the mantle—apparently straight down. Where they break, they make a window through which warmer mantle can flow upward; this would also affect surface topography. A technology called seismic imaging offers soft-focus but increasingly definitive visions of slabs down there. It's kind of like a CAT scan of the earth's interior.

Since no one can ever see the mantle more directly than through seismic imaging, geologists' speculations regarding the force that drives the plates are a bit like the proverbial description of an elephant by several seeing-impaired persons. This force must involve convection—hotter stuff rising and cooler stuff sinking. The mantle convects because radioactive decay heats the earth internally. Where slabs sink, they drive the downward limb of a convection cell; mantle heat helps to drive the upward limb, and between them a horizontal limb drives the overlying plate.

Oceanic crust and continental crust share the same plate, but only the oceanic crust is dense enough to subduct. After a plate's oceanic portion has subducted under a continental edge for some time, its continental crust may arrive at the trench. Then subduction ceases there (and soon begins somewhere else) and the two continents collide, crumpling and pushing up great mountain ranges like the Himalayas.

When an island chain on an oceanic plate meets an overriding continental edge, it may achieve a similar result, suturing to the continent, forcing subduction to jump to its outboard side, and raising mountains. The island chain is now a new terrane accreted to the edge of the continent. In this fashion, the continents have grown for billions of years.

Most of North America breaks down into terranes all with different accretion ages. The ones under most of our Rockies have been together for more than a billion years as part of the proto-continent Laurentia, which at times has been part of various supercontinents. Terranes described as "exotic" accreted much more recently (less than 400 million years ago) and were never part of Laurentia. They underlie the Pacific coast states and province, and enter our range in Washington, Oregon, and a westernmost bit of Idaho.

When an oceanic plateau—a patch of extra-thick basaltic crust—hits a subduction trench, it also resists subducting. If it subducts, it will likely do so less steeply than normal, perhaps even so shallowly that it grinds along against the underside of the overriding continent for a ways. This flat subduction is more speculative than terrane accretion, since it takes place far below the surface, but you're going to hear a lot more about it, because it is thought to be the key to the Rocky Mountains.

Picture accretion as a roiling pot of thick soup with bits of foam appearing out of the rising part, floating off to the side, coalescing with other bits of foam, and accreting to a mass of foam along one edge. The mantle is not actually a liquid like that soup. It is mainly solid, but semifluid, flowing very, very slowly, as many solids do. (Solid glaciers flow at a glacial pace; window glass flows a little bit, becoming wavy over a century or more.)

Rock can of course get hot enough to melt into a liquid—magma or lava—and the earth

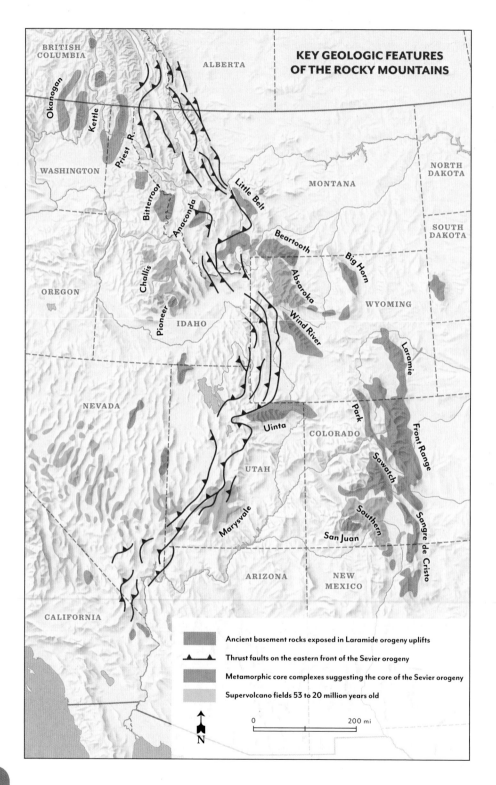

KEY GEOLOGIC FEATURES
OF THE ROCKY MOUNTAINS

BRITISH
COLUMBIA

ALBERTA

Okanogan

Kettle

Priest R.

Bitterroot

Anaconda

Little Belt

MONTANA

NORTH
DAKOTA

WASHINGTON

Challis

Pioneer

IDAHO

Beartooth

Absaroka

Wind River

Big Horn

SOUTH
DAKOTA

WYOMING

OREGON

NEVADA

Uinta

Laramie

Park

Front Range

COLORADO

UTAH

Sawatch

Sangre de Cristo

Marysvale

Southern

San Juan

CALIFORNIA

ARIZONA

NEW
MEXICO

Ancient basement rocks exposed in Laramide orogeny uplifts

Thrust faults on the eastern front of the Sevier orogeny

Metamorphic core complexes suggesting the core of the Sevier orogeny

Supervolcano fields 53 to 20 million years old

0   200 mi

N

is generally hotter the deeper you go, but along with depth comes higher pressure that raises the rock's melting point. Therefore, most mantle fails to melt, because it's under so much pressure. The even hotter layer below the mantle, the outer core, is molten. But from there on up, most mantle is solid rock, and flows at a slower-than-glacial pace.

VOLCANOES AND PLUTONS

Parallel to a subduction zone there typically lies a line of volcanoes called an arc, either offshore (creating islands) or up to 200 miles inland. (Not all arcs look arcuate on a map.) Surprisingly, what melts rock at a subduction arc is thought to be the addition not of heat but of water—seawater carried into the depths within the oceanic slab and squeezed out into adjacent minerals (the chemicals rocks are made of), combining with them into new minerals, some of which have lower melting points.

(Mantle material melts for a different reason under mid-ocean ridges: its melting point drops there because, as it convects upward, the pressure on it drops.)

The melting under the volcanic arc is partial: just the minerals with lower melting points melt. The term of art, "crystal mush," describes magma as crystals mixed with melt in a wide range of ratios. As pulses of magma rise, gravity differentiates them further, selecting less-dense minerals to rise. At the same time, the primitive mantle magma evolves on its way up by melting and then incorporating crustal minerals it meets. The resulting chemistry determines which kind of rocks, and volcanoes, the magma will produce. We classify magmas on a scale from pale to dark, with paleness defined largely as a higher percentage of silica, the chemical in quartz crystals.

Darker magmas are more fluid, and more likely to reach the surface, erupting as a volcano. Pale magmas are extremely viscous, like cold peanut butter, and they have a higher melting point, so it takes more heat to get them to erupt. If they reach the surface, they can build lava domes, or they can explode, blasting copious ash and tuff all over and leaving a caldera (a broad hole) at the eruption site.

More commonly, pale magmas fail to reach the surface. About 90 percent of magma instead crystallizes slowly into a pluton, a deep mass of intrusive rock. This can be either an entire magma body that never surfaced or a portion of a magma body that got left behind. The biggest plutons, called batholiths, grow over millions of years in increments of thin layers.

A mountain range may result, due to the thickening or deepening of continental crust along the line of magma formation. The thickened line floats higher than surrounding areas of continental crust. Overlying volcanoes and other rocks erode off the top, exposing the plutons, which persist for millions of years as a granitic mountain range.

This flotation principle, called isostasy, was a brilliant nineteenth-century breakthrough in geologic thought. It works well for ranges—Himalayas, Alps, Appalachians—that thickened where two continents collided and crumpled together. The range as a whole can be buoyant either because its crust is thicker than adjacent crust or because it contains more of those lighter rocks—or both. It turns out, though, that many mountain ranges don't fit that model. Neither thick crust nor low-density crust are sufficient to give the Rockies their elevation today, 45 million years after the events that crumpled them into mountains. It's got to be something else.

Those original Rocky Mountain orogenies (mountain-building events) are themselves puzzling. Ordinary subduction—that is, with no terrane or plateau arriving—produces a volcanic arc and often also a coast range of scraped-off sediments, but the latter tends to be modest in size, and close to the subduction trench. Something seriously scrunched the Rockies, far from any trench.

But let's step back to the beginning.

ARCHAEAN CRATONS

From 4 to 1.6 billion years ago

The oldest, most stable areas of a continental plate are cratons. Barely a third of the way through Earth's history, some granites of the Wind River and Beartooth Ranges crystalized as part of an early craton. Some of its rocks formed as early as 3.96 billion years ago, the era when asteroids bombarded the earth, ranking them close to the oldest rocks in the world. That craton (the Wyoming) underlies an area that runs east to the Black Hills, south to near Cheyenne, and north and west well into Montana and Idaho.

Wind River Range gneisses metamorphosed in one of the earliest known subduction events, 2.7 billion years ago. Subduction was different back then, in the Archean Eon. Earth was not far along in the process of sorting elements—raising lighter ones to the surface, sinking heavier ones to the core. Life consisted of bacteria and archaea. Some of them got their energy from sulfates, and produced methane, giving us the first

climatic greenhouse. Without that, Earth's water would have all been ice, and photosynthetic life would have been impossible. Photosynthetic cyanobacteria were at work releasing oxygen for several hundred million years before there was enough oxygen for oxygen-breathing life to come into being. At first, all the free oxygen was nabbed by oxygen-hungry compounds in volcanic gases. Indeed, 70 percent of the earth's mineral types could not exist until there was free oxygen thanks to cyanobacteria.

Plate tectonics were part and parcel of the development of life on Earth. Elements erupted into the atmosphere and the ocean, where microbes recombined them; then subduction returned them to the crust and mantle, metamorphism re-sorted them, volcanism again ejected just some of them, and so on in a great sorting cycle until at last there was an atmosphere, a climate, and an ocean that could nurture complex life. That, in turn, continued to change the biosphere's chemistry. We had one era that produced nearly all of the iron ore, one that produced most coal, and others that produced petroleum, red rocks, or giant insects.

Wyoming joined Montana—a somewhat younger cratonic block called the Medicine Hat—1.81 billion years ago, in the Proterozoic Eon. (No, cratons did not actually anticipate state lines.) Together, they joined cratons east and north of them to form Laurentia—the early iteration of the North American continent—which was in the process of joining most of the world's other continents in the supercontinent Nuna.

(Geologists have not entirely settled on the name "Nuna"; some call it Columbia. Evidence of supercontinents before that is murky, but everyone agrees on two global ones since then: Rodinia and later Pangaea. Gondwana and Laurasia [Laurentia plus Asia] were supercontinents that each held

Archaean gneiss in foreground and background

about half of the world's land before they joined as Pangaea.)

Beginning 1.7 billion years ago, accretions from the south brought Colorado and New Mexico, but these were not coherent, old cratons. Instead, the Yavapai and Mazatzal provinces comprised smaller terranes of young material—volcanic arcs either on islands or onshore. As Nuna broke up, terrane accretions and mountain-building in this area continued, producing nearly all the granitic rocks in Colorado's high mountains, mainly either 1.4 billion or 1.1 billion years ago. By 1 billion years ago, the continents reassembled in a new supercontinent, Rodinia. The final step in assembling Rodinia was a huge, possibly globe-spanning collisional orogeny; southern cratons (now in Africa and South America) bashing into the southern side of Laurentia were just part of it. While the core of that Grenville orogeny crosses Texas and northernmost Mexico, its energy ramified as far as British Columbia, as reflected metamorphically in a few scattered outcrops.

A cliff through Belt sedimentary layers

THE BELT SUPERGROUP

From 1.6 billion to 400 million years ago
For about 70 million years, our part of the supercontinent Nuna contained a huge, continually subsiding basin filled with water. Rivers washed voluminous sediments down into this sea while it kept sinking. The beds of sediment reached 11 miles thick—a phenomenal rate of sedimentation, though some say offshore sediment fans of the Indus and Ganges Rivers today are in the same league. There was no life on land to hold the soil in place. During the early phase, starting 1.47 billion years ago, the bulk of the sediments apparently came from mountains that are not any part of North America today. They came from the other side, the craton that drifted far away when Nuna broke up;

one school of thought sees it in Tasmania and Antarctica today, another in Siberia.

The rocks those sediment beds turned into are the Belt Supergroup.[2] It forms most of Glacier National Park and crops up throughout western Montana, northern Idaho, a bit of Washington, and north into Canada.

After 70 million years of geologically rapid motion, for 1.3 billion years the basin lay still, becoming a strong, rigid block that suffered very little folding or faulting. When eventually the Sevier orogeny built the Rockies, younger sedimentary formations broke up on countless thrust faults. The Belt moved on a thrust fault too, but at least within Glacier National Park it did not break up, instead sliding in one giant horizontal layer. As you

2. *Supergroups form when star rocks join up. Sorry, no: they are simply groups of related rock formations. This one was the Belt Formation—named for Montana's Big Belt Mountains—until geologists saw that it was too huge and varied to treat as a single formation.*

Stromatolite

can see at a glance anywhere in the park, it is still generally horizontal and right side up.

Other parts of the Belt saw action. The Idaho Cobalt Belt, near Salmon, is the richest known US source of cobalt, which is in high demand for batteries as the world decarbonizes energy. The cobalt apparently rose to the earth's surface in ancient arc volcanoes, was eroded and deposited as grains in Belt sediments, and was concentrated later during episodes of magma and metamorphism.

Belt rocks display astonishingly gorgeous and strange textures. Though too old to bear fossils of macroscopic animals, they do contain stromatolites—fossilized, round colonies of cyanobacteria, the creatures that gave us our oxygen. Additionally, you'll find:

- Parallel ripples formed under tides
- Cross-current ripples
- Contrasting colors of very fine beds, around a millimeter thick
- Ordinary mud cracks where mud shrank as it dried out in the air
- Mysterious crinkle-cracks that pinch out at their ends rather than meeting each other
- "Molar-tooth structures"—crumply textures punctuated with small, bright-white, crooked shards consisting of calcite injected into the mud where CO_2

from decomposing bacteria bubbled up through the mud
- Ooids (**oh**-oids, meaning egglike)—very round, sand-sized limestone grains of calcite precipitated concentrically around nuclei
- Pillow basalt from lava erupting underwater
- Pahoehoe basalt lava that flowed above the water level

The subsequent Paleozoic Era, beginning 600 million years ago, left a lot of fossils (but not dinosaurs; those came later). For a long, lonely time, our region remained generally quiescent and submerged, near Laurentia's western edge. Our abundant Paleozoic sedimentary rocks originated variously—well offshore on the continental shelf, or as lake bottoms, mudflats, or beaches.

PANGAEA, THE ANCESTRAL ROCKIES, AND TERRANES
From 400 to 140 million years ago

During the assembly, life, and breakup of the most recent supercontinent, Pangaea, our region built mountains once again. Enough time has passed to wear those mountains down to plains, but their rocks remain—or at least their sands.

North America's western flank, no longer quiescent, became the active plate margin that it remains to this day. For some time, our side of the Pacific was a sea of major, mountainous island chains riding a maze of small tectonic plates subducting in several directions. A modern analog would be Indonesia, the Philippines, and Japan. Our islands began accreting to the continent 400 million years ago. Early ones probably overrode a subducting continental shelf, producing the Antler Mountains in Nevada and terranes now found in Oregon's Wallowas and Washington's Selkirks. Meanwhile, Montana and Wyoming remained quiet and low, often submerged.

At 340 million years ago, there was a pause in terrane accretion, and mountain-

Maroon Formation beds derived from erosion of Ancestral Rockies, Colorado

building moved to our southern flank. The large Amazonian Craton collided with southeastern Laurentia, completing Pangaea. This raised a mountain range across Oklahoma, paralleling the earlier one from 1 billion years ago; this time, the orogeny extended into Colorado as the Ancestral Rockies, centered on a pair of ranges aligned southeast to northeast, one through Pikes Peak and one through the Uncompahgre Plateau.

Geologists find no volcanic rocks or granitic rocks stemming from this event, a rather baffling absence. We do have massive sedimentary formations, including Denver's Red Rocks and the purple rocks of the Maroon Bells, from sediments eroded from the high Ancestrals. During some periods, the sediments settled in shallow seawater, indicating that these Rockies were big islands. Somewhat later, a Sahara-like erg, or sand-dune sea, developed from southwest Wyoming to northern Arizona, eventually petrifying into Utah's Wave sandstones. Some sands in those dunes originated in the Appalachians and were carried here by continent-spanning rivers.

That was the last collision from the south. The final terrane to hit California from the west, called Wrangellia, did not stay there. It currently makes up most of Vancouver Island and is smeared out northward all the way to the Wrangell Mountains in Alaska. The specifics of where and when Wrangellia was before that are controversial; smearing out northward was just the final step in a lot of northward movement along the coast. It impacted California by 100 million years ago; this collision drove the Sevier orogeny— birthing the modern Rockies.

THE SEVIER MOUNTAINS
From 125 to 65 million years ago
The characteristic look of the Sevier orogeny, as seen from space, is narrow, parallel mountain ranges running north to south, sometimes curved. We see them in central Montana, extreme southwestern Wyoming, all along the Alberta-British Columbia border, and in central Utah near the headwaters

GEOLOGIC TIME

4.7 billion years ago	Earth formed.
4 billion	The oldest rocks solidified from magma. Oceans formed.
3.96 billion	The oldest crystals in the Rockies solidified from magma.
3.85 billion	A possible origin of life in the form of primitive bacteria appeared.
3+ billion	Cyanobacteria began releasing free oxygen.
2.3 billion	Oxygen became abundant enough to oxidize most elements.
1.81 billion	Wyoming and other cratons formed the continent Laurentia.
1.4 billion	The Belt basin collected sediments for the Belt rocks.
580 million	The oldest macroscopic, multicelled animal fossils appeared.
470 million	Fungi and algae moved onto land; first plants move soon after.
245 million	The Permian Extinction occurred, the most severe ever.
225 million	Dinosaurs became dominant on land. Pangea rifted.
180 million	Most of our region was still flat and shallowly submerged.
100 million	Flowering plants came to predominate over spore plants.
140 million	Thrust faulting began raising the Canadian Rockies.
125–65 million	Sevier orogeny raised the US Rockies.
75–45 million	The Laramide orogeny raised the east-central Rockies.
65 million	The Cretaceous Extinction deleted dinosaurs, except for birds.
53–23 million	Huge volcanoes blanketed several Rocky Mountain regions.
17 million	The Basin and Range Extension accelerated.
16.6 million	The Yellowstone Hotspot began (if not earlier). The Columbia Basalt floods began.
3.5 million	A long cooling trend was underway, climaxing in the Pleistocene Ice Ages.
2 million	Huckleberry Ridge eruption occurred, Yellowstone's largest.
1.8 million	Ancestral hominids shaped stone tools.
1,293,000	Mesa Falls eruption at Yellowstone occurred.
630,000	Lava Creek eruption at Yellowstone occurred.

HUMAN TIME	
130,000 BCE	Anatomically modern humans appeared.
115,000	The most recent ice age glacial stage began.
30,000	Possible first human migration from Asia to Alaska occurred.
16–14,500	Last major advance of most alpine glaciers here occurred.
15–13,000	Purcell and Flathead Lobes of Cordilleran Ice Sheet were at their maximum. Missoula (or Bretz) flooded. Early Americans lived in Nipéhe village on Salmon River, Idaho.
11,500	The ice cap was almost completely gone from Yellowstone.
11,200	Hunters with Clovis spear points and spear throwers appeared; many large mammals died out.
7,000–4,000	Holocene Climatic Optimum occurred—about as warm as the twentieth century.
5,700	Mount Mazama (Oregon) erupted, leaving a marker of that date in soils.
2,000	Colorado's youngest volcano, Dotsero, erupted.
0	Youngest basalt flow at Craters of the Moon, Idaho, occurred.
730 CE	Oldest living whitebark pine germinated in Idaho.
800–1250	Medieval Climate Anomaly (warm period) occurred.
1600s	European diseases began killing Indigenous Americans in huge numbers; Plains groups got horses; European honeybees reached the plains.
1785	Fur trade began between Northwest Coast, China, and Europe.
1792	A Hudson's Bay Company trader reached the Canadian Rockies.
1805	Lewis and Clark crossed the US Rockies.
1820	Post–Ice Age glaciers reached their greatest extent ("Little Ice Age"). Fur trade was going strong in the Rockies.
1887±	The Plains bison population reached its low: 541 animals.
1959	Hebgen Lake earthquake, our largest recorded (7.5), occurred.
1983	Borah Peak (Idaho) earthquake, magnitude 7.3, occurred.
Today	Glaciers are almost gone. Human-caused extinctions are accelerating. The Rockies are still rising.

Aerial view of Montana Overthrust Belt

of the river they are named for. The Sevier front never touched Colorado or New Mexico. All of this territory had for a long time been basins accumulating sediments and turning them into rock. When subjected to strong eastward compression, these layered rocks broke along parallel fault lines, with each piece pushing partway on top of the piece to its east. Geologists call this process thrusting, on thrust faults; the mountain ranges are called the Overthrust Belt.

The Farallon oceanic plate commenced subducting eastward under the North American plate 175 million years ago. That instigated a voluminous volcanic arc, whose legacy today is enormous granitic batholiths in Idaho, Baja California, coastal British Columbia, and the Sierra Nevada. As Farallon subduction delivered its final terrane, Wrangellia, the relative motion of the two plates began to speed up because a new part of the Atlantic was now spreading. Eastward compression intensified, instigating thrusting along the Sevier mountain front.

This front line advanced gradually from west to east. The overthrust belt ranges are just the youngest, easternmost of the Sevier mountains. This was a huge range, its axis running just west of the Nevada-Utah state

line and north through the growing Idaho Batholiths.

All that core portion of it was "soon" (in geologic time) to be broken down.

THE LARAMIDE MOUNTAINS
Peaking between 75 and 50 million years ago

The Farallon plate's next special delivery, after Wrangellia, was an oceanic plateau.

Far away on the other side of the Pacific there's an oceanic plateau, the Shatsky Rise, a patch of thicker crust produced at a mid-ocean ridge. It's thicker because for a few million years magma surfaced much faster than usual. Since mid-ocean ridges are spreading centers where new basalt comes out and splits in two directions, the Shatsky is presumed to be one half of a matched pair, with its twin moving east on the Farallon plate. That spot on the Farallon plate subducted under Southern California 90 million years ago. Thus began what geologists call the "flat slab."

Plateaus subducting are common globally, and plateau slabs are generally understood to descend at a shallower angle, due to slightly greater buoyancy. According to one hypothesis on the Rockies, Shatsky's twin so resisted sinking that the slab all around it turned horizontal and slid north-northeast right against the underside of the lithosphere overriding it. Picture its progress like a mouse moving through the gut of a snake. It reached the broad, low sedimentary basin behind the Sevier mountains 75 million years ago, and raised new mountains there of a whole new style: the Laramide orogeny.

Laramide faulting was (like the Sevier) thrust faulting, meaning it resulted from compression, or shortening of distance from one side to the other. While the amount of shortening was less than in the Sevier, a key difference is that these faults ran deeper. Sevier faults broke up thick sedimentary beds that covered the Rocky Mountain region, and those sedimentary rocks are most of what you'll see today in a Sevier range. In contrast, Laramide faults broke the basement underneath those beds—the granitic and metamorphic rocks that date back to Archaean times. Those rocks are a lot of what you'll see in a Laramide range today. Curiously, if you could have climbed a Laramide mountain when it was rising 65 million years ago, you might have seen only sedimentary rocks: those thick layers didn't just vaporize when they rose; they eroded away over time.

The Laramide uplifts don't line up as a continuous mountain chain, but are scattered, with oddly diverse orientations—northwest to southeast is typical, but the Front Range runs north-south and the Uintas run east-west. Elsewhere in the world, basement-involved faulting is unusual but not unheard of. It intuitively fits the idea that the faulting resulted from a force tugging the lithosphere's underside. Laramide faults in cross-section are long and shallow, pushing the sedimentary beds above them into a broad arch. Indeed, most Laramide uplifts are better described as arches than as fault blocks. That does not mean they were modest: total vertical motion on the Wind River Fault was at least 8 miles.

Volcanoes are strikingly absent from the Sevier and Laramide orogenies, with one exception: a swath developed crossing Colorado from Cortez to Boulder. It's called the Colorado Mineral Belt because it's where nearly all of the state's valuable minerals were mined—gold, silver, copper, molybdenum, and on down the preciousness scale to lead and zinc. These metals are pretty scarce in the earth's crust. They get concentrated enough for mining where magma from the mantle brings them up into the crust, then water-based brines circulate under pressure intense enough to dissolve metals, and rise to where the pressure is lower, precipitating the metals out again in veins.

Calculations suggest that the mineral belt developed along an edge of the Shatsky plateau at a time when that was just getting dense enough to break off and sink into the mantle. This may have initiated a tear in the slab, inviting hot mantle to rise toward the surface and erupt as the volcanoes of the Mineral Belt. Later, San Juan volcanism added further megadonations of ore. When the region rose, the volcanoes eroded off the top, bringing metal-rich veins to the surface.

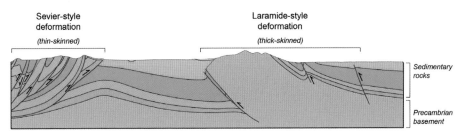

Wyoming and Salt River Ranges (Sevier) face Wind River Range (Laramide) across the Green River Valley.

Hogback, Stonewall, and Flatiron

Hogbacks—sharp ridges consisting of a steeply tilted sedimentary rock bed, and running for miles and miles—are common in the Rockies. They are created by differential erosion of weaker sedimentary beds, with the strongest bed left outstanding, and they occur where arches and troughs tilted rock strata and then erosion had time to create a plain for them to stand up from. Find them in almost any intermountain lowland in the arid West.

One in particular, the Dakota Hogback, puts a neat border around the foothills of many big Laramide ranges, mainly on the east side, but sometimes west of the mountain range as well. The Dakota is so long that it seems absurd to call it a single hogback; however, it's true that every instance of it is made of Dakota sandstone, a single bed that settled in the interior seaway 150 million years ago, preserving countless Jurassic fossils.

Finding gaps in the Dakota Hogback was a serious challenge for nineteenth-century travelers. Interstate 70 demanded a massive roadcut through it at the edge of Denver. Highway 34 approaching Rocky Mountain National Park encounters an extra fold and has to cross the Dakota three times, exploiting gaps. Colorado 12 finds a nifty gap near the town of Stonewall, where the Dakota is tilted straight up, earning its local name—the Stonewall.

Hogbacks at front of Sangre de Cristo Mountains, New Mexico

Formations called flatirons are similar erosion-resistant tilted beds in the foothills. These rich red strata are a little thicker than the Dakota, and a little older, placing them underneath the Dakota and thus closer to the core of the mountain range.

Hogbacks and flatirons are the broken-off edges of the sedimentary beds that once covered the arches like blankets, then got eroded off the tops but remain layered like blankets over the surrounding lowlands. They help us see the Laramide ranges as gigantic arches cored with granites and metamorphic rocks.

From western Colorado, Shatsky's twin ground northeastward to Wyoming where, shockingly, it may have gotten stuck—glued, as it were, to the underside of the North American plate. Recent seismic imagery shows a flat-lying mass of colder material there, suggesting Shatsky's size and shape. The complex timing and regional variation of Rocky Mountain uplifts seem to require that some crazy-sounding things happened to pieces of both the flat-subducting lithosphere and the overriding North American lithosphere. Basement-involved uplifts occurred earlier and over a wider area than the past mapped locations of the Shatsky twin. (That mismatch is a problem for the flat-slab hypothesis. A persistent minority of geologists attributes the Laramide as well as the Sevier orogeny to the dragged-out Wrangellia terrane collision, rather than to flat subduction.)

The flat Farallon slab seems to have shaved off, from below, some of the thickness of the lithosphere, while also hydrating it by adding water. Remember, volcanic arcs like the Cascades are caused by seawater getting squeezed out of a descending slab and into an overriding plate, altering the minerals there into ones with lower melting points. They melt, rise, and erupt. A flat slab would also lose its water into the overriding plate, but it would be spread out over a broad area and time span. The entire area's lower lithosphere would see minerals changing into other minerals, predominantly less dense ones. For example, garnet alters to mica. Less dense means more buoyant: this broad area of the lithosphere would want to float higher. So hydration is one contributor to the Rockies' elevation.

The Sevier orogeny preceded the Laramide, with some overlap. The Sevier was the bigger, higher mountain range. The Laramide uplifts were not as high during their active phase as they are today, and their feet were barely if at all above sea level: when they began to rise, eastern Wyoming and Colorado were awash in the Cretaceous interior seaway, where dinosaurs splashed. They owe their taller status today to later events.

SUPERVOLCANOES
From 53 to 23 million years ago

As the Laramide orogeny waned, volcanism returned, beginning in interior British Columbia and slowly working its way south. The Absaroka Mountains, on the northern and eastern sides of Yellowstone, are erosional mountains carved into volcanic material that erupted beginning 53 million years ago. The similar-sized Challis volcanic field began erupting concurrently in the eastern part of the Idaho Batholiths. These fields likely included supervolcanoes, meaning explosive eruptions many times larger than any eruptions in written history. Their chemistry was at the pale (rhyolite) end of

"Tent rocks" of volcanic tuff in Colorado

Snowshoe Mountain in the bowl of the Creede Caldera

the scale, and produced more ash and tuff than flowing lava.

Still-larger supervolcanoes built Colorado's eastern San Juan Mountains and West Elk Mountains. A single eruption 28.2 million years ago was one of the biggest- known volcanic events anywhere, leaving 1,000 cubic miles of welded tuff spread out 300 feet deep over southwestern Colorado, and then collapsing into a 47-mile by 22-mile caldera named La Garita. The San Juans alone contain around twenty-one calderas and around five supervolcanoes, defined as at least 1,000 cubic kilometers of lava in one eruption. Only a few of these are still conspicuous as big, round features on a satellite image. One is the Creede Caldera, where the Rio Grande neatly arcs the northern half of the caldera's moat, surrounding Snowshoe Mountain, a broad, round volcanic dome. The dome and the Creede Caldera itself were built by resurgent eruptions in the heart of the vast La Garita caldera, helping to overwrite La Garita as a visible feature of today's landscape. In the eastern half of the San Juans, nearly all rock is volcanic rock from that period. Most peaks don't show you where volcanoes were; they're simply high points that remain after millions of years of erosion.

San Juan volcanism was a second major source of Colorado's mineral wealth. Its magmas superheated water, which in turn dissolved minerals, carried them up through caldera-related faults and fractures, and deposited them in veins that were exploited at mining towns like Creede.

Westward across southern Utah and Nevada, similar supervolcanic fields including forty-eight calderas developed around the same time. One of the biggest, the Marysvale, built Utah's Tushar Mountains. Still more were in southern Arizona and across a vast region of Mexico, the Sierra Madre Occidental.

This volcanic flare-up was part of a stunning reversal in the plate-tectonic regime, thought to result from the flat slab at last turning downward, tearing in places, and falling into the mantle. After tens of millions of years of the entire West being under east-west compression, raising many mountains but very few volcanoes, it flipped into east-west stretching, with a lot of volcanism.

All slabs gradually get heavier, and eventually sink into the mantle. Slab rocks are under greater pressure than they were before they subducted, because now they have another plate on top of them. Gradually this metamorphoses them: it recombines their elements into new minerals denser, on average, than the old minerals. If it's a flat slab flush against the plate over it, it weighs the plate down as it densifies. This happened to Wyoming and Colorado prior to and during the Laramide orogeny. But when a slab delaminates and sinks, the plate above it rebounds while hotter mantle material flows in to fill the space. That's one part of what makes the inland West high. The flat slab's plunge brought hot mantle upward, and that in turn produced rising magma. The entire region's lithosphere heated up, and it remains hotter than elsewhere in North America today. Paradoxically, lithospheric heating both raised the regions where and when it was active, and weakened the lithosphere, setting the stage for a collapse.

POST-OROGENIC STRETCHING

The past 50 million years

Big mountain ranges collapse under their own weight. When geologists say "collapse," they're describing extension, or horizontal stretching, of the lithosphere and crust. The crust thins and the region loses average elevation—a fact in no way contradicted by fault-block mountains rising as part of this process. These "normal faults" raise ranges while dropping the intervening valleys about four times as far. Extensional normal faults are usually "listric," meaning the fault lines arc belowground, tilting the raised blocks so that they resemble fallen dominoes in cross-section. Under them, the lowest crust layer stretches like taffy.

Extension of the Sevier region began by 50 million years ago, progressing from British Columbia southward, but curiously it did not reduce the average elevation by much until later. Initially, it was expressed as scattered metamorphic core complexes, and later—after the supervolcano flare-up heated the lithosphere, weakening it—as normal faults. Many geologists think that between the high Sierra Nevada arc and

the front-line Sevier mountains we had a "Nevadaplano" for millions of years, a very high plateau analogous to Bolivia's Altiplano in the heart of the Andes. Both in Bolivia and here, they attribute the high plateau to the end of a flat-slab era.

Other geologists doubt there was a high plateau or very much extension and lowering prior to 17 million years ago, when a new force came into play. At that point, the Farallon plate had almost entirely subducted. It was gone from Baja and Southern California, leaving that part of North America in direct contact with the Pacific plate, which doesn't subduct there. Instead, Baja and Southern California are actually on

Satellite view of Basin-and-Range fault block ranges. Idaho–Montana state line indicated in red on right side of image for reference.

Listric faults in cross-section

the Pacific plate and slide northwest along a plate-boundary fault you may have heard of— the San Andreas.

Paralleling the San Andreas, there's a whole zone of active faults farther east, moving nearly all of California northwest, pushing western Oregon north and dragging Nevada along for the ride. A huge block comprising Oregon and large parts of Nevada, Utah, Idaho, and Washington is rotating clockwise, as confirmed by GPS devices placed all across the block. It breaks away at the Teton and Wasatch Ranges, two of the most active fault blocks in the West. Not far away, active faults of the Madison and Lost River Ranges produced the two strongest earthquakes south of Alaska, north of Mexico, and outside of California since 1906: Hebgen Lake in 1959 and Borah Peak in 1983. Idaho's Sawtooths had one almost as big in 2020. In sum, a new plate-boundary relationship is the additional force driving extension today, raising our fastest-rising mountains.

The Great Basin (nearly all of Nevada, plus parts of Oregon, Idaho, and Utah) falls within the Basin and Range province, a textbook case of extension: range after parallel range of lofty fault-block mountains, with big, flat basins separating them. The basins are flat because they filled up with sediments eroding off the mountains as they rose, and the region is too arid for its rivers to carry the sediments out of the basin and out to sea. Basin and Range faulting extends north to raise at least six parallel ranges in Idaho and Montana, from the Sawtooths to the Madison Range. They were separated from their Nevada kin when Yellowstone Hotspot volcanoes demolished intervening fault-block ranges, leaving the Snake River Plain in their place.

THE RIO GRANDE RIFT

The past 25 million years

Extension took a dramatic form in New Mexico and southern Colorado: a valley opened up, miles wide and hundreds of miles long, sinking between opposing normal faults. Named the Rio Grande Rift after the river it captured, it crosses New Mexico from north to south. It may extend as far north as the Wyoming state line. It contains the headwaters of the Arkansas River, where it split open and sank the very axis of the Sawatch Laramide uplift, and it broke up Laramide uplifts in the Sangre de Cristo Mountains. Those two mountain blocks that it raised on its flanking faults contain Colorado's and New Mexico's highest peaks today. The modern Sangre de Cristos are a hodgepodge: ancient basement rocks appear atop some ridges, while much younger sedimentary rocks appear on others.

The rift appears to be a member of the global family of continental rifts, interpreted as potential beginnings of spreading centers that could split continents apart, opening new ocean basins. However, rifting here slackened 8 million years ago. When surface motions are checked with GPS arrays, the flanking mountains are getting a bit farther apart, but no faster than many other parts of the Basin and Range. Earthquake activity continues, but again, it seems less intense than in the Basin and Range. This may be a failing rift.

Geologists also look for heat in active rifts. In northern New Mexico, no heat runs up and down the rift, but they find a line of heat

Metamorphic Core Complexes

Producing metamorphic rocks generally requires deep burial; the ones we can see are those that have risen back up from the depths. Common settings for high-grade metamorphism include arc terranes and sutures between plates that collided: subduction performed the deep burial, and intense mountain-building brought some of the altered rocks back up. The Wallowas, an island arc, have metamorphic marble. Idaho, located near the same suture, also has plenty of metamorphics.

The cores of our Laramide ranges, appropriately called basement rocks, were deep for a very long time. Any rock that makes it to the ripe old age of two or three billion is likely to incur some metamorphism during that span, and these are no exception: they are mixtures of granitic and metamorphic rocks, mainly gneiss.

In contrast, sedimentary rocks of Wyoming, Colorado, and central Montana formed later, in shallow Paleozoic and Mesozoic seas, and were never near a plate boundary. Few of them ever got a chance to metamorphose, so—outside of the Laramide uplifts—sedimentary rocks far outnumber metasedimentary rocks at the surface. The Belt Supergroup, being older and thicker, widely experienced mild metamorphism, producing slate, argillite, and greenstone. In Colorado, very old quartzites and high-grade metamorphic rocks are scattered around, remnants of long-gone mountain ranges.

A less common but classic setting for metamorphism arose from the core of the Sevier orogeny. The post-orogenic collapse of the Sevier mountains, 50 million years ago, entailed big, shallow-angle extensional faults that thinned the crust regionally, then rapidly drew to the surface massive domes of very deep rocks metamorphosed under the intense compression of the orogeny. The domes prove durable, and are still standing as mountain ranges today. Called metamorphic core complexes, they scatter in a broken line from interior British Columbia to northern Mexico. Within our range we have the Pioneers in Idaho, the Bitterroots on the Idaho-Montana border, the Anacondas in Montana, the Selkirks in northern Idaho and Washington, and the Kettle and Okanogan domes in Washington. Look in any of them for taffylike swirls of migmatite from the earth's mid-crust.

Geologists know the temperature and pressure that produce each metamorphic mineral, and from these they calculate the depth where rocks metamorphosed. For core complexes, the minimum standard is "tens of kilometers" deep.

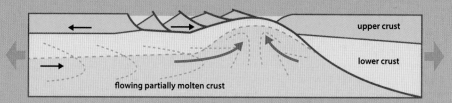

A metamorphic core complex rising in a context of crustal stretching

crossing it—the Jemez Lineament, a line of volcanoes mostly from the past 10 million years. The biggest of those volcanoes comprises the Jemez Mountains; its center collapsed, forming the Valles Caldera, in two climactic eruptions 1.61 and 1.25 million years ago. A new cone grew within the caldera, which remains more or less active. In scale, the Valles compares with Crater Lake: both are smaller than Yellowstone or La Garita.

Any causal connection between the rift and the Jemez volcano remains unclear. The Jemez Lineament follows the line of a suture where Mazatzal terranes accreted to the Yavapai province 1.6 billion years ago. The suture persists as a line of weakness in the crust that rising magma exploits.

THE YELLOWSTONE HOTSPOT

The past 17 million years

Can you name the biggest non-extinct volcano in the Americas? Try Yellowstone. The Yellowstone volcano consists of three overlapping calderas half the size of the park, topping a broad swell. The calderas are basins where the volcano collapsed into its own emptied magma chamber at the end of an eruptive sequence. These supervolcano eruptions happened 2.3, 1.3, and 0.6 million years ago. At that rate, we should be due for another one in 100,000 years, and there is no reason to doubt that it will happen at some point. The heat is obviously still there, as are some interesting up-and-down motions of the caldera floor in recent years. (These motions as well as ground tremors are monitored, and should give us plenty of warning if a magma body rises to the surface.) Yellowstone may be Earth's hottest large caldera.

Yellowstone's eruptions were explosions. Rather than piling their lava up to make a mountain, they blasted it into the sky. The finer particles darkened skies and chilled weather for years, if not decades, while coarser ash blanketed two-thirds of the United States. Geologists use the three big ash layers to date sediments far and wide.

Yellowstone's swell raises the entire area by about 1,200 feet, enough to give Yellowstone a huge ice cap during the ice ages. (The swell is not a volcano: it's caused by upward pressure deep in the earth's crust, not by lava piling up on the surface.)

Yellowstone volcanism aligns with a series of progressively older volcanic centers across Idaho's Snake River Plain. Most geologists consider this a "hotspot track"—the surface expression of a plume that rises from a point in the deep mantle while the crust drifts over it, so that it appears to move in a line. The earliest undisputed trace of the Yellowstone Hotspot is a group of calderas 17 million years old near McDermitt, Oregon. Most geologists today would also include the Steens and Columbia River basalts, a series of flood basalts that began at the same time, just south of the Wallowas. Many geologists think hotspots as a rule start out as basalt floods—vast provinces of flat basalt flows. A hotspot starting at McDermitt would run in a nice, slightly curving line; one starting with the flood basalts would not, as their source vents were well to the north. That's a problem, but it can perhaps be explained by some of the magma getting deflected by the subducting slab, or by the continent's edge.

A few geologists trace Yellowstone still further back, to vast basalts now located in Oregon and Washington's coast ranges. This basalt forms an accreted terrane, Siletzia, born 56 million years ago as a chain of hotspot seamounts off the coast.

The first hotspot lists were purely oceanic—the Hawaiian Islands being textbook—but starting with Yellowstone, continental hotspots have been proposed. The sharp bend in the Hawaii hotspot track was long seen as proof of a sharp shift in plate motions 43 million years ago; hotspots were thought to remain fixed in place in the mantle, offering a frame of reference to help map out ancient plate movements. As more data

Hot Springs, Geysers, and Mudpots

Yellowstone has dozens of hot springs and mudpots, and a majority of the world's geysers (a majority that's increasing, as geothermal wells kill off some of the competition).

Almost anywhere, temperature increases with depth below the earth's surface. No volcanoes or magma are required. At 2 miles down, for example, seeping groundwater commonly heats to 200°F. Hot, it tends to rise. To reach the surface, it just needs broken or porous rock to let it through quickly, before it cools down or dilutes. The Rockies abound in hot springs near deep faults, which provide pathways of broken rock.

Yellowstone, a huge caldera of thoroughly fractured rock, takes "geothermal" to a higher level: it has magma (interlaced in solid rock) just 3 miles down; 460°F rock is just 1,500 feet down; and 400°F water is close to the surface. Under a geyser, even 400°F water doesn't boil, due to high pressure, as long as the slender vent is plugged with crusty mineral. Pressure builds until it bursts those crusts, reducing the pressure, instantly boiling the superheated water, shooting more water and steam out, and so on in a chain reaction that perpetuates itself for the seconds or minutes it takes to shoot out thousands of gallons of water.

Old Faithful often takes only eighty minutes or so to rebuild its pressure, but most geysers take much longer. While the pressure is down, minerals precipitate in the neck of the vent, making a new, but weak, plug. Sooner or later, they may plug the vent permanently, or at least until earthquakes crack them. Yellowstone obliges with frequent local tremors and less frequent Basin and Range quakes like the 1959 Hebgen Lake quake, which reactivated 160 geysers and turned a few into continuous spouters. On the other hand, a prolonged drought in the 1300s apparently dried up Old Faithful for many years.

Mudpots occur where sulphurous steam seeps up without water. The sulphuric acid breaks the rock down, and steam saturates and bubbles the resulting silt.

Mineral-rich hot water provides habitat for colorful, highly specialized organisms that hold as much magic for scientists and pharmaceutical companies as they do for sightseers. (See p. 492.)

Castle Geyser in Yellowstone National Park

comes in, it looks like that plate-motion shift never happened; the plume changed course in the mantle. Today, few geophysicists maintain that plumes are perfectly fixed or, for that matter, that they all originate at the core-mantle boundary. Debate rages on as to how far down in the mantle the Yellowstone Hotspot originates.

THE ROCKIES RISE AGAIN

The past 10 million years

The orogenies that created the Rockies completed their work 45 million years ago. The mountains have not just frittered away ever since.

Even before plate tectonics transformed the debate, geologists for decades told tales of the Rocky Mountains falling and rising again. In one view—one you may still encounter on an interpretive sign—the summit flats up on several ranges like the Beartooths, the Wind Rivers, and the Colorado Front Range were called remnant surfaces from the early Great Plains, which then rose 12,000 feet locally.

More than a century on, geologists still debate how high the Rockies were at various points along the way. All agree that there has been post-Laramide uplift. The mountains wore down somewhat in their first 10 billion years, and rose later, at least once.

For evidence of uplift, look along the eastern face of the Laramie Mountains and Front ranges for a large, gently sloping bench, typically at around 9,000 feet. This Rocky Mountain erosion surface, or pediment, developed as a plain at the foot of the range millions of years ago. Pediments are all over Colorado and Wyoming. They don't come labeled with the dates and elevations of their formation, but geologists agree that most are higher today than when they formed.

(Summit flats are different; they can form right on top of mountain ranges. Ice freezing and thawing in cracks breaks rock down into soil, producing gentle surface topography broken in places by picturesque tors—scattered small bedrock outcrops 10 to 40 feet high. River canyons and glacial valleys carve their way into ranges from the margins, while leaving central summit flats intact.)

There's decent agreement on some extremely broad uplift within the past 10 million years in particular, plus a handful of more focused regional uplifts. By extremely broad, I mean that it raises our entire Rocky Mountain region *plus* western parts of the Great Plains—but not by a lot. Maybe 1,000 feet.

Broad uplift may be explained, in part or in whole, as continuing effects of the flat slab. The Great Plains and adjacent Rockies were tugged down by the sinking slab under them, flooding them with shallow Cretaceous seas and letting sediments more than 2 miles thick accumulate on the sinking seafloor. When the Laramide ranges rose, the basins between them remained low. After the flat slab sank, the area rebounded as mantle material rose around the falling slab's edges. Some of it melted and rose toward the surface, and chemical changes made the part that's left behind less dense. Hydration (alteration to lighter-weight minerals) contributed, and could take a long time to play out.

Smaller-scale post-Laramide uplift processes include:

- The Aspen and San Juan anomalies, seismically imaged patches of especially hot mantle—likely where a piece of lithosphere delaminated and fell away
- The Yellowstone Hotspot swell
- The volcanic Jemez Lineament and the Rio Grande Rift
- The Wallowa fault block, likely an after-effect of pouring out the vast Columbia River Basalts from that location
- Isostatic adjustment following ice age glaciation (see p. 499)
- Isostatic adjustment after the Colorado River changed course and carved the Grand Canyon to near sea level, leading to massive erosion in many areas upstream from there

None of the above uplift forces except post-glacial adjustment reached as far north as Glacier National Park. Several Montana ranges may have stayed as high as they are now—getting buoyancy from mantle flow—until about 34 million years ago, and then subsided a little.

ICE AGES

The final, very crucial step in shaping our Rocky Mountains had less to do with plates or slabs. About 2.6 million years ago, Earth entered the Pleistocene Ice Ages. Repeatedly, about half of North America was blanketed by two huge ice sheets similar to those blanketing Greenland and Antarctica today. One lay on either side of the Canadian Rockies. The western one (the Cordilleran) covered parts of Alaska, the Yukon, and British Columbia, and entered the United States briefly at its last maximum, 17,000 years ago. It reached the Columbia River in northeast Washington, and the Mission and Swan valleys in Montana.

Farther south—all the way to New Mexico—each big mountain group accumulated its own cluster of valley glaciers, which gave our region a facelift (including a bit of an uplift). Wherever you see a U-shaped valley today, picture a thick glacier flowing for at least 1,000 years. Broad mountain groups—Yellowstone, the Wallowas, and the San Juans—acquired broad central ice caps with valley glaciers emanating from them.

Glaciers occur wherever an average year brings more new snow than can melt: snow accumulates and slowly compacts into ice. Eventually, the ice gets so thick and heavy that it flows slowly downhill until it reaches an elevation warm enough to melt it as fast as it arrives. This flowing ice is a glacier, a mechanism that balances the snow's "mass budget." Ideally, the rate of flow is equal to both the excess snow accumulation in the upper part of the glacier and the excess melting in the lower part; this ideal glacier would neither advance nor retreat.

A U-shaped valley with lakes in Wyoming

Few glaciers are so stable. Instead, the elevation where the glacier terminates in a melting snout advances and retreats (drops and rises) in response to climate trends. Retreating glaciers don't turn around and flow back uphill; they simply melt away at the bottom faster than they arrive there. A glacier "stagnates" after shrinking to where it no longer has enough mass and slope to keep flowing. This is what we have seen in recent decades due to the warming climate. We'll be lucky if any of of our glaciers last to 2050.

The work glaciers did on our mountains will endure in the form of U-shaped valleys, cirques, moraines, and an abundance of lakes. (Lakes are inherently temporary: a depression that gets enough rain or snow to become a lake will keep filling until it overflows, and then that outlet stream will keep cutting downward until it empties the lake. In recent times, glaciers have been Earth's chief mechanism for creating new lakes.)

Glacier ice has a consistency utterly unlike that of a snowbank. Beneath a firn of last year's snow, what was once snow is recrystallized into coarse, nubbly granules with hardly any air space. Eventually that granular texture will grade into massive blue ice,

Moraines, Cirques, and Tarns

You're gazing at a beautiful blue lake on the edge of a valley or plain, at the foot of a defile down the front of a great mountain range. The lake is one in a row of similar lakes. Each of them formed behind a natural dam called a "terminal moraine." (If it has a man-made dam instead, you're at the wrong lake; try another.)

A terminal moraine is typically an arc-shaped heap of rocks of all sizes, dumped across a valley by a glacier in pause mode, before it started to retreat. If a valley has a series of several "recessional moraines," each records a minor re-advance during a long retreat. The terminal moraine—farthest downvalley—marks the last glacial maximum, around twenty-three thousand years ago. The youngest one dates from the third Little Ice Age maximum, in the mid-1800s. (No moraine remains from the two previous little maximae; the 1800s advance wiped them out.)

Don't picture a bulldozer; picture a mobile conveyor belt running in place for a while before backing up; as it retreats, it continues to dump its load. So does a retreating glacier, spreading rocks across the valley floor as a "ground moraine," not so much a heap as a bouldery texture all over the valley floor. It may be pocked with pits or tiny lakes. Pits form where big ice blocks are deposited among rocks, and then melt. Many moraine boulders have one or more flattened and striated sides that were ground against bedrock while this boulder was on the glacier's sole.

We also have "rock glaciers," which move like a glacier but are like ground moraines in consisting of rocks of mixed sizes in a valley. Underneath, the rocks are held in a great mass of ice and frozen mud that has flow characteristics similar to glacial ice.

A "lateral moraine" is a terrace on the valley flank, parallel to the valley. It, too, was left behind by a thinning, narrowing glacier. The glacier's edges carry more than their share of rocks because rocks fall from adjacent cliffs.

Having expended a liter of sweat since departing from that morainal lake we discussed above, you're now admiring a small subalpine lake tucked in among cliffs of an amphitheater-shaped hanging valley—a tarn, in a cirque.

though a microscope will still show a texture more granular than, say, frozen lake ice.

Our glaciers are "warm" glaciers at close to 32°F throughout, in contrast to cold Arctic glaciers. Warm glaciers, like ice skaters, glide on a film of pressure-melted water. They pour around bedrock outcrops by melting under pressure against the knobs' upstream sides and refreezing against the downstream sides (repaying the heat debt incurred by the change of state from solid to liquid). They erode rock ferociously, not because either the water layer or the ice is abrasive, but because sand, pebbles, and boulders are gripped and ground along over the substrate. It's the sediment load that does the grinding in both glacier and river erosion—comparable, respectively, to a power belt-sander and a sandstorm. The glacier's grit leaves parallel grooves or "striations" across bedrock outcrops.

Because glaciers carve faster than streams do, on average, the ice ages sped up the rate of erosion and of isostatic uplift (the principle by which the crust layer floats atop the mantle layer). But the size of this effect is controversial, at least for small alpine glaciers, which do move rock, but not very far. If the

Glacially carved valley heads can all be called cirques. The word *cirque* (which means both "circle" and "circus" in French) refers to the half-bowl shape passed down from Greek amphitheaters to European circuses. Typically, the lip of the cirque, just above the dropoff, is eroded down to bedrock and striated with parallel scratches. The interior floor (near the tarn, if there is one) collects fine sediment, and may be marshy: the tarn is filling up and becoming a marsh, and perhaps later a meadow. *Tarn* is Scandinavian for a small lake.

Where the glaciers of two adjacent cirques grow until they almost touch, they leave a saw-edged ridge, an "arête" (a-**rett**). Where the heads of cirque glaciers on three or four sides of a mountain erode back near each other, they carve a "horn," as in Matterhorn. Abundant horns, arêtes, and cirques earned Glacier National Park its name.

LEFT: *A large cirque in Montana* RIGHT: *Rock glacier showing glacier-like curving lines*

rock they moved still sits in moraines a few miles downvalley, it still weighs the region down, and isostasy won't be able to lift much.

Here's how erosion can make mountaintops higher: think of the mountain range as an iceberg. Famously, only the tip of the iceberg—10 percent of its mass—floats above the waterline. Imagine a cylindrical iceberg 100 feet tall and at least that wide, like a giant hockey puck; 10 feet (10 percent) shows above the water. If you melt off those top 10 feet, the berg, now 90 feet tall, rises 9 feet to keep its top 10 percent (9 feet) above water. Now suppose instead you carve those top 10 feet into a perfect cone, like sharpening a pencil. This time you've removed two-thirds as much mass as in the first case, so the berg rises two-thirds as far—6 feet instead of 9 feet; but since the tip remains 100 feet above the base, all 6 feet are a gain in its peak height.

Similarly, when an ice age inflicts alpine glaciers on a plateaulike mountain range, they can carve deep, U-shaped valleys fairly close together without reducing the highest ridges by much. They have removed significant mass from the range as a whole, so it floats higher isostatically. The *average* elevation across the range is lower than before (erosion is working to level the range), but the *peak* elevations are higher. Local relief is greater.

This works even without glaciers to speed it up: suppose a broadly arched mountain range sits at isostatic equilibrium for millions of years, and then the climate changes from arid to wet and stormy, carving canyons and then wide valleys until the high surface remains only as narrow ridges. Erosion has removed a huge weight of rock. Provided that rivers carry most of that mass off to distant lowlands or seas, isostatic compensation will raise the ridges to new elevations.

In the real world, this peak-raising effect may be modest, but the effect of glacial erosion making slopes steeper and some peaks peakier is undeniably dramatic and alluring.

By fourteen thousand years ago, all ice was in full rout, the glacial stage over. By nine thousand years ago, most alpine glaciers were gone from the US Rockies. Some small ones came back during modest cooling in the last four thousand years; they maxed out in the 1800s, the so-called Little Ice Age. Wyoming's Wind River Range holds the biggest glaciers in the US Rockies today. Glaciers south of the Winds are small, and soon to vanish.

GIANT ICE AGE LAKES AND MEGAFLOODS

When there's light snow on the slopes above Missoula, you can see dozens of wave-cut terraces, showing that the town site was once 1,000 feet underwater. Not so very long ago, a lake with about as much water as Lake Erie and Lake Ontario put together filled the Clark Fork Valley, stretching from Idaho eastward to near Gold Creek on I-90, south in the Bitterroot Valley to near Sula, and northeast to near Kalispell.

Glacial Lake Missoula formed again and again, each time the Cordilleran Ice Sheet in Canada advanced far enough south to dam the Clark Fork near Sandpoint, Idaho. The ice dam either floated or burst each time the lake rose high enough, at intervals of several decades, and the entire lake would drain within a few days, ripping across eastern Washington and out to sea through the Columbia River Gorge. These Missoula floods were faster—and forty-six times larger, by one calculation—than any river flood in written history. (I decline to compare them with the refilling of the Black Sea, escaped by Noah in Genesis.) The flow rate was equal to all the world's present-day rivers combined, the power and turbulence almost unimaginable. Satellite imagery of eastern Washington helps you see how floods anastomosed in broad channels, leaving half the state looking like a sandcastle swept by a wave. Just south of Markle Pass, Montana, cur-

Slight terraces showing multiple Lake Missoula shore levels, Missoula, Montana

rent ripples stand 35 feet high and 300 feet from crest to crest; they formed on the lake bottom while it was draining.

There were at least one hundred of these floods between 21,000 and 13,400 years ago, plus an additional series of them during at least one earlier Pleistocene ice age. Indigenous people must have witnessed Missoula floods, as there's evidence that they lived in Idaho's mountains 15,575 years ago. The successive lakes, their floods, and their life spans diminished gradually toward the end, because the ice-sheet lobe shrank as the Ice Age waned, requiring less water depth to float it.

Sudden floods due to cyclical ice-dam failure are called *jökulhlaups* in Iceland, where small jökulhlaups occur to this day. Alaskans see them too, but probably can't pronounce them.

Then there's Glacial Lake Bonneville, our name for the Great Salt Lake on ice age steroids. Since that lake does not drain to the sea, it rises and falls depending on rates of precipitation and evaporation. During the ice ages, there was more precipitation than in recent times, and slower evaporation; the lake grew huge, peaking eighteen thousand years ago and then releasing a megaflood down the Snake and Columbia

Rivers. Compared to the biggest Missoula floods, it released twice as much water but at only one-tenth the flow rate, because the release was less sudden. There was no ice dam; there was simply Red Rock Pass. The lake rose to where it began draining over the pass, whose surface was an easily eroded alluvial fan. Once the overflow began, it took a few weeks to cut 370 feet down to bedrock, tanking the lake level by that amount. (It fell lower over the next eighteen thousand years of drier climate, and is falling dangerously today, with humans diverting water for our own use.) You can see wave-cut terraces from Lake Bonneville's maximum level on many Wasatch slopes.

Idaho's Big Wood River and Big Lost River basins held small ice-dammed lakes, and also witnessed floods.

Sedimentary Rocks

Rocks are classed into three groups that reflect which kind of reprocessing they underwent most recently:

- Sedimentary rocks were compacted and/or chemically cemented after settling as small particles (like mud).
- Igneous rocks solidified from molten rock (magma or lava).

- Metamorphic rocks recrystallized (without melting) from other rocks under intense heat and pressure.

LIMESTONE AND DOLOMITE

Limestone's abundance belies what an exceptional rock it is. First off, it's the most abundant rock (together with its close sibling dolomite) that typically originates through the work of living organisms. (Coal is a distant second.) Various kinds of plants, animals, and bacteria separate out the mineral calcite (a.k.a. lime—$CaCO_3$) from solution in seawater. Animals typically use calcite to make protective shells. You see a lot of those shells—trilobites, crinoid segments, weird tubes, and so forth—in some Paleozoic limestones. In others, and in all Proterozoic (older) limestones, there are no macroscopic fossils. Their calcite was produced by those amazing creatures to whom we owe everything we are today: cyanobacteria. (It was they that freed oxygen, making animal life possible, and fixed nitrogen, allowing plants to evolve.) Some calcite is made by free-floating cyanobacteria and settles to the seafloor after they die; some is made in situ on the seafloor by cyanobacterial mats and stromatolites. As lime mud turns into limestone, the microscopic needles that the cyanobacteria made—as well as the much larger forms that animals made—mostly break down, and the calcite recrystallizes, wiping out any fossil textures.

Second, carbonate rocks provide the world's largest long-term carbon sink—bigger than fossil fuels. If it hadn't been for their sequestration of carbon, the greenhouse climate would long ago have become and remained so hot that large animals probably never would have evolved. This is one of the chief self-regulating cycles described by the Gaia hypothesis: when CO_2 levels rise, calcite critters thrive, sucking up carbon and sequestering it in limestone, reducing CO_2 levels. Unfortunately, this works at a geologic pace; it can't speed up enough to save us from our industrial-era carbon emissions.

Third, carbonate rocks are the most abundant rocks that make plant ecologists sit up and pay attention. Some plants cannot grow on soil derived from limestone, a few only grow on it, and many occur on both calcareous and non-calcareous soils but show a clear preference. These highly alkaline soils have improved availability of nitrogen and, of course, calcium, but low availability of phosphorus and some trace elements. Lodgepole pine was found to grow 36 to 50 percent slower on calcareous than on acidic soils. Rocky Mountain alpine plants widely turn to mycorrhizal fungi as the only way to find usable phosphorus in calcareous soils.

Lime's effect is patchy: rain leaches lime downward, creating lime horizons 6 to 20 inches down depending on water-flow patterns. Shallow-rooted lime-hating plants can find happiness where this horizon is deep enough. On alpine tundra, limestone inflicts physical hardships more critical than the chemical ones: it breaks

Patterns formed of alternating limestone and dolostone Belt rocks

down into fine chips that are more prone to both drought and frost heaving than other soils.

Finally, limestones dissolve—very slowly, of course, but with dramatic results that include underground rivers, sinkholes, the world's long cave systems, stalagmites and stalactites, and tower-shaped hills in southern China. Pure water cannot dissolve limestone, but groundwater easily picks up enough CO_2 to make a weak carbonic acid, which can.

Of the two carbonate rocks, limestone is the one with at least 50 percent calcite content. If the mineral dolomite—$CaMg(Co_3)_2$—predominates along with a lot of calcite, you have dolomite, the rock. (There is a case for reserving "dolomite" for the mineral while calling the rock "dolostone," but "dolomite" is still widely used for both.) The test is a drop of hydrochloric acid, which effervesces in reaction with the alkaline calcite; on dolomite, it effervesces only weakly. Many sedimentary geologists carry a squirt bottle of HCl; others, not wanting to be seen in shredded shirts, avoid the acid test by calling both rocks just "carbonate rock."

Limestones get dolomitized by magnesium-rich brine percolating through them, replacing some of the calcium atoms with magnesium. One common setting would be shallow Proterozoic seas where cyanobacteria thrived for centuries, creating a limestone seafloor. Then some climate or drainage shift reduced the influx of fresh water, and evaporation concentrated the salts while cyanobacteria kept on consuming the calcium, turning the sea to high-magnesium, low-calcium brine. Dolomitization obliterates fossils and yields a porous, "vuggy" rock. (Vugs are visible holes.)

Limestone formations commonly have a bad case of the stripes. Dark-gray limestone bands alternate with brown shale bands, or with pinkish-buff dolomite bands. The stripes may originate from cyclic changes

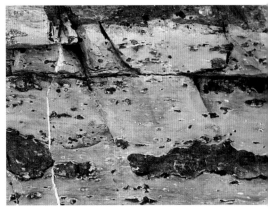

Madison limestone, Montana

in sea level. Limestone formed as "carbonate platforms" in vast, shallow seas—rapidly when the seas were shallow, then slowing and getting overwhelmed with mud (shale) when they were a little deeper. What could have caused slow, regular sea level cycles? During the Carboniferous Period, a huge ice cap waxed and waned on the supercontinent Gondwana in long cycles, which may have caused sea level cycles here, as well as cyclic coal formation almost worldwide. CO_2 levels in the air were likely involved, since CO_2 controls the rate at which calcite forms and dissolves. It's challenging to figure out a great cyclical mechanism among climate, CO_2, sea level, possibly tectonic movements, and the flourishing of different forms of life.

Though one kind of very pure limestone is chalk, most limestone in the Rockies is hard—hard enough to be an overhang on a cliff. Calcite is white, chalk is white, but limestone strata in the Rockies have weathered to dark-gray bands. The aptly named Leadville limestone crops out in many parts of the Colorado Rockies. In central Montana, the pale-gray mid-Paleozoic Madison limestone provides an erosion-resistant backbone for several ranges. Throughout Glacier National Park, massive lower cliffs are dark-gray Altyn carbonates: the Altyn Formation forms the base of the Lewis thrust sheet and is the park's oldest rock, at around 1.5 billion years.

Travertine being created in Canary Hot Springs, Yellowstone National Park

In Proterozoic limestones in Glacier, look for patches of whorls about 6 to 20 inches across—ancient stromatolites. Geologists long assumed they had non-biological origins, and came up with many hypotheses. They turned out to be a kind of cyanobacterial colony still found in a few small, saline, tropical bays. They were the dominant life-form for billions of years (much longer than dinosaurs) but became uncommon after creatures evolved to eat them. They thrive today only in water too salty or too hot for those creatures.

TRAVERTINE

Rocks of calcium carbonate can also form without biological help. If the mineral precipitates to the seafloor, it becomes limestone, but if it precipitates from water dripping inside a cave (as a stalactite, etc.) or gushing from a hot spring, it takes a denser form called travertine. In a cave, the drip becomes supersaturated with dissolved carbonate, precipitating carbonate as the water evaporates. In a hot spring, water becomes supersaturated when it reaches the outside world and cools, reducing the amount of minerals

it can hold in solution. The amazing terraces of Yellowstone's Mammoth Hot Springs are travertine. The water cools abruptly at the exact moment when it becomes a thin film going over the "dam" of the terrace, so new travertine forms right there and builds the dam higher. Algae sometimes get in the way of precipitating a dense travertine, yielding porous tufa. Long-defunct similar springs produced travertine that's quarried near Gardiner, Montana, and sliced into facing stone with lacy-patterned holes. Hanging Lake near Glenwood Springs, Colorado, exists thanks to a natural travertine dam. Its water is not hot, but is loaded with calcium and bicarbonate.

QUARTZ AND CHERT

Quartz is the most familiar "rock crystal"—transparent (though sometimes tinted), with six generally unequal sides and a six-faceted point. It's the only stone in this chapter that forms single crystals—because it's a mineral, rather than a rock.[3]

Quartz is silicon dioxide (SiO_2, also called silica) with too few impurities to disrupt its crystal form. It is harder than steel (i.e., too

3. *A mineral is a single chemical compound, or continuum of compounds, with one characteristic crystal shape. A rock, on the other hand, is a mixture of minerals whose proportions to each other vary only within limits that define that kind of rock. Its constituent minerals may form visible crystals or not.*

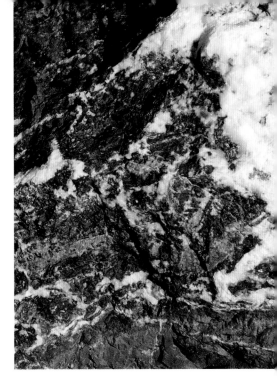

hard to scratch with a knife), lightweight and light-colored, and abundant as a rock ingredient; it is in nearly every light-to-medium-colored rock you see, except limestone and marble. Its abundance at the earth's surface probably results from its light weight—it has risen among heavier materials while powerful forces stirred the earth around it for some four billion years. The heavy metals nickel and iron gravitated to the earth's core; they are still the most abundant elements in the earth as a whole, whereas silica outweighs them near the surface.

Large, free-sided quartz crystals are uncommon, but they do catch the eye. They form in cavities in rocks through which groundwater rich in dissolved silica has seeped for a long time. Silica precipitates out as "rock crystals" on the cavity lining, like sugar crystallizing on a string as "rock candy" (which is named after quartz crystals). Veins of quartz became gleams in a prospector's eye, since lucrative metals often occur in or near quartz veins where pressurized subterranean brines delivered silica and metals together.

Nearly pure quartz may occur in an opaque, cryptocrystalline form, meaning the crystals are too small to see even with a microscope. This rock, chert, resembles porcelain, whereas crystalline quartz is transparent like glass. That's no coincidence: ground-up silica rock is the main ingredient in both glass and porcelain. Both porcelain and a hard chert called Arkansas novaculite can whet a fine edge on a knife. Like glass and obsidian, chert chips with an even, shallowly concave fracture. Its dark form, flint, was second to obsidian as arrowhead material.

Cherts form in several ways. The most common is similar to limestone—crumbled marine shell material settling to the ocean floor or precipitating there from solution in seawater. Whereas most seashells are made of calcite, the mineral in limestone, other organisms—diatoms, radiolarians, and

TOP: *Quartz veins* BOTTOM: *Chert*

sponges—make a sort of shell of silica. They are the main source of chert.

Oceanic plate subduction tends to destroy massive chert beds rather than toss them up onto continents, and sure enough, few are found here. More common are small nodules of chert within limestone formations.

Don't infer that silica dissolves easily in water just because quartz and chert have histories as silica solutions. The silica in those solutions dissolved under great pressure deep in the earth. At surface pressures, the insolubility, hardness, and chemical stability of silica are what make quartz more durable than most rocks in a stream, ultimately yielding the high quartz content of beaches, sandstones, and quartzite. Pressurized, superheated water in the Yellowstone caldera picks up silica as it flows through high-silica rhyolite, then precipitates it around geysers as a porcelainlike glaze called sinter, or geyserite. Hot water rising instead through carbonate strata yields the lime counterpart, travertine.

SANDSTONE

Sandstone is a broad term for sedimentary rock made up of compacted, sand-sized (0.06-to-2-millimeter) particles cemented together with water-soluble minerals like silica or calcite. The sand grains may almost all be quartz. It's puzzling that beds or beaches of nearly pure quartz sand ever collect, since the waters that bring them are eroding all the various rocks of a watershed. This is because quartz is both hard and chemically stable, tending to break down to sand-sized particles and then stop breaking down. In contrast, feldspars and other abundant min-

erals weather chemically, and end up as finer silt or clay particles. Ocean waves collect the sand into beaches, while washing the finer silt and mud particles out to sea. Rivers also separate sand from silt.

Utah, Colorado, and southern Wyoming are famously rich in red sandstones. These are quartz-rich, but with enough iron oxide mixed in to dye them red. Many Utah sandstones and some Colorado ones display obvious wind-blown dune textures: they are literally petrified sand dunes. Ancient sedimentary beds in the Rockies typically produced shale or limestone if they were underwater, and sandstone if they were beaches or dunes. Sandy and limey phases alternated many times, up until the middle of the Laramide orogeny.

RIGHT: *Microlaminated sandstone and mudstone of the Belt* BOTTOM: *Sandstone in the Dakota Hogback, Colorado*

CONGLOMERATE

Typical conglomerates are petrified pebble beaches and river bars. By definition, they're sedimentary rocks composed of at least 50 percent erosion-rounded stones. They require soluble minerals as glue, as well as fine rock particles to fill in the spaces. Well-cemented conglomerates will break straight through pebbles and interstices alike—a fine sight. Weaker ones break through the cementing matrix only, leaving the pebbles sticking out like from a concrete paver. Some metaconglomerates were compressed during metamorphism, ovalizing the pebbles all in one direction. Montana's Gravelly Range is named for its exposed conglomerates.

BRECCIA

Breccia consists of unrounded rock pieces bound up in a fine-grained matrix other than volcanic tuff. The distinction from conglomerate's erosion-rounded pieces may seem trifling, but on further thought it is puzzling, since only eroded rocks are normally deposited as sediments. There turn out to be several brecciating processes, each involving breakage, the root meaning of the Italian word *breccia* (**bretch**-ia).

Sedimentary breccia is often a turbidite—shaley shards ripped from deep-sea sediments by a sort of submarine landslide, and redeposited in a jumble within a matrix of mud. Terrestrial landslides in shaley, slaty, or limestone terrain can also end up compacted into breccia.

Volcanic breccia can be the skin of a slow lava flow that shatters and then slips back into the flow. It can be the edge of a magma body that incorporates broken chunks of surrounding rock; the magma needs to be cool enough to leave sharp, unmelted edges on the chunks.

Tectonic breccia is a mix of coarse and fine rock fragments broken by shearing forces within a fault, and then compacted; sometimes you can visually match the broken edges of nearby fragments.

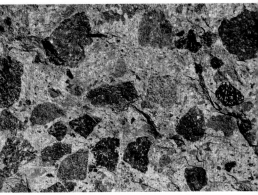

TOP: *Conglomerate above a layer of Maroon Formation sandstone* **BOTTOM:** *Volcanic breccia*

There is a lot of breccia out there, but much of it looks like an obscure mess. It takes rock saws and polishers to reveal the splendid breccia used as architectural facings.

SHALE, MUDSTONE, AND ARGILLITE

Throughout the Overthrust Belt, layers of soft, weak shale provided surfaces where the earth could break into thrust blocks and sheets. While playing this critical role, shale also tended to crumble or erode away where exposed.

Shale originates as clayey river mud carried out to sea and piled up to such depths that, through thousands or millions of years under the weight of subsequent layers, it compacts into more or less solid rock. Most

TOP: *Large boulder of current-rippled Belt argillite*
BOTTOM: *Cross-rippled Belt rock*

Proterozoic Belt mudstones break on their original bedding planes, which often show mud-cracks or perfect ripples, as if small waves had rippled a shallow pond floor just last week. "Ladderback" criss-crossed ripples, for example, may have been imposed in one direction by occasional storms strong enough to stir fairly deep water, then stabilized by algae, then imposed in a second direction by a tsunami from an earthquake. Squiggly crackle-textured pavements owe their striking texture to the shock of quakes themselves.

The Proterozoic was an eon of few or no animals. Younger mudstones don't often show good ripple marks, even if they formed in shallow seas, because worms, mollusks, and other animals burrowed through the muds before they fossilized, messing up any ripples.

Mildly metamorphosed shale is argillite. Grinnell Formation rocks are the most colorful in Glacier National Park thanks to their iron-rich argillite, either green or red depending on whether there was much oxygen around to rust the iron during deposition. That formation is 1.5 billion years old. Vast red shales were again laid down—probably on arid coastal plains at the downwind edge of Pangea—between 270 and 210 million years ago in Montana and Wyoming. Oxygen-rich epochs that allowed giant insects to evolve (p. 500) also made red beds. But most Colorado and Utah red rocks are sandstones.

Intrusive Igneous Rocks

Igneous rocks are divided into two textural classes—fine-grained and coarse-grained—and graded by chemical (or mineral)

sediments need water-soluble minerals to cement them into sedimentary rocks, but clay particles are so minute and flaky that they become shale through compaction alone, or with only minor amounts of cement. You can break shale in your hands, breathe on it and smell clay, and scrape it with your knife and spit on the scrapings to mix up some clay mud. Shale is generally gray, and breaks into flattish leaves along its bedding planes. Mudstone is similar, but doesn't break into thin leaves due to its different mineral content. Where we find shales and mudstones, we may find less- or more-processed beds nearby—dense clays that lay on the ocean floor too briefly and shallowly to get compacted into rock, and others that got buried long and deeply, metamorphosing into slate or phyllite.

composition. High-silica rocks are "light," low-silica rocks "dark." Each "darkness" grade on the compositional scale can occur with either fine or coarse texture. The two textures tend to correlate with two origins; for our purposes we will assume they always correlate as follows:

- Volcanic, from lava that erupted upon the continental surface or the ocean floor, congealing quickly to produce fine crystals, or
- Intrusive, from magma that solidified into rock somewhere beneath the surface, and cooled slowly, producing coarse texture

DIORITE AND GRANITE

Granite has high name recognition, partly for its role as the heavy in rock-climbing thrillers, and partly for its ubiquity in cemeteries and kitchen counters whose polished surfaces teach us the texture of granitic intrusive rocks—interlocked coarse crystals of varicolored minerals. Quartz is the white-to-buff, translucent component; feldspars are salmon to pale gray; pale, glittering flakes are muscovite mica; glittering black flakes are biotite, or "black mica"; hornblende and pyroxene are duller blacks.

The word *granitic* refers to those granular crystals, and the word *intrusive* describes their likely origin. While some magma (molten rock) rises and extrudes onto the earth's surface through a volcano, other magma merely intrudes among subsurface rocks and solidifies there, at depth. This may be an entire magmatic event that failed to reach the surface, or it may be a remnant left behind in the plumbing. In either case, if it cools in a large mass, it changes from a liquid to a solid state very slowly, allowing its atoms plenty of time to bond within larger and larger mineral crystals. Large crystal size defines the granitic class of igneous rocks.

Intruded masses are called plutons; the biggest (over 100 square kilometers, or 38 square miles, of surface exposure) are batho-liths ("deep rocks"). The Idaho Batholith was long thought of as a single humongous intrusion, but further study found dozens of distinct plutons. In fact, not much of what was once common knowledge about batholiths goes unquestioned. Textbooks have illustrated them as bottomless, capital-dome-shaped cross-sections. These were said to take millions of years to rise and to crystallize, but no one could figure out how they made so much room for themselves, even over millions of years. More likely they are structured like big clusters of honey mushrooms.

Rock climbers and kitchen designers call any sort of coarse-grained igneous rock granite, but geologists do not. True granites here include the pink granite of the Pikes Peak batholith and the whitish granite of the Wind River Range. For coarse igneous rocks that are not quite granite, hedging with the word "granitic" is vague but correct. Diorite, a common granitic rock containing less than 20 percent quartz, is salt-and-pepper speckled. Geologists say diorite is darker than granite, with granodiorite (abundant in Idaho) in between, but by "darker" they really mean "contains less quartz." A granite can look darker than a diorite if the granite has more quartz but a lot less feldspar and more blackish minerals.

Granite, Sawtooth Mountains, Idaho

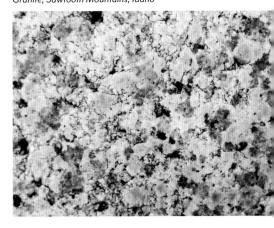

Diorite is the coarse-grained equivalent of fine-grained andesite. Lava of diorite's chemical proportions could also erupt and crystallize as andesite; diorite often represents the roots of andesitic volcanoes. A diorite pluton related to Challis/Absaroka volcanism forms the core of the Crazy Mountains.

Granite is equivalent to rhyolite; these magmas are so viscous that 90 percent of them freeze in their own plumbing to become plutons rather than volcanoes.

If you can cram two more coarse igneous "-ites" into your brain, try these oddballs:

- Pegmatite is granitic rock with extra-large crystals (greater than 1 centimeter) of any mineral content; it is most often found as pale dikes, or swirls in migmatite (p. 538); and it is beloved by designers (even those who take it for granite).
- Syenite is a rare, pinkish, all-pale granitoid rock, made up almost purely of feldspars: it's a foiditoid. (I'm not making this up!) Find it in Montana in the northern Crazies or on Skalkaho Mountain in the Sapphires. In Colorado, you can see the combination plate—pegmatitic syenite—on Deckers Road within the Hayman burn.

Countless peaks attest to granite's durability, yet heaps of coarse, sharp sand show its proneness to "decomposing." Granite boulders round themselves as they decompose. When mica crystals in the granite get wet, they can alter into clay minerals, expanding as they do so and thus crumbling the near-surface part of the boulder. Broad granite exposures may exfoliate in sheets parallel to the surface, as heat expansion and contraction split the plutons in patterns.

PERIDOTITE

Peridotite (per-**id**-o-tight) rocks are apparently pieces brought up from the mantle. They are largely restricted to environs of deep faulting within mountain ranges, because they're too heavy to reach the sur-

Peridotite in the Stillwater Complex, Montana

face without exceptionally deep and fast rock movement. The western Idaho shear zone—where the Wallowa terrane slid north along the old continental edge before suturing—is such an environment. Look for patches of blackish-green peridotite, flecked with black crystals, in roadcuts north of Riggins.

Peridotite is defined by a 40-plus percent proportion of the yellow-green mineral olivine, the predominant mineral in the mantle and thus in the earth as a whole. It has no fine-grained (volcanic) equivalent because volcanoes aren't hot enough to melt it. Solid peridotite rocks carried in molten lava occasionally shoot out of volcanoes.

Peridotite gets concentrated in an unusual feature called a layered mafic intrusion, thought to be dark magma that failed to reach the surface in a flood basalt. The layers form as different minerals crystalize at slowly dropping temperature points, and settle to the bottom of the magma pool. Like several other layered mafic intrusions, the Stillwater Complex on the north edge of the Beartooths is rich in chromium, palladium, and platinum ores. It was emplaced 2.7 billion years ago during the birth of the Wyoming Craton.

Gem-quality crystals of olivine are known as peridot. Peridotite and its relatives

TOP: *The dark Purcell Sill outlined with limestone it bleached* RIGHT: *Gabbro*

weather into reddish "serpentine" soils difficult for most plants to grow in.

GABBRO AND DIABASE

Magma of the same mineral composition as basalt that crystallizes slowly underground without reaching the surface and erupting becomes gabbro, a dark rock with a coarse texture like granite. While basalt is the most abundant volcanic rock, gabbro is a rather rare intrusive rock. The impressive Sneffels pluton in Colorado's San Juans includes gabbro, as does Montana's Crazy Mountains pluton.

Basalt is very fluid, and doesn't tend to plug up plumbing. So it's rare for basaltic magma to form large plutons, but common for it to crack rock formations when forcing its way up, filling the crack, and then solidifying in it. If the magma took advantage of a weak plane between sedimentary beds, it's a "sill." If it cut across bedding planes (or if there are no bedding planes), it's a "dike." Both make lines or stripes across a rock face.

The Purcell Sill was injected horizontally between two beds of limestone through much of Glacier National Park. The magma's heat vaporized the organic impurities that make

Siyeh limestone gray, leaving dramatic white outlines above and below the sill. The magma cooled fast enough to make a fine-grained black rock that you can call either basalt or diabase (the name for basalt when it's in a sill or dike, making it intrusive rather than volcanic). Dikes branch off from the Purcell Sill in many places in the Belt Supergroup.

The stripe up the middle of Mount Moran's face, the Black Dike, is the most photogenic of several 1.2-billion-year-old dikes in the Tetons. Dikes decorate the ancient Wyoming craton in the Wind Rivers and Beartooths as well—clues of a likely rift zone, since lava explores many cracks on its way to erupting as a new spreading center.

The Chief Joseph Dike Swarm in the Wallowas is a batch of some 1,800 dikes that supplied the lava for the Columbia River Basalt floods. Look for dark stripes through pale granite—for example, by looking to the southwest from the south end of Wallowa Lake. If you get to see one up close, note how its heat altered the granite, partially melting it and mixing with it a bit. Unmelted crystals drift a few inches into the basalt.

Dikes may be several yards wide, and miles long. The word "dike" applies to either large or small crack fillings, so long as they are an igneous mixture, showing that they solidified from magma. The other, very different mode of crack-filling is veins, which form by precipitating from solution in a water-based fluid that flows through the crack.

Recent pahoehoe lava at Craters of the Moon

Volcanic Igneous Rocks

The familiar word for volcanic igneous rock is "lava." Other than a few of the youngest basalt flows, most of ours don't look like lava flows anymore, but they were once.

ANDESITE

Named after the world's highest volcanoes, the Andes, andesite is better than other lavas at building up tall, layered stratovolcanoes. Andesite lavas make rough gray, greenish, or sometimes reddish-brown rocks. Often they are speckled with crystals up to ¾ inch across, scattered throughout a fine matrix otherwise lacking crystals of visible size. These were already crystallized in the molten magma when it erupted. With close examination and a little practice, you can easily tell them from the shards of noncrystalline, whole rock jumbled up in tuff. It can be hard, though, to tell andesite from lavas like basalt and dacite.[4] Color descriptions are unavoidably vague and not very useful for ID; andesite is "medium-dark"—between dark basalt and light dacite and rhyolite—but the actual tints overlap.

BASALT

The most abundant rock in the earth's crust as a whole, basalt is dark lava. From black, it ranges down to light gray, and may be altered to greenish or reddish. Its surface is drab, without conspicuous crystals or other features except, frequently, bubbles called vesicles. These show that the lava rose from the depths full of dissolved gases. Just as uncapping a bottle of soda releases bubbles of CO_2 that, under pressure, had been dissolved invisibly in the liquid, so lavas often foam up as they near the surface. Vesicles abound near the original top surface of a lava flow. In old basalts, they have often become solid polka dots by filling with water-soluble minerals such as calcite.

Polygonal (four-, five-, or six-sided) basalt columns, often neatly vertical at one level and splayed out at another, result from shrinkage

4. *The technical definitions are ranges on a graph of bulk chemistry; positive IDs often require specimens to be pulverized and analyzed in a lab. Light color correlates with silica (SiO_2) content—around 50 percent in basalt, 60 percent in andesite, and over 70 percent in rhyolite. Most SiO_2 in magma forms silicate (SiO_4) tetrahedrons that bind to other elements to make feldspar and other minerals; only the SiO_2 excess over about 55 percent stays purer, as quartz crystals. Thus andesite has few quartz crystals, and basalt has none.*

during cooling of large flows. See them on Table Mountain above Golden, Colorado—lava flows of the Colorado Mineral Belt.

Basalt lava erupts in several styles. One of them, the basalt flood, does not build a mountain, being so liquid that it flows out flat. The Columbia River Basalt is a vast series of floods that covered much of eastern Oregon and Washington between 17 and 12 million years ago. See it on Hells Canyon's walls, or raised high in the western part of the Wallowas. Much younger, less voluminous Snake River Basalts cover the Snake River Plain in Idaho, notably at the Craters of the Moon, the largest young lava field in the continental United States. Its latest eruption came two thousand years ago.

Basalt flows are widespread in the Southern Rockies—not just at Basalt, Colorado, which has a basalt-capped mesa. West of there, basalt around 11 million years old caps the Grand Mesa. Colorado's huge San Juan volcanic flare-up produced almost no basalt at all during its main phases, but began erupting basalts as it waned, 25 million years ago. Since then, the Rio Grande Rift and Jemez Lineament have carried on making basalt.

Seafloor basalt is far and away the most voluminous lava in the earth's crust. When basalt erupts underwater, it chills and solidifies quickly, in small batches that we see as glassy-rinded, blob-shaped "pillow" basalt forms, 1 to 4 feet across. A blob squeezes out of the seafloor and freezes, then another blob squeezes out next to it, and so on. You can see pillow and other oozy flow shapes in Glacier National Park's Purcell Lava formation. It crops out at Boulder Pass and Granite Park (a misnomer referring to its basalt). Having been around for 1.4 billion years, much Purcell Lava has metamorphosed to sea-green greenstone.

RHYOLITE AND TUFF

Rhyolite lavas typically become pale, pinkish-tan rocks, fairly easy to tell from other lavas but sometimes confused with tuff, which is lava—often rhyolite—that exploded rather than flowed. Though named for the Greek word for "flow," rhyolite is among the least fluid lavas. Lava viscosity increases dramatically with silica content, and flowing rhyolite may reach a thousand times the viscosity of basalt, squeezing out of the ground like dried-up peanut butter. It so resists flowing that it may plug up a vent forever, ending activity at that volcano.

Rhyolite magma is often rich in gases that it can contain only as long as it remains under pressure. On nearing the surface, the magma literally explodes. The Yellowstone volcano did that three times in the last 2 billion years, each time blasting mountains of pumice, ash, rocks, and gases into the air with hundreds of times the explosive force Mount St. Helens released in 1980. Ashfall from Yellowstone's second-biggest eruption is seen in soil profiles from Los Angeles to Galveston. As the Yellowstone Hotspot marched across Idaho for 15 million years, it left an immense trough, the Snake River Plain. Basalt flows that cover much of the

Tuff in Tusher Mountains, Utah

plain are just a thin skin over rhyolite-tuff layers thousands of feet thick.

Lava blasted into fine smithereens is called ash. Amalgamated into solid rock, it becomes tuff. If ash settles back to Earth at modest temperatures, decades pass before enough groundwater full of dissolved silica percolates through to cement it into an ashfall tuff. If, on the other hand, ash is carried by hot, whirling volcanic gas—a "glowing avalanche," or *nuee ardente*—its own heat quickly cements it into a welded tuff. Ashfall tuff is often crumbly and vulnerable to erosion; it resembles sandstone. Welded tuff resembles lava. Tuff and lava are easiest to tell apart if there are inclusions: in lava, these are crystals, often of a contrasting color; in tuff, they're irregular fragments about the same color as the rest of the tuff itself.

The biggest volcanic episodes in our Rocky Mountain story—San Juan, Challis, Absaroka, and Marysvale—mainly produced rhyolite and tuff.

OBSIDIAN

A few rhyolite lavas with hardly any gas or water in them erupt quietly, and then cool without organizing their ions into mineral crystals. This yields volcanic glass, called obsidian. It is typically black or dark red-brown, even though chemically it is a "pale"

Obsidian arrowhead

lava. Like other glass, it breaks in concentric arcs. It was so prized for knapping into arrowheads and other extremely sharp tools that Yellowstone obsidian was traded all the way to the East Coast, even long before horses were available for hauling. A 17-inch-long obsidian blade was found in Ohio. Yellowstone's Obsidian Cliff was the preeminent source due not only to its size but also to its purity and range of colors.

The Crow wear red obsidian arrowheads for spiritual protection.

Metamorphic Rocks

Metamorphic rocks were altered by intense heat and pressure while deeply buried, a process that recombined their atoms into new minerals. Each mineral requires a particular degree of heat and pressure; geologists rank the results from low- to high-grade metamorphics. Low-grade metamorphism is rife in our Proterozoic and early Paleozoic sediments, turning limestone to marble, shale to argillite or slate, sandstone to quartzite, and occasionally reaching schist grade, but not gneiss, a high-grade metamorphic.

SLATE AND PHYLLITE

Shale and its metamorphic product, slate, share dark-gray colors (varying to reds and greens in slates and argillites) and platy textures that break into flat pieces. But the flat cleavage of slate was never the flat bedding plane of its parent rock, shale. Bedding planes started out horizontal, as layers of sediment settling out from water. Slate cleavage planes, in contrast, are perpendicular to the direction of pressure that metamorphosed the rock. There may have been two or more such directions at different times, yielding slate that fractures into slender "pencils." Some slates metamorphosed from mudstone, and never had distinct bedding planes; in others, the bedding has faded. Commonly, though, the parent shale's bedding planes remain as

LEFT: *Layers of slate broken and intruded by granite magma* RIGHT: *Quartzite*

streaks or bands in slate, visible but no longer causing a plane of weakness.

Slate gets a satiny luster from microscopic crystals of mica and other minerals. Metamorphism aligned these all parallel, creating the cleavage plane and the shine. Higher-pressure metamorphism of the same rock produces larger crystals (just barely visible without a lens); a stronger, glittery, often wavy gloss; and less tendency to split. Though sold as roofing slate, that rock is properly phyllite. Slate is shatter-prone and weak in landforms, phyllite slightly less so.

QUARTZITE

Sandstones can be more than 90 percent pure silica if both the sand grains and the cement binding them together are mostly silica. Such a rock—quartzite—is strikingly different from ordinary sandstone. It has virtually no pore spaces between the grains, and when it breaks, it breaks right through the grains rather than between them. You can't scrape sand off the surface with a knife, and you can scarcely see sand with the naked eye, but you can see a kind of sugary texture. The color is often pale gray, pinkish, or sometimes blackened on exposed surfaces. In skilled Paleolithic hands, it was fashioned into sharp tools, mainly when obsidian and flint were unavailable.

The mineral quartz resists not only physical and chemical breakdown, but also metamorphic alteration: a 90-percent-quartz sandstone after metamorphism is still 90 percent quartz. Quartzite actually names two kinds of rocks—sedimentary orthoquartzite and metamorphic metaquartzite—that are almost identical to the naked eye. Magnified, metamorphosed quartz crystals are a bit larger, more polygonal in shape, and perhaps elongated in one direction. (I'm placing quartzite among the metamorphics because our quartzite beds tend to be ancient and metamorphosed.)

The highest parts of Wyoming's Medicine Bow Range are an exceptionally white quartzite that is 2.2 billion years old. Colorado's Needle Mountains above the Animas River have ancient quartzite beds more than 1 mile thick and exceptionally pure—99 percent silica. Such ultra-pure quartzite beds formed in the shallow water of several basins around mountains that arose from the Wyoming-Yavapai collision, 1.7 billion years ago. Similar beds turn up in the Park, Front, and Sawatch Ranges and in New Mexico—and also Wisconsin, Ontario, South Dakota, and California. All of them formed

LEFT: *Schist outcrop in Idaho* RIGHT: *Gneiss with small pegmatite dike*

within a half-billion-year period. They don't seem to have formed anywhere since then; all are at least a billion years old, suggesting that our Earth no longer refines quartz sand as effectively. What changed? Oceans were more acidic back then, and continents were likely windier without plants on them—and wind sorts and rounds sand grains.

SCHIST

With increasing heat and pressure, parallel layers in metamorphic rocks may get crimped, crumpled, or curved, while the rocks' strength increases. At the same time, the flaky crystals (mainly mica) grow larger as "slaty cleavage" turns into "schistose foliation," or is eventually replaced by coarser crystals that may show "gneissose banding." Though "schist" derives from the Greek for "split" (as in schizophrenia), schists aren't nearly as easy to split as slates. Schistose texture is often dulled and obscured by weathering or rust, but in unweathered specimens the parallel mica flakes glimmer. Black mica gives many schists a charcoal gray hue, often streaked with white quartz layers. Some schists have scattered, reddish, smeared garnet crystals ¼ inch or larger, with foliation circling them in almond shapes.

Metamorphism makes new minerals by recombining atoms into new molecules. Migration of atoms between molecules is extremely slow unless there is water flowing between the rock particles—but there usually is. Geologists say "fluids," rather than water, since this water carries so many dissolved minerals that it's a strong brine—and so hot that it's commonly a brine vapor. Each mineral product requires not only the right grade of heat and pressure but also the right elements.

Low-grade metamorphism (greenschist grade, with characteristic green minerals) turns basalt to greenstone. Very high-pressure, relatively low-temperature metamorphism yields blueschist. In blueschist terrain, you may find glittery, pure-black amphibolite rocks with aligned fine crystals.

GNEISS

Gneiss ("nice") excels as mountain material: not only is it hard, but it has few microfractures to admit water and invite frost splitting. Gneiss falls between schist and granitic rocks in appearance, and often in location. Some gneiss started out as shale or sandstone and was metamorphosed by heat and pressure past the phyllite and schist stages,

Skarn above a layer of marble

nearly to the point of melting. Other gneiss was granitic before metamorphism. Gneisses of sedimentary and igneous parentage look similar in the field. Coarse gneiss may overlap fine-grained granitics in crystal size; if the rock is gneiss (as opposed to granite), the grains will be at least slightly flattened and aligned. The plane of flattening may show as light and dark bands, often contorted. Gneiss fragments may break along planes of glittery mica flakes—planes that are basically little bands of schist. For the rock to be gneiss (as opposed to schist), fat mineral grains must outweigh flaky ones overall.

Heat and pressure beyond gneiss grade can melt the rock into granite, or it can get just partway to melting, producing either showy mixtures called migmatites, or subtle ones with shifting areas of aligned and nonaligned crystals.

Gneiss and granite make up the bulk of continental crust, the ancient "basement." To illustrate the scale with a metaphor borrowed from the Roadside Geology series by Alt and Hyndman, if the earth were a house, the basement would be the clapboard siding, and the world's sedimentary rocks would be the paint—which has flaked off, exposing clapboard in most Laramide uplifts.

HORNFELS AND SKARN

If you ever notice that you have just left an area of granitic rocks, you may be in the vicinity of contact metamorphic rock baked by close contact with hot, intrusive magma. That gives it a high metamorphic grade in terms of heat, but low in terms of pressure. Some such rocks are altered by the addition of volatile elements that emanated from the magma. Skarn, for one, is baked from limestone or dolomite with the addition of silica. It's typically found sandwiched between granite (magma that brought the silica) and

marble—the same limestone and dolomite metamorphosed without the silica.

Hornfels means "horn rock" in German, suggesting that it's one tough rock, regardless of whether the name refers to animal horns or to the Matterhorn and its ilk. It is a fine-grained crystalline rock, often dark gray to black, altered from shale and its kin by heat. The easiest way to tell it from basalt is by its location near granitics.

On Cathedral Peak, south of Aspen, you can see a transition into metamorphism. Part of the peak is unmetamorphosed, reddish Maroon Formation sedimentary rock, and part is altered to gray hornfels. Hot magma nearby baked the color out, but left the maroon beds still visible as layers. Colorado's Spanish Peaks are a pair of thin, vertical plutons flanked by hornfels still showing, again, the bedding planes of the sediments they were baked from. In the heart of Montana's Crazy Mountains, magma intruded soft, weak mudstones, baking them granite-hard.

MARBLE

Marble, the metamorphic derivative of either limestone or dolomite, is widespread in the central Wallowas (photo on p. 29), in Montana sedimentary beds of both the Paleozoic and Proterozoic ages, and in Colorado. Marble, Colorado, is home to the Yule Quarry, America's most illustrious marble quarry. The Lincoln Memorial is pure white because limestone around Marble

LEFT: *Painted Wall, Black Canyon of the Gunnison in Colorado* RIGHT: *Migmatite*

was nearly pure calcite. Strong marbled colors come from other rock types mixed in with the limestone or dolomite. (Don't expect marble in the field to display the marbling and luster of polished slabs.) Yule marble was Leadville Formation limestone baked by hot magma—the pluton that stands today as Treasure Mountain—intruding nearby. Such contact metamorphism is a frequent origin for marble.

MIGMATITE

View magnificent migmatites (from the Greek for "mix") on cirque walls above Kane Lake, Idaho; in many parts of the Front Range, the Wind Rivers, and the Beartooths; and on the walls of the National Pegmatite/ Migmatite Gallery (a.k.a. the Black Canyon of the Gunnison). The canyon's rocks may not be all that different from Proterozoic rocks deep in other Laramide uplifts, but they are much better displayed, thanks to a sharp increase in the incision rate of the Gunnison River 2.5 million years ago. That rate—about 2 inches per century—is the fastest measured in the region, easily beating the Grand Canyon. Inflicting rapid cutting upon tough metamorphic rock was the recipe that yielded high, vertical cliffs crowding a stunningly narrow canyon. Accelerated cutting resulted from further fortuitous combina-

tions: recent uplift of western Colorado; the Colorado River dropping its downstream sections by cutting the Grand Canyon and then capturing the Gunnison drainage; and ice ages accelerating the delivery of water full of tumbling rocks.

Migmatites make our most luscious rock slabs, and our best aid for visualizing continental tectonics. It's one thing to say that when tectonic plates scrunch, the solid rock deep down can flow—like solid glacier ice only even slower. It's another to look at a massive chunk of very solid rock that obviously flowed like swirly cake batter full of nuts, marshmallows, and fudge—yet did so mostly without melting.

This migmatite is what was going on deep in the heart of an orogeny. Even right before our eyes it retains its mystery. Edges between the granitic and the metamorphosed parts are sharp, so we know the mass as a whole didn't melt. Nearby magma bodies injected dikes, whose magma cooled very slowly into pegmatite with large crystals. Or maybe the dikes filled with magma that oozed out of the solid rock like melting fat oozing from a steak searing in a pan; crumples were superimposed on earlier folds; cross-cutting cracks filled up with silica, making quartz veins—at least that's how I think it happened. I never got to see the work in progress.

Acknowledgments

I thank my wife, Sabrina, for her unstinting support without which this book would have remained a fantasy. Yes, it's an Acknowledgments cliche; it just happens to be true. As does Sabrina.

I thank my unflappable backpacking partner who once taught me, "You can't touch the real sky, but the trees can, right, Daddy?" and her brother who named his hover fly friend Jerry and said, each time a hover fly came to land on him, that it was Jerry back for another visit. I thank the backpacking brothers Wilson, who never met a stromatolite that bored them, and my brother, Pagen, who never ceases finding inspiration in the infinite forms our world presents.

I thank the capturers of exquisite images of nature. They transformed this book from merely fascinating to a feast for the eyes. Find these photographers on page 540. Especially I thank Patrick Alexander, Ginnie Skilton, Richard Droker, Matt Lavin, Nick Dean, Marli Miller, and Mike Ashbee.

I thank dozens of ecologists who were utterly generous with their time in helping me make this book as accurate as I could manage. Diana Tomback, Diana Six, Colin Maher, Craig Allen, Ellis Margolis, and Tova Spector showed me around their study sites in the Rockies. I thank scientists who spent time in conversation straightening out details: Candace Galen, Tom Rodhouse, Johanna Varner, Embere Hall, Andrew Larson, Alina Cansler, Krysten Schuler, Connie Millar, Kim Davis, Camille Stephens-Rumann, Dan Gavin, Paula Fornwalt, Bryan Harvey, Sharon Hood, and Jimmy Kagan. I thank those who did the same via correspondence: David Giblin, Stephen Meyer, Christopher Marsh, Jonathan Coop, and Karl Karlstrom. I thank experts who made suggestions to improve my textual accuracy (don't blame them for my errors that undoubtedly slipped in!): Gene Humphreys, Bruce McCune, Michael Beug, Becky Bolinger, Justine Karst, and Jonathan Thompson. Similar improvements to 2003's *Rocky Mountain Natural History* persist in the current edition; thanks go to Walt Fertig, Larry Evans, Trevor Goward, Stephen Johnston, Peter Lesica, Jeff Lockwood, Jim O'Connor, Dennis Paulson, Bob Pyle, Roger Rosentreter, and Steve Sharnoff.

Finally I thank the couriers of the final stage—the editorial polishing works at Mountaineers Books. (Can you believe Apple and Microsoft *still* don't play nicely together? And what a torpedo that can launch?) You persevered. Thank you, Kate Rogers, Laura Shauger, Mary Metz, Jen Grable, Erin Greb, Matt Samet, Sarah Breeding, Callie Stoker-Graham, and the Carbon Footprint Reduction Committee.

Photo Credits

Photos and illustrations are used with permission, unless noted as CC-BY, in the public domain (PD), or from a government agency. CC-BY indicates images used under Creative Commons Licenses 2.0 through 4.0, requiring attribution. The more than 360 images not credited below are by author Daniel Mathews.

Ian Adams, CC-BY: dense spikemoss

Patrick Alexander: adonis blazingstar, alpine mertensia, alpine primrose, alpine spring-parsley, ballhead sandwort, blue-eyed-grass, Boulder blackberry, Butte candle, canary violet, cheatgrass, Colorado ragwort, curl-leaf mountain-mahogany, cutleaf anemone, Dalmatian toadflax, darkthroat shooting star, dunhead sedge, dwarf lupine, Fee's lip fern, Fendler's lip fern, forget-me-not, fourwing saltbush , fuzzytongue penstemon, golden draba, hairy arnica, hairy clematis, Hall's penstemon, Hayden's aster, Hooker's onion, Idaho fescue, Lambert's locoweed, leafy spurge, longleaf phlox, matted buckwheat, Mertens rush, miner's candle, needle-and-thread, netskin onion, pale mountain-dandelion, pearly everlasting, plateau striped whiptail, prairie rattlesnake, purple reedgrass, rockjasmine, Rocky Mountain fringed gentian, sand lily, sharpleaf twinpod, silverleaf phacelia, spike woodrush, spiny hopsage, toadlily, Utah sweetvetch, whiplash saxifrage, woolly hawkweed

Bryn Armstrong: trail-plant

Michael Ashbee: American dipper, belted kingfisher, bighorn sheep, brown creeper, bufflehead, cherry-faced meadowhawk, elk, green-winged teal, grizzly bear, Hudsonian whiteface, lazuli bunting, lynx, mayfly, northern pygmy owl, northern saw-whet owl, rock wren, snowshoe hare, sora, Townsend's warbler, vivid dancer, yellow-headed blackbird, spreadwing emerging

David Bird, PD: black moss

Justin Baldwin: white fir

Andrew Block: eyed sphinx moth

Nick Block, CC-BY: Mormon fritillary

Matt Bowser, CC-BY: fuzzy-foot morel

John Brew: bead lily, purple saxifrage, turpentine biscuitroot, cliffrose, pipsissewa

Matt Berger, CC-BY: alpine collomia, American thorow-wax, Mat rock-spiraea, smooth cliffbrake, Spiny milkvetch, Telesonix

J. R. Cagle, CC-BY: dark wood nymph, orange-tip, Rocky Mountain dotted blue

Caleb Catto, CC-BY: alpine bluegrass, red-backed vole

Jamie Chavez: American goldfinch, Lorquin's admiral, Vaux's swift, western tiger swallowtail

James Cheshire, CC-BY: northern black currant

Terri Clements: aspen bracket, fuzzy truffle, shoestring root rot or honey mushroom, snowbank false-morel, summer oyster mushroom

Jordan Collins: frayed-cap moss

Whitney Cranshaw, Colorado State University, Bugwood.org: mountain pine beetle

Dick Culbert, CC-BY: hedgehog mushroom

Nick Dean: American pipit, anise swallowtail, bald eagle, black-capped chickadee, blue-eyed darner, Bullock's oriole, Canada goose, clodius parnassian, common merganser, common raven, common yellowthroat, harlequin duck, hoary marmot, horse fly, killdeer, mallard, margined white, mariposa copper, mosquito, mountain bluebird, mourning cloak, mourning dove, northern pintail, Pacific wren, paddle-tailed darner, peregrine falcon, pied-billed grebe, pine siskin, red crossbill, ring-billed gull, ruby-crowned kinglet, song sparrow, Steller's jay, ten-lined June beetle, tundra swan, violet-green swallow, warbling vireo, western tanager, yellow warbler, yellow pine chipmunk

J. Dreier (66dodge): thirteen-lined ground squirrel

Richard Droker: alpine jelly cone, American parsley fern, bracken, bristly haircap moss, brown helmets, common liverwort, dyer's polypore, forking bone lichen, goldspeck lichen, great blue heron, hooded sunburst lichen, hooded tube lichen, Iceland-moss, ladder lichen, lungwort, map lichen, mealy pixie goblet, membranous dog lichen, orange chocolate-chip lichen, peat moss, perforated rock tripe, pixie goblet, Purcell Sill, ragbag lichen, running clubmoss, sandpaper stippleback, snakewort, snow lichen, spraypaint lichen, sword fern, turkey tail, white-lined sphinx moth, Wilson's snipe, witch's hair, woolly foam lichen

Patricia Drukker-Brammall, courtesy of Pam Schofield: drawings of alpine haircap moss, big redstem moss, common haircap moss, crisping cushion moss, erect-fruited iris moss, false haircap moss, forked broom moss. juniper haircap moss, paw-wort

Ryan Durand, CC-BY: alpine lady fern, fairy fingers, palm tree moss, pipecleaner moss

Eric Eaton: drawings of black fire beetle, black fly, no-see-um, striped ambrosia beetle, western pine beetle

Kris Elkin: drawings of cat and dog prints

Nolan Exe, CC-BY: smallflower miterwort, tofieldia

Jon Farmer, iStock: lenticular clouds

5 Acre Geographic: brown elfin, common ringlet, common wood nymph, great spangled fritillary, hydaspe fritillary, lupine blue, pine white, purplish copper, western pine elfin

John Flannery, CC: black swallowtail

flickr.com/photos/owlp: Cooper's hawk, moss, golden buprestid, thymeleaf speedwell, snakefly, yellow starry fen

Richard B. Forbes: long-tailed vole

Judy Gallagher, CC-BY: hover fly, smooth sumac

John Game, CC-BY: leopard lily, star moss

Kirk Gardner, CC-BY: masked shrew

Laura Gaudette, CC-BY: Gillette's checkerspot, hedgehog fly, ponderous borer

Ellyne Geurts, PD: northern spur-throat grasshopper

Brad Geyer: Dinwoody Glacier

Rick and Susie Graetz: Chinook wall

Tim Hagan: heartleaf pyrola, pussypaws, redstem saxifrage, white mountain-heather

Robb Hannawacker: Weidemeyer's admiral

Jason Headley: paw-wort

Scott Hecker, CC-BY: red-backed vole

Alex Heyman: golden currant, spineless horsebrush

Hendrik Hollander, PD: painting of Linnaeus

Jason Hollinger, CC-BY: black-and-gold tile lichen, brittle prickly pear, creeping juniper, dog lichen, green rock posy, green starburst lichen, pitted beard lichen, powdered beard lichen, tumbleweed rock-shield

Dick E. Hoskins: camas (both), lightning

David W. Inouye, Biology Image Library: broad-tailed hummingbird

Robert Ivens: western water shrew

Sarah C. Jackson: Wyoming kittentails

Jeanne R. Janish, courtesy of University of Washington Press: drawings of blue wildrye, Brewer's cliff-brake, brittle fern, brook saxifrage, buckbean, Columbian needlegrass, cow-parsnip, coyote willow, curly sedge, elk sedge, ground-cedar, heartleaf twayblade, lady fern, northwestern twayblade, Oregon woodsia, Ross's sedge, rough scouring-rush, spike fescue, tufted hairgrass, Scouler willow, slender cinquefoil, sticky cinquefoil, sheep fescue, sword fern, twisted-stalk, western poison ivy, white bog orchid

Willis Linn Jepson: drawings of red elderberry, poison hemlock, water hemlock

Braden J. Judson: bigleaf rosette moss

Ben Keen, CC-BY: long-toed salamander

C. K. Kelly, CC-BY: Douglas-fir, pandora pine moth

Ken Kertell: Colorado hairstreak, Sheridan's green hairstreak

Iris Kilpatrick: sharp-shinned hawk

Jerry Kirkhart, CC-BY: loggerhead shrike, sandhill crane

Russ Kleinman, Western New Mexico University Department of Natural Sciences and Dale A. Zimmerman Herbarium: fringed puccoon, hoar moss, muttongrass, New Mexico checkermallow, showy goldeneye, stemless townsendia

Russ Kleinman and Karen Blisard: Baltic rush, rolled-leaf moss

Russ Kleinman and Richard Felger: Scouler willow

klukoff: yellow coral fungus

Ron Knight, CC-BY: Bohemian waxwing, brown-capped rosy-finch, dark-eyed junco, horned lark, western grebe, white-breasted nuthatch

Brandon Lally, CC-BY: American mink, golden-crowned kinglet

Matt Langemeier: alpine kittentails

milkweed, sharptooth angelica, silverberry, sticky currant, three-toothed mitrewort, Utah honeysuckle, waxleaf penstemon, western mountain-ash, white virgin's bower, woolly mullein, wormleaf stonecrop, yellow dryad

Susan Stacy: death cap

Barbara Stafford: big shaggy moss

Peter Stevens, CC-BY: Pacific treefrog

Matt Strieby: western white pine, metamorphic core complex (adapted from Donna L. Whitney)

Nigel Tate: Bullock's oriole, mountain bluebird, Pacific wren, ruby-crowned kinglet, violet-green swallow, warbling vireo, western tanager, Wilson's warbler, yellow warbler

Alberto Tabernero: Wyoming ground squirrel

Glen Tepke: black swift

Joe Thompson CC-BY: lilac-bordered copper

Joseph Tomelleri: paintings of arctic grayling, brook trout, brown trout, bull trout, cutthroat trout, sockeye salmon, longnose dace, longnose sucker, mountain whitefish, rainbow trout, redside shiner, sculpin, smallmouth bass

Ken-Ichi Ueda, CC-BY: littleleaf mock-orange, elm sawfly

Chloe and Trevor Van Loon, CC-BY: American moonwort, water strider

Doug Waylett, courtesy of Charlene Jutra: alpine bistort, arctic harebell, bastard toadflax, bog blueberry, buckbean, Canadian white violet, chocolate tips, Colorado rock-shield, common alpine, crown coral fungus, deer fly, dwarf sweetroot, elegant sunburst lichen, fire-rim tortoiseshell, fool's huckleberry, glacier lily, greenish pyrola, hairy golden-aster, inky cap, knight's-plume, ladies-tresses, limestone columbine, long-beak willow, low larkspur, mint-leaf bee balm, naked miterwort, nobo bluet, northern twayblade, one-flowered wintergreen, one-leaf bog orchid, pinesap, pleated gentian, prairie coneflower, prickly rose, police car moth, reindeer lichen, round-leaf bog orchid, ruffled freckle pelt, scarlet globemallow, serviceberry, showy aster, showy locoweed, shy wallflower, sibbaldia, snow cinquefoil, spiny phlox, sulfur dust lichen, swamp gooseberry, wavyleaf thistle, western white, yampah, yellow lady's-slipper, yellow owl's-clover

Andrey Zharkikh, CC-BY: alpine buttercup, Colorado blue spruce, dwarf mountain groundsel, goosefoot violet, one-flower daisy, Richardson's geranium, Rocky Mountain wood tick, wapato

Krzysztof Ziarnek, CC-BY: gray aster

National Park Service, PD: Curtis Akin: Grand Prismatic Spring; **Sherman F. Denton:** lake trout; **Addy Falgous:** travertine; **Neal Herbert:** American beaver, Canada jay, gray wolf, red fox, ruffed grouse, western meadowlark; **Jacob W. Frank:** American red squirrel, badger, balsamroot on p. 15, bracted lousewort, Castle Geyser, Columbia spotted frog, common goldeneye, common merganser young, coyote, darkwoods violet, evening grosbeak, Hallett Peak, Lamar River Valley, Pacific marten, pale tiger swallowtail, pointed-tip cat's ears, pronghorn, Saint Mary Lake, short-tailed weasel, spotted sandpiper, tree swallow, trumpeter swan, typical fair weather cumulus, virga, white-tailed deer, white-throated swift, yellow-bellied marmot, yellow-rumped warbler; **Glacier National Park:** northern river otter, western white pine cone, Townsend's solitaire, varied thrush; **Eric Johnston:** black bear; **Don Loarie CC-BY:** bumblebee, Emma's dancer, orange sulphur; **Daniel A. Leifheit:** boreal owl; **Mesa Verde National Park:** collared lizard; **Jim Peaco:** great gray owl. great horned owl, white-crowned sparrow; **Diana Renkin:** mountain goat; **Ann Schonlau:** long-tailed weasel, mountain cottontail, pika; **Dan Stahler:** wolf pack; **Yellowstone National Park:** a low-key fire; **Zion National Park:** ringtail

US Forest Service, PD: Unattributed: grand fir, Lochsa River, Rocky Mountain maple, subalpine fir, swamp gooseberry; **Ellen Blonder:** drawing of western jumping mouse; **Sheri Hagwood:** whiplash willow; **Dave Powell:** lodgepole pine pollen cone

US Fish and Wildlife Service, PD: Unattributed: black-billed magpie, northern leopard frog; **Tom Koerner:** red-tailed hawk, Swainson's hawk, northern harrier (both), white-tailed prairie dog

LANDSAT, PD: Front Range fire scars, satellite view of Basin-and-Range

Wyoming State Geological Survey, PD: diagram of Sevier and Laramide mountains

Glossary

Note: Italic in definitions indicates glossary terms. Refer to their definitions to learn more.

abdomen The hindmost section of an adult insect, enclosing reproductive and digestive systems.

adipose fin A soft, fleshy bump on the back of some fish right in front of the tail fin.

alevin A hatchling fish, especially a salmon or trout in the stage when it is still attached to and nourished by an egg yolk sac.

allelopathy Deposition in the soil, by plants, of chemicals that discourage growth of competing plant or fungal *species*.

alpine The elevation zone too high for tree *species* to grow numerously in upright tree form.

alpine timberline A broad, high-elevation belt where cold and snow severely inhibit tree growth. Ranges from tree line, the highest elevation (locally) where trees can grow upright, up to scrub line, the highest elevation where *conifer species* can grow even as *krummholz*.

alternate Arranged with only one leaf at any given distance along a stem. Contrast with *opposite* or *whorled*.

anadromous Born in fresh water, growing for an extended period at sea, and returning to fresh water to breed.

angle of repose The steepest slope angle that a given loose sediment is able to hold.

anther A plant *stamen* tip, bearing the *pollen* grains.

archaea Single-celled organisms once seen as primitive bacteria, but now seen as one of three domains—the very top taxonomic rank, with bacteria being a second domain.

Archaean Eon A very early time in geologic history (p. 504), from 4 to 2.5 billion years ago, when the earliest life-forms developed.

ash Fine particles of *volcanic rock* (usually non-crystalline) formed when *magma* is blasted out of a volcano in a fine spray.

aspect The compass direction a slope faces.

awn A stiff, hairlike extension of the tip of a *bract* in the *floret* of a grass.

axil The point on a stem where a leaf or a branch attaches.

basal Describes leaves attached to a plant at ground level as opposed to higher.

batholith A large formation of *intrusive igneous rock* exposed at the surface over a large area—technically at least 100 square kilometers.

biomass Total living matter, usually expressed as a measure of dry weight (or sometimes volume) per unit area.

bisexual With functioning *stamens* and *pistils* in the same flower.

bloom A pale, powdery coating on a surface.

Boreal Of the globe-encircling northerly belt dominated by *conifer* forests, transitional between Arctic and *Temperate* climate.

bract A modified (usually smaller) leaf, often attached just below a flower or *inflorescence*.

broadleaf A term for all trees and shrubs other than *conifers*. A few have leaves narrow enough to be confused with conifer needles.

call Any vocalization common to birds of a given *species*. Compare *song*.

calyx The outer *whorl* or circle of parts of most flowers—the *sepals* spoken of collectively, or a ring of sepal-like lobes that would be sepals if they were separate all the way to their bases. Compare *corolla*.

canopy In a forest, the collective mass of branches and foliage of the taller trees.

cap The spreading top portion of a typical mushroom, supported by a stem and bearing *gills*, tubes, or spines on its lower surface.

capsule A seedpod; technically a nonfleshy *fruit* that splits to release seeds. In mosses, the *spore*-containing organ.

catkin An *inflorescence* type on certain trees and shrubs, consisting of a dense *spike* of minute, dry, *petal*-less flowers. Our *species* with catkins make distinct male and female catkins.

cirque A mountain valley given a characteristic amphitheater shape by glacial erosion.

clade A group of *species* or other taxa sharing a single common ancestor. From the Greek for "branch," it's a branch of the evolutionary tree of life.

class A *taxonomic* group broader than an *order* or *family* and narrower than a (plant) division or (animal) phylum. Examples: mammals, birds, spiders, insects.

climax A hypothetical condition of stability in which all *successional* changes resulting from plant community growth have taken place, so that further changes can only follow destructive disturbances.

cloaca An orifice (in birds, reptiles, fishes, etc.) through which the urinary, reproductive, and gastrointestinal tracts all discharge.

clone (*noun*) A group of individuals that are genetically identical because they arose by cloning. (*verb*) To reproduce vegetative or other asexual reproduction from a single progenitor.

composite A member of the plant *family* Asteraceae (formerly Compositae), characterized by composite flower *heads* that resemble single flowers but are actually well-organized *inflorescences* of tiny flowers. Examples: daisies, thistles, dandelions.

compound leaf A structure of three or more leaflets on stalklike ribs, resembling separate leaves on a branchlet except that it terminates in a leaflet rather than a flower, a bud, a growing shoot, and so forth. It grows from a node on a stem, but its leaflets do not each grow from nodes of their own.

congener A member of a *species* in the same *genus*.

conifer A tree or shrub within a large group generally characterized by needlelike or scalelike leaves and often *woody* "cones" bearing the seeds.

conspecific An individual of the same *species*.

convect To move upward and downward due to temperature contrasting with that of other fluids in the same system: hot air rises, cold air sinks.

Cordillera The entire mountain system running from Alaska to Tierra del Fuego, encompassing the Rocky Mountains and all other mountains in the western parts of the Americas. In Spanish, *cordillera* means "mountain range."

corolla The *whorl* or circle of a typical flower's parts lying second from the outside; that is, the *petals* collectively, or else a ring of lobes that would be petals if they were separate all the way to their bases. Compare *calyx*.

cosymbiont A symbiotic partner of any given organism.

crevasse A large crack in a glacier, expressing flow stresses.

crown The leafy (top) part of a tree.

crust The surface layer of the earth, averaging about 6 miles thick under the oceans and 20 miles thick in the continents, defined primarily by having lower-density rock than the *mantle* and core areas underneath it.

crystal The structure of many chemical compounds (including all *minerals*) in which the atoms position themselves in a particular geometry that repeats indefinitely; usually visible in broken surfaces of the material.

deciduous A tree or shrub that sheds its leaves annually; that is, not *evergreen*.

dicot A member of *class* Magnoliopsida, which includes all *broadleaf* trees and shrubs and most terrestrial *herbs* except lilies, orchids, and grasslike plants (the *monocots*). Short for "dicotyledon."

dike A homogeneous body of rock that is long and deep but not thick—typically an *igneous rock* body solidified from fluid *magma* intruded into a *fault* or crack.

disjunct Separated—said of a local population of a *species* distant from its other populations.

disk flower Tiny, often dry and drab-colored flowers making up a dense circle (disk) in the center of a *composite* flower *head*, in the *family* Asteraceae. The eye of a daisy is one style of disk.

disturbance An abrupt change in a plant community—for example, logging, major windstorms, floods, or fires.

DNA Deoxyribonucleic acid, the basic molecule that makes up genes, which are passed from parent to offspring and control the form taken by each living cell. Researchers can analyze DNA to determine the inherited lineage of an organism.

doghair stand An extremely dense patch of small (growth-stunted) trees.

dominant A tree whose crown is part of the highest canopy layer in a stand; or more generally, a plant *species* covering more ground than others within a layer (*herb*, low shrub, tall shrub, tree) of a plant community.

duff Matted, partly decayed litter on the forest floor.

elliptic A leaf shape widest at the middle and tapered at both ends, broader than *lanceolate* but less broad than oval.

endophyte An organism (usually a fungus) living harmlessly or beneficially within a plant—an extremely common and widespread but poorly studied *symbiosis.*

entire A leaf or leaflet margin that lacks teeth, scallops, or lobes.

epiphyte An organism that lives attached to and supported by a host organism without directly exchanging any substances with the host and typically without causing it major harm. Lichens, mosses, ferns, and flowering plants can be epiphytes on trees.

evergreen Bearing relatively heavy leaves that normally keep their form, color, and function on the plant for at least twelve months and often a few years. Opposed to *deciduous;* compare *persistent.*

family A *taxonomic* group broader than a *genus* or *species*, and narrower than an *order* or *class*. Latin names for families end in *-aceae* if they are plants and *-idae* if they are animals.

fascicle A bundle, especially the characteristic cluster of one to five pine needles sheathed at their base in tiny, dry, membranous *bracts.*

fault A fracture in rock or the earth (or the location of that fracture) where relative earth movement on the two sides has taken place; may be of any size.

filament The stalk portion of a stamen.

flora All plant *species*, collectively. Also, a reference work cataloguing those species.

floret A small, simple flower within a compact inflorescence such as the *spikelet* of a grass.

fruit Among the seed plants, an *ovary* wall matured into a seed-bearing structure. Also, more broadly, any seed-bearing or *spore*-bearing structure.

fry Young fish (both singular and plural). Among salmon and trout, the fry stage follows the *alevin* stage and precedes the *parr* stage.

fused Unseparated; fused *petals* (or *sepals*) form a single *corolla (calyx)* ring or tube, usually with lobes at the front edge that can be counted, like petals or sepals, for identification purposes.

gall A growth induced on a plant by insects, fungi, or bacteria. Many galls serve as protective homes for insect larvae.

genus Taxonomic group broader than a *species* and narrower than a *family*; the first of the two words in a scientific name, or binomial, of an organism. Plural: *genera.*

gills On aquatic animals, respiratory organs where oxygen suspended in the water passes into the animal. On mushrooms, *spore*-bearing organs, either

paper-thin or wrinklelike, radiating on the underside of the *cap.*

gland An organ that secretes a fluid. Glands are common on plant surfaces in the form of dots or microscopic bulbs on the ends of short hairs, leading to the description of the leaf or *sepal* as "sticky-hairy."

granitic rocks Coarse-grained *igneous rocks*; see *intrusive* rocks.

head A tight, compact *inflorescence.*

herbaceous Not *woody*; the aboveground parts normally die or wither at the end of the growing season. (Some low, soft-stemmed plants classified as *herbs* may persist through some mild winters and into summer.)

herb layer All the *herbaceous* plants and low shrubs of comparable size, as a group comprising a structural element of a plant community.

high-grade *Metamorphic rocks* altered by relatively high degrees of heat and pressure.

hybrid An offspring of parents of different *species.*

hyphae Minute tubular filaments that make up the body (mycelium) of a fungus. Singular: *hypha.*

igneous rocks Rocks that reached their present *mineral* composition and texture by cooling and solidifying from a fluid *magma*; may be either *volcanic* or *intrusive.*

inflorescence A cluster of flowers from one stem; the cluster's shape or structure.

instar A stage in insect growth and development separated from the next instar by molting.

intergrade To vary along a continuum.

introduced Not thought to have lived in a given area (the Rocky Mountains, in this book) before Euro-Americans arrived; opposed to *native.*

intrusive Igneous rocks that solidified underground, without erupting onto the earth's surface. I sometimes substitute the loose term *granitic.*

invasive *Introduced* and spreading quickly.

involucral bracts Small leaves in a *whorl* immediately beneath an *inflorescence* (such as a flower *head* in the *composite family*).

irregular A flower made of *petals* (or of *sepals*) that are conspicuously not all alike in size and/or shape; many are bilaterally symmetrical.

isostasy Earth's tendency to even out its surface anomalies by raising (floating) the areas of *crust* that exert less gravity because they are thinner or less dense.

juvenile plumage Feathers (and their color pattern) of birds old enough to fly, but not yet entering their first breeding season; the two sexes typically resemble each other at this stage.

krummholz Dwarfed, ground-hugging shrubby growth, near *alpine timberline*, of *conifer* species that grow as upright trees elsewhere under more moderate conditions. German for "crooked wood." Some Canadians prefer the term "kruppelholz."

lanceolate A leaf that is tapered to a point at each end, usually narrower than *elliptic* and broader than *linear*.

landlocked Fish that complete their life cycle in fresh water, where some barrier (usually a dam) blocks the *anadromous*, or seagoing, life cycle the *species* is capable of.

larva An insect, amphibian, or other animal in a youthful form that differs strongly from the adult form. Caterpillars are the larvae of moths and butterflies, maggots of flies, and tadpoles of frogs.

lava Rock that flows, or once flowed, across the earth in a melted state in a volcanic eruption; compare *magma*.

layering Vegetative or clonal reproduction where new roots and stems grow from branches in contact with the earth.

leader The topmost central shoot of a plant.

leaflet An apparent leaf that is actually a member of a *compound* leaf.

leafstalk A narrowed stalk portion of a leaf, distinguished from the leaf "blade." (Synonym of "petiole.")

linear Of a leaf that is very narrow and almost uniform in width over its whole length.

lip On an orchid, the lowest, largest *petal*, specialized as a landing platform for an insect pollinator.

lobe A convexity in a leaf, *calyx*, or *corolla* outline.

low-grade Altered by relatively mild degrees of heat and pressure (*metamorphic rocks*).

lumpers Taxonomists inclined to reduce the numbers of *species, genera*, and so forth, by recognizing any given distinction at a lower *taxonomic* level. Opposed to *splitters*.

magma Subsurface rock in a melted, liquid state. If and when it reaches the surface, it erupts either as flowing *lava* or into the air as *pyroclastic* fragments.

mantle The major portion of the earth's interior below the *crust* and above the core.

matrix Fine-grained material in some rocks in which much larger grains, crystals, or contrasting pieces of rock are embedded.

meta- A prefix used in naming *metamorphic rocks* in terms of the parent rock they remineralized from. Examples: metaconglomerate, metasedimentary.

metamorphic rocks Rocks whose *mineral* composition and texture resulted from remineralization by great heat and/or pressure deep in the earth.

metamorphism The heat-and-pressure process that makes *metamorphic rocks*.

metamorphosis The process by which *larval* animals transform into adults.

milt fish semen.

midden Organic material heaped by an animal. Both zoologists and anthropologists use this Old English word, originally defined as a dung heap. Animal middens may include dung, but commonly include stored food.

mineral Any natural, non-living earth substance with a well-defined chemical formula and a characteristic crystalline structure. Rocks are mixtures of minerals.

monocot A member of the *class* Liliopsida, generally characterized by parallel-veined leaves and flower parts in threes or sixes. No monocots in the Rocky Mountain region are *woody*. Short for "monocotyledon"; compare *dicot*.

moraine A usually elongated heap or hill of rock debris lying where it was deposited by a glacier when it retreated from that margin.

mutualism *symbiosis* from which both partners benefit.

mycology The scientific study of fungi.

mycophagy The eating of fungi; the collecting of wild fungi for food.

mycorrhiza A tiny organ formed jointly by plant roots and fungi that passes substances between plant and fungus. Plural: "mycorrhizae" is preferred in the United States, "mycorrhizas" in Britain.

naiad An aquatic *larva* in insects that undergo several larval *instars* and no *pupa*.

native Species thought to have lived in a given area (the Rockies, in this book) before Euro-Americans arrived. Compare *introduced*.

nature The universe as it was preceding or is outside of human management; a hypothetical construct useful in some contexts but misleading in others, since the effects of Indigenous people are under-recognized.

nectar A sugary liquid secreted in some flowers, attracting pollinators.

niche The role of a *species* within an ecological community, in terms especially of how it exploits the habitat and other community members to meet its life requirements.

oblanceolate A leaf shape tapered to the leafstalk and more round at the tip, broadest past the midpoint; narrower than *obovate*.

obovate A leaf shape roughly oval and broadest past the midpoint, often tapered to the leafstalk; broader than *oblanceolate*.

offset A short propagative shoot from the side of a plant base, or a small bulb from the side of a plant's bulb.

old-growth Never-logged forest containing many trees older than, let's say, two hundred years. It's debated.

opposite Leaves attached in pairs along a stem. The stem terminates in a bud, flower(s), or a growing shoot. If it terminates in a leaf or a *tendril*, these are not opposite leaves, but leaflets of *compound* leaves that may be either opposite or *alternate*.

order A *taxonomic* group broader than a *family*, *genus* or *species*, and narrower than a *class*; for example, rodents, carnivores, owls, and woodpeckers—but no longer used for plant groups.

orogeny A mountain-building event in geologic history.

ovary The egg-producing organ; in most flowers, an enlarged basal part of the *pistil*.

palmately lobed Shaped (like a maple leaf) with the leaf outline indented between main veins that radiate from one point at the leaf base.

panicle A many-flowered *inflorescence* with a main stem and branches, at least some of which are branched again. Compare *raceme, spike*.

parasite An organism that draws nutrients out of an unrelated organism, more or less harming the host without ingesting the host itself.

parr marks Vertical blotches on the sides of *anadromous* salmon and trout of "parr" age, when they are actively feeding in fresh water before migrating to sea.

pelvic fins Side-by-side fins midway along a fish's belly.

persistent Leaves that tend to stay on the stem through fall and winter, but lacking the heavy weight and gloss characteristic of true *evergreen* leaves.

petal Part of a flower that is typically tender and showy, in the inner of two or more *whorls*. If there is only one whorl, its members are *sepals*, no matter how showy.

pheromone A chemical emitted by an animal (or a plant, some say) that stimulates some behavior in others of the same *species*.

photobiont The photosynthesizing (alga or cyanobacteria) partner in a lichen.

photosynthesis The synthesis of carbohydrates out of simpler molecules, using energy from sunlight; the basic function of chlorophyllous or green parts of plants and algae.

phylum A broad *taxonomic* group, ranking just below the kingdom. Animal phyla include chordates and mollusks. Plant taxonomists have replaced phyla with divisions.

pinna The fern part branching directly off of the main fern stalk or leaf axis; if the pinna is again branched, the ultimate leaflets are *pinnules*. Plural: pinnae.

pinnately compound A leaf composed of five, seven, nine, or more leaflets attached (except for the last one) in opposite pairs along a leaf axis. If they are attached to mini-axes paired in turn along a central axis, the leaf is pinnately twice compound.

pioneer A *species* growing on recently disturbed (e.g., burned, clearcut, deglaciated, or volcanically deposited) terrain.

pistil The female organ of a flower, including the *ovary* and ovules and any *styles* or *stigmas* that catch *pollen*.

pitch Sticky, aromatic resin (water-insoluble) in *conifers*, that defends them from insect attack. Contrast *sap*.

plankton Aquatic organisms that drift with the currents, lacking effective powers of either locomotion or attachment.

plate tectonics The dynamics of Earth's *crust* in terms of the relative motion of often continent-sized plates of crust drifting across an effectively fluid layer underneath. The major unifying theory dominating geophysics since the 1960s, when it was known as "continental drift."

pollen Dustlike male reproductive cells (pollen grains) borne on the *anthers* of a flower, each capable, if carried to the *stigma* of a *conspecific* flower by wind, animal pollinators, or other agents, of fertilizing a female reproductive cell.

pore A minute hole, especially one allowing substances to pass between an organism and its environment.

propagate To reproduce, either sexually or not.

propagule In lichens and mosses, any multi-celled structure more or less specialized to break off and grow into an organism independent of (but genetically identical to) the one it grew on; a means of asexual reproduction or cloning.

prostrate Growing more or less flat upon the substrate.

pupa An insect in its immobile life stage in which a *larva* metamorphoses into an adult. Plural: pupae. Verb: pupate.

pyroclastic Composed of rock fragments formed by midair solidification of *lava* during an explosive volcanic eruption. Volcanic *ash* is loose, fine pyroclastic material; tuff (a rock) is consolidated pyroclastic material.

raceme A slender *inflorescence* with each flower borne on an unbranched side stalk from the central axis. Compare *panicle, spike.*

ray flower One of many small asymmetrical flowers radially arranged in a *composite* flower *head*, typically *petal*-like and strap-shaped for most of its length but tubular at its base, often enclosing a *pistil*. Only in the *family* Asteraceae. Compare *disk flower.*

regular With all of its *sepals* and with all of its *petals* (if any) essentially alike in size, shape, and spatial relationship to the others.

relief The vertical component of distance between high and low points on the earth's surface.

resin blisters Horizontally elongated blisters, initially full of liquid *pitch*, on the bark of young true fir trees.

rhizine A tough, threadlike appendage on certain lichens, serving as a holdfast, not a vessel.

rhizome A rootstalk—a horizontal stem just beneath the soil surface connecting several aboveground stems; sometimes thickened, storing starches.

rhizosphere The layer of soil permeated and affected by roots and *hyphae*.

runner A stem that trails across the ground surface, capable of producing new root crowns and upright stems—a mode of asexual reproduction.

rut The annual period of sexual excitement and activity in certain mammals.

sap A watery fluid within plant vessels, carrying materials between parts of the plant. Contrast *pitch*.

saprobe A fungus or bacterium that absorbs its carbohydrate nutrition from dead organisms; synonym of "saprotroph." (The old term "saprophyte" has been abandoned. Nongreen plants that get their carbohydrates indirectly from live plants via mycorrhizae were once called saprophytes; this error persists in some literature; see p. 154.)

scat Feces. (Wildlife biologists' slang derived by back-formation from "scatology.")

schistosity A *metamorphic rock* texture in which *minerals* are arranged in layers of aligned crystal grains.

scree Loose rock debris lying at or near its *angle of repose* upon or at the foot of a steeper rock face

from which it broke off. (It implies gravel and small rocks, not boulders, which would be *talus*.)

sedimentary Rocks formed by slow compaction and/or cementation of particles previously deposited by wind or water currents or by chemical precipitation in water.

sepal A modified leaf within the outermost *whorl* (the *calyx*) of a flower's parts. Typically, sepals are green and leaflike and enclose a concentric whorl of petals, but in some *species* they are showy and petal-like. Compare *involucral bracts, tepal, petal.*

serotinous Remaining closed indefinitely. Serotinous pine cones open and release seeds in response to heat after being on the tree for years.

simple Not *compound.*

smolt A young salmon or trout when first migrating to sea, typically losing its *parr marks* and becoming silvery. (From the same Old English word as "smelt.")

song A relatively long and variegated bird *call*, generally identifiable to *species*; heard mainly in the *order* Passeriformes, sometimes called "songbirds."

sorus A *spore*-bearing clump visible as a raised dot, line, crescent, or circle on a fern leaf. Plural: sori.

spathe A specialized leaf forming a hood over, or a wrapper around, an *inflorescence.*

species Basic unit in classification of living things; problematic to define in a way that works equally well for all types of living things. A scientific name is a binomial in which the first word is the genus and the second word is the species within that genus.

spike An *inflorescence* of flowers attached directly (i.e., without a stalk for each flower) to a central stalk. Compare *raceme, panicle.*

spikelet In the *inflorescence* of a grass, a compact group of one to several *florets* and associated scalelike *bracts*, on a single axis branching off of the central stem.

splitters Taxonomists inclined to increase the numbers of *species, genera*, etc., by recognizing any given distinction at a higher *taxonomic* level. Opposed to *lumpers.*

spore A single cell specialized for growing into a new multicelled individual, among the lower plants and fungi. Sometimes viewed as a seed counterpart, since it travels; but a fern or moss spore is not the sexually produced stage in the life cycle, making it more comparable to a *pollen* grain.

spreading Borne in a position perpendicular to the stalk or branch. Contrasts with erect or pendent.

spur A hollow, closed-ended extension from a *corolla* or *calyx*, often bearing *nectar*.

stamen A male organ of a flower, typically consisting of a *pollen*-bearing *anther* at the tip of a slender *filament*.

steppe An unforested, arid plant community type dominated by sparse grasses and/or shrubs. Ecologists reserve "desert" for sites so barren that no real plant community exists—for example, active dunes, and saline or alkaline dry lakebeds.

stigma The *pollen*-receptive tip of the *pistil* of a flower.

stomata Minute *pores* in leaf surfaces for the *transpiration* of gases. Sometimes Anglicized to "stomates." Singular: stoma.

style The stalklike part of a flower's *pistil*, supporting the *stigma* and conveying male sexual cells from it to the *ovary*.

subalpine Of or in the highest elevation zone where *conifers* growing in tree form are numerous; below the *alpine* zone where tree *species* are absent, dwarfed, or prostrate.

subduction The process in which the edge of a *plate* of the earth's *crust* bends and sinks underneath the edge of an adjacent plate, and returns to the underlying *mantle*.

subshrub A perennial plant with a persistent, somewhat *woody* base, thus intermediate between an herb and a shrub.

subspecies A *taxonomic* rank narrower than a *species* and, according to some, slightly broader than a variety.

succession The gradual change in the *species* composition of a plant community that results from soil-biotic interactions internal to the community.

symbiosis Intimate association of unlike organisms for the benefit of at least one of them. Symbioses can be mutualistic, with both or all partners benefitting; parasitic, where one benefits and one suffers; or commensal, where neither partner suffers substantially, and little or no material passes between them. The same symbiosis can shift between those categories over time.

talus Rocks and boulders lying at the foot of a rock face from which they broke off. Compare *scree*.

taxon A *species*, *genus*, or other unit of classification. Plural: taxa.

taxonomy The scientific naming and classifying of organisms, particularly within the Linnaean system (*species*, *genus*, etc.), which uses Latin and Latinized names and endeavors to reflect lines of evolutionary descent.

tectonic Related to large-scale geologic deformation such as folding, *faulting*, volcanism, and *subduction*.

Temperate Zone The climatically moderate parts of the world, lying between the Tropics and the Arctic, and including the entire US Rocky Mountains.

tendril A slender organ that supports a climbing plant by coiling or twining around something.

tepal A *petal* or *sepal* on a flower (such as some lilies) whose petals and sepals are identical to each other.

terminal At an end, such as a growing tip of a plant or the lower end of a glacier.

terrane A geologically mapped area whose rock formations have a history utterly unlike that of adjacent terranes, suggesting that at some time they were far from each other.

thorax The middle section of many adult insect bodies, bearing legs and wings, if present.

transpiration The emission of water vapor into the air from plants.

turf A top layer of soil permeated by a dense, cohesive mat of grass and/or sedge roots.

umbel An umbrella-shaped *inflorescence* with many flower stalks branching from one point. Characteristic of the *family* Apiaceae.

veil A membrane extending from the edge of the *cap* to the stem, in certain mushrooms when immature, soon rupturing and often persisting as a ring around the stem; also The "universal veil" in most *Amanita* mushrooms when they are young buttons, an additional, outer membrane extending from the edge of the cap to the underground base of the mushroom, soon rupturing and often visibly persisting as a cuplike growth at the mushroom's base.

vesicle A small hole in a *volcanic rock*, originating as a gas bubble in *lava*; vesicles characterize all pumice and much basalt and andesite.

volcanic rock An *igneous rock* that solidified from fluid *lava* in a volcanic eruption. Compare *intrusive*.

whorled Arrangement of three or more leaves (or other parts) around one point along a stem or axis.

woody Reinforced with fibrous tissue so as to remain rigid and functional from one year to the next. Woody plants are trees or shrubs. Opposed to *herbaceous*.

zooplankton Non-plant, drifting aquatic plankton, ranging from one-celled organisms up to jellyfish and insect *larvae*. Singular: zooplankter.

Recommended Reading

Many scientific findings mentioned in this guide (but not quoted) are cited in an online bibliography at https://raveneditions.com/bibliographies.

DIRECT QUOTATIONS

All quotations from Meriwether Lewis are from *The Journals of Lewis and Clark*. Bernard DeVoto (Boston, MA: Houghton Mifflin, 1953).

All quotations from David Douglas are from *Journal Kept by David Douglas 1823–1827*. (New York: Antiquarian Press, reprinted 1959).

How to Use This Guide: Names

p. 13 "Oregon's top natural heritage botanist": Kagan, James. "Plant Conservation in Oregon over the Last Half Century." Online address to Institute for Natural Resources, Corvallis, Oregon, January 4, 2023. media.oregonstate.edu/media/t/1_2 jwxwte9/209417243. Accessed January 30, 2023.

p. 13 "This is a necessary": Dorn, Robert D. *Vascular Plants of Wyoming*. 3rd ed. Cheyenne, WY: Mountain West Publishing, 2001.

p. 13 "folk taxonomy": Cotterill, F. P. D., et al. "Why One Century of Phenetics Is Enough." *Systematic Biology* 63, no. 5 (2014): 819–32.

p. 14 "It may be more fruitful": Money, Nicholas P. "Against the Naming of Fungi." *Fungal Biology* 117 (2013): 463–65.

Chapter 1

p. 18 "This version of ecoregion boundaries" https://gaftp.epa.gov/EPADataCommons/ORD /Ecoregions/cec_na/NA_LEVEL_III.pdf. Accessed September 1, 2023.

p. 26 "This may be because the rich": Thompson, Jonathan P. 2023. "The West: Land of Income Extremes." *The Land Desk* blog. landdesk.org/p /the-west-land-of-income-extremes. Accessed March 24, 2023.

p. 29 "highest commercial value of any species, wherever found": Rockwell, F. I. "The White Pines of Montana and Idaho—Their Distribution, Quality, and Uses. *Forestry Quarterly* 9 (1911): 219–31.

Chapter 3

p. 51 "Our drive of forenoon": H. K. Hines and R. Ketcham quoted in Peters, Harold J. *Seven Months to Oregon: 1853*. Tooele, UT: Patrice Press, 2008.

Chapter 4

p. 90 "inhaled the smoke": Maximilian, Prince of Wied and Neuwied, quoted in McKelvey, Susan Delano. *Botanical Exploration of the Trans-Mississippi West, 1790–1850*. Corvallis, OR: Oregon State University Press, 1991.

Chapter 5

p. 142 "the hospitality the Jesuits showed to me was scant": Karl A. Geyer quoted in McKelvey, ibid.

p. 173 "small, fibrous, woody, and tough": Harrington, H. D. *Manual of the Plants of Colorado*. Denver: Sage Press, 1954.

p. 175 "Really big in Oregon": Gorham, John, and Liz Crain. *Toro Bravo*. San Francisco: McSweeney's, 2013.

p. 186 "aggression, hyperactivity, a stiff and clumsy gait": Cornell College of Agriculture and Life Sciences. "Swainsonine Poisoning." poisonousplants.ansci.cornell.edu/locoweed /swain2.html#loco

p. 192 "When the boat touches the shore": Henry Marie Brackenridge, *Views of Louisiana*. Quoted in McKelvey, op cit.

p. 223 "It is most effective": Moore, Michael. *Medicinal Plants of the Pacific West*. Santa Fe: Museum of New Mexico Press, 1993.

p. 228 "Early in the spring, when livestock": Holmgren, Noel H., Patricia K. Holmgren, Arthur Cronquist, et al. 1977–2012. *Intermountain Flora*. NY Botanical Garden.

Chapter 6

p. 246 "My prospects for the first 8 days": Letter from Augustus Fendler to Georg Engelmann, July 26, 1849. biodiversitylibrary.org /item/139836#page/18/mode/1up. Accessed January 2, 2024.

Chapter 7

p. 260 "Occasionally a semiporous barrier": Campbell, Alsie, and Jerry F. Franklin. *Riparian Vegetation in Oregon's Western Cascade Mountains*. Bulletin 14, Coniferous Biome. Seattle: University of Washington Press, 1979.

Chapter 8

p. 264 "It was as shocking as putting a pizza": Wang, L. 2001. "Fungi Slay Insects and Feed Host Plants." *Science News,* April 7, 2001. This study was: John N. Klironomos and Miranda M. Hart. "Animal Nitrogen Swap for Plant Carbon." *Nature* 410 (2001): 65152.

p. 269 "odor like dirty gym socks": Bessette, Alan. *Mushrooms of Northeast North America*. Syracuse: Syracuse University Press, 1997.

p. 277 "a fun puzzle for scientists": Raudabaugh, D. B., et al. "Where Are They Hiding? Testing the Body Snatchers Hypothesis in Pyrophilous Fungi." *Fungal Ecology,* 43 (2020): 100870.

p. 278 "We recommend that it be cooked well": Trudell, Steve, and Joe Ammirati. *Mushrooms of the Pacific Northwest*. Portland, OR: Timber Press, 2009.

p. 293 "Their diet was varied with burnt bones": Seton, Ernest Thompson. *The Book of Woodcraft*. New York: Garden City, 1912.

Chapter 9

p. 293 "Their diet was varied with burnt bones": Seton, Ernest Thompson. *The Book of Woodcraft,* p. 247. New York: Garden City, 1912.

p. 299 "I personally think [the Indigenous peoples": Goward, Trevor. "Twelve Readings on the Lichen Thallus: VI. Reassembly." 2009. www.waysoflichenment.net/essays /readings_6_reassembly.pdf. Accessed October 28, 2023.

Chapter 10

p. 312 "In the morning Mr N and myself": Townsend, John Kirk. *Narrative of a Journey across the Rocky Mountains to the Columbia River*, 1839. Reprinted 1999. Corvallis: Oregon State University Press.

p. 336 "At night they are so fearless": Townsend, John Kirk. *Narrative of a Journey across the Rocky Mountains to the Columbia River,* 1839. Reprinted 1999. Corvallis: Oregon State University Press.

p. 361 "Saw large herds of buffalo": Townsend, John Kirk. *Narrative of a Journey across the Rocky Mountains to the Columbia River,* 1839. Reprinted 1999. Corvallis: Oregon State University Press.

Chapter 11

p. 383 "bird of rebirth": Welch, Lew. *Ring of Bone: Collected Poems*. Expanded Edition. San Francisco: City Lights, 2012.

p. 377 "Strainers of aerial plankton": Attributed to Evelyn Bull.

p. 379 "Trans-Pacific migrants": Piersma, Theunis, et al. "The Pacific as The World's Greatest Theater of Bird Migration: Extreme Flights Spark Questions about Physiological Capabilities, Behavior, and the Evolution of Migratory Pathways." *Ornithology*, 139, no. 2 (2022): ukab086.

p. 412 "An astonishing recording": Kroodsma, Donald E. *The Singing Life of Birds*. Boston: Houghton Mifflin, 2005.

p. 409 "Birds don't do it": Sly and Robbie (reggae duo) with Bootsy Collins. *Rhythm Killers*. 1987.

Chapter 15

p. 449 "hosts the greatest": James, David, and David Nunnallee. *Life Histories of Cascadia Butterflies*. Corvallis: Oregon State University Press, 2011.

p. 452 "How to Not Get Bit" data comes mostly via the Environmental Working Group. ewg.org /consumer-guides/ewgs-advice-avoiding-bug -bites. Updated 2018; accessed January 22, 2024.

p. 473 "Plants are really just very slow animals." Schultz, Jack C. "Biochemical Ecology: How Plants Fight Dirty." *Nature* 416 (2002): 267. https://doi .org/10.1038/416267a.

Chapter 16

p. 493 "could have provided cradles for the emergence of life": Farmer, Jack. "Hydrothermal Systems: Doorways to Early Biosphere Evolution." *GSA Today* 10, no. 7 (2000): 1–9.

p. 494 "the only microorganism we know to smile": Upcroft, Jacqui, and Peter Upcroft. "My Favorite Cell: Giardia." *BioEssays* 20, no. 3 (1998): 256–63.

p. 494 "education efforts": Welch, Timothy E. "Risk of Giardiasis from Consumption of Wilderness Water in North America: A Systematic Review of Epidemiologic Data." *International Journal of Infectious Diseases* 4, no. 2 (2000): 100–103.

Chapter 17

p. 496 "a theory for the origin of mountains with the origin of mountains left out": Widely attributed to James Dwight Dana.

AUTHORITIES FOR COMMON AND SCIENTIFIC NAMES IN THIS GUIDE

Note: All websites were accessed in April 2023.

Ackerfield, Jennifer. *A Flora of Colorado.* 2nd ed. Fort Worth, TX: BRIT Press, 2022.

Brodo, Irwin M., and Sylvia and Stephen Sharnoff. *Lichens of North America.* New Haven, CT: Yale University Press, 2001.

Checklist of North American Birds. American Ornithological Society. checklist.american ornithology.org/taxa. 2023.

Colorado Herpetological Society. coloherps.org.

Cook, Thea, Stephen C. Meyers, and Scott Sundberg, eds. *Oregon Plant Atlas, Photo Gallery, and Vascular Plant Checklist.* oregonflora.org. Version 2.1. 2023.

Flora of North America Editorial Committee. 1993–. *Flora of North America North of Mexico.* 19 volumes to date, of 30 planned. New York: Oxford University Press. efloras.org/flora_page .aspx?flora_id=1.

Giblin, D. E., and B. S. Legler, eds. WTU Image Collection: Vascular Plants, MacroFungi, and Lichenized Fungi of Washington State. University of WA Herbarium. burkeherbarium .org/imagecollection. 2003+.

Lesica, Peter. *Manual of Montana Vascular Plants.* 2nd ed. Fort Worth, TX: BRIT Press, 2022.

Lotts, Kelly, and Thomas Naberhaus. Butterflies and Moths of North America. butterflies andmoths.org. 2023.

Mammal Diversity Database, American Society of Mammalogists. mammaldiversity.org.

McCune, Bruce, and Linda Geiser. *Macrolichens of the Pacific Northwest.* 3rd ed. Corvallis: Oregon State University Press, 2024.

Stevens, P. F. (2001 onward). Angiosperm Phylogeny Website. Version 14, July 2017 [updated since]. www.mobot.org/MOBOT /research/APweb/

WEBSITES

A bibliography of scientific papers reflected in this book is posted at https://raveneditions.com /bibliographies. All websites were accessed in April 2023.

Audubon Guide to North American Birds. Many song and call recordings. audubon .org/bird-guide.

Biota of North America Program. Maps of plant species presence by county. bonap.net /NAPA/Genus/Traditional/County.

Breeding Bird Survey. mbr-pwrc.usgs.gov.

BugGuide. Insect identification. bugguide.net.

Consortium of Lichen Herbaria. lichenportal .org/portal.

earth.nullschool.net. Rendering of atmosphere in the present, recent past, and near future.

Enlichenment (Trevor Goward). waysoflichen ment.net/ways/home.

GOES-East Sector View: Northern Rockies. Satellite views updated every five minutes, or animated; great for seeing where it's cloudy or smoky. star.nesdis.noaa.gov/goes/sector .php?sat=G16&sector=nr.

iNaturalist. inaturalist.org/home.

Inciweb. Fire status information. inciweb .wildfire.gov.

Lewis and Clark Trail Heritage Foundation. lew- is-clark.org. More than 2300 pages of all things Lewis and Clark.

MODIS Today. Images from the latest satellite pass. ge.ssec.wisc.edu/modis-today.

Mushroom Expert (Michael Kuo). mushroom expert.com.

Mushroom Observer. mushroomobserver.org.

Native American Ethnobotany Database (Daniel Moerman). naeb.brit.org.

North American Bumblebees. bumblebee.org.

PLANTS Database, National Plant Data Center. plants.usda.gov.

RealClimate. Climate science from climate scien- tists. realclimate.org.

Southwest Colorado Wildflowers (Al Schneider). swcoloradowildflowers.com.

Spire Measure (David Metzler and Edward Earl). Analysis of mountain topography. peaklist.org/spire/ index.html.

Stephen Sharnoff's Lichens. sharnoffphotos.com /lichens/lichens_home_index.html.

The Gymnosperm Database (Chris Earle). conifers.org.

Xerces Society for Invertebrate Conservation. xerces.org.

BOOKS

Allred, Kelly, and Eugene M. Jercinovic. *Flora Neomexicana III: An Illustrated Identification Manual*, 2nd ed. Self-published: lulu.com, 2020.

Armstrong, David M. *Mammals of Colorado*, 2nd ed. Boulder: University Press of Colorado, 2011.

Arno, Stephen F., and Carl E. Fiedler. *Douglas Fir: The Story of the West's Most Remarkable Tree*. Seattle: Mountaineers Books, 2020.

Arno, Stephen F., and Ramona P. Hammerly. *Northwest Trees*, 2nd ed. Seattle: Mountaineers Books, 2007.

——. *Timberline: Mountain and Arctic Forest Frontiers*. Seattle: Mountaineers Books, 1984.

Arora, David. *Mushrooms Demystified*. Berkeley, CA: Ten Speed Press, 1986.

Aubry, Keith B., et al. (eds.). *Biology and Conservation of Martens, Sables, and Fishers*. Ithaca, NY: Comstock Publishing Associates, 2012.

Behnke, Robert J. *Trout and Salmon of North America*. Illustrated by Joseph R. Tomelleri. New York: Free Press, 2002.

Beug, Michael W., A. E. Bessette, and A. R. Bessette. *Ascomycete Fungi of North America*. Austin: University of Texas Press, 2014.

Birkhead, Tim. *The Magpies*. London: T. & A. D. Poyser, 1991.

Brodo, Irwin M., and Sylvia and Stephen Sharnoff. *Lichens of North America*. New Haven, CT: Yale University Press, 2001.

Chadwick, Douglas H. *The Wolverine Way*. Ventura, CA: Patagonia, 2010.

Chittka, Lars. *The Mind of a Bee*. Princeton, NJ: Princeton University Press, 2022.

Cripps, Cathy L., Vera Evenson, and Michael Kuo. *The Essential Guide to Rocky Mountain Mushrooms by Habitat*. Urbana: University of Illinois Press, 2016.

Dalton, David A. *The Natural World of Lewis and Clark*. Columbia: University of Missouri Press, 2008.

Denver Botanic Gardens. *Wildflowers of the Rocky Mountain Region*. Portland, OR: Timber Press, 2018.

Dorn, Robert D. *Vascular Plants of Wyoming*. Cheyenne: Mountain West, 2001.

Eaton, Eric R. *Insectpedia*. Princeton: Princeton University Press, 2022.

Eaton, Eric R., and Kenn Kaufman. *Kaufman Field Guide to Insects of North America*. New York: Houghton Mifflin Harcourt, 2007.

Elbroch, Mark, and Kurt Rinehart. *Behavior of North American Mammals*. New York: Houghton Mifflin Harcourt, 2011.

Evenson, Vera Stucky, and Denver Botanic Gardens. *Mushrooms of the Rocky Mountain Region*. Portland, OR: Timber Press, 2015.

Farrell, Justin. *Billionaire Wilderness: The Ultra-wealthy and the Remaking of the American West*. Princeton. Princeton University Press, 2020.

Farjon, Aljos. *A Natural History of Conifers*. Portland, OR: Timber Press, 2008.

Geiger, Rudolf, et al. *The Climate Near the Ground*, 7th ed. Lanham, MD: Rowman & Littlefield, 2009.

Gillett, J. D. *Mosquitoes*. London: Weidenfeld and Nicolson, 1971.

Goldfarb, Ben. *Crossings: How Road Ecology Is Shaping the Future of Our Planet*. New York: W. W. Norton, 2023.

Halfpenny, James C. *Scats and Tracks of the Rocky Mountains*, 3rd ed. Helena, MT: Falcon Press, 2015.

Hansman, Heather. *Powder Days: Ski Bums, Ski Towns and the Future of Chasing Snow*. Toronto: Hanover Square, 2021.

Heinrich, Bernd. *Winter World*. New York: HarperCollins, 2003.

Hitchcock, C. Leo, et al. *Vascular Plants of the Pacific Northwest*, in 5 volumes. Seattle: University of Washington Press, 1955–1969. Source of the Jeanne R. Janish drawings in this book.

Hitchcock, C. Leo, et al. *Flora of the Pacific Northwest*, 2nd ed. Seattle: University of Washington Press, 2018.

Holmgren, Noel H., Patricia K. Holmgren, Arthur Cronquist, et al. *Intermountain Flora*, in 8 volumes. New York: NY Botanical Garden, 1977–2012.

Hyndman, Donald W., and Robert C. Thomas. *Roadside Geology of Montana*, 2nd ed. Missoula, MT: Mountain Press, 2020.

James, David G., and David Nunnallee. *Life Histories of Cascadia Butterflies*. Corvallis: Oregon State University Press, 2011.

Lanner, Ronald M. *The Piñon Pine*. Reno: University of Nevada Press, 1981.

Lewis, Meriwether, and William Clark. *The Journals of Lewis and Clark*. Edited by Bernard DeVoto. Boston: Houghton Mifflin, 1953.

Link, Paul, Shawn Willsey, and Keegan Schmidt. *Roadside Geology of Idaho*, 2nd ed. Missoula, MT: Mountain Press, 2021.

Lücking, Robert, and Toby Spribille. *The Lives of Lichens: A Natural History*. Princeton, NJ: Princeton University Press, 2024.

Mathews, Daniel. *Trees in Trouble: Wildfires, Infestations, and Climate Change*. Berkeley, CA: Counterpoint, 2020.

McCune, Bruce. *Microlichens of the Pacific Northwest*. Corvallis, OR: Wild Blueberry Media, 2017.

McCune, Bruce, and Linda Geiser. *Macrolichens of the Pacific Northwest*, 3rd ed. Corvallis: Oregon State University Press, 2024.

McCune, Bruce, and Trevor Goward. *Macrolichens of the Northern Rocky Mountains*. Eureka, CA: Mad River, 1995.

McKelvey, Susan Delano. *Botanical Exploration of the Trans-Mississippi West, 1790–1850*. Corvallis: Oregon State University Press, 1991.

Miller, Marli B. *Roadside Geology of Oregon*, 2nd ed. Missoula: Mountain Press, 2014.

Miller, Marli B., and Darrel S. Cowan. *Roadside Geology of Washington*, 2nd ed. Missoula: Mountain Press, 2017.

Paulson, Dennis. *Dragonflies and Damselflies of the West*. Princeton: Princeton University Press, 2009.

Pyle, Robert Michael. *Butterflies of the Pacific Northwest*. Portland, OR: Timber Press, 2018.

Robbins, Chandler S., Bertel Bruun, and Herbert S. Zim. *Birds of North America: Golden Field Guide*, 2nd ed. New York: St. Martins Press, 2001.

Russo, Ron. *Field Guide to Plant Galls of California and Other Western States*. Berkeley: University of California Press, 2006.

Schofield, W. B. *Field Guide to Liverwort Genera of Pacific North America*. Seattle: University of Washington Press, 2002. Source of the liverwort drawings in this book.

———. *Some Common Mosses of British Columbia*. Victoria: Royal BC Museum, 1992. Source of some moss drawings in this book.

Sibley, David Allen. *The Sibley Guide to Bird Life and Behavior*. New York: Knopf, 2009.

Smith, Douglas W., Daniel R. Stahler, and Daniel R. MacNulty, eds. *Yellowstone Wolves*. Chicago: University of Chicago Press, 2020.

Tilford, G. L. *Edible and Medicinal Plants of the West*. Missoula, MT: Mountain Press, 1997.

Tomback, Diana F., Stephen F. Arno, and Robert E. Keane. *Whitebark Pine Communities: Ecology and Restoration*. Washington, DC: Island Press, 2001.

Turner, Nancy J. *Food Plants of Interior First Peoples*. Vancouver: University of British Columbia Press, 1997.

———. *Plant Technology of First Peoples of British Columbia*. Vancouver: University of British Columbia Press, 1988.

Van Pelt, Robert. *Identifying Old Trees and Forests in Eastern Washington*. Olympia: Washington State Department of Natural Resources, 2008.

Weber, William, and Ronald C. Wittmann. *Bryophytes of Colorado: Mosses, Liverworts, and Hornworts*. Santa Fe, NM: Pilgrim's Progress, 2007.

———. *Colorado Flora: Eastern Slope*, 4th ed. Boulder: University Press of Colorado, 2012.

———. *Colorado Flora: Western Slope*, 4th ed. Boulder: University Press of Colorado, 2012.

Weidensaul, Scott. *A World on the Wing: The Global Odyssey of Migratory Birds*. New York: W. W. Norton, 2021.

Whiteman, Noah. *Most Delicious Poison: The Story of Nature's Toxins—From Spices to Vices*. New York: Little, Brown Spark, 2023.

Williams, Felicie, and Halka Chronic. *Roadside Geology of Colorado*, 3rd ed. Missoula, MT: Mountain Press, 2014.

Williams, Paul, et al. *Bumblebees of North America*. Princeton, NJ: Princeton University Press, 2014.

Williams, Roger Lawrence. *A Region of Astonishing Beauty: The Botanical Exploration of the Rocky Mountains*. Lanham, MD: Roberts Rinehart, 2003.

Winkler, Daniel. *Fruits of the Forest: A Field Guide to Pacific Northwest Edible Mushrooms*. Seattle: Skipstone, 2023.

Index

Note: If you search for a species here and don't find it, the author may have treated it as an alternate name, i.e., not the currently preferred name. With few exceptions, space allows only preferred names to be listed below.

A

abbreviations and symbols, 16
Abert's squirrel, 321
Abies concolor, 67–68
Abies grandis, 53, 68–70
Abies lasiocarpa, 65–67, 70
accipiter, 388–89
Accipiter cooperii, 388
Accipiter gentilis, 388
Acer glabrum, 89
Acer negundo, 89–90
acknowledgments, 539
Aconitum columbianum, 191
Actaea rubra, 240–41
Actitis macularia, 380–81
Adejeania vexatrix, 454–55
Adelges cooleyi, 456–57
Adelges piceae, 457
Adelges tsugae, 457
Adenocaulon bicolor, 168–69
Adiantum aleuticum, 248–49
admiral (butterfly), 482
Adonis blazingstar, 215
Aechmophorus occidentalis, 374
Aedes spp., 452–53
Aegolius acadicus, 393
Aegolius funereus, 393
Aeronautes saxatilis, 376
Aeshna juncea, 469
Aeshna palmata, 468
African cheetah, 364
Agastache urticifolia, 184
Agelaius phoeniceus, 420
Aglais milberti, 483–84

Agoseris aurantiaca, 171
Agoseris glauca, 171
Agropyron cristatum, 127
Agrostis stolonifera, 129
Agulla bicolor, 465
air, rising and sinking, 32–33
Airbnb, 26
Alberta penstemon, 181
Alces alces, 358–60
alder, 86–87
alderleaf mountain-mahogany, 98
Alectoria sarmentosa, 299
Aleuria aurantia, 281–82
Allium acuminatum, 143
Allium brevistylum, 142–43
Allium cernuum, 142
Allium geyeri, 143
Allium schoenoprasum, 143
Allium textile, 143
Allocetraria madreporiformis, 300
Allotropa virgata, 153–54
Alnus alnobetula, 86
Alnus incana, 86
alp lily, 144
alpine avens, 216
alpine bistort, 224
alpine bluegrass, 129–30
alpine buttercup, 219
alpine clover, 189
alpine collomia, 195
alpine forget-me-not, 202
alpine grasshopper, 465–66
alpine haircap moss, 254
alpine jelly cone, 281
alpine kittentails, 184
alpine lady fern, 249
alpine laurel, 119
alpine meadow butterweed, 161
alpine mertensia, 201

alpine plants, snow-depth specialties, 196
alpine primrose, 198
alpine springbeauty, 222
alpine spring-parsley, 226
alpine sunflower, 162
alpine timothy, 128
alpine waxy cap, 268
alpine willow-herb, 229–30
alpine wintergreen, 120
alpinegold, 162
Amanita muscaria, 266
Amanita pantherinoides, 266
Amanita phalloides, 267
Ambystoma macrodactylum, 432–33
Ambystoma mavortium, 432
Amelanchier alnifolia, 96–97
amenity migration, 26–27
American alpine speedwell, 231
American beaver, 325–26
American bistort, 224
American crow, 402
American dipper, 413
American goldfinch, 418–19
American kestrel, 399
American mink, 343–44
American moonwort, 250
American Ornithological Society (AOS), 367
American parsley fern, 247
American pipit, 415–16
American red squirrel, 321
American robin, 414
American thorow-wax, 226
American three-toed woodpecker, 396
Ammirati, Joe, 278
amphibians, 431–35. *See also specific species*
global decline of, 431–32

Anaphalis margaritacea, 168
Anas acuta, 367
Anas crecca, 368
Anas platyrhynchos, 367
Anaxyrus boreas, 434
Anaxyrus woodhousii, 434
ancestral Rockies, 503–04
andesite, 532
Androloma maccullochii, 473–74
Androsace chamaejasme, 197
Androsace montana, 198
Androsace septentrionalis, 197
anemone, 239–40
Anemone drummondii, 240
Anemone multifida, 240
Anemone narcissiflora, 240
Anemone occidentalis, 239–40
Anemone patens, 239
angelica, 229
Angelica arguta, 229
Angelica pinnata, 229
anise swallowtail, 475–76
Antennaria anaphaloides, 167
Antennaria lanata, 167
Antennaria media, 167
Antennaria parvifolia, 167
Antennaria rosea, 167
Antennaria umbrinella, 167
Anthocharis julia, 479–80
Anthocharis stella, 479–80
Anthoxanthum hirtum, 129
Anthus rubescens, 415–16
Anticlea elegans, 147
Anticlea occidentalis, 147
Antigone canadensis, 379–80
Antilocapra americana, 364–65
Aphyllon fasciculatum, 155
Aphyllon purpureum, 155
Apioperdon pyriforme, 283–84
Apis mellifera, 458
Apocynum androsaemifolium, 204
Aquarius remigis, 457
Aquila chrysaetos, 385–86
Aquilegia coerulea, 220
Aquilegia flavescens, 220
Aquilegia formosa, 220
Aquilegia jonesii, 220
Aquilegia saximontana, 220
Aralia nudicaulis, 205
Arceuthobium, 175–76
archaea, 492
Archaean cratons, 500–01
arctic aster, 158
arctic blue, 488

arctic fritillary, 486
arctic gentian, 237
arctic grayling, 446
arctic harebell, 199
arctic sagewort, 170
arctic sandwort, 212–13
Arctostaphylos uva-ursi, 119–20
Ardea herodias, 383
Argentina anserina, 218–19
Argia emma, 470
Argia vivida, 470
argillite, 528
Argynnis cybele, 486
Argynnis hydaspe, 485
Argynnis mormonia, 485
Armillaria solidipes, 268–69
arnica, 159
Arnica cordifolia, 159
Arnica latifolia, 159
Arnica mollis, 159
Arnica rydbergii, 159
Arora, David, 266, 274
arrowleaf balsamroot, 163
arrowleaf groundsel, 160
Artemisia cana, 112
Artemisia dracunculus, 170
Artemisia ludoviciana, 170
Artemisia michauxiana, 170
Artemisia norvegica, 170
Artemisia scopulorum, 170
Artemisia tridentata, 112–13
Artomyces pyxidatus, 282–83
Asarum caudatum, 241, 242
Ascaphus montanus, 435
Asclepias speciosa, 204–05
Aspen, Colorado, 26, 27
aspen. *See* quaking aspen
aspen bracket, 279–80
aspen leaf miner, 472
aspen oyster mushroom, 271
Aspidoscelis velox, 426
aster, 157–58
Astragalus kentrophyta var.
    tegetarius, 186–87
Astragalus purshii, 187
Athene cunicularia, 392–93
Athyrium distentifolium, 249
Athyrium filix-femina, 249
Atriplex canescens, 88–89
Auricularia americana, 281
avalanches, 40

B
badger, 349–50
Baker's bluebells, 201

bald eagle, 386–87
baldhip rose, 101
ballhead sandwort, 212
ballhead waterleaf, 199
balsam woolly adelgid, 457
Balsamorhiza incana, 163
Balsamorhiza sagittata, 163
balsamroot, 163
Baltic rush, 137
Bambi (movie), 49
baneberry, 240–41
banner clouds, 35
Barbey's larkspur, 193
Barbilophozia lycopodioides, 261
Barclay willow, 84
bark beetle, 462–63
barred owl, 392
Barrow's goldeneye, 370
basalt, 532–33
Bassariscus astutus, 338
bastard toadflax, 203–04
bat, 307–08
bead lily, 143–44
beaked moss, 258
bear, 319
bear safety, 341
beardlip penstemon, 183
beargrass, 125, 147–48
bear's head, 283
bedstraw sphinx moth, 474
bees and wasps, 457–60
beetles, 460–65
bellflower, 199
Belt supergroup, 501–02
belted kingfisher, 394
Bering, Vitus, 404–05
besseya kittentails, 183–84
Bessey's locoweed, 186
Betula glandulosa, 87
Betula occidentalis, 87
Betula papyrifera, 87
Beug, Micheal, 271
Bierstadt, Albert, 217
big brown bat, 507
big redstem moss, 259
big sagebrush, 112–13
big shaggy moss, 258–59
big-head groundsel, 160
bighorn sheep, 360–61
bigleaf lupine, 188
bigleaf rosette moss, 256
biological crusts (biocrusts), 288
biomass, 43–44
birch, 87–88
birchleaf spiraea, 100

bird flu H5N1, 344
birds, 366–423. *See also specific species*
    extra-pair copulations, 409
    migration, 379
    recycled brains, 406
    soaring wings, 386
    speed wings, 398
Birds of North America (Robbins), 17
biscuitroot, 225
bison, 361–63
bistort, 224
Bistorta bistortoides, 224
Bistorta vivipara, 224
bitterbrush, 102–03
bitterroot, 242–43
black alpine sedge, 133–34
black bear, 59, 338–40
black cottonwood, 81–82
black fire beetle, 465
black fly, 453–54
black hawthorne, 99
black huckleberry, 109
black locust, 114
black morel, 276
black moss, 255
black pine, 52
black rosy-finch, 416
black swallowtail, 476
black swift, 376
black twinberry, 91
black-and-gold tile lichen, 289–90
black-backed woodpecker, 395
black-billed magpie, 405–07
black-capped chickadee, 407
blackneck garter snake, 427
blackroot sedge, 133
black-veined forester, 473–74
blanketflower, 161
blazingstar, 215
blister rust, 57
blistered rock tripe, 294
blue (butterfly), 488–89
blue elderberry, 92–93
blue flax, 205
blue grama, 128
Blue Mountains, 28–29
blue virgin's clematis, 94
blue wildrye, 127
bluebells-of-Scotland, 199, 201
blueberry, 109–10
bluebunch wheatgrass, 127
blue-eyed darner, 469

blue-eyed Mary, 181
blue-eyed-grass, 149
blue-staining slippery jack, 275–76
bluejoint reedgrass, 130
blueleaf cinquefoil, 218
blueleaf strawberry, 215
blushing stippleback, 296
bobcat, 350–51
Boechera lyallii, 233
bog blueberry, 109–10
bog orchid, 149–50
bog saxifrage, 207–08
Bohemian waxwing, 411
boletus, 274–75
Boletus barrowsii, 274–75
Boletus edulis, 274–75
Boletus rubriceps, 274
Boloria chariclea, 486
Bombus appositus, 459
Bombus huntii, 459
Bombus morrisoni, 459
Bombus spp., 458–60
Bombycilla cedrorum, 411
Bombycilla garrulus, 411
Bonasa umbellus, 373–74
Bonneville shooting star, 198
boreal chorus frog, 433
boreal owl, 393
Bos bison, 361–63
Botany of Northwest America (Lyall), 62
Botrychium neolunaria, 250
bottlebrush squirreltail, 126
boulder raspberry, 99
Bouteloua dactyloides, 128
Bouteloua gracilis, 128
box elder, 89–90
bracken, 249
bracted lousewort, 177
Branta canadensis, 371
breccia, 527
Brewer's cliff-brake, 246
Brewer's mitrewort, 206
Brickellia grandiflora, 168
bristly beard lichen, 298
bristly haircap moss, 254
brittle fern, 247
brittle prickly pear, 118
broadleaf arnica, 159
broadleaf evergreen shrubs, 114–24
broadlip twayblade, 151
broad-tailed hummingbird, 377
Bromus carinatus, 131

Bromus tectorum, 131
bronze bells, 147
brook saxifrage, 207
brook trout, 442–43
broomrape, 155
brown creeper, 412
brown elfin, 490
brown helmets, 303
brown tile lichen, 289
brown trout, 443–44
brown-capped rosy-finch, 416
brown-eyed wolf lichen, 301
Bruner's spur-throat grasshopper, 466
Bryoria fremontii, 298
Bryoria fuscescens, 298
Bubo virginianus, 391
Bucephala albeola, 370
Bucephala clangula, 370
Bucephala islandica, 370
buckbean, 203
buckwheat, 237–38
buffalograss, 128
bufflehead, 369
bugs, true, 456–57
bull trout, 441–42
bullfrog, 435
Bullock's oriole, 420–21
bullsnake, 429
bumblebee, 455, 458–60
bunchberry, 235
Bupleurum americanum, 226
Buprestis aurulenta, 461
Burke, Joseph, 142, 220
burrowing owl, 392–93
busky-tailed woodrat, 326–27
Buteo jamaicensis, 389–90
Buteo swainsoni, 390
Butte candle, 203
buttercup, 219
butterflies. *See also specific species*
    and moths, 471–91
    puddling, 487
    word origin, 479
butterweed, 160–61
buzzard, 385

C

Calamagrostis canadensis, 130
Calamagrostis purpurascens, 129
Calamagrostis rubescens, 129
Calbovista subsculpta, 284
calcite, 523
California corn lily, 148

California gull, 382
California tortoiseshell, 483
calliope hummingbird, 377
Callophrys augustinus, 490
Callophrys eryphon, 489
Callophrys mossii, 490
Callophrys sheridanii, 490
Callospermophilus lateralis, 315–16
Calochortus apiculatus, 141
Calochortus elegans, 141
Calochortus eurycarpus, 141
Calochortus gunnisonii, 140–41
Calochortus nuttallii, 141
Caltha chionophila, 239
Calvitimela armeniaca, 289–90
Calypso bulbosa, 152
calypso orchid, 152
Calyptridium umbellatum, 232
camas, 145–46
Camassia quamash, 145–46
Campanula rotundifolia, 199
Campanula uniflora, 199
Campbell, Alsie, 260
Campylium stellatum, 258
Canada goose, 371
Canada jay, 404–05
Canadian buffaloberry, 93–94
Canadian gooseberry, 103–04
Canadian white violet, 192
canary violet, 191
Candelariella vitellina, 289–90
Canis latrans, 334–36
Canis lupus, 336–38
Cansler, Alina, 69
Cantharellus roseocanus, 272
Cantharellus subalbidus, 272
carbon dioxide, 41–42, 44
Cardamine cordifolia, 233
Cardellina pusilla, 422
Carex aquatilis, 132
Carex elynoides, 133
Carex geyeri, 134
Carex heteroneura, 133
Carex mertensii, 134
Carex nigricans, 133–34
Carex phaeocephala, 132–33
Carex raynoldsii, 134
Carex rossii, 133
Carex rupestris, 133
Carex scopulorum, 132
Carex spectabilis, 134
Carpodacus, 417
Carraway, Leslie, 506
Carson, Kit, 297

cascara, 108
Cassin's finch, 417
Cassiope mertensiana, 120–21
Cassiope tetragona, 121
Castilleja flava, 178
Castilleja hispida, 178
Castilleja integra, 178
Castilleja liniarifolia, 178
Castilleja miniata, 178
Castilleja occidentalis, 178
Castilleja pulchella, 178
Castilleja rhexiifolia, 178
Castor canadensis, 325–26
Catastomus catastomus, 446–47
Cathartes aura, 383–85
Catharus guttatus, 414
Catharus ustulatus, 414
cattail, 131–32
Ceanothus sanguineus, 107–08
Ceanothus velutinus, 115–16
cedar waxwing, 411
Centaurea stoebe, 168
Cerastium arvense, 213
Cercocarpus ledifolius, 115
Cercocarpus montanus, 98
Cercyonis oetus, 480
Cercyonis pegala, 480
Certhia americana, 412
Cervus canadensis, 356–58
Cetraria ericetorum, 300
Cetraria islandica, 300–01
Cetraria juniperina, 301
Cetraria nivalis, 301
Chaenactis douglasii, 166–67
Chaetura vauxi, 376
Chamaenerion angustifolium, 230
Chamaenerion latifolium, 230
Chamisso, Adelbert von, 200
chanterelle, 272–73
Charadrius vociferus, 381
Charina bottae, 427
cheatgrass, 131
checkerspot, 484–85
cheetah, 364
cherry-faced meadowhawk, 469–70
chert, 525–26
chickadee, 406, 407
chickenlike fowl, 372–74
chicken of the woods, 280
Chief Joseph Dike Swarm, 532
Chimaphila umbellata, 119
Chinook wall, Chinook arch (clouds), 35

Chinook winds, 31, 38–39
Chionophila jamesii, 184
chipmunk, 313–15
chipping sparrow, 419
Chlamydomonas nivalis, 493–94
Chlosyne acastus, 484
chocolate-tips, 225
chokecherry, 97
Chordeiles minor, 375–76
chorus frog, 433–34
chronic wasting disease, 355
Chrysemys picta, 430
Chrysops spp., 454
Chrysothrix chlorina, 291
chryxus arctic, 480–81
Cicuta douglasii, 228
Cicuta maculata, 228
Cimbex americanus, 460
Cinclus mexicanus, 413
cinquefoil, 216, 218
Circotettix rabula, 466
Circus hudsonius, 387–88
cirrus clouds, 33
Cirsium eatonii, 169
Cirsium griseum, 169
Cirsium scariosum, 169
Cirsium undulatum, 169
cladonia, 303–04
Cladonia arbuscula, 302
Cladonia cariosa, 303
Cladonia chlorophaea, 303
Cladonia coniocraea, 303
Cladonia deformis, 304
Cladonia pyxidata, 303
Cladonia rangiferina, 302
Cladonia sulphurina, 303
Cladonia verticillata, 303
claret cup cactus, 118
Clark, William, 64, 241, 349
Clarkia pulchella, 231
Clark's nutcracker, 56, 58–59, 60, 402–03
Claytonia lanceolata, 222
Claytonia megarhiza, 222
Claytonia parvifolia, 222
clematis, 94–95
Clematis columbiana, 94
Clematis hirsutissima, 232
Clematis ligusticifolia, 94
Clematis occidentalis, 94
Clethrionomys gapperi, 329–30
cliff cinquefoil, 218
cliff fendlerbush, 95–96
cliff swallow, 408
cliff-brake, 246

cliffbush, 95
cliffrose, 103
Climacium dendroides, 259–60
climate feedback and plants, 43–44
climate change
  and biological crusts, 288
  long-term, 41–42
  recent, 42–43
  trees and, 53
  weeds and, 174
climograms, 37–38
Clintonia uniflora, 143–44
Clodius parnassian, 477
clouds
  rain, 33–34
  types of, 35
cloudcaps, 35
clouded sulphur, 478
clubmoss, 250–51
clustered broomrape, 155
Coccothraustes vespertinus, 416
Coenonympha tullia, 480
Coeur d'Alene salamander, 433
coiled-beak lousewort, 177
Colaptes auratus ssp. *cafer*, 397
Colias alexandra, 479
Colias eurytheme, 478–79
Colias philodice, 478
collared lizard, 425–26
Collinsia parviflora, 181
Collomia debilis, 195
Coloradia pandora, 474
Colorado blue columbine, 220
Colorado chipmunk, 314
Colorado Flora (Weber), 16
Colorado hairstreak, 491
Colorado ragwort, 160
Colorado rock-shield, 296–97
Coluber constrictor, 427–28
Columba livia, 375
Columbia River basalt lava, 28
Columbia Mountains, 29–30
Columbia spotted frog, 434
Columbian ground squirrel, 316–17
Columbian needlegrass, 126
columbine, 219–21
Comandra umbellata, 203–04
common alpine, 481
common garter snake, 427
common goldeneye, 370
common haircap moss, 254
common juniper, 78
common liverwort, 260

common loon, 382–83
common merganser, 371–72
common nighthawk, 375–76
common raven, 401–02
common ringlet, 480
common snowberry, 92
common twinpod, 234
common wood nymph, 480
common yellowthroat, 422
Confucius, 165
conglomerate, 527
conifers, 50–78. *See also specific species*
  with bunched needles, 50–52, 54–63
  with single needles, 63–74
  that grow as low shrubs, 77–78
  trees with tiny, scalelike leaves, 74–77
Conium maculatum, 228
Conocephalum salebrosum, 260–61
Contopus cooperi, 399–400
convergent ladybug, 460–61
Cooley spruce gall adelgid, 456–57
copper (butterfly), 487–88
Cooper's hawk, 388
Coprinopsis atramentarius, 270
Coprinus comatus, 270
coprophagy, 310
Corallorhiza maculata, 153
Corallorhiza mertensiana, 153
Corallorhiza striata, 153
coralroot, 152–53
corkscrew trees, 59
corn lily, 148
Cornus canadensis, 235
Cornus sericea, 90–91
Corthylio calendula, 410
Corvus brachyrhynchos, 402
Corvus corax, 401–02
Corydalis aurea, 191
cottongrass, 134–35
cottonwood morel, 277
cottonwoods. See black or plains cottonwood
Cottus punctulatus, 448
Cottus semiscaber, 448
cougar, 351–52
COVID-19, 344
cow-parsnip, 228–29
coyote, 334–36
coyote willow, 84
cranes, 379–80

Crataegus chrysocarpa, 98–99
Crataegus douglasii, 99
Crataegus rivularis, 98–99
creeping juniper, 77–78
creeping Oregon-grape, 117–18
Crepis nana, 172
Crepis occidentalis, 172
crested wheatgrass, 127
crisping cushion moss, 256
Crotalus atrox, 430
Crotalus oreganus, 429
Crotalus oreganus concolor, 430
Crotalus viridis, 429
Crotaphytus collaris, 425–26
crow, 402
crown coral fungus, 282–83
Cryptogramma acrostichoides, 247
Culex pipiens, 452
Culicoides spp., 453
cumulus clouds, 34–36
curl-leaf mountain-mahogany, 115
curly sedge, 133
curlyhead goldenweed, 164
currant, 103–05
cushion buckwheat, 238
cushion phlox, 195–96
Cusick's speedwell, 231
cutleaf anemone, 240
cutleaf coneflower, 166
cutleaf daisy, 156–57
cutthroat trout, 438–40
Cyanocitta stelleri, 405
Cygnus buccinator, 371
Cygnus columbianus, 371
Cymopterus alpinus, 226
Cymopterus nivalis, 226
Cymopterus terebinthinus, 226
Cynomys leucurus, 324–25
Cypripedium montanum, 152
Cypripedium parviflorum, 152
Cypseloides niger, 376
Cystopteris fragilis, 247

D
daisy, 156–57
dalmatian toadflax, 180–81
damselflies, 470–71
dancer (insect), 470–71
dandelion, 171
dark pussytoes, 167
dark wood nymph, 480
dark-eyed junco, 419–20
darkthroat shooting star, 198

darkwoods violet, 192
darner, 468–69
Darwin, Charles, 13
Dasiphora fruiticosa, 100
datil yucca, 116–17
death cap, 267
death-stalk suillus, 276
deciduous trees, 79–114
deer, 352–56
    chronic wasting disease, 355
    horns and antlers, 358
    pheromones, 353
deer fly, 454
deer mouse, 327
Delphinium barbeyi, 193
Delphinium bicolor, 193
Delphinium nuttallianum, 193
Delphinium occidentale, 193
Dendragapus obscurus, 373
Dendroctonus brevicomis, 462
Dendroctonus ponderosae, 462
Dendroctonus pseudotsugae, 462
Dendroctonus rufipennis, 462
dense spikemoss, 250
dense-flowered dock, 238
Dermacentor andersoni, 495
Dermatocarpon miniatum, 296
Dermatocarpon reticulatum, 296
Deschampsia cespitosa, 130
devil's club, 106–07
diabase, 531–32
diamondleaf saxifrage, 208
Dicentra uniflora, 190
dicots, 153–243
Dicranum scoparium, 256
diorite, 529–30
Diphasiastrum complanatum, 250
Diplacus nanus, 180
Distichum capillaceum, 255
DNA analysis and names, 13
Dodecatheon, 198
Doellingeria engelmannii, 157–58
dog lichen, 291–92
dogwood. *See* red osier dogwood
dolomite, 523–24
dotted-stalk suillus, 276
Doty, Sharon, 82
Douglas, David, 64, 65, 99, 100, 111, 120, 121, 136, 146, 154
Douglas-fir, 20, 25, 49, 53, 63–65
Douglas-fir beetle, 462
Douglasia, 198
Douglas's triteleia, 145
downy oatgrass, 129

downy woodpecker, 395
draba, 233–34
Draba aurea, 234
Draba oligosperma, 234
dragonflies, 467–71
Drummond, Thomas, 65, 136
Drummond's rush, 136
dryad, 121–23
Dryas drummondii, 122
Dryas hookeriana, 122
Drymocallis arguta, 218
Drymocallis glandulosa, 218
Dryobates pubescens, 395
Dryobates villosus, 395
Dryocopus pileatus, 396–97
Dryopteris expansa, 248
Dryopteris filix-mas, 248
duck, 367–72
dunhead sedge, 132–33
dusky butterweed, 161
dusky grouse, 373
dusky horsehair, 298
dusky shrew, 506
dust lichen, 290–91
dusty maiden, 166–67
dwarf alpine hawksbeard, 172
dwarf blueberry, 109
dwarf clover, 189
dwarf lupine, 188
dwarf mistletoe, 175–76
dwarf mountain groundsel, 160
dwarf primrose, 198
dwarf purple monkeyflower, 180
dwarf sweetroot, 228
dyer's polypore, 280

E
eagles, 385–90
Earle, Chris, 78
early cinquefoil, 216, 218
eastern fence lizard, 425
eastern white pine, 56–57
Echinocereus triglochidiatus, 118–19
Echinodontium tinctorium, 73
edible horsehair, 298
Elaeagnus commutata, 108–09
elderberry, 92–93
elegant cat's ears, 141
elegant sunburst lichen, 293
elephant's head, 176
elfin, 489–90
elk, 356–58
elk sedge, 134
elk thistle, 169

elm sawfly, 460
Elymus elymoides, 126
Elymus glaucus, 127
emerald spreadwing, 470
Emma's dancer, 470
Empidonax traillii, 400
Enallagma annexum, 471
Enallagma boreale, 471
Engelmann, Georg, 70, 142
Engelmann spruce, 20, 25, 47
Engelmann's aster, 157–58
entire paintbrush, 178
Ephemeroptera spp., 467
Epilobium anagallidifolium, 229–30
Eptesicus fuscus, 507
Equisetum arvense, 251
Equisetum hyemale, 252
Equisetum laevigatum, 251–52
Equisetum sylvaticum, 251
Erebia epipsodea, 481
erect-fruited iris moss, 255
Eremogone capillaris, 212
Eremogone congesta, 212
Eremogone fendleri, 212
Eremogone obtusiloba, 212–13
Eremophila alpestris, 407–08
Erethizon dorsatum, 332–33
Ericameria nauseosa, 113
Ericameria suffruticosa, 113
Erigeron compositus, 156
Erigeron glacialis, 156–57
Erigeron grandiflorus, 157
Erigeron humilis, 157
Erigeron lanatus, 157
Erigeron speciosus, 156
Eriocoma nelsonii, 126
Eriogonum caespitosum, 238
Eriogonum flavum, 238
Eriogonum ovalifolium, 238
Eriogonum umbellatum, 237
Eriophorum angustifolium, 134–35
Eriophyllum lanatum, 162
Eritrichium nanum, 202
Erysimum capitatum, 235
Erysimum inconspicuum, 235
Erythranthe guttata, 179
Erythranthe lewisii, 179
Erythranthe primuloides, 180
Erythranthe tilingii, 179–80
Erythronium grandiflorum, 140
Eschscholtz, Johann Friedrich von, 200
Eucephalus, 177–78

Euphilotes ancilla, 489
Euphorbia esula, 173–74
Euphydryas gilletti, 484
Eurhynchiastrum pulchellum, 258
European honeybee, 458
Eurybia conspicua, 158
Eurybia glauca, 158
Eurybia sibirica, 158
evening grosbeak, 416
evening-primrose, 230–31
exploding beard lichen, 298
extremophiles, 492–93
eyed sphinx moth, 475

F
fairy bells, 138
Falco mexicanus, 399
Falco peregrinus, 397–99
Falco rusticolus, 398
Falco sparverius, 399
falcons, 384, 397–99
false haircap moss, 254–55
false-Solomon's-seal, 139
Farmer, Jack, 493
featherleaf kittentails, 232
Fee's lip fern, 245–46
Felis, 350
fellfield anemone, 240
Fendler, Augustus, 246
Fendlera rupicola, 95–96
Fendler's lip fern, 246
Fendler's pennycress, 233
Fendler's sandwort, 212
Fendler's waterleaf, 199
fernleaf lousewort, 177
ferns, 244–50
ferret, 325
Festuca brachyphylla, 130
Festuca idahoensis, 130
Festuca saximontana, 130
few-seeded draba, 234
field chickweed, 213
field crescent, 485
field dog lichen, 291
field horsetail, 251
field locoweed, 186
fir engraver, 463–64
fir trees, 48. *See also specific species*
fire(s)
   in Colorado, 21
   Las Conchas Fire, 86
   in New Mexico, 21

in the Rockies, 46–49
   succession after, 48
firecracker penstemon, 183
fire-rim tortoiseshell, 483–84
fireweed, 230
fisher, 347
fishes, 436–48. *See also specific species*
   civilization vs., 437
   lateral line on, 439
   whirling disease, 440
five fingers, 300
five-stamen mitrewort, 206
five-veined little sunflower, 163
fleabane, 156–57
flies, 449–56
Flora of the Pacific Northwest (Hitchcock), 106
flowering trees and shrubs, 79–124. *See also specific species*
fly amanita, 266–67
fog, 41
Fomitopsis mounceae, 278–79
fool's huckleberry, 110–11
Forest Service and fire suppression, 49
forest succession, 48
forget-me-not, 202
forked broom moss, 256
forking bone lichen, 300
four-angled mountain-heather, 121
four-nerve daisy, 162
fourwing saltbush, 88–89
foxtail barley, 126
fractocumulus, fractostratus clouds, 35
Fragaria vesca, 215
Fragaria virginiana, 215
fragrant bedstraw, 235–36
fragrant sumac, 105
Frangula purshiana, 108
Frank, Albert B., 285
Franklin, Ken, 398
Franklin, Sir John, 136, 447
Frasera speciosa, 236
frayed-cap moss, 255
freckle pelt, 292
Frémont, John C., 224, 297
fringed puccoon, 202–03
Fritillaria atropurpurea, 144–45
Fritillaria pudica, 144
frogs, 433–35
frost riving, 20

funeral bell, 267
fungal species, naming, 14, 15
fungal web and herbs, 154
fungi, 262–85. *See also specific species*
   generally, 262–63
   and roots, symbiosis between, 263–64, 266
   rodents and, 330
fuzzy truffle, 284–85
fuzzy-foot morel, 277
fuzzytongue penstemon, 183

G
gabbro, 531–32
Gaillardia aristata, 161
Galerina marginata, 267
Galium boreale, 235
Galium triflorum, 235
Gallinago delicata, 381
galls, 71, 456
Gambel oak, 85–86
garter snakes, 426–27
Gaultheria humifusa, 120
Gavia immer, 382–83
Geiser, Linda, 287
gelifluction, 20
gemmed puffball, 283
gentian, 236–37
Gentiana affinis, 237
Gentiana algida, 237
Gentiana calycosa, 237
Gentiana prostrata, 237
Gentianopsis thermalis, 236
geologic time (fig.), 504
geology of Rocky Mountains, 496–538
Geopora cooperi, 284–85
Geopyxis carbonaria, 282
Geothlypis trichas, 422
geranium, 210
Geranium richardsonii, 210
Geranium viscosissimum, 210
Gerard, John, 206
Geum rossii, 216
Geum triflorum, 215–16
Geyer, Karl Andreas, 84, 142, 220, 297
Geyer willow, 84
Geyer's onion, 143
giant ice age lakes, 520–21
giant wildrye, 127
giardia, 494–95
Gillette's checkerspot, 484
ginger, wild, 241, 242

Glacial Lake Bonneville, 521
glacier lily, 140
Glacier National Park, 30, 31, 531, 533
Glaucidium gnoma, 393–94
Glaucomys sabrinus, 320–21
globeflower, 240
globemallow, 221
glossary, 543–51
gneiss, 536–37
Gnophaela vermiculata, 473
gold cobblestone lichen, 289
golden banner, 190
golden buprestid, 461
golden currant, 104
golden draba, 234
golden eagle, 385–86
golden smoke, 191
golden-crowned kinglet, 409–10
goldeneye, 370
golden-mantled ground squirrel, 315–16
goldenrod, 164–65
goldenweed, 164
goldspeck lichen, 289–90
Gondwana, 523
Goodyera oblongifolia, 150–51
gooseberry, 103–05
goosefoot violet, 191
gopher snake, 429
gopher teeth, 315
Gordon, Alexander, 142
Gorham, John, 175
Goward, Trevor, 299
grand fir, 49, 53, 68–70
granite, 529–30
grasses, 126
grasshopper, 465–67
grasslike herbs, 125–37. *See also specific species*
grass-of-Parnassus, 206
Gravity Recovery and Climate Experiment (GRACE) satellites, 45
Gray, Asa, 217
gray aster, 158
gray reindeer lichen, 302
gray thistle, 169
gray wolf, 336–38
gray-crowned rosy-finch, 416
Grayia spinosa, 88
grayleaf willow, 84
greasewood, 88
Great Basin bristlecone, 60
Great Basin gopher snake, 429

great blue heron, 383
great gray owl, 391–92
great horned owl, 391
great spangled fritillary, 486
greater fritillary, 485–86
grebes, 374–75
green alder, 86
green corn lily, 148
green rock posy, 296
green starburst lichen, 296
greenhouse gases, 42–43
greenish pyrola, 213
green-winged teal, 368
Grey, Zane, 113
Grimmia montana, 255
grizzled rock tripe, 294
grizzly bear, 59, 340–42
ground-cedar, 250
groundsel, 159–60
grouseberry, 110
Guepiniopsis alpina, 281
guide, how to use this, 10–17
Gulo gulo, 348–49
Gunnison, John (lieutenaut), 217
Gunnison's mariposa lily, 140–41
Gymnocarpium dryopteris, 248
Gymnorhinus cyanocephalus, 403–04
gyrfalcon, 398
Gyromitra montana, 278

H
Haemorhous cassinii, 417
haircap mosses, 254
hairy arnica, 159
hairy clematis, 232
hairy golden-aster, 161–62
hairy woodpecker, 395
Haliaeetus leucocephalus, 386–87
Hall, Embere, 311
Hall's penstemon, 182
hardhack, 100
harlequin duck, 368–69
Harrington, H. D., 173
harsh paintbrush, 178
Hart, Miranda, 264
hawks, 384, 385–90
hawksbeard, 172
hawkweed, 171–72
hawthorn, 98–99
Hayden, Ferdinand Vandiveer, 217
Hayden's alpine clover, 189
Hayden's aster, 158
heartleaf arnica, 159

heartleaf bittercress, 233
heartleaf pyrola, 213
heartleaf twayblade, 151
heather. *See* mountain-heather
heather vole, 330
hedgehog fly, 454–55
hedgehog mushroom, 273
Hedwigia ciliata, 258
Hedysarum boreale, 187
Hedysarum occidentale, 187
Hedysarum sulphurescens, 187
Heinrich, Bernd, 402
Helianthella quinquenervis, 162–63
Helianthella uniflora, 162–63
Heliomeris multiflora, 162
hemlock woolly adelgid, 457
Heracleum maximum, 228–29
Hericium abietis, 283
hermit thrush, 414
herons, 383
Herpotrichia nigra, 66
Herrickia, 178
Hesperostipa comata, 126–27
Heterotheca villosa, 161–62
Heuchera cylindrica, 209
hibernation and torpor, 318–19
Hieracium albiflorum, 172
Hieracium aurantiacum, 172
Hieracium scouleri, 172
Hieracium triste, 172
Hines, Harvey Kimball, 51
Hippodamia convergens, 460–61
Histrionicus histrionicus, 368–69
Hitchcock, Dr. Leo, 17, 106
hoar moss, 258
hoary balsamroot, 163
hoary comma, 484
hoary marmot, 318
hollow suillus, 275
Holm's Rocky Mountain sedge, 132
Holodiscus discolor, 101–02
honey mushroom, 268–69
honeybees, 458
honeysuckle. *See* Utah honeysuckle
Hood, Robert, 447
hooded sunburst lichen, 293–94
hooded tube lichen, 300
Hooker, Sir Joseph, 122
Hooker, William Jackson, 65, 122, 142
Hooker's fairy bells, 138
Hooker's onion, 143

hookspur violet, 192
hopsage, 88
Hordeum jubatum, 126
horsehair lichen, 298–99
horned lark, 407–08
hornfels, 537
horse fly, 454
horsetails, 251–52
hover fly, 455–56
huckleberry and blueberry, 109–10
Hudsonian whiteface, 468
Hulsea algida, 162
human time (fig.), 505
hummingbird, 377–78
Hybomitra spp., 454–55
hydaspe fritillary, 485
Hydnum repandum, 273
Hydrophyllum capitatum, 199
Hydrophyllum fendleri, 199
Hygrophorus pudorinus, 268
Hygrophorus subalpinus, 268
Hyles gallii, 474
Hyles lineata, 474–75
Hylocomium splendens, 259
Hymenoloma crispulum, 256
Hymenoxys grandiflora, 162
Hypaurotis crysalus, 491
Hypericum perforatum, 214
Hypericum scouleri, 214
Hypnum revolutum, 260
Hypogymnia imshaugii, 300
Hypogymnia occidentalis, 300
Hypogymnia physodes, 300
Hypomyces lactifluorum, 270–71

I
ice ages, 517–20
Iceland-moss, 300–01
Icmadophila ericetorum, 290
Icterus bullockii, 420–21
Ictidomys tridecemlineatus, 316
Idaho batholiths, 25–28
Idaho fescue, 130
igneous rocks
   intrusive, 528–32
   volcanic, 532–34
Iliamna rivularis, 221
Indian paint fungus, 73
Indian pipe, 154
inky cap, 270
insecticides, 459
insects, 449–91. *See also specific species*
   in decline, 450–51

how not to get bit, 452
International Space Station, 293
intrusive igneous rocks, 528–32
invasive species, 44, 46
Ipomopsis aggregata, 194–95
Iris missouriensis, 149
isostasy, 499
Ives, Joseph Christmas, 217
ivesia, 216
Ivesia gordonii, 216
Ixoreus naevius, 413–14

J
Jackson Hole, 26. 27
Jacob's ladder, 194
James, David, 449
James, Edwin, 95, 202
Jamesia americana, 95
Jefferson, Thomas, 453
Jökulhlaups, 521
Junco hyemalis, 419–20
Juncus balticus, 137
Juncus drummondii, 136
Juncus mertensianus, 137
Juncus parryi, 136–37
juniper, 76–77
juniper haircap moss, 254
Juniperus communis, 78
Juniperus horizontalis, 77–78
Juniperus monosperma, 76–77
Juniperus occidentalis, 76–77
Juniperus osteosperma, 76–77
Juniperus scopulorum, 76–77

K
Kalmia microphylla, 119
Ketcham, Rebecca, 51
killdeer, 381
King, Clarence (captain), 217
king boletus, 274
kingfishers, 394
kinglet, 409–10
king's crown, 210
kinnickinnick, 119–20
kit fox, 334
Klironomos, John, 264
knight's-plume, 259
kokanee salmon, 444–45
krummholz, 55, 63, 66, 68, 78

L
lace fern, 245
ladder lichen, 303
ladies' tresses, 151
lady fern, 249

ladybug, 460–61
lady's slipper, 152
Laetiporus conifericola, 280–81
Lagapus leucura, 372–73
lake trout, 443
Lambert's locoweed, 185
lambstongue groundsel, 160
Lampropeltis gentilis, 428
lanceleaf stonecrop, 211
Lanius ludovicianus, 400–01
Lanzwert's sweetpea, 190
Laramide Mountains, 506–07, 509
larch trees, 47, 49, 51
Larix lyallii, 62–63
Larix occidentalis, 55, 61–62
larkspur, 193
Larus californicus, 382
Larus delawarensis, 382
Lashchinsky, Nikolai, 80
Lathyrus lanszwertii, 190
lattice tube lichen, 300
Lazuli bunting, 423
leafy aster, 157
leafy spurge, 173–74
least chipmunk, 314
Lecidea atrobrunnea, 289
Lemmiscus curtatus, 330–31
lemon sagewort, 170
lenticular clouds, 33, 35
leopard lily, 144–45
Lepraria neglecta, 291
Leptosiphon nuttallii, 195
Lepus americanus, 313
Lepus townsendii, 311–12
Lesica, Peter, 14
Lestes dryas, 470
Letharia columbiana, 301
Letharia lupina, 301
Leucocrinum montanum, 145
Leucopoa kingii, 129
Leucorrhinia hudsonica, 468
Leucosticte atrata, 416
Leucosticte australis, 416
Leucosticte tephrocotis, 416
Lewis, Meriwether, 64, 88, 96, 97, 102, 146, 222, 241, 242, 243, 423
Lewisia pygmaea, 242
Lewisia rediviva, 242–43
Lewis's mock-orange, 96
Leymus cinereus, 127
lichens, 253, 286–304
   generally, 286–88
   vagrant, 295

Lichens of North America (Brodo and Sharnoff), 15
Life Histories of Cascadia Butterflies (James and Nunnallee), 449
lightning and thunder, 36–37
Ligusticum filicinum, 227
Ligusticum porteri, 227
lilac-bordered copper, 487
Lilium philadelphicum, 144
limber pine, 25, 31, 55, 59–60
Limenitis lorquini, 482
Limenitis weidemeyerii, 482
limestone, 522–24
limestone columbine, 220
Linaria dalmatica, 180–81
Linnaea borealis, 124
Linnaean binomials, 12–13
Linnaeus, Carl, 13, 123, 124, 231, 383, 476
Linum lewisii, 205
lip fern, 245–46
Lithobates pipiens, 434–35
Lithophragma glabrum, 209–10
Lithospermum incisum, 202
Lithospermum ruderale, 202
lithosphere, 496
little sunflower, 162–63
littleleaf angelica, 229
littleleaf cinquefoil, 218
littleleaf miner's-lettuce, 222
littleleaf mock-orange, 96
littleleaf pussytoes, 167
Listera, 151
liverworts, 260–61
lizards, 424–26
Lloydia serotina, 144
Lobaria pulmonaria, 292–93
lobster mushroom, 271
Lockwood, Jeffrey, 466
locoweed, 185–86
locusts, 466
lodgepole pine, 20, 25, 47, 48, 54–55
loggerhead shrike, 400–01
Lomatium cous, 225
Lomatium dissectum, 225
Lomatium sandbergii, 225
Lomatium triternatum, 225
Long, Stephen, 95
long-beak willow, 84
longleaf phlox, 196
longnose dace, 447
longnose sucker, 446–47
long-tailed vole, 328

long-tailed weasel, 344
long-toed salamander, 432–33
Lonicera involucrata, 91
Lonicera utahensis, 91
Lontra canadensis, 345–46
loon, 367, 382–83
Lorquin's admiral, 482
lousewort, 176–77
lovage, 226–27
lovely paintbrush, 178
low daisy, 157
low huckleberry, 109
low larkspur, 193
Loxia curvirostra, 417–18
lungwort, 292–93
lupine, 187–89
lupine blue, 488
Lupinus argenteus, 187–88
Lupinus lepidus, 188
Lupinus polyphyllus, 188
Lupinus sericeus, 188
Lütke, Fyodor (count), 120
Luzula parviflora, 135
Luzula spicata, 135
Lyall, David, 62
Lyall's goldenweed, 164
Lyall's phacelia, 200–01
Lyall's rockcress, 233
Lycaena helloides, 487
Lycaena mariposa, 487–88
Lycaena nivalis, 487
Lycoperdon perlatum, 283
Lycopodium clavatum, 251
lynx, 350
Lynx canadensis, 350
Lynx rufus, 350–51
Lysichiton americanus, 137–38

M
MacGillvray's warbler, 422
Mackenzie, Alexander, 241
magnetoreception, 379
Mahonia repens, 117–18
Maianthemum racemosum, 139
Maianthemum stellatum, 139
Malacosoma californica, 472–73
male fern, 248
mallard, 367
mallow ninebark, 101
mammals, 305–65. *See also specific species*
Manual of Montana Vascular Plants (Lesica), 14
Manual of the Flowering Plants of California (Jepson), 17

map lichen, 288–89
maple. *See* Rocky Mountain maple
maps
  Rocky Mountain geology, 498
  Rocky Mountains ecoregions, 6
marble, 537–38
Marchantia polymorpha, 260
margined white, 477–78
mariposa copper, 487–88
mariposa lily, 140–42
marmot, 317–20, 323
Marmota caligata, 317
Marmota flaviventris, 317–18
marshmarigold, 239
Martes caurina, 346–47
Martin, Jon, 320
masked shrew, 506
mat rock-spiraea, 123–24
Mathews, Daniel, 575
matted buckwheat, 238
Maximilian, Prince of Wied, 90
mayfly, 467
McIlvaine, Charles (captain) 265
McLoughlin, John, 142
meadow death camas, 146–47
meadow vole, 328–29
mealy pixie goblet, 303
Mech, David, 337
Megaceryle alcyon, 394
megafloods, 520–21
Melanophila acuminata, 465
Melanoplus alpinus, 465–66
Melanoplus borealis, 465
Melanoplus bruneri, 466
Melilotis albus, 190
Melilotis officinalis, 189–90
Melospiza melodia, 419
membranous dog lichen, 291–92
Mentzelia laevicaulis, 215
Mentzelia multiflora, 215
Menyanthes trifoliata, 203
Menzies, Archibald, 64, 110, 111
Mephitis mephitis, 342
Mergus merganser, 371–72
Mertens, Franz Karl, 120
Mertens, Karl Heinrich, 120
Mertens rush, 137
Mertens sedge, 134
Mertensia bakeri, 201
Mertensia ciliata, 201
Mertensia paniculata, 201
Mertensia tweedyi, 201
mesoclimate, 39–40
metamorphic rocks, 534–38

Mexican jay, 56
Micranthes ferruginea, 207
Micranthes lyallii, 207
Micranthes occidentalis, 207
Micranthes odontoloma, 207
Micranthes oregana, 207–08
Micranthes rhomboidia, 208
microclimate, 40–41
Micropterus dolomieu, 446
Microtus drummondii, 328
Microtus longicaudus, 328
Microtus montanus, 328
Microtus richardsoni, 329
Middle Rockies, 24–25
midget faded rattlesnake, 430
migmatite, 538
milksnake, 428
miner's candle, 203
minnow, 447–48
mint-leaf bee balm, 185
Missouri goldenrod, 164–65
Mitella breweri, 206
Mitella nuda, 207
Mitella pentandra, 206
Mitella stauropetala, 206
Mitella trifida, 206
mitrewort, 206–07
mock-orange, 96
Monarda fistulosa, 185
Moneses uniflora, 214
monkeyflower, 179–80
Monochamus scutellatus, 464–65
Monotropa hypopitys, 154–55
Monotropa uniflora, 154
monkshood, 191
Monsoon, North American, 37
Montana overthrust belt, 30–31
montane vole, 328
Montia chamissoi, 223
monument plant, 236
Moore, Michael, 223
moose, 358–60
mooseberry, 93
Morchella americana, 277
Morchella elata, 276
Morchella populiphila, 277
Morchella tomentosa, 277
morel, 276–78
Mormon fritillary, 485
mosquito, 449–53
moss gentian, 237
moss-campion, 211
mosses, 253–60
Moss's elfin, 489
moths. *See also specific species*

and butterflies, 471–91
mountain alder, 86
mountain bluebird, 415
mountain bog gentian, 237
mountain brome, 131
mountain chickadee, 407
mountain cottontail, 311
mountain death camas, 147
mountain emerald, 468
mountain goat, 363–64
mountain hemlock, 72–73
mountain hollyhock, 221
mountain kittentails, 232
mountain lady's-slipper, 152
mountain monkeyflower, 179–80
mountain parnassian, 477
mountain pine beetle, 53, 55,
    462, 463
mountain sweetroot, 228–29
mountain whitefish, 445–46
mountain-ash, 98
mountain-dandelion, 171
mountain-heather, 120–22
mountain-sorrel, 191
The Mountaineers' suggestions
    for carbon-footprint
    reduction, 23
mourning cloak, 483
mourning dove, 375
mudstone, 528
mule deer, 354–56
mule's ears, 163–64
mullein, woolly, 184
Munro's globemallow, 221
mushroom worker's lung, 272
mushrooms, poisonous, 265
muskrat, 331–32
Mustela richardsoni, 344
muttongrass, 130
Myadestes townsendi, 415
mycorrhizal symbiosis, 154, 322
Myosotis asiatica, 202
Myotis ciliolabrum, 307–08
Myriopteris fendleri, 246
Myriopteris gracilis, 245–46
Myriopteris gracillima, 245
Myxobolus cerebralis, 440

N

naked broomrape, 155
naked mitrewort, 207
names, common and scientific,
    10–11, 12–17
Napoleon, 453
narcissus anemone, 240

narrowleaf cottonwood, 81–82
narrowleaf desert-parsley, 225
needle-and-thread, 126–27
Neogale frenata, 344
Neogale vison, 343–44
Neophasia menapia, 477
Neotamias amoenus, 314
Neotamias minimus, 314
Neotamias quadrivittatus, 314
Neotamias umbrinus, 314
Neotoma cinerea, 326–27
Neottia banksiana, 151–52
Neottia borealis, 151
Neottia convallarioides, 151
Neottia cordata, 151
netskin onion, 143
netted rock tripe, 294
nettle-leaf giant hyssop, 184
New Manual of Botany (Gray), 17
New Mexico checkermallow, 221
New World vultures, 383–85
Newberry, John Strong, 217
nightjars, 375–76
nobo bluet, 471
Noccaea fendleri, 233
nodding onion, 142
Nootka rose, 101
North American Monsoon, 37
northern bedstraw, 235
northern black currant, 104
northern blue, 488
northern bog violet, 192
northern flying squirrel, 320–21
northern goshawk, 388
northern harrier, 387–88
northern holly fern, 245
northern leopard frog, 434–35
northern pintail, 367
northern pocket gopher, 323–24
northern pygmy owl, 393–94
northern red bank conk, 278–79
northern red-shafted flicker, 397
northern river otter, 345–46
northern rubber boa, 427
northern saw-whet owl, 393
northern spur-throat
    grasshopper, 465
northern sweetgrass, 129
northern twayblade, 151
northwestern twayblade, 151–52
Norvell, Lorelei, 269
Norwood, William, 148
no-see-um, 453
Nucifraga columbiana, 402–03
Nunnallee, David, 449

Nuphar polysepala, 239
nuthatch, 411–12
Nuttall, Thomas, 192, 312
Nuttall's larkspur, 193
Nuttall's linanthus, 195
Nymphalis antiopa, 483
Nymphalis californica, 483

O
oak fern, 248
obsidian, 534
ocean-spray, 101–02
Ochotona princeps, 309–11
Odocoileus hemionus, 354–56
Odocoileus virginianus, 352–54
Oeder's lousewort, 177
Oeneis chryxus, 480–81
Oenothera caespitosa, 230
Oenothera elata, 231
old-man's-beard, 297–98
olive-sided flycatcher, 399–400
On the Origin of Species
    (Darwin), 13
Oncorhynchus clarkii, 438–40
Oncorhynchus mykiss, 440–41
Oncorhyncus nerka, 444–45
Ondatra zibethicus, 331–32
one-flower daisy, 157
one-flower little sunflower,
    162–63
one-flowered wintergreen, 214
one-leaf bog orchid, 150
one-sided pyrola, 213
Opheodrys vernalis, 428–29
Oplopanax horridus, 106–07
Oporornis talmiei, 422
Opuntia fragilis, 118
Opuntia polyacantha, 118
orange hawkweed, 172
orange sulphur, 478–79
orange-peel fungus, 281–82
orange-tip, 479–80
Oreamnos americanus, 363–64
Oregon woodsia, 247
Oregon-boxwood, 116
oreocarya, 203
Oreocarya glomerata, 203
Oreocarya virgata, 203
Oreostemma alpigenum var.
    haydenii, 158
Orthilia secunda, 213
Orthocarpus luteus, 179
oshá, 227
Osmorhiza berteroi, 227–28
Osmorhiza depauperata, 228

Osmorhiza occidentalis, 228
osprey, 385
Ötzi, the Bronze Age man, 279
oval-leaf huckleberry, 109
Ovis canadensis, 360–61
owls, 384, 390–94
Oxyria digyna, 191
Oxytropis besseyi, 186
Oxytropis campestris, 186
Oxytropis lambertii, 185
Oxytropis sericea, 186
Oxytropis splendens, 185
oyster mushroom, 271–72
Ozomelis, 205

P
Pacific marten, 346–47
Pacific rattlesnake, 429
Pacific treefrog, 433
Pacific trillium, 140
Pacific wren, 412–13
Packera cana, 161
Packera pseudaurea, 160
Packera streptanthifolia, 161
Packera subnuda, 161
paclitaxel, 74
paddle-tailed darner, 468
paintbrush, 177–79
painted lady, 481–82
painted turtle, 430
pale tiger swallowtail, 475
palm-tree moss, 259–60
Pandion haliaetus, 385
Pando grove, 80
pandora pine moth, 474
Pangaea, 503, 528
paper birch, 87
Papilio eurymedon, 475
Papilio multicaudata, 475
Papilio polyxenes, 476
Papilio rutulus, 475
Papilio zelicaon, 475–76
Parmeliopsis ambigua, 296
Parnassia fimbriata, 206
parnassian, 476–77
Parnassius clodius, 477
Parnassius smintheus, 477
Parry, Charles C., 212
Parry's alpine clover, 189
Parry's primrose, 198
Parry's rush, 136–37
Parry's townsendia, 159
Passerina amoena, 423
paw print identification, 349
paw-wort, 261

Paxistima myrsinites, 116
pear puffball, 283–84
pearly everlasting, 168
pearly pussytoes, 167
peat moss, 256–58
Pectiantia, 205
Pedicularis bracteosa, 176
Pedicularis contorta, 176
Pedicularis cystopteridifolia, 176
Pedicularis groenlandica, 176
Pedicularis oederi, 176
Pedicularis racemosa, 176
Pedicularis sudetica, 176
pegmatite, 530
Pekania pennanti, 347
Pellaea breweri, 246
Pellaea glabella, 247
Peltigera aphthosa, 292
Peltigera canina, 291
Peltigera leucophlebia, 292
Peltigera membranacea, 291–92
Peltigera rufescens, 291–92
Penicillium, 279
penstemon, 181–83
Penstemon albertinus, 181
Penstemon barbatus, 183
Penstemon confertus, 183
Penstemon cyaneus, 181
Penstemon eatonii, 183
Penstemon ellipticus, 183
Penstemon eriantherus, 183
Penstemon fruticosus, 182
Penstemon hallii, 182
Penstemon nitidus, 181
Penstemon procerus, 181–82
Penstemon rydbergii, 182
Penstemon strictus, 182
Penstemon whippleanus, 182
perching birds, 399–423
peregrine falcon, 397–99
perforated rock tripe, 294
Perideridia gairdneri, 227
peridotite, 530–31
Perisoreus canadensis, 404–05
permafrost, 20
Peromyscus maniculatus, 327
Peromyscus truei, 327–28
Petrophytum caespitosum,
    123–24
Petrochelidon pyrrhonota, 408
phacelia, 200–01
Phacelia hastata, 200
Phacelia linearis, 201
Phacelia lyallii, 200–01
Phacelia sericea, 200

Phaeolus schweinitzii, 280
Phellinus tremulae, 279–80
Phenacomys intermedius, 330
pheromones, deer, 353
Philadelphus lewisii, 96
Philadelphus microphyllus, 96
Phleum alpinum, 128
phlox, 195–96
Phlox diffusa, 196
Phlox hoodii, 196
Phlox longifolia, 196
Phlox multflora, 196
Phlox pulvinata, 195–96
Phyciodes pulchella, 485
phyllite, 535
Phyllocnistis populiella, 472
Phyllodoce empetriformis, 121
Phyllodoce glanduliflora, 121
Phylloxera bug, 72
Physaria acutifolia, 234
Physaria didymocarpa, 234
Physaria floribunda, 234
Physocarpus malvaceus, 101
Phytophthora, 74
Pica hudsonia, 405–07
Picea engelmannii, 70–72
Picea glauca, 71–72
Picea pungens, 70–71
Picoides arcticus, 395
Picoides dorsalis, 396
pied-billed grebe, 374–75
Pieris marginalis, 477–78
pigeons, 375
pika, 309–11
Pike, Zebulon (captain), 95
pileated woodpecker, 396–97
Pinchot, Gifford, 48
pine beetles, 57
pine siskin, 418
pine trees, 47, 49
pine white, 477
pinedrops, 153–54
pinegrass, 129
pinesap, 154–55
pink mountain-heather, 121
piñon-juniper woodland, 20, 22
piñon mouse, 327–28
piñon pine (two-leaf), 55–56
Pinus albicaulis, 57–59
Pinus aristata, 60–61
Pinus contorta, 54–55
Pinus edulis, 55–56
Pinus flexilis, 59–60
Pinus longaeva, 60
Pinus monophylla, 56

Pinus monticola, 55, 56–57
Pinus ponderosa, 47, 48, 49,
    51–52, 54
Pinus strobiformis, 60
Pinus strobus, 56–57
pinyon jay, 56, 403–04
pipecleaner moss, 258
pipsissewa, 119
Piranga ludoviciana, 423
pitch (resin), 50
pitted beard lichen, 298
Pituophis catenifer ssp.
    deserticola, 429
Pituophis catenifer ssp. sayi, 429
pixie goblet, 303
plains cottonwood, 82
plains prickly pear, 118
plant and climate feedback,
    43–44
plantainleaf buttercup, 219
Platanthera dilatata, 149–50
Platanthera obtusata, 150
Platanthera orbiculata, 150
Platanthera stricta, 150
plate tectonics, 496–99
plateau fence lizard, 425
plateau striped whiptail, 426
Platismatia glauca, 293
pleated gentian, 237
Plebejus glandon, 488–89
Plebejus icarioides, 488
Plebejus idas, 488
Pleopsidium flavum, 289
Plethodon idahoensis, 433
Pleurotus populinus, 271
Pleurotus pulmonarius, 271
Pleurozium schreberi, 259
plutons, 499
Poa alpina, 129–30
Poa fendleriana, 130
Podilymbus podiceps, 374–75
Poecile atricapilla, 407
Poecile gambeli, 407
pointed-tip cat's ears, 141
point-tip twinpod, 234
poison hemlock, 228
poisonous mushrooms, 265
Polemonium foliosissimum, 194
Polemonium pulcherrimum, 194
Polemonium viscosum, 194
police car moth, 473
Polygonia gracilis, 484
polymerase chain reactions
    (PCRs), 492
Polyphylla decemlineata, 461

Polypodium hesperium, 244–45
Polystichum lonchitis, 245
Polystichum munitum, 245
Polytrichastrum alpinum, 254
Polytrichum commune, 254
Polytrichum juniperinum, 254
Polytrichum piliferum, 254
ponderosa pine, 20, 55
ponderous borer, 461–62
Pontia occidentalis, 478
Populus angustifolia, 81–82
Populus deltoides, 82
Populus tremuloides, 79–81
Populus trichocarpa, 81–82
porcupines, 315, 332–33
Porzana carolina, 381
post-orogenic stretching, 511–12
Potentilla brevifolia, 218
Potentilla concinna, 216, 218
Potentilla glaucophylla, 218
Potentilla gracilis, 218
Potentilla nivea, 218
powdered beard lichen, 297
powderhorn lichen, 303
prairie coneflower, 165–66
prairie crocus, 239
prairie dog, 323
prairie falcon, 399
prairie lizard, 425
prairie rattlesnake, 429
prairie smoke, 215–16
prickly pear, 118
prickly rose, 101
primrose, 198
primrose monkeyflower, 180
Primula angustifolia, 198
Primula conjugens, 198–99
Primula parryi, 198
Primula pauciflora, 198–99
pronghorn, 364–65
pronouncing scientific Latin
    names, 16–17
Prosartes hookeri, 138
Prosartes trachycarpa, 138
Prosopium williamsoni, 445–46
Prunella vulgaris, 185
Prunus virginiana, 97
Pseudacris maculata, 433
Pseudacris regilla, 433
Pseudoroegneria spicata, 127
Pseudotsuga menziesii, 53, 63–65
Pteridium aquilinum, 249
Pterospora andromedea, 153–54
Pterourus, 475–476
Ptilium crista-castrensis, 259

puccoon, 202
puffball, 283–84
Pulsatilla, 241–42
Puma concolor, 351–52
purple hairgrass, 130
purple monkeyflower, 179
purple reedgrass, 129
purple salsify, 173
purple saxifrage, 208–09
purplish copper, 487
Pursh, Friedrich, 102, 243
Purshia stansburyana, 103
Purshia tridentata, 102–03
pussypaws, 232
pussytoes, 167
pygmy bitterroot, 242
pygmy goldenweed, 164
pygmy nuthatch, 411
pygmy shrew, 506
pygmyflower, 197
pyrodiversity, 47
pyrola, 213–14
Pyrola asarifolia, 213
Pyrola chlorantha, 213
Pyrola picta, 213
Pyrrocoma crocea, 164

Q

quaking aspen, 20, 48, 79–81
quartz, 524–26
quartzite, 535–36
Queen Alexandra sulphur, 479
queen's crown, 210–11
Quercus gambelii, 85–86

R

rabbits, 308–13
racer, 427–28
Racomitrium canescens, 255
Rafinesque, Constantine, 165–66
ragbag lichen, 293
ragged robin, 231
rain, 33–34
rainbow chanterelle, 272
rainbow trout, 440–41
Ramaria rasilospora, 282
Rana catesbeiana, 435
Rana luteiventris, 434
Ranunculus adoneus, 219
Ranunculus alismifolius, 219
Ranunculus eschscholtzii, 219
Ranunculus glaberrimus, 219
raptors reconfigured, 384
raspberry. *See* boulder raspberry, wild raspberry

Ratibida columnifera, 165
rattlesnake, 429–30
rattlesnake-plantain, 150–51
raven, 401–02
Raynolds, William F., 217
Raynolds sedge, 134
reading, recommended, 553–58
red columbine, 220
red crossbill, 417–18
red elderberry, 92
red fox, 333–34
red gooseberry, 103
red hawthorne, 98–99
red osier dogwood, 90–91
red squirrel, 59, 321–23
red-backed vole, 329–30
red-breasted nuthatch, 411
red-crowned sulphur cup, 304
red-napped sapsucker, 397
redside shiner, 447
redstem, 107–08
redstem saxifrage, 207
red-tailed hawk, 389–90
redtop, 129
red-winged blackbird, 420
Regulus satrapa, 409–10
reindeer lichen, 302
reptiles, 424–30
resin birch, 87
resin (pitch), 50
Rhinichthys cataractae, 447–48
Rhionaeschna multicolor, 469
Rhizocarpon geographicum, 288–89
Rhizomnium magnifolium, 256
Rhizoplaca melanophthalma, 296
Rhodiola integrifolium, 210
Rhodiola rhodanthum, 210–11
rhododendron. see white rhododendron
Rhododendron albiflorum, 111–12
Rhododendron columbianum, 114–15
Rhododendron menziesii, 110–11
Rhus aromatica, 105
Rhus glabra, 105
rhyolite, 530, 533–34
Rhytidiadelphus triquetrus, 258–59
Rhytidiopsis robusta, 258
Ribes aureum, 104
Ribes cereum, 104
Ribes hudsonianum, 104
Ribes inerme, 104
Ribes lacustre, 103

Ribes laxiflorum, 104
Ribes montigenum, 103
Ribes oxyacanthoides, 103–04
Ribes viscosissimum, 104
Ribes wolfii, 104
Richardonius balteatus, 447–48
Richardson, John (sir), 136, 447
Richardson's geranium, 210
Richardson's ground squirrel, 317
ring-billed gull, 382
ringtail, 338
Rio Grande rift, 512, 514
river beauty, 230
river hawthorne, 98–99
Robinia neomexicana, 114
rock clematis, 94
rock pigeon, 375
rock tripe, 294–95
rock willow, 84
rock wren, 413
rockjasmine, 197
rocks. *See also specific rock, mineral*
   intrusive igneous, 528–32
   metamorphic, 534–38
   sedimentary, 521–28
   volcanic igneous, 532–34
rockworm, 304
rocky ledge penstemon, 183
Rocky Mountain bristlecone pine, 60–61
Rocky Mountain butterweed, 161
Rocky Mountain columbine, 220
Rocky Mountain dotted blue, 489
Rocky Mountain Douglas-fir, 64
Rocky Mountain elderberry, 93
Rocky Mountain fringed gentian, 236
Rocky Mountain goldenrod, 164–65
Rocky Mountain iris, 149
Rocky Mountain lousewort, 176
Rocky Mountain maple, 89
Rocky Mountain penstemon, 182
Rocky Mountain phlox, 196
Rocky Mountain sagewort, 170
Rocky Mountain spotted fever, 495
Rocky Mountain tailed frog, 435
Rocky Mountain wood tick, 495
Rocky Mountain woodsia, 248
Rocky Mountains
   Blue Mountains, 28–29
   climate, 39–46
   Columbia Mountains, 29–30

economic inequity, 26–27
fire, 46–49
geology, 496–538
Idaho batholiths, 25–28
map, 6
Middle Rockies, 24–25
Montana overthrust belt, 30–31
rise again, 516–17
Southern Rockies, 18–21
Utah Rockies, 22, 24
in a warmer future, 44–46
weather, 32–39
rodents, 313–33. *See also specific species*
and fungi, 330
rolled-leaf moss, 260
Romanzoffia sitchensis, 200
roots and fungi, symbiosis between, 263–64, 266
Rosa acicularis, 101
Rosa gymnocarpa, 101
Rosa nutkana, 101
Rosa woodsii, 100
Rosentreter, Roger, 295
Ross's sedge, 133
rosy paintbrush, 178
rosy pussytoes, 167
rosy waxy cap, 268
rosy-finch, 416
rotor clouds, 35
rough-fruit fairy bells, 138
rough-scouring-rush, 252
roundleaf alumroot, 209
round-leaf bog orchid, 150
rubber rabbitbrush, 113
Rubel, William, 266
Rubus deliciosus, 99
Rubus idaeus, 99
Rubus parviflorus, 99
ruby boletus, 274
ruby-crowned kinglet, 410
Rudbeckia laciniata, 166
Rudbeckia occidentalis, 166
ruffed grouse, 373–74
ruffled freckle pelt, 292
Rumex densiflorus, 238
Rumiantzev, Nikolai, 200
running clubmoss, 251
Rusavskia elegans, 293
rush, 135–37
Russula brevipes, 270–71
rusty-hair saxifrage, 207
Rydberg's arnica, 159
Rydberg's penstemon, 182

S

sagebrush buttercup, 219
sagebrush checkerspot, 484
sagebrush lizard, 425
sagebrush sooty hairstreak, 490–91
sagebrush vole, 330–31
Sagittaria cuneata, 138
salamanders, 432–33
Salix barclayi, 84
Salix bebbiana, 84
Salix brachycarpa, 84–85
Salix cascadensis, 85
Salix exigua, 84
Salix geyeriana, 84
Salix glauca, 84
Salix lasiandra var. *caudata*, 83–84
Salix nivalis, 85
Salix petrophila, 85
Salix pseudomonticola, 84
Salix scouleriana, 83–85
Salix vestita, 84
Salmo trutta, 443–44
Salpinctes obsoletus, 413
salsify, 172–73
saltcedar, 108
Salvelinus confluentus, 441–42
Salvelinus fontinalis, 442–43
Salvelinus namaycush, 443
Sambucus cerulea, 92
Sambucus racemosa, 92, 93
sand lily, 145
Sandberg's biscuitroot, 225
sandhill crane, 379–80
sandpaper stippleback, 296
sandstone, 526
sandwort, 212–13
Sarcobatus vermiculatus, 88
Sarcodon imbricatus, 273–74
sarsaparilla, 205–06
Satyrium semiluna, 490–91
Saxifraga austromontana, 208
Saxifraga cespitosa, 208
Saxifraga flagellaris, 208
Saxifraga oppositifolia, 208–09
saxifrage
    mat-forming species, 208–09
    tall species, 207–08
Say, Thomas, 95
scarlet globemallow, 221
scarlet paintbrush, 178
Sceloporus graciosus, 425
Sceloporus tristichus, 425
schist, 536

scientific and common names, 10–11, 12–17
Sciurus aberti, 321
Scolytus ventralis, 463–64
scorched-earth cup, 282
Scouler willow, 83
Scouler's hawkweed, 172
scouring rush, 251–52
scrub jay, 56
sculpin, 448
sedge darner, 469
sedges, 125, 132
sedimentary rocks, 521–28
Sedum lanceolatum, 211
Sedum stenopetalum, 211
sego lily, 141
Selaginella densa, 250
Selasphorus calliope, 377
Selasphorus platycercus, 377
self-heal, 185
Senecio fremontii, 160
Senecio integerrimus, 160
Senecio megacephalus, 160
Senecio soldanella, 160
Senecio triangularis, 160
serviceberry, 96–97
serviceberry willow, 84
Seton, Ernest Thompson, 294–95
Setophaga coronata, 421
Setophaga petechia, 421–22
Setophaga townsendi, 422
Sevier Mountains, 503, 506
sex vs. survival, underground, 323
shaggy mane, 270
shale, 527–28
sharpleaf twinpod, 234
sharptooth angelica, 229
sheep fescue, 130
Shepherdia canadensis, 93–94
Sheridan's green hairstreak, 490
shingled hedgehog, 273–74
shoestring root rot, 268–69
shooting star, 198–99
shorebirds, 380–82
short-fruit willow, 84–85
short-stalk suillus, 276
short-stemmed russula, 270–71
short-styled onion, 142–43
short-tailed weasel, 344
showy aster, 158
showy fleabane, 156
showy goldeneye, 162
showy Jacob's ladder, 194
showy locoweed, 185

showy milkweed, 204–05
showy monocots, 137–53
showy sedge, 134
shrew, 506–07
shrubby cinquefoil, 100
shrubby penstemon, 182
shrubs. *See also specific species*
    broadleaf evergreen, 114–24
    flowering deciduous, 79–114
shy wallflower, 235
Sialia currucoides, 415
sibbaldia, 205
Sibbaldia procumbens, 205
sickletop lousewort, 176
Sidalcea neomexicana, 221
Silene acaulis, 211
Silene parryi, 211–12
silky lupine, 188
silky phacelia, 200
silverberry, 108–09
silverleaf phacelia, 200
silverweed, 218–19
silvery lupine, 187–88
Simard, Suzanne, 264
Simulium, 453–54
single goldenbush, 113
single-leaf piñon pines, 56
Sisyrinchium idahoense, 149
Sisyrinchium montanum, 149
Sitka mist maiden, 200
Sitka mountain-ash, 98
Sitka valerian, 223
Sitta canadensis, 411
Sitta carolinensis, 411
Sitta pygmaea, 411
skarn, 537
skunk, 342–43
skunk-cabbage, 137–38
sky pilot, 194
skyrocket, 194–95
slate, 534–35
slender bog orchid, 150
slender cinquefoil, 218
smallflower mitrewort, 206
small-flowered penstemon,
    181–82
small-flowered woodland star,
    209–10
small-flowered woodrush, 135
smallmouth bass, 446
smelowskia, 233
Smelowskia americana, 233
Smerinthus astarte, 475
Smokey Bear period, 49
smooth cliff-brake, 247

smooth green snake, 428–29
smooth sumac, 105
smooth-fruit sedge, 133
smooth-scouring-rush, 251–52
smoothstem blazingstar, 215
snakefly, 465
snakes, 426–30
snakewort, 260–61
snow buttercup, 219
snow cinquefoil, 218
snow lichen, 301
snow mold, 66
snowbank false-morel, 278
snowberry, 91–92
snowbrush, 115–16
snow-depth specialties, 196
snowline spring-parsley, 226
snowlover, 184
snowshoe hare, 313
soapweed yucca, 116–17
sockeye salmon, 444–45
Socrates, 228
Solidago lepida, 164
Solidago missouriensis, 164–65
Solidago multiradiata, 164–65
Solidago simplex, 165
Solorina crocea, 294
Somatochlora semicircularis, 468
song sparrow, 419
songbirds, 399–423. *See also*
    specific species
sora, 381
Sorbus scopulina, 98
Sorbus sitchensis, 98
Sorex cinereus, 506
Sorex hoyi, 506
Sorex monticola, 506
Sorex navigator, 506
Sorex vagrans, 506
sources, recommended reading,
    553–58
Southern Rockies, 18–21
southwestern red squirrel, 321
southwestern white pine, 60
sparrow, 419
speedwell, 231
Speyeria, 485–86
Sphaeralcea coccinea, 221
Sphaeralcea munroana, 221
sphagnum moss, 257
Sphyrapicus nuchalis, 397
spike fescue, 129
spike woodrush, 135
spikemosses, 250–51
Spilogale gracilis, 342

spineless horsebrush, 113–14
Spinulum annotinum, 250
Spinus pinus, 418
Spinus tristis, 418–19
spiny milkvetch, 186–87
spiny phlox, 196
spiny wood fern, 248
spiraea, 100
Spiraea douglasii, 100
Spiraea lucida, 100
Spiraea splendens, 100
Spiranthes romanzoffiana, 151
Spizella passerina, 419
spotted knapweed, 168
spotted sandpiper, 380–81
spotted saxifrage, 208
spraypaint lichen, 290
spreading dogbane, 204
spreading phlox, 196
Spribille, Toby, 299
springbeauty, 222–23
spring-parsley, 226
spruce, 70–72
spruce beetle, 72, 462
stairstep moss, 259
Stansbury, Captain Howard, 217
star moss, 256
starry false-Solomon's-seal, 139
steelhead trout, 440–41
steer's head, 190
Steller, Georg, 404–05, 444
Steller's jay, 56, 405
stemless townsendia, 159
Stereocaulon tomentosum, 302
Stewart, Sir William D., 142
sticky cinquefoil, 218
sticky currant, 104
sticky geranium, 210
sticky goldenrod, 165
stiff clubmoss, 250
stinging nettle, 174–75
stippleback, 295–96
St.-John's-wort, 214–15
stonecrop, 211
storm warnings, 33
strawberry, 215
streambank butterweed, 160
Streptopus amplexifolius, 139
striped ambrosia beetle, 464
striped coralroot, 153
striped skunk, 342
Strix nebulosa, 391–92
Strix varia, 392
Sturnella neglecta, 421
subalpine daisy, 156–57

subalpine fir, 20, 25, 47, 49, 65–67, 70
subalpine larch, 62–63
subalpine spiraea, 100
Suillus, 275–76
Suillus ampliporus, 275
Suillus brevipes, 276
Suillus lakei, 275
Suillus subalpinus, 276
Suillus tomentosus, 275–76
sulfur dust lichen, 291
sulphur cup, 303
sulphur (insect), 478–79
sulfur-flower, 237
sumac, 105–06
summer oyster mushroom, 271
supervolcanoes, 509–11
Swainson's hawk, 390
Swainson's thrush, 414
swallow, 408–09
swallowtail, 475–76
swamp gooseberry, 103
swan, 370–71
sweet rockjasmine, 197
sweet-clover, 189–90
sweetroot, 227–28
sweetvetch, 187
Swertia perennis, 205
swift, 376–77
swift fox, 334
sword fern, 245
syenite, 530
Sylvilagus nuttallii, 311
symbols and abbreviations, 16
Sympetrum internum, 469–70
Symphoricarpos albus, 92
Symphoricarpos occidentalis, 92
Symphoricarpos rotundifolius, 91–92
Symphyotrichum foliaceum, 157
synthyris kittentails, 231–32
Syntrichia ruralis, 256
Syrphidae, 456

T

Tabanus spp., 454
Tachycineta bicolor, 408
Tachycineta thalassina, 408
Taft, William (president), 48
tall bluebells, 201
tall cinquefoil, 218
tall evening-primrose, 231
tall fringed bluebells, 201
Tamarix chinensis, 108
Tamiasciurus fremonti, 321

Tamiasciurus hudsonicus, 321
Taraxacum officinale, 171
tarragon, 170
tasselflower, 168
Taxidea taxus, 349–50
Taxus brevifolia, 73-74
telesonix, 209
Telesonix heucheriformis, 209
ten-lined June beetle, 461
tent caterpillars, 97, 472–73
Tephroseris lindstroemii, 161
terranes, 503–04
Tetradymia canescens, 113–14
Tetraneuris acaulis, 162
Thalictrum occidentale, 232–33
Thamnolia subuliformis, 304
Thamnophis cyrtopsis, 427
Thamnophis elegans, 427
Thamnophis sirtalis, 427
Tharsalea, 487
thermals and cumulus clouds, 34–36
thermophiles, 492
Thermopsis rhombifolia, 190
Thermus aquaticus, 492
thimbleberry, 99
thirteen-lined ground squirrel, 316
thistle, 169–70
Thomomys talpoides, 323–24
threadleaf phacelia, 201
threadleaf sandwort, 212
three-toothed mitrewort, 206
Thuja plicata, 75
thunder and lightning, 36–37
Thymallus arcticus, 446
thymeleaf speedwell, 231
tiarella, 209
Tiarella trifoliata, 209
tick, 495
Tilford, Gregory, 165
timberline and tree line, 68–69
time
    geologic (fig.), 504
    human (fig.), 505
Timmia austriaca, 254–55
toadlily, 223
toads, 434
tobacco-root, 223
tofieldia, 139–40
Tonestus lyallii, 164
Tonestus pygmaeus, 164
torpor and hibernation, 318–19
towering Jacob's ladder, 194
towhead baby, 239–40

Townsend, John Kirk, 112, 312, 361
townsendia, 158–59
Townsendia exscapa, 159
Townsendia montana, 159
Townsendia parryi, 159
Townsend's solitaire, 415
Townsend's warbler, 422
Toxicodendron radicans, 106
Toxicoscordion venenosum, 146–47
Tragopogon dubius, 173
Tragopogon porrifolius, 173
trailing black currant, 104
trail-plant, 168–69
Trametes versicolor, 279
trapper's tea, 114–15
travertine, 524
tree ear, 281
tree line and timberline, 68–69
tree swallow, 408
trees. *See also specific species*
    and climate change, 53
    conifers, 50–78
    deciduous, 79–114
    broadleaf evergreens, 114–24
Tremella mesenterica, 281
Triantha occidentalis, 139–40
Trichocnemis spiculatus, 461–62
Tricholoma murrillianum, 269–70
Trifolium dasyphyllum, 189
Trifolium haydenii, 189
Trifolium nanum, 189
Trifolium parryi, 189
Trillium ovatum, 140
Trisetum spicatum, 129
Triteleia grandiflora, 145
Troglodytes pacificus, 412–13
Trollius albiflorus, 240
trout, 441–42
Trudell, Steve, 278
true bugs, 456–57
truffles, 265, 284–85, 329–30
trumpeter swan, 371
Trypodendron lineatum, 464
Tsuga heterophylla, 55, 73
Tsuga mertensiana, 72–73
tube lichen, 300
Tucker, Vance, 398
tuff, 533–34
tufted alpine saxifrage, 208
tufted evening-primrose, 230
tumbleweed rock-shield, 297
tundra, patterns on, 20–21

tundra swan, 371
Turbinellus floccosus, 273
Turdus migratorius, 414
turkey tail, 279
turkey vulture, 383–85
Turner, Nancy, 97, 239
turpentine biscuitroot, 226
turted hairgrass, 130
turtles, 430
twayblade, 151–52
twinflower, 124
twinpod, 234
twisted-stalk, 139
two-tailed swallowtail, 475
Typha latifolia, 131–32

U

Uinta chipmunk, 314
Uinta clover, 189
Uinta ground squirrel, 316
umber pussytoes, 167
Umbilicaria decussata, 294
Umbilicaria hyperborea, 294
Umbilicaria torrefacta, 294
Umbilicaria vellea, 294
Urocitellus armatus, 316
Urocitellus columbianus, 316–17
Urocitellus elegans, 316
Urocitellus richardsonii, 317
Ursus americanus, 338–40
Ursus arcos, 340–42
Urtica gracilis, 174–75
Usnea cavernosa, 298
Usnea hirta, 298
Usnea intermedia, 298
Usnea lapponica, 297
Utah honeysuckle, 91
Utah Rockies, 22, 24
Utah sweetvetch, 187

V

V fingers, 300
Vaccinium cespitosum, 109
Vaccinium membranaceum, 109
Vaccinium myrtillus, 109
Vaccinium ovalifolium, 109
Vaccinium scoparium, 110
Vaccinium uliginosum, 109–10
Vahlodea atropurpurea, 130
valerian, 223–24
Valeriana edulis, 223
Valeriana occidentalis, 223
Valeriana sitchensis, 223
Vancouver, George, 111
Vanessa cardui, 481–82

varied thrush, 413–14
Vaux, family, 377
Vaux's swift, 376
Veratrum viride, 148
Verbascum thapsus, 184
Veronica besseya, 184
Veronica cusickii, 231
Veronica dissecta, 232
Veronica missurica, 232
Veronica serpyllifolia, 231
Veronica wormskjoldii, 231
Veronica wyomingensis, 183–84
Vespertilionidae family, 507–08
Viburnum edule, 93
Viola adunca, 192
Viola canadensis, 192
Viola nephrophylla, 192
Viola nuttallii, 191
Viola orbiculata, 192
Viola purpurea, 191
violet, 191–92
violet-green swallow, 408
Vireo gilvus, 400
virga, 34
vivid dancer, 470
volcanic igneous rocks, 532–34
volcanoes, 499
Vulpes macrotis, 334
Vulpes velox, 334
Vulpes vulpes, 333–34

W

Walker, Skip, 196
wallflower, 234–35
wandering shrew, 506
wapato, 138
warbler, 421–22
warbling vireo, 400
warted giant puffball, 284
wasps and bees, 457–60
water hemlock, 228
water sedge, 132
water strider, 457
water vole, 329
waterleaf, 199
watermelon snow, 493–94
Watson, Sereno, 217
wavyleaf thistle, 169
wax currant, 104
waxleaf penstemon, 181
waxwing, 410–11
waxy cap, 268
weasel, 344–45
Weber, Dr. William, 16

*Webster's Third New
    International Dictionary*, 17
weeds, invasive species, 174
Weidemeyer's admiral, 482
Welch, Lew, 383
West Nile virus, 402, 407, 453
western balsam bark beetle, 67
western Canada goldenrod, 164
western coneflower, 166
western coralroot, 153
western diamondback
    rattlesnake, 430
western grebe, 374
western hemlock, 55, 73
western jumping mouse, 328
western larch, 55
western larch, 47, 61–62
western larkspur, 193
western maidenhair fern, 248–49
western matsutake, 269–70
western meadow vole, 328
western meadowlark, 421
western meadow-rue, 232–33
western mountain-ash, 98
western paintbrush, 178
western painted suillus, 275
western panther amanita,
    266–67
western pine beetle, 462, 463
western pine elfin, 489
western poison ivy, 106
western polypody, 244–45
western redcedar, 75
western saxifrage, 207
western spotted skunk, 342
western springbeauty, 222
western spruce budworm, 53
western St.-John's-wort, 214
western sweetroot, 228
western sweetvetch, 187
western tanager, 423
western tent caterpillar moth,
    472–73
western terrestrial garter snake,
    427
western tiger salamander, 432
western tiger swallowtail, 475
western toad, 434
western valerian, 223
western wallflower, 235
western water shrew, 506
western white, 478
western white pines, 56–57
western wormwood, 170
western yew, 73–74

whiplash willow, 83–84
Whipple, Amiel Weeks, 217
Whipple's penstemon, 182
white bog orchid, 149–50
white catchfly, 211–12
white chanterelle, 272
white clover, 190
white dryad, 122
white fir, 67–68
white hawkweed, 172
white king boletus, 274–75
white locoweed, 186
white mariposa lily, 141
white mountain-heather, 120–21
white mule's-ears, 163
white pine blister rust, 29, 56, 61
white rhododendron, 111–12
white spruce, 71–72
white virgin's bower, 94
whitebark pine, 25, 47, 57–59
white-breasted nuthatch, 411
white-crowned sparrow, 419
white-lined sphinx moth, 474–75
white-spotted pine sawyer, 464–65
whitestem gooseberry, 104
white-tailed deer, 352–54
white-tailed jackrabbit, 311–12
white-tailed prairie dog, 324–25
white-tailed ptarmigan, 372–73
white-throated swift, 376
white-veined pyrola, 213
wild chive, 143
wild ginger, 241, 242
wild onion, 142–43
wild raspberry, 99
wild sarsaparilla, 205–06
wildfire. See fire
wildflowers, 125
wildrose, 100–01
Williamson, Robert S., 217
willow flycatcher, 400
willows, 82–86
Wilson's snipe, 381

Wilson's warbler, 422
wintergreen, 214
witch's hair, 299
witch's butter, 281
Wolf, John, 104
wolf lichen, 301–02
wolf's currant, 104
wolverine, 348–49
wood horsetail, 251
wood lily, 144
wood nymph, 480
wood strawberry, 215
Woodhouse's toad, 434
woodpeckers, 394–97
woodrush, 135
Woods rose, 100
woodsia, 247
Woodsia oregana, 247
Woodsia scopulina, 248
woolly butterweed, 161
woolly chanterelle, 273
woolly daisy, 157
woolly foam lichen, 302
woolly hawkweed, 172
woolly mullein, 184
woolly pussytoes, 167
woolly sunflower, 162
woollypod milkvetch, 187
wormleaf stonecrop, 211
wormwood, 170–71
wrangler grasshopper, 466
Wyeth, Nathaniel, 192, 312
Wyethia amplexicaulis, 163
Wyethia helianthoides, 163
Wyoming ground squirrel, 316
Wyoming paintbrush, 178
Wyoming townsendia, 159

X
Xanthomendoza fallax, 293–94
Xanthocephalus xanthocephalus, 420
Xanthoparmelia chlorochroa, 297

Xanthoparmelia coloradoensis, 296–97
Xerophyllum tenax, 147–48

Y
yampah, 227
yellow bell, 144
yellow buckwheat, 238
yellow columbine, 220
yellow coral fungus, 282
yellow dryad, 122
yellow fever, 453
yellow lady's-slipper, 152
yellow monkeyflower, 179
yellow morel, 277
yellow mountain-heather, 121
yellow mule's-ears, 163
yellow owl's-clover, 179
yellow paintbrush, 178
yellow penstemon, 183
yellow pine chipmunk, 314
yellow ruffle lichen, 301
yellow salsify, 173
yellow starry fen moss, 258
yellow sweetvetch, 187
yellow warbler, 421
yellow-bellied marmot, 317–18
yellow-headed blackbird, 420
yellow-pond-lily, 239
yellow-rumped warbler, 421
Yellowstone hotspot, 25, 514–16, 533–34
Yellowstone plateau, 24–25
yew bark, 74
yucca, 116–17
Yucca baccata, 116
Yucca glauca, 116–17

Z
Zapus princeps, 328
Zenaida macroura, 375
zoned dust lichen, 291
Zonotrichia leucophrys, 419

About the Author

Stephen Weil

DANIEL MATHEWS comes from a line of botanically trained forebears, who taught him names of trailside plants in his childhood. Informed by thousands of scientific papers as well as six decades on and off hiking trails in the West, his writing dives deeper into science than other field guides, while also being more "linguistically alive."

Mathews is the author of *Trees in Trouble, Rocky Mountain Natural History,* and *Cascadia Revealed,* and contributing author of *National Audubon Society Field Guide to the Rocky Mountain States* and *National Wildlife Federation Field Guide to Trees of North America.*

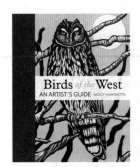

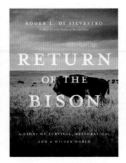
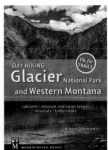